Lecture Notes in Computer Scier

T0238597

Commenced Publication in 1973
Founding and Former Series Editors:
Gerhard Goos, Juris Hartmanis, and Jan van Leeuwen

Hong Gao   Lipyeow Lim   Wei Wang
Chuan Li   Lei Chen (Eds.)

Web-Age
Information Management

13th International Conference, WAIM 2012
Harbin, China, August 18-20, 2012
Proceedings

 Springer

Volume Editors

Hong Gao
Harbin Institute of Technology
Harbin 150001, Heilongjiang, China
E-mail: honggao@hit.edu.cn

Lipyeow Lim
University of Hawaii
Honolulu 96822, HI, USA
E-mail: lipyeow@hawaii.edu

Wei Wang
Fudan University
Shanghai 200433, China
E-mail: weiwang1@fudan.edu.cn

Chuan Li
Sichuan University
Chengdu 610064, Sichuan, China
E-mail: lcharles@scu.edu.cn

Lei Chen
Hong Kong University of Science and Technology
Kowloon, Hong Kong, China
E-mail: leichen@cse.ust.hk

ISSN 0302-9743   e-ISSN 1611-3349
ISBN 978-3-642-32280-8   e-ISBN 978-3-642-32281-5
DOI 10.1007/978-3-642-32281-5
Springer Heidelberg Dordrecht London New York

Library of Congress Control Number: 2012943037

CR Subject Classification (1998): H.2.4, H.2.7-8, H.3.3-5, F.2.2, H.2, H.4, C.2, H.5, G.2.2, I.5.3

LNCS Sublibrary: SL 3 – Information Systems and Application, incl. Internet/Web and HCI

Typesetting: Camera-ready by author, data conversion by Scientific Publishing Services, Chennai, India

Printed on acid-free paper

Springer is part of Springer Science+Business Media (www.springer.com)

Preface

This volume contains the proceedings of the 13th International Conference on Web-Age Information Management (WAIM), held August 18–20, 2012, in Harbin, China. WAIM is a leading international conference on research, development, and applications of Web technologies, database systems and software engineering. WAIM is based in the Asia-Pacific region, and previous WAIM conferences were held in Shanghai (2000), Xi'an (2001), Beijing (2002), Chengdu (2003), Dalian (2004), Hangzhou (2005), Hong Kong (2006), Huangshan (2007), Zhangjiajie (2008), Suzhou (2009), Jiuzhaigou (2010), and Wuhan (2011). As the 13th event in the increasingly popular series, WAIM 2012, which was organized and supported by Harbin Institute of Technology, attracted outstanding researchers from all over the world to Harbin, China. In particular, this year WAIM and Microsoft Research Asia jointly sponsored a database summer school, which was collocated with WAIM.

This high-quality program would not have been possible without the authors who chose WAIM as a venue for their publications. Out of 178 submissions from various countries and regions, we selected 32 full papers and 10 short papers for publication. The acceptance rate for regular paper was 18% while the overall acceptance rate is 23.6% including short papers. The contributed papers address a wide range of topics such as spatial databases, query processing, XML and Web data, graph and uncertain data, distributed computing, information extraction and integration, data warehousing and data mining, similarity search, wireless sensor networks, social networks, data security, etc. We are grateful to our distinguished keynote speakers Wenfei Fan, Jiawei Han and Rakesh Agrawal for contributing to the event's success.

A conference like WAIM can only succeed as a team effort. We want to thank the Program Committee members and the reviewers for their invaluable efforts. Special thanks to the local Organizing Committee headed by Jizhou Luo. Many thanks also go to our Workshop Co-chairs (Xiaochun Yang and Hongzhi Wang), Publication Co-chairs (Lei Chen and Chuan Li), Publicity Co-chairs (Weiyi Meng and Guohui Li), Industrial chairs (Mukesh Mohania and Xiaoxin Wu), Panel Co-chairs (Jian Pei and Zhaonian Zou), and Finance Co-chairs (Howard Leung and Shengfei Shi). Last but not least, we wish to express our gratitude for the hard work of our webmaster (Hai Cao), and for our sponsors who generously supported the smooth running of our conference.

We hope that you enjoy reading the proceedings of WAIM 2012.

August 2012

Hong Gao
Lipyeow Lim
Wei Wang

Conference Organization

General Co-chairs

Jianzhong Li   Harbin Institute of Technology, China
Qing Li   City University of Hong Kong

Program Committee Co-chairs

Hong Gao   Harbin Institute of Technology, China
Lipyeow Lim   University of Hawaii, USA
Wei Wang   Fudan University, China

Workshop Co-chairs

Xiaochun Yang   Northeast University, China
Hongzhi Wang   Harbin Institute of Technology, China

Panel Co-chairs

Jian Pei   Simon Fraser University, USA
Zhaonian Zou   Harbin Institute of Technology, China

Industry Co-chairs

Mukesh Mohania   IBM, India
Xiaoxin Wu   Huawei, China

Publication Co-chairs

Lei Chen   Hong Kong University of Science and Technology, Hong Kong
Chuan Li   Sichuan University, China

Publicity Co-chairs

Weiyi Meng   Binghamton University, USA
Guohui Li   Huazhong University of Science and Technology, China

Local Organization Chair

Jizhou Luo   Harbin Institute of Technology, China

Finance Co-chairs

Howard Leung   City University of Hong Kong
Shengfei Shi   Harbin Institute of Technology, China

Steering Committee Liaison

Xiaofeng Meng   Renmin University, China

CCF DBS Liaison

Zhiyong Peng   Wuhan University, China

Microsoft Summer School Liaison

Haixun Wang   Microsoft Research Asia, China

Program Committee

Alfredo Cuzzocrea   University of Calabria, Italy
Aoying Zhou   East China Normal University, China
Bin Cui   Peking University, China
Carson K. Leung   University of Manitoba, Canada
Chengkai Li   University of Texas at Arlington, USA
Chuan Li   Sichuan University, China
Donghui Zhang   Microsoft Jim Gray Systems Lab, USA
Feifei Li   Florida State University, USA
Ge Yu   Northeast University, China
Giovanna Guerrini   Università di Genova, Italy
Heng Tao Shen   University of Queensland, Australia
Hiroaki Ohshima   Kyoto University, Japan
Hong Chen   Chinese University of Hong Kong, Hong Kong
Hongzhi Wang   Harbin Institute of Technology, China
Hong Gao   Harbin Institute of Technology, China
Hwanjo Yu   Pohang University of Science and Technology, Korea
Jeffrey Yu   Chinese University of Hong Kong
Jianyong Wang   Tsinghua University, China
Jianzhong Li   Harbin Institute of Technology, China
JianliangXu   Hong Kong Baptist University, Hong Kong
Jimmy Huang   York University, Canada
Jizhou Luo   Harbin Institute of Technology, China

Jun Gao   Peking University, China
Johann Gamper   Free University of Bozen-Bolzano, Italy
Lei Duan   Sichuan University, China
Lei Chen   Hong Kong University of Science and Technology,
                         Hong Kong
Lei Zou   Peking University, China
Ling Feng   Tsinghua University, China
Ning Jing   National University of Defense Technology, China
Ning Ruan   Kent State University, USA
Ning Yang   Sichuan University, China
Peng Wang   Fudan University, China
Raymond Ng   University of British Columbia, Canada
Shuai Ma   University of Edinburgh, UK
Shuigeng Zhou   Fudan University, China
Shuming Shi   Microsoft Research Asia, China
Tao Li   Florida International University, USA
Toshiyuki Amagasa   University of Tsukuba, Japan
Wei Wang   University of New South Wales, Australia
Weiyi Meng   State University of New York at Binghamton, USA
Xiangliang Zhang   King Abdullah University of Science and Technology,
                         Saudi Arabia
Xiaofeng Meng   Renmin University of China
Xin Dong   AT&T Research, USA
Xingquan Zhu   University of Technology, Sydney
Xintao Wu   University of North Carolina at Charlotte, USA
Xuemin Lin   University of New South Wales, Australia
Xiaofang Zhou   University of Queensland, Australia
Xiaochun Yang   Northeastern University, China
Yanghua Xiao   Fudan University, China
Yan Jia   National University of Defense Technology, China
Yang-Sae Moon   Kangwon National University, Korea
YaokaiFeng   Kyushu University, Japan
Yi Cai   City University of Hong Kong
Ke Yi   Hong Kong University of Science and Technology,
                         Hong Kong
Yoshiharu Ishikawa   Nagoya University, Japan
Yunjun Gao   Zhejianing University, China
Yuqing Wu   Indiana University at Bloomington, USA
Zhanhuai Li   Northwestern Polytechnical University, China
Zhaonian Zou   Harbin Institute of Technology, China
Zhiyong Peng   Wuhan University, China
Zongmin Ma   Northeastern University, China

Table of Contents

Keynotes

Session 1: Wireless Sensor Networks

Session 2: Data Warehouse and Data Mining

Session 3: Query Proceeding (1)

Session 4: Query Proceeding (2)

Session 5: Spatial Database

Session 6: Similarity Search and Queries

Session 7: XML and Web Data

Session 8: Graph and Uncertain Data

Session 9: Distributed Computing

Session 10: Data Security and Management

Session 11: I nformation Extraction and Integration

Session 12: Social Networks and Modern Web Services

Data Quality: Theory and Practice

Wenfei Fan*

University of Edinburgh and Harbin Institute of Technology

Abstract. Real-life data are often dirty: inconsistent, inaccurate, incomplete, stale and duplicated. Dirty data have been a longstanding issue, and the prevalent use of Internet has been increasing the risks, in an unprecedented scale, of creating and propagating dirty data. Dirty data are reported to cost US industry billions of dollars each year. There is no reason to believe that the scale of the problem is any different in any other society that depends on information technology. With these comes the need for improving *data quality*, a topic as important as traditional data management tasks for coping with *the quantity* of the data.

We aim to provide an overview of recent advances in the area of data quality, from theory to practical techniques. We promote a conditional dependency theory for capturing data inconsistencies, a new form of dynamic constraints for data deduplication, a theory of relative information completeness for characterizing incomplete data, and a data currency model for answering queries with current values from possibly stale data in the absence of reliable timestamps. We also discuss techniques for automatically discovering data quality rules, detecting errors in real-life data, and for correcting errors with performance guarantees.

1 Data Quality: An Overview

Traditional database systems typically focus on *the quantity of data*, to support the creation, maintenance and use of large volumes of data. But such a database system may not find correct answers to our queries if the data in the database are "dirty", *i.e.*, when the data do not properly represent the real world entities to which they refer.

To illustrate this, let us consider an employee relation residing in a database of a company, specified by the following schema:

employee (FN, LN, CC, AC, phn, street, city, zip, salary, status)

Here each tuple specifies an employee's name (first name FN and last name LN), office phone (country code CC, area code AC, phone phn), office address (street, city, zip code), salary and marital status. An instance D_0 of the employee schema is shown in Figure 1.

---

* Fan is supported in part by EPSRC EP/J015377/1, the RSE-NSFC Joint Project Scheme, the 973 Program 2012CB316200 and NSFC 61133002 of China.

H. Gao et al. (Eds.): WAIM 2012, LNCS 7418, pp. 1–16, 2012.

	FN	LN	CC	AC	phn	street	city	zip	salary	status
t_1:	Mike	Clark	44	131	null	Mayfield	NYC	EH4 8LE	60k	single
t_2:	Rick	Stark	44	131	3456789	Crichton	NYC	EH4 8LE	96k	married
t_3:	Joe	Brady	01	908	7966899	Mtn Ave	NYC	NJ 07974	90k	married
t_4:	Mary	Smith	01	908	7966899	Mtn Ave	MH	NJ 07974	50k	single
t_5:	Mary	Luth	01	908	7966899	Mtn Ave	MH	NJ 07974	50k	married
t_6:	Mary	Luth	44	131	3456789	Mayfield	EDI	EH4 8LE	80k	married

Fig. 1. An employee instance

Consider the following queries posted on relation D_0.

(1) Query Q_1 is to find the number of employees working in the NYC office (New York City). The answer to Q_1 in D_0 is 3, by counting tuples t_1, t_2 and t_3. However, the answer may not be correct, for the following reasons. First, the data in D_0 are *inconsistent*. Indeed, the CC and AC values of t_1, t_2 and t_3 have conflicts with their corresponding city attributes: when CC = 44 and AC = 131, the city should be Edinburgh (EDI) in the UK, rather than NYC; and similarly, when CC = 01 and AC = 908, city should be Murray Hill (MH) in the US. It is thus likely that NYC is not the true city value of t_1, t_2 and t_3. Second, the data in D_0 may be *incomplete* for employees working in NYC. That is, some tuples representing employees working in NYC may be *missing* from D_0. Hence we cannot trust 3 to be the answer to Q_1.

(2) Query Q_2 is to find the number of distinct employees with FN = Mary. In D_0 the answer to Q_2 is 3, by enumerating tuples t_4, t_5 and t_6. Nevertheless, the chances are that t_4, t_5 and t_6 actually refer to the same person: all these tuples were once the true values of Mary, but some have become obsolete. Hence the correct answer to Q_2 may be 1 instead of 3.

(3) Query Q_3 is to find Mary's current salary and current last name, provided that we know that t_4, t_5 and t_6 refer to the same person. Simply evaluating Q_3 on D_0 will get us that salary is either 50k or 80k, and that LN is either Smith or Luth. However, it does not tell us whether Mary's current salary is 50k, and whether her current last name is Smith. Indeed, reliable timestamps for t_4, t_5 and t_6 may not be available, as commonly found in practice, and hence, we cannot tell which of 50k or 80k is more current; similarly for LN.

This example tells us that when the data are dirty, we cannot expect a database system to answer our queries correctly, no matter what capacity it provides to accommodate large data and how efficient it processes our queries.

Unfortunately, real-life data are often *dirty*: inconsistent, duplicated, inaccurate, incomplete and/or out of date. Indeed, enterprises typically find data error rates of approximately 1%–5%, and for some companies it is above 30% [41]. In most data warehouse projects, data cleaning accounts for 30%-80% of the development time and budget [43], for improving the quality of the data rather than developing the systems. When it comes to incomplete information, it is estimated that "pieces of information perceived as being needed for clinical decisions were missing from 13.6% to 81% of the time" [38]. When data currency is concerned, it is known that "2% of records in a customer file become obsolete in one month" [14]. That is, in a database of 500 000 customer records, 10 000 records may go stale per month, 120 000 records per year, and within two years about 50% of all the records may be obsolete.

Why do we care about dirty data? Data quality has become one of the most pressing challenges to data management. It is reported that dirty data cost US businesses 600 billion dollars annually [14], and that erroneously priced data in retail databases alone cost US consumers $2.5 billion each year [16]. While these indicate the daunting cost of dirty data in the US, there is no reason to believe that the scale of the problem is any different in any other society that is dependent on information technology. Dirty data have been a longstanding issue for decades, and the prevalent use of Internet has been increasing the risks, in an unprecedented scale, of creating and propagating dirty data.

These highlight the need for *data quality management*, to improve *the quality* of the data in our databases such that the data consistently, accurately, completely and uniquely represent the real-world entities to which they refer.

Data quality management is at least as important as traditional data management tasks for coping with *the quantity of data*. There has been increasing demand in industries for developing data-quality management systems, aiming to effectively detect and correct errors in the data, and thus to add accuracy and value to business processes. Indeed, the market for data-quality tools is growing at 16% annually, way above the 7% average forecast for other IT segments [34]. As an example, data quality tools deliver "an overall business value of more than 600 million GBP" each year at BT [40]. Data quality management is also a critical part of big data management, master data management (MDM) [37], customer relationship management (CRM), enterprise resource planning (ERP) and supply chain management (SCM), among other things.

This paper aims to highlight several central technical issues in connection with data quality, and to provide an overview of recent advances in data quality management. We present five important issues of data quality (Section 2), and outline a rule-based approach to cleaning dirty data (Section 3). Finally, we identify some open research problems associated with data quality (Section 4).

The presentation is informal, to incite curiosity in the study of data quality. We opt for breadth rather than depth in the presentation: important results and techniques are briefly mentioned, but the details are omitted. A survey of detailed data quality management techniques is beyond the scope of this paper, and a number of related papers are not referenced due to space constraints. We refer the interested reader to papers in which the results were presented for more detailed presentation of the results and techniques. In particular, we encourage the reader to consult [3,4,9,17,21] for recent surveys on data quality management. In fact a large part of this paper is taken from [21].

2   Central Issues of Data Quality

We highlight five central issues in connection with data quality: data consistency, data deduplication, data accuracy, information completeness and data currency.

2.1   Data Consistency

Data consistency refers to the validity and integrity of data representing real-world entities. It aims to detect inconsistencies or conflicts in the data. In a relational database, inconsistencies may exist within a single tuple, between different tuples in the same table, and between tuples across different relations.

As an example, consider tuples t_1, t_2 and t_3 in Figure 1. There are conflicts within each of these tuples, as well as inconsistencies between different tuples.

(1) It is known that in the UK (when $CC = 44$), if the area code is 131, then the city should be Edinburgh (EDI). In tuple t_1, however, $CC = 44$ and $AC = 131$, but city $\neq$ EDI. That is, there exist inconsistencies between the values of the CC, AC and city attributes of t_1; similarly for tuple t_2. These tell us that tuples t_1 and t_2 are erroneous.

(2) Similarly, in the US ($CC = 01$), if the area code is 908, the city should be Murray Hill (MH). Nevertheless, $CC = 01$ and $AC = 908$ in tuple t_3, whereas its city is not MH. This indicates that tuple t_3 is not quite correct.

(3) It is also known that in the UK, zip code uniquely determines street. That is, for any two tuples that refer to employees in the UK, if they share the same zip code, then they should have the same value in their street attributes. However, while $t_1[CC] = t_2[CC] = 44$ and $t_1[zip] = t_2[zip]$, $t_1[street] \neq t_2[street]$. Hence there are conflicts between t_1 and t_2.

Inconsistencies in the data are typically identified as violations of *data dependencies* (*a.k.a.* integrity constraints [1]). Errors in a single relation can be detected by intrarelation constraints, while errors across different relations can be identified by interrelation constraints.

Unfortunately, traditional dependencies such as functional dependencies (FDs) and inclusion dependencies (INDs) fall short of catching inconsistencies commonly found in real-life data, such as the errors in tuples t_1, t_2 and t_3 above. This is not surprising: the traditional dependencies were developed for schema design, rather than for improving data quality.

To remedy the limitations of traditional dependencies in data quality management, conditional functional dependencies (CFDs [23]) and conditional inclusion dependencies (CINDs [7]) have recently been proposed, which extend FDs and INDs, respectively, by specifying patterns of semantically related data values. It has been shown that conditional dependencies are capable of capturing common data inconsistencies that FDs and INDs fail to detect. For example, the inconsistencies in t_1–t_3 given above can be detected by CFDs.

A theory of conditional dependencies is already in place, as an extension of classical dependency theory. More specifically, the satisfiability problem, implication problem, finite axiomatizability and dependency propagation have been studied for conditional dependencies, from the complexity to inference systems to algorithms. We refer the interested reader to [7,6,23,31] for details.

2.2 Data Deduplication

Data deduplication aims to identify tuples in one or more relations that refer to the same real-world entity. It is also known as entity resolution, duplicate detection, record matching, record linkage, merge-purge, database hardening, and object identification (for data with complex structures).

For example, consider tuples t_4, t_5 and t_6 in Figure 1. To answer query Q_2 given earlier, we want to know whether these tuples refer to the same employee. The answer is affirmative if, for instance, there exists another relation which indicates that Mary Smith and Mary Luth have the same email account.

The need for studying data deduplication is evident: for data cleaning it is needed to eliminate duplicate records; for data integration it is to collate and fuse information about the same entity from multiple data sources; and for master data management it helps us identify links between input tuples and master data. The need is also highlighted by payment card fraud, which cost $4.84 billion worldwide in 2006 [42]. In fraud detection it is a routine process to cross-check whether a credit card user is the legitimate card holder. As another example, there was a recent effort to match records on licensed airplane pilots with records on individuals receiving disability benefits from the US Social Security Administration. The finding was quite surprising: there were forty pilots whose records turned up in both databases (cf. [36]).

No matter how important it is, data deduplication is nontrivial. Indeed, tuples pertaining to the same object may have different representations in various data sources with different schemas. Moreover, the data sources may contain errors. These make it hard, if not impossible, to match a pair of tuples by simply checking whether their attributes pairwise equal. Worse still, it is often too costly to compare and examine every pair of tuples from large data sources.

Data deduplication is perhaps the most extensively studied data quality problem. A variety of approaches have been proposed: probabilistic, learning-based, distance-based, and rule-based (see [15,36,39] for recent surveys).

We promote a dependency-based approach for detecting duplicates, which allows us to capture the interaction between data deduplication and other aspects of data quality in a uniform logical framework. To this end a new form of dependencies, referred to as *matching dependencies*, has been proposed for data deduplication [18]. These dependencies help us decide *what attributes to compare* and *how to compare these attributes* when matching tuples. They allow us to deduce alternative attributes to inspect such that when matching cannot be done by comparing attributes that contain errors, we may still find matches by using other, more reliable attributes.

In contrast to traditional dependencies that we are familiar with such as FDs and INDs, matching dependencies are dynamic constraints: they tell us what data have to be updated as a consequence of record matching. A dynamic constraint theory has been developed for matching dependencies, from deduction analysis to finite axiomatizability to inference algorithms (see [18] for details).

2.3 Data Accuracy

Data accuracy refers to the closeness of values in a database to the true values of the entities that the data in the database represent. Consider, for example, a person schema:

person (FN, LN, age, height, status)

where each tuple specifies the name (FN, LN), age, height and marital status of a person. An instance of person is shown below, in which s_0 presents the "true" information for Mike.

	FN	LN	age	height	status
s_0:	Mike	Clark	14	1.70	single
s_1:	M.	Clark	14	1.69	married
s_2:	Mike	Clark	45	1.60	single

Given these, we can conclude that the values of $s_1[\text{age, height}]$ are more accurate than $s_2[\text{age, height}]$, as they are closer to the true values for Mike, while $s_2[\text{FN, status}]$ are more accurate than $s_1[\text{FN, status}]$. It is more challenging, however, to determine the relative accuracy of s_1 and s_2 when the reference s_0 is unknown, as commonly found in practice. In this setting, it is still possible to find that for certain attributes, the values in one tuple are more accurate than another by an analysis of the semantics of the data, as follows.

(1) Suppose that we know that Mike is still going to middle school. From this, we can conclude that $s_1[\text{age}]$ is more accurate than $s_2[\text{age}]$. That is, $s_1[\text{age}]$ is closer to Mike's true age value than $s_2[\text{age}]$, although Mike's true age may not be known. Indeed, it is unlikely that students in a middle school are 45 years old. Moreover, from the age value ($s_1[\text{age}]$), we may deduce that $s_2[\text{status}]$ may be more accurate than $s_1[\text{status}]$.

(2) If we know that $s_1[\text{height}]$ and $s_2[\text{height}]$ were once correct, then we may conclude that $s_1[\text{height}]$ is more accurate than $s_2[\text{height}]$, since the height of a person is typically monotonically increasing, at least when the person is young.

2.4 Information Completeness

Information completeness concerns whether our database has complete information to answer our queries. Given a database D and a query Q, we want to know whether Q can be completely answered by using only the data in D. If the information in D is incomplete, one can hardly expect its answer to Q to be accurate or even correct.

In practice our databases often do not have sufficient information for our tasks at hand. For instance, the value of $t_1[\text{phn}]$ in relation D_0 of Figure 1 is missing, as indicated by null. Worse still, tuples representing employees may also be missing from D_0. As we have seen earlier, for query Q_1 given above, if some tuples representing employees in the NYC office are missing from D_0, then the answer to Q_1 in D_0 may not be correct. Incomplete information introduces serious problems to enterprises: it routinely leads to misleading analytical results and biased decisions, and accounts for loss of revenues, credibility and customers.

How should we cope with incomplete information? Traditional work on information completeness adopts either the Closed World Assumption (CWA) or the Open World Assumption (OWA), stated as follows (see, *e.g.*, [1]).

- The CWA assumes that a database has collected all the tuples representing real-world entities, but some *attribute values* of the tuples may be *missing*.

- The OWA assumes that in addition to missing values, some *tuples* representing real-world entities may also be *missing*. That is, our database may only be a proper subset of the set of tuples that represent real-world entities.

Database textbooks typically tell us that the world is closed: all the real-world entities of our interest are assumed already represented by tuples residing in our database. After all, database theory is typically developed under the CWA, which is the basis of negation in our queries: a fact is viewed as false unless it can be proved from explicitly stated facts in our database.

Unfortunately, in practice one often finds that not only attribute values but also tuples are missing from our database. That is, the CWA is often too strong to hold in the real world. On the other hand, the OWA is too weak: under the OWA, we can expect few sensible queries to find complete answers.

The situation is not as bad as it seems. In the real world, neither the CWA nor the OWA is quite appropriate in emerging applications such as master data management. In other words, real-life databases are *neither* entirely closed-world *nor* entirely open-world. Indeed, an enterprise nowadays typically maintains *master data* (*a.k.a. reference data*), a single repository of high-quality data that provides various applications with a synchronized, consistent view of the core business entities of the enterprise (see, *e.g.*, [37], for master data management). The master data contain complete information about the enterprise in certain categories, *e.g.*, employees, departments, projects, and equipment. Master data can be regarded as a closed-world database for the core business entities of the enterprise. Meanwhile a number of other databases may be in use in the enterprise for, *e.g.*, sales, project control and customer support. On one hand, the information in these databases may not be complete, *e.g.*, some sale transaction records may be missing. On the other hand, certain parts of the databases are *constrained by* the master data, *e.g.*, employees and projects. In other words, these databases are *partially closed*. The good news is that we often find that partially closed databases have complete information to answer our queries at hand.

To rectify the limitations of the CWA and the OWA, a theory of relative information completeness has been proposed [20,19], to specify partially closed databases *w.r.t.* available master data. In addition, several fundamental problems in connection with relative completeness have been studied, to determine whether our database has complete information to answer our query, and when the database is incomplete for our tasks at hand, to decide what additional data should be included in our database to meet our requests. The complexity bounds of these problems have been established for various query languages.

2.5 Data Currency

Data currency is also known as *timeliness*. It aims to identify the current values of entities represented by tuples in a database that may contain stale data, and to answer queries with the current values.

The question of data currency would be trivial if all data values carried valid timestamps. In practice, however, one often finds that timestamps are unavailable or imprecise [46]. Add to this the complication that data values are often copied or imported from other sources [12,13], which may not support a uniform scheme of timestamps. These make it challenging to identify the "latest" values of entities from the data in our database.

For example, recall query Q_3 and the employee relation D_0 of Figure 1 given earlier. Assume that tuples t_4, t_5 and t_6 are found pertaining to the same employee Mary by data deduplication. As remarked earlier, in the absence of reliable timestamps, the answer to Q_3 in D_0 does not tell us whether Mary's current salary is 50k or 80k, and whether her current last name is Smith or Luth.

Not all is lost. In practice it is often possible to deduce currency orders from the semantics of the data, as illustrated below.

(1) While we do not have timestamps associated with Mary's salary, we know that the salary of each employee in the company does *not* decrease, as commonly found in the real world. This tells us that $t_6[\mathsf{salary}]$ is more current than $t_4[\mathsf{salary}]$ and $t_5[\mathsf{salary}]$. Hence we may conclude that Mary's current salary is 80k.

(2) We know that the marital status can only change from single to married and from married to divorced; but not from married to single. In addition, employee tuples with the most current marital status also contain the most current last name. Therefore, $t_6[\mathsf{LN}] = t_5[\mathsf{LN}]$ is more current than $t_4[\mathsf{LN}]$. From these we can infer that Mary's current last name is Luth.

A data currency model has recently been proposed in [26], which allows us to specify and deduce data currency when temporal information is only partly known or not available at all. Moreover, a notion of certain current query answers is introduced there, to answer queries with current values of entities derived from a possibly stale database. In this model the complexity bounds of fundamental problems associated with data currency have been established, for identifying the current value of an entity in a database in the absence of reliable timestamps, answering queries with current values, and for deciding what data should be imported from other sources in order to answer query with current values. We encourage the interested reader to consult [26] for more detailed presentation.

2.6   Interactions between Data Quality Issues

To improve data quality we often need to deal with each and every of the five central issues given above. Moreover, there issues interact with each other, as illustrated below.

As we have seen earlier, tuples t_1, t_2 and t_3 in the relation D_0 of Figure 1 are inconsistent. We show how data deduplication may help us resolve the inconsistencies. Suppose that the company maintains a master relation for its offices, consisting of consistent, complete and current information about the address and phone number of each office. The master relation is specified by schema:

office (CC, AC, phn, street, city, zip),

and is denoted by D_m, given as follows:

	CC	AC	phn	street	city	zip
t_{m1}:	44	131	3456789	Mayfield	EDI	EH4 8LE
t_{m2}:	01	908	7966899	Mtn Ave	MH	NJ 07974

Then we may "clean" t_1, t_2 and t_3 by leveraging the interaction between data deduplication and data repairing processes (for data consistency) as follows.

(1) If the values of the CC, AC attributes of these tuples are confirmed accurate, we can safely update their city attributes by letting $t_1[\text{city}] = t_2[\text{city}] := \text{EDI}$, and $t_3[\text{city}] := \text{MH}$, for reasons remarked earlier. This yields t'_1, t'_2 and t'_3, which differ from t_1, t_2 and t_3, respectively, only in their city attribute values.

(2) We know that if an employee tuple $t \in D_0$ and an office tuple $t_m \in D_m$ agree on their address (street, city, zip), then the two tuples "match", i.e., they refer to the same address. Hence, we can update $t[\text{CC, AC, phn}]$ by taking the corresponding master values from t_m. This allows us to change $t'_2[\text{street}]$ to $t_{m1}[\text{street}]$. That is, we repair $t'_2[\text{street}]$ by matching t'_2 and t_{m1}. This leads to tuple t''_2, which differs from t'_2 only in the street attribute.

(3) We also know that for employee tuples t_1 and t_2, if they have the same address, then they should have the same phn value. In light of this, we can augment $t'_1[\text{phn}]$ by letting $t'_1[\text{phn}] := t''_2[\text{phn}]$, and obtain a new tuple t''_1.

One can readily verify that t''_1, t''_2 and t'_3 are consistent. In the process above, we "interleave" operations for resolving conflicts (steps 1 and 3) and operations for detecting duplicates (step 2). On one hand, conflict resolution helps deduplication: step 2 can be conducted only after $t_2[\text{city}]$ is corrected. On the other hand, deduplication also helps us resolve conflicts: $t'_1[\text{phn}]$ is enriched only after $t'_2[\text{street}]$ is fixed via matching.

There are various interactions between data quality issues, including but not limited to the following.

− Data currency can be improved if more temporal information can be obtained in the process for improving information completeness.

- To determine the current values of an entity, we need to identify tuples pertaining to the same entity, via data deduplication. For instance, to find Mary's LN in relation D_0 of Figure 1, we have to ask whether tuples t_4, t_5 and t_6 refer to the same person.
- To resolve conflicts in tuples representing an entity, we have to determine whether the information about the entity is complete, and only if so, we can find the true value of the entity from the data available in our database.

These suggest that a practical data quality management system should provide functionality to deal with each and every of five central issues given above, and moreover, leverage the interactions between these issues to improve data quality. There has been preliminary work on the interaction between data deduplication and data repairing [27], as illustrated by the example above.

3   Improving Data Quality

Real-life data are often dirty, and dirty data are costly. In light of these, effective techniques have to be in place to improve the quality of our data. But how?

Errors in Real-Life Data. To answer this question, we first classify errors typically found in the real world. There are two types of errors, namely, syntactic errors and semantic errors, as illustrated below.

(1) Syntactic errors: violations of domain constraints by the values in our database. For example, name = 1.23 is a syntactic error if the domain of attribute name is string, whereas the value is numeric. Another example is age = 250 when the range of attribute age is [0, 120].

(2) Semantic errors: discrepancies between the values in our database and the true values of the entities that our data intend to represent. All the examples we have seen in the previous sections are semantic errors, related to data consistency, deduplication, accuracy, currency and information completeness.

While syntactic errors are relatively easy to catch, it is far more challenging to detect and correct semantic errors. Below we focus on semantic errors.

Dependencies as Data Quality Rules. A central question concerns how we can tell whether our data have semantic errors, *i.e.,* whether the data are dirty or clean? To this end, we need data quality rules to detect semantic errors in our data and fix those errors. But what data quality rules should we adopt?

A natural idea is to use data dependencies (*a.k.a.* integrity constraints). Dependency theory is almost as old as relational databases themselves. Since Codd [10] introduced functional dependencies, a variety of dependency languages, defined as various classes of first-order (FO) logic sentences, have been developed. There are good reasons to believe that dependencies should play an important role in data quality management systems. Indeed, dependencies specify a fundamental part of the semantics of data, in a declarative way, such that errors emerge as violations of the dependencies. Furthermore, inference systems, implication analysis and profiling methods for dependencies have shown promise as a systematic method for reasoning about the semantics of the data. These help us deduce and discover rules for improving data quality, among other things. In addition, all the five central aspects of data quality – data consistency, deduplication, accuracy, currency and information completeness – can be specified in terms of data dependencies. This allows us to treat various data quality issues in a uniform logical framework, in which we can study their interactions.

Nevertheless, to make practical use of dependencies in data quality management, classical dependency theory has to be extended. Traditional dependencies were developed to improve *the quality of schema* via normalization, and to optimize queries and prevent invalid updates (see, *e.g.*, [1]). To *improve the quality of the data*, we need new forms of dependencies, such as conditional dependencies by specifying patterns of semantically related data values to capture data inconsistencies [23,7], matching dependencies by supporting similarity predicates to accommodate data errors in record matching [18], containment constraints by enforcing containment of certain information about core business entities in master data to reason about information completeness [20,19], and currency constraints by incorporating temporal orders to determine data currency [26].

Care must be taken when designing dependency languages for improving data quality. Among other things, we need to balance the tradeoff between expressive power and complexity, and revisit classical problems for dependencies such as the satisfiability, implication and finite axiomatizability analyses.

Improve Data Quality with Rules. After we come up with the "right" dependency languages for specifying data quality rules, the next question is how to effectively use these rules to improve data quality? In a nutshell, a rule-based data quality management system should provide the following functionality.

Discovering Data Quality Rules. To use dependencies as data quality rules, it is necessary to have efficient techniques in place that can *automatically discover* dependencies from data. Indeed, it is unrealistic to rely solely on human experts to design data quality rules via an expensive and long manual process, or count on business rules that have been accumulated. This suggests that we learn informative and interesting data quality rules from (possibly dirty) data, and prune away trivial and insignificant rules based on a threshold specified by users.

More specifically, given a database instance D, the *profiling problem* is to find a *minimal cover* of all dependencies (*e.g.*, CFDs, CINDs, matching dependencies) that hold on D, *i.e.*, a non-redundant set of dependencies that is logically equivalent to the set of all dependencies that hold on D.

To find data quality rules, several algorithms have been developed for discovering CFDs [8,24,35] and matching dependencies [44].

Validating Data Quality Rules. A given set Σ of dependencies, either automatically discovered or manually designed by domain experts, may be dirty itself. In light of this we have to identify "consistent" dependencies from Σ, *i.e.*, those rules that make sense, to be used as data quality rules. Moreover, we need to deduce new rules and to remove redundancies from Σ, via the implication or deduction analysis of those dependencies in Σ.

This problem is, however, nontrivial. It is already NP-complete to decide whether a given set of CFDs is satisfiable [23], and it becomes undecidable for CFDs and CINDs taken together [7]. Nevertheless, there has been an approximation algorithm for extracting a set S' of consistent rules from a set S of possibly inconsistent CFDs, while guaranteeing that S' is within a constant bound of the maximum consistent subset of S (see [23] for details).

Detecting Errors. After a validated set of data quality rules is identified, the next question concerns how to effectively catch errors in a database by using these rules. Given a set Σ of data quality rules and a database D, we want to *detect inconsistencies* in D, *i.e.*, to find all tuples in D that violate some rule in Σ. When it comes to relative information completeness, we want to decide whether D has complete information to answer an input query Q, among other things.

We have shown that for a centralized database D, given a set Σ of CFDs and CINDs, a fixed number of SQL queries can be *automatically* generated such that, when being evaluated against D, the queries return all and only those tuples in D that violate Σ [23]. That is, we can effectively detect inconsistencies by leveraging existing facility of commercial relational database systems.

In practice a database is often fragmented, vertically or horizontally, and is distributed across different sites. In this setting, inconsistency detection becomes nontrivial: it necessarily requires certain data to be shipped from one site to another. In this setting, error detection with minimum data shipment or minimum response time becomes NP-complete [25], and the SQL-based techniques for detecting violations of conditional dependencies no longer work. Nevertheless, effective batch algorithms [25] and incremental algorithms [29] have been developed for detecting errors in distributed data, with certain performance guarantees.

Data Imputation. After the errors are detected, we want to automatically localize the errors, fix the errors and make the data consistent. We also need to identify tuples that refer to the same entity, and for each entity, determine its latest and most accurate values from the data in our database. When some data are missing, we need to decide what data we should import and where to import from, so that we will have sufficient information for tasks at hand. As remarked earlier, these should be carried out by capitalizing on the interactions between processes for improving various aspects of data quality, as illustrated in Section 2.6.

As another example, let us consider *data repairing* for improving data consistency. Given a set Σ of dependencies and an instance D of a database schema $\mathcal{R}$, it is to find a candidate *repair* of D, *i.e.*, an instance D' of $\mathcal{R}$ such that D' satisfies Σ and D' *minimally differs* from the original database D [2]. This is the method that US national statistical agencies, among others, have been practicing for decades for cleaning census data [33,36]. The data repairing problem is, nevertheless, highly nontrivial: it is NP-complete even when a fixed set of FDs or a fixed set of INDs is used as data quality rules [5], even for centralized databases. In light of these, several heuristic algorithms have been developed, to effectively repair data by employing FDs and INDs [5], CFDs [11,45], CFDs and matching dependencies [27] as data quality rules. A functional prototype system [22] has also shown promises as an effective tool for repairing data in industry.

The data repairing methods mentioned above are essentially heuristic: while they improve the consistency of the data, they do not guarantee to find correct fixes for each error detected, *i.e.*, they do not warrant a precision and recall of 100%. Worse still, they may introduce new errors when trying to repair the data. In light of these, they are not accurate enough to repair critical data such as medical data, in which a minor error may have disastrous consequences. This highlights the quest for effective methods to find *certain fixes* that are guaranteed correct. Such a method has recently be proposed in [28]. While it may not be able to fix all the errors in our database, it guarantees that whenever it updates a data item, it correctly fixes an error without introducing any new error.

4   Conclusion

Data quality is widely perceived as one of the most important issues for information systems. In particular, the need for studying data quality is evident in big data management, for which two central issues of equivalent importance concern how to cope with *the quantity* of the data and *the quality* of the data.

The study of data quality management has raised as many questions as it has answered. It is a rich source of questions and vitality for database researchers. However, data quality research lags behind the demands in industry. A number of open questions need to be settled. Below we address some of the open issues.

Data Accuracy. Previous work on data quality has mostly focused on data consistency and data deduplication. In contrast, the study of data accuracy is still in its infancy. One of the most pressing issues concerns how to determine whether one value is more accurate than another in the absence of reference data. This calls for the development of models, quantitative metrics, and effective methods for determining the relative accuracy of data.

Information Completeness. Our understanding of this issue is still rudimentary. While the theory of relative information completeness [20,19] circumvents the limitations of the CWA and the OWA and allows us to determine whether a database has complete information to answer our query, effective metrics and algorithms are not yet in place for us to conduct the evaluation in practice.

Data Currency. The study of data currency has not yet reached the maturity. The results in this area are mostly theoretical: a model for specifying data currency, and complexity bounds for reasoning about the currency of data [26]. Among other things, effective methods for evaluating the currency of data in our databases and for deriving current values from stale data are yet to be developed.

Interaction between Various Issues of Data Quality. As remarked earlier, there is an intimate connection between data repairing and data deduplication [27]. Similarly, various interactions naturally arise when we attempt to improve the five central aspects of data quality: information completeness is intimately related to data currency and consistency, and so is data currency to data consistency and accuracy. These interactions require a full treatment.

Repairing Distributed Data. Already hard to repair data in a centralized database, it is far more challenging to efficiently fix errors in distributed data. This is, however, a topic of great interest to the study of big data, which are typically partitioned and distributed. As remarked earlier, data quality is a central aspect of big data management, and hence, effective and scalable repairing methods for distributed data have to be studied.

The Quality of Complex Data. Data quality issues are on an even larger scale for data on the Web, *e.g.*, XML data and social graphs. Already hard for relational data, error detection and repairing are far more challenging for data with complex structures. In the context of XML, for example, the constraints involved and their interaction with XML Schema are far more intriguing than their relational counterparts, even for static analysis [30,32], let alone for data repairing. In this setting data quality remains by and large unexplored. Another issue concerns object identification, *i.e.*, to identify complex objects that refer to the same real-world entity, when the objects do not have a regular structure. This is critical not only to data quality, but also to Web page clustering, schema matching, pattern recognition, and spam detection, among other things.

References

1. Abiteboul, S., Hull, R., Vianu, V.: Foundations of Databases. Addison-Wesley (1995)
2. Arenas, M., Bertossi, L.E., Chomicki, J.: Consistent query answers in inconsistent databases. In: PODS (1999)
3. Batini, C., Scannapieco, M.: Data Quality: Concepts, Methodologies and Techniques. Springer (2006)
4. Bertossi, L.: Database Repairing and Consistent Query Answering. Morgan & Claypool Publishers (2011)
5. Bohannon, P., Fan, W., Flaster, M., Rastogi, R.: A cost-based model and effective heuristic for repairing constraints by value modification. In: SIGMOD (2005)
6. Bravo, L., Fan, W., Geerts, F., Ma, S.: Increasing the expressivity of conditional functional dependencies without extra complexity. In: ICDE (2008)
7. Bravo, L., Fan, W., Ma, S.: Extending dependencies with conditions. In: VLDB (2007)
8. Chiang, F., Miller, R.: Discovering data quality rules. In: VLDB (2008)
9. Chomicki, J.: Consistent Query Answering: Five Easy Pieces. In: Schwentick, T., Suciu, D. (eds.) ICDT 2007. LNCS, vol. 4353, pp. 1–17. Springer, Heidelberg (2006)
10. Codd, E.F.: Relational completeness of data base sublanguages. In: Data Base Systems: Courant Computer Science Symposia Series 6. Prentice-Hall (1972)
11. Cong, G., Fan, W., Geerts, F., Jia, X., Ma, S.: Improving data quality: Consistency and accuracy. In: VLDB (2007)
12. Dong, X.L., Berti-Equille, L., Srivastava, D.: Integrating conflicting data: The role of source dependence. In: VLDB (2009)

13. Dong, X.L., Berti-Equille, L., Srivastava, D.: Truth discovery and copying detection in a dynamic world. In: VLDB (2009)
14. Eckerson, W.W.: Data quality and the bottom line: Achieving business success through a commitment to high quality data. The Data Warehousing Institute (2002)
15. Elmagarmid, A.K., Ipeirotis, P.G., Verykios, V.S.: Duplicate record detection: A survey. TKDE 19(1) (2007)
16. English, L.: Plain English on data quality: Information quality management: The next frontier. DM Review Magazine (April 2000)
17. Fan, W.: Dependencies revisited for improving data quality. In: PODS (2008)
18. Fan, W., Gao, H., Jia, X., Li, J., Ma, S.: Dynamic constraints for record matching. VLDB J. 20(4), 495–520 (2011)
19. Fan, W., Geerts, F.: Capturing missing tuples and missing values. In: PODS, pp. 169–178 (2010)
20. Fan, W., Geerts, F.: Relative information completeness. TODS 35(4) (2010)
21. Fan, W., Geerts, F.: Foundations of Data Quality Management. Morgan & Claypool Publishers (2012)
22. Fan, W., Geerts, F., Jia, X.: Semandaq: A data quality system based on conditional functional dependencies. In: VLDB, demo (2008)
23. Fan, W., Geerts, F., Jia, X., Kementsietsidis, A.: Conditional functional dependencies for capturing data inconsistencies. TODS 33(1) (2008)
24. Fan, W., Geerts, F., Li, J., Xiong, M.: Discovering conditional functional dependencies. TKDE 23(5), 683–698 (2011)
25. Fan, W., Geerts, F., Ma, S., Müller, H.: Detecting inconsistencies in distributed data. In: ICDE, pp. 64–75 (2010)
26. Fan, W., Geerts, F., Wijsen, J.: Determining the currency of data. TODS (to appear)
27. Fan, W., Li, J., Ma, S., Tang, N., Yu, W.: Interaction between record matching and data repairing. In: SIGMOD (2011)
28. Fan, W., Li, J., Ma, S., Tang, N., Yu, W.: Towards certain fixes with editing rules and master data. VLDB J. 21(2), 213–238 (2012)
29. Fan, W., Li, J., Tang, N., Yu, W.: Incremental detection of inconsistencies in distributed data. In: ICDE (2012)
30. Fan, W., Libkin, L.: On XML integrity constraints in the presence of DTDs. J. ACM 49(3), 368–406 (2002)
31. Fan, W., Ma, S., Hu, Y., Liu, J., Wu, Y.: Propagating functional dependencies with conditions. In: VLDB, pp. 391–407 (2008)
32. Fan, W., Siméon, J.: Integrity constraints for XML. JCSS 66(1), 256–293 (2003)
33. Fellegi, I., Holt, D.: A systematic approach to automatic edit and imputation. J. American Statistical Association 71(353), 17–35 (1976)
34. Gartner. Forecast: Enterprise software markets, worldwide, 2008-2015, 2011 update. Technical report, Gartner (2011)
35. Golab, L., Karloff, H., Korn, F., Srivastava, D., Yu, B.: On generating near-optimal tableaux for conditional functional dependencies. In: VLDB (2008)
36. Herzog, T.N., Scheuren, F.J., Winkler, W.E.: Data Quality and Record Linkage Techniques. Springer (2009)
37. Loshin, D.: Master Data Management. Knowledge Integrity, Inc. (2009)
38. Miller, D.W., et al.: Missing prenatal records at a birth center: A communication problem quantified. In: AMIA Annu. Symp. Proc. (2005)
39. Naumann, F., Herschel, M.: An Introduction to Duplicate Detection. Morgan & Claypool Publishers (2010)

40. Otto, B., Weber, K.: From health checks to the seven sisters: The data quality journey at BT (September 2009), BT TR-BE HSG/CC CDQ/8
41. Redman, T.: The impact of poor data quality on the typical enterprise. Commun. ACM 2, 79–82 (1998)
42. SAS (2006), http://www.sas.com/industry/fsi/fraud/
43. Shilakes, C.C., Tylman, J.: Enterprise information portals. Technical report. Merrill Lynch, Inc., New York (November 1998)
44. Song, S., Chen, L.: Discovering matching dependencies. In: CIKM (2009)
45. Yakout, M., Elmagarmid, A.K., Neville, J., Ouzzani, M.: GDR: a system for guided data repair. In: SIGMOD (2010)
46. Zhang, H., Diao, Y., Immerman, N.: Recognizing patterns in streams with imprecise timestamps. In: VLDB (2010)

Construction of Web-Based, Service-Oriented Information Networks:
A Data Mining Perspective
(Abstract)

Jiawei Han

Department of Computer Science
University of Illinois at Urbana-Champaign
Urbana, IL 61801, U.S.A.
hanj@cs.uiuc.edu
https://www.cs.uiuc.edu/homes/hanj

Abstract. Mining directly on the existing networks formed by explicit webpage links on the World-Wide Web may not be so fruitful due to the diversity and semantic heterogeneity of such web-links. However, construction of service-oriented, semi-structured information networks from the Web and mining on such networks may lead to many exciting discoveries of useful information on the Web. This talk will discuss this direction and its associated research opportunities.

The World-Wide Web can be viewed as a gigantic information network, where webpages are the nodes of the network, and links connecting those pages form an intertwined, gigantic network. However, due to the unstructured nature of such a network and semantic heterogeneity of web-links, it is difficult to mine interesting knowledge from such a network except for finding authoritative pages and hubs. Alternatively, one can also view that Web is a gigantic repository of multiple information sources, such as universities, governments, companies, news, services, sales of commodities, and so on. An interesting problem is whether this view may provide any new functions for web-based information services, and if it does, whether one can construct such kind of semi-structured information networks automatically or semi-automatically from the Web, and whether one can use such new kind of networks to derive interesting new information and expand web services.

In this talk, we take this alternative view and examine the following issues: (1) what are the potential benefits if one can construct service-oriented, semi-structured information networks from the World-Wide Web and perform data mining on them, (2) whether it is possible to construct such kind of service-oriented, semi-structured information networks from the World-Wide Web automatically or semi-automatically, and (3) research problems for constructing and mining Web-Based, service-oriented, semi-structured information networks.

This view is motivated from our recent work on (1) mining semi-structured heterogeneous information networks, and (2) discovery of entity Web pages and their corresponding semantic structures from parallel path structures.

H. Gao et al. (Eds.): WAIM 2012, LNCS 7418, pp. 17–19, 2012.

First, real world physical and abstract data objects are interconnected, forming gigantic, interconnected networks. By structuring these data objects into multiple types, such networks become *semi-structured heterogeneous information networks*. Most real world applications that handle big data, including interconnected social media and social networks, scientific, engineering, or medical information systems, online e-commerce systems, and most database systems, can be structured into heterogeneous information networks. For example, in a medical care network, objects of multiple types, such as patients, doctors, diseases, medication, and links such as visits, diagnosis, and treatments are intertwined together, providing rich information and forming heterogeneous information networks. Effective analysis of large-scale heterogeneous information networks poses an interesting but critical challenge. Our recent studies show that the semi-structured heterogeneous information network model leverages the rich semantics of typed nodes and links in a network and can uncover surprisingly rich knowledge from interconnected data. This heterogeneous network modeling will lead to the discovery of a set of new principles and methodologies for mining interconnected data. The examples to be used in this discussion include (1) meta path-based similarity search, (2) rank-based clustering, (3) rank-based classification, (4) meta path-based link/relationship prediction, (5) relation strength-aware mining, as well as a few other recent developments.

Second, it is not easy to automatically or semi-automatically construct *service-oriented, semi-structured, heterogeneous information networks* from the WWW. However, with the enormous size and diversity of WWW, it is impossible to construct such information networks manually. Recently, there are progresses on finding entity-pages and mining web structural information using the structural and relational information on the Web. Specifically, given a Web site and an entity-page (e.g., department and faculty member homepage) it is possible to find all or almost all of the entity-pages of the same type (e.g., all faculty members in the department) by growing parallel paths through the web graph and DOM trees. By further developing such methodologies, it is possible that one can construct *service-oriented, semi-structured, heterogeneous information networks* from the WWW for many critical services. By integrating methodologies for construction and mining of such web-based information networks, the quality of both construction and mining of such information networks can be progressively and mutually enhanced.

Finally, we point out some open research problems and promising research directions and hope that the construction and mining of Web-based, service-oriented, semi-structured heterogeneous information networks will become an interesting frontier in the research into Web-aged information management systems.

References

1. Brin, S., Page, L.: The anatomy of a large-scale hypertextual web search engine. In: Proc. 7th Int. World Wide Web Conf. (WWW 1998), Brisbane, Australia, pp. 107–117 (April 1998)

2. Ji, M., Han, J., Danilevsky, M.: Ranking-based classification of heterogeneous information networks. In: Proc. 2011 ACM SIGKDD Int. Conf. on Knowledge Discovery and Data Mining (KDD 2011), San Diego, CA (August 2011)
3. Ji, M., Sun, Y., Danilevsky, M., Han, J., Gao, J.: Graph regularized transductive classification on heterogeneous information networks. In: Proc. 2010 European Conf. Machine Learning and Principles and Practice of Knowledge Discovery in Databases (ECMLPKDD 2010), Barcelona, Spain (September 2010)
4. Kleinberg, J.M.: Authoritative sources in a hyperlinked environment. J. ACM 46, 604–632 (1999)
5. Sun, Y., Aggarwal, C.C., Han, J.: Relation strength-aware clustering of heterogeneous information networks with incomplete attributes. PVLDB 5, 394–405 (2012)
6. Sun, Y., Barber, R., Gupta, M., Aggarwal, C., Han, J.: Co-author relationship prediction in heterogeneous bibliographic networks. In: Proc. 2011 Int. Conf. Advances in Social Network Analysis and Mining (ASONAM 2011), Kaohsiung, Taiwan (July 2011)
7. Sun, Y., Han, J.: Mining Heterogeneous Information Networks: Principles and Methodologies. Morgan & Claypool Publishers (2012)
8. Sun, Y., Han, J., Yan, X., Yu, P.S., Wu, T.: PathSim: Meta path-based top-k similarity search in heterogeneous information networks. In: Proc. 2011 Int. Conf. Very Large Data Bases (VLDB 2011), Seattle, WA (August 2011)
9. Sun, Y., Han, J., Zhao, P., Yin, Z., Cheng, H., Wu, T.: RankClus: Integrating clustering with ranking for heterogeneous information network analysis. In: Proc. 2009 Int. Conf. Extending Data Base Technology (EDBT 2009), Saint-Petersburg, Russia (March 2009)
10. Sun, Y., Yu, Y., Han, J.: Ranking-based clustering of heterogeneous information networks with star network schema. In: Proc. 2009 ACM SIGKDD Int. Conf. Knowledge Discovery and Data Mining (KDD 2009), Paris, France (June 2009)
11. Wang, C., Han, J., Jia, Y., Tang, J., Zhang, D., Yu, Y., Guo, J.: Mining advisor-advisee relationships from research publication networks. In: Proc. 2010 ACM SIGKDD Conf. Knowledge Discovery and Data Mining (KDD 2010), Washington D.C. (July 2010)
12. Weninger, T., Danilevsky, M., Fumarola, F., Hailpern, J., Han, J., Ji, M., Johnston, T.J., Kallumadi, S., Kim, H., Li, Z., McCloskey, D., Sun, Y., TeGrotenhuis, N.E., Wang, C., Yu, X.: Winacs: Construction and analysis of web-based computer science information networks. In: Proc. 2011 ACM SIGMOD Int. Conf. Management of Data (SIGMOD 2011) (system demo), Athens, Greece (June 2011)
13. Weninger, T., Fumarola, F., Lin, C.X., Barber, R., Han, J., Malerba, D.: Growing parallel paths for entity-page discovery. In: Proc. 2011 Int. World Wide Web Conf. (WWW 2011), Hyderabad, India (March 2011)

Electronic Textbooks and Data Mining

Rakesh Agrawal, Sreenivas Gollapudi, Anitha Kannan, and Krishnaram Kenthapadi

Search Labs, Microsoft Research,
Mountain View, CA, USA

Abstract. Education is known to be the key determinant of economic growth and prosperity [8,12]. While the issues in devising a high-quality educational system are multi-faceted and complex, textbooks are acknowledged to be the educational input most consistently associated with gains in student learning [11]. They are the primary conduits for delivering content knowledge to the students and the teachers base their lesson plans primarily on the material given in textbooks [7].

With the emergence of abundant online content, cloud computing, and electronic reading devices, textbooks are poised for transformative changes. Notwithstanding understandable misgivings (e.g. Gutenberg Elegies [6]), textbooks cannot escape what Walter Ong calls 'the technologizing of the word' [9]. The electronic format comes naturally to the current generation of 'digital natives' [10]. Inspired by the emergence of this new medium for "printing" and "distributing" textbooks, we present our early explorations into developing a data mining based approach for enhancing the quality of electronic textbooks. Specifically, we first describe a diagnostic tool for authors and educators to algorithmically identify deficiencies in textbooks. We then discuss techniques for algorithmically augmenting different sections of a book with links to selective content mined from the Web.

Our tool for diagnosing deficiencies consists of two components. Abstracting from the education literature, we identify the following properties of good textbooks: (1) *Focus* : Each section explains few concepts, (2) *Unity*: For every concept, there is a unique section that best explains the concept, and (3) *Sequentiality*: Concepts are discussed in a sequential fashion so that a concept is explained prior to occurrences of this concept or any related concept. Further, the tie for precedence in presentation between two mutually related concepts is broken in favor of the more significant of the two. The first component provides an assessment of the extent to which these properties are followed in a textbook and quantifies the comprehension load that a textbook imposes on the reader due to non-sequential presentation of concepts [1,2]. The second component identifies sections that are not written well and can benefit from further exposition. We propose a probabilistic decision model for this purpose, which is based on the syntactic complexity of writing and the notion of the dispersion of key concepts mentioned in the section [4].

For augmenting a section of a textbook, we first identify the set of key concept phrases contained in a section. Using these phrases, we find web articles that represent the central concepts presented in the section and endow the section with links to them [5]. We also describe techniques for finding images that are most relevant to a section of the textbook, while respecting the constraint that the same image is not repeated in different sections of the same chapter. We pose this problem of matching images to sections in a textbook chapter as an optimization problem and present an efficient algorithm for solving it [3].

H. Gao et al. (Eds.): WAIM 2012, LNCS 7418, pp. 20–21, 2012.

We finally provide the results of applying the proposed techniques to a corpus of widely-used, high school textbooks published by the National Council of Educational Research and Training (NCERT), India. We consider books from grades IX–XII, covering four broad subject areas, namely, Sciences, Social Sciences, Commerce, and Mathematics. The preliminary results are encouraging and indicate that developing technological approaches to embellishing textbooks could be a promising direction for research.

References

1. Agrawal, R., Chakraborty, S., Gollapudi, S., Kannan, A., Kenthapadi, K.: Empowering authors to diagnose comprehension burden in textbooks. In: KDD (2012)
2. Agrawal, R., Chakraborty, S., Gollapudi, S., Kannan, A., Kenthapadi, K.: Quality of textbooks: An empirical study. In: ACM DEV (2012)
3. Agrawal, R., Gollapudi, S., Kannan, A., Kenthapadi, K.: Enriching textbooks with images. In: CIKM (2011)
4. Agrawal, R., Gollapudi, S., Kannan, A., Kenthapadi, K.: Identifying enrichment candidates in textbooks. In: WWW (2011)
5. Agrawal, R., Gollapudi, S., Kenthapadi, K., Srivastava, N., Velu, R.: Enriching textbooks through data mining. In: ACM DEV (2010)
6. Birkerts, S.: The Gutenberg Elegies: The Fate of Reading in an Electronic Age. Faber & Faber (2006)
7. Gillies, J., Quijada, J.: Opportunity to learn: A high impact strategy for improving educational outcomes in developing countries. USAID Educational Quality Improvement Program, EQUIP2 (2008)
8. Hanushek, E.A., Woessmann, L.: The role of education quality for economic growth. Policy Research Department Working Paper 4122. World Bank (2007)
9. Ong, W.J.: Orality & Literacy: The Technologizing of the Word. Methuen (1982)
10. Prensky, M.: Digital natives, digital immigrants. On the Horizon 9(5) (2001)
11. Verspoor, A., Wu, K.B.: Textbooks and educational development. Technical report. World Bank (1990)
12. World-Bank. Knowledge for Development: World Development Report: 1998/99. Oxford University Press (1999)

Topology-Based Data Compression
in Wireless Sensor Networks

Shangfeng Mo[1,2,3], Hong Chen[1,2,*], and Yinglong Li[1,2,3]

[1] Key Laboratory of Data Engineering and Knowledge Engineering of MOE,
Renmin University of China, Beijing 100872, China
[2] School of Information, Renmin University of China, Beijing 100872, China
[3] Hunan University of Science and Technology, Xiangtan 411201, China
moshangfengxy@yahoo.com.cn, chong@ruc.edu.cn,
liyinglong518@126.com

Abstract. In this paper, we address the problem of Data Compression which is critical in wireless sensor networks. We proposed a novel Topology-based Data Compression (TDC) algorithm for wireless sensor networks. We utilize the topological structure of routing tree to reduce the transmission of message packets. We analyzed the differences and relations between our algorithm and other compression algorithms. Extensive experiments are conducted to evaluate the performance of the proposed TDC approach by using two kinds of data sets: real data set and synthetic data set. The results show that the TDC algorithm substantially outperforms Non-compression algorithm in terms of packets transmitted.

Keywords: WSNs, topology-based, data compression.

1 Introduction

With the development of microelectronics, embedded computing and wireless communication technology, sensor hardware manufacturing ability is also improved. Low-power, inexpensive sensor nodes can be integrated with information collection, data processing and wireless communication functions [1]. Wireless sensor networks (WSNs) composed by a large number of sensor nodes deployed in the monitoring regions, are used to collect and process information of perceived objects.

The sensor nodes are usually battery-powered and deployed in harsh physical environments. It is usually impossible to replace the batteries and the nodes. So the goal of data compression in wireless sensor networks (WSNs) is to reduce the energy consumption and prolong the network lifetime. Compare with the calculation, the communication between sensor nodes consume much more energy [2]. So the key problem of saving the energy consumption is to reduce the amount of data transmission.

---

* Corresponding author.

H. Gao et al. (Eds.): WAIM 2012, LNCS 7418, pp. 22–34, 2012.

In this paper, we focus on Topology-based Data Compression algorithm (TDC) for wireless sensor networks. Our TDC algorithm will minimize the amount of data during transmission, and the sink can obtain the accurate information.

Our contributions are summarized as follows:

(1) We utilize the topological structure of routing tree to reduce the transmission of common information including node id, sequence number, etc. Whether all nodes sending their data to the sink or parts of nodes sending their data to the sink, our compression algorithm TDC is basically better than the Non-compression Algorithm.

(2) We analyzed the differences and relations between our compression algorithm TDC and other compression algorithms, such as wavelet compression algorithms [3] [4], Huffman encoding algorithms [5] [6], etc.

(3) Extensive experiments are conducted to evaluate the performance of the proposed TDC approach by using two kinds of data sets: real data set and synthetic data set. The results provide a number of insightful observations and show that TDC algorithm substantially outperforms Non-compression algorithm in terms of packets transmitted under various network configurations.

The remainder of this paper is organized as follows. Section 2 summarizes related work. Section 3 introduces some assumptions. Section 4 describes the proposed TDC scheme in details. Section 5 presents experimental results. Section 6 concludes the paper.

2   Related Work

There are many compression algorithms in wireless sensor networks, such as wavelet compression algorithms [3] [4], Huffman encoding algorithms [5] [6], etc.

Wavelet compression algorithms [3] [4] utilize the wavelet transformation to achieve the data compression. The goal is to achieve large compression ratio and maintain a certain degree of accuracy. Wavelet compression algorithms will obtain a good approximation ability and high compression ratio. But the resulting values of the sink obtained are approximate and not completely accurate, which is not the same as our compression algorithm. Our compression algorithm is to reduce the amount of transmission of useless data, and the data which the sink obtained are accurate.

The core idea of Huffman encoding algorithms [5] [6] are to encode the data of sensor nodes using the shorter codes, as the sensor nodes are farther away from the base station or the data occur more frequently. Our compression algorithm is to reduce the amount of transmission of useless data. Therefore, there are no conflict between Huffman encoding algorithms and our compression algorithm. The normalized data based on our compression algorithm can be encoded using Huffman encoding algorithms, which will further increase the compression ratio.

3   Preliminaries

In this paper, there are N sensor nodes randomly deployed in an area, which constitute a network by self-organized manner. The sensor nodes sample the data periodically.

Each sampling period is called a round. Sink node continuously requests the data in every sampling period. The i-th sensor node is denoted by si and the corresponding sensor nodes set $S = \{s1, s2, ..., sn\}$, $|S| = N$. Each si has the maximum communication radius R.

We make the following assumptions:

1. All ordinary sensor nodes are homogeneous and have the same capabilities. When all nodes are deployed, they will be stationary, and each one has a unique identification (ID).
2. There is only one sink (base station), and the sink node can be recharged.
3. Links are symmetric. If node si can communicate with node sj, node sj can also communicate with node si.
4. The energy resource of ordinary sensor nodes is highly-limited and unreplenished.

4   The TDC Scheme

In recent years, there are many routing algorithms appeared in the wireless sensor networks (WSNs). In general, the routing protocols are classified into two types: flat routing protocols and hierarchical routing protocols. TAG [7] and HEAR [8] belong to the flat routing protocols; LEACH [9], EEUC [10], HEED [11] and CCEDG [12] belong to the hierarchical routing protocols. Flat routing protocols are tree-based routing approach mostly, and they route data to the sink node through a multi-hop network. In hierarchical routing protocols, sensor nodes construct clusters for routing and then data transmission occurs as two steps, i.e., intra-cluster routing and inter-cluster routing [13].

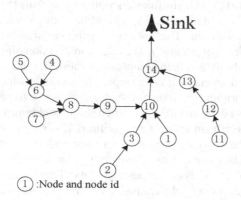

Fig. 1. An example of TAG routing tree

In TAG [7] protocol, as shown in Figure 1, firstly, sink sk broadcasts a **building** message to build the tree. The node si which received the **building** message may be the child of the sender sk. Then the child node si will transfer the **building** message to its neighbors, and so on. The child node si also selects an optimal parent node to send back the **response** message for building the routing tree. We will describe our Data

Compression Algorithm based on the topology of TAG routing protocol, and it is also effective to other topologies: flat routing protocols and hierarchical routing protocols.

4.1 Overview of TDC Algorithm

The sensory data usually consists of two parts. One part is common information, such as node id, date and time, sequence number or epoch, etc. Another part is the sensory attributes. For example, in the real dataset [14], the common information include node id, date and time, epoch; and the sensory attributes include temperature, humidity, light, etc.

In the traditional data transmission process, the data is transmitted along the routing tree in turn. As shown in Figure 1, if node 4 wants to send its data to the sink, its data will be sent to its parent node 6 first, then node 8, 9, 10, 14, and finally the sink. In this process, the useful information is the sensory attributes, and the common information transmitted to the sink only make the sink identifying the data belong to. If a node only transmits its sensory attributes and a small number of other auxiliary information to the sink, the sink can also identify the owner of the sensory attributes based on the auxiliary information. At the same time, the bytes which the auxiliary information occupied are less than the common information. This method will reduce amount of packets transmitted and save the energy consumption.

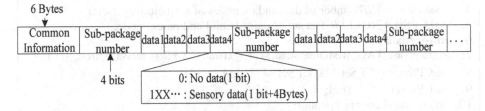

Fig. 2. The data package of a node

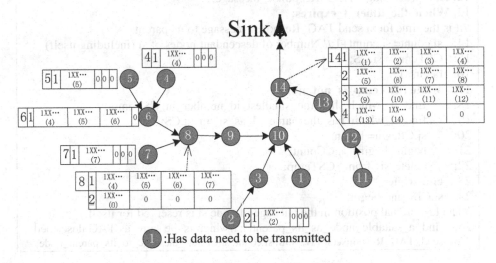

Fig. 3. All nodes need to send their data to the sink

As shown in Figure 2, there is a data package of a node *si*. The data of descendant nodes will be encapsulated into the package. Every *m* (*m*=4) data of descendant nodes forms a sub-package, and each sub-package has a sub-package number. If the descendant node *sj* sends a data to the sink, the data will be denoted as 1XX··· (1 bit + sizeof(attribute values)), otherwise denoted as 0.

As shown in Figure 3, all nodes need to send their data to the sink. The leaf nodes in the tree, such as node 4, 5, 7, etc, have no descendant node. There is only one sub-package in the data package and the sub-package number starts with 1. The intermediate nodes in the tree, such as node 6, 8, 9 etc, have one or more than one descendant nodes. The intermediate node 8 has two child nodes 6 and 7, and the node 6 has two child nodes 4 and 5. The number of descendant nodes of node 8 (including node 8) is 5. The order of the descendant nodes in the node 8 package is: 4, 5, 6, 7 and 8.

4.2   The Detailed TDC Algorithm

We will describe the constructing process of our Topology-based Data Compression (TDC) algorithm in this subsection.

```
1:  initialize
2:    si.PN = NULL;// Parent Node of si;
3:    si.CS = NULL;//Child nodes set of si is null;
4:    si.count = 0; //Number of descendant nodes of si (including itself);
5:    si.CPSet = NULL; // The set of pair of child and parent node.
6:  end-initialize
7:  on_receiving_TAG_Response(si. id, sj.id, sj.count, sj. CPSet)_message_from_sj ( );
8:    si.CPSet = si.CPSet ∪ sj.CPSet ∪ { sj.id, si. id};
9:    si.CS= si.CS ∪ sj.id;
10:   si.count= si.count+ sj.count;
11:   sj.CCount = sj.count;   //save the number of descendant nodes of child node sj.
12: end_processing_ TAG_Response_message;
13: When_the_timer_t_expires;
//it is the time for si send TAG_Response message to its parent.
14:   si.count= si.count+1;// Number of descendant nodes of si (including itself)
15:   CSTemp = si.CS;
16:   begin=1;
17:   while (CSTemp_is_not_null)
18:       find_sj_which_ has_the_smallest_id_number_in_CSTemp;
19:       find_sj_which_has_the_same_id_as_sj_in_si.CS;
20:       sj.CBegin= begin;
21:       begin= begin+ sj.CCount;
22:       delete_sj_from_ CSTemp;
23:   end-while
24:   si.CBegin= begin;
//The last ordinal position in the package of node si is reserved for itself.
25:   find_a_suitable_node_as_its_parent(); //which is the same as TAG described.
26:   send_TAG_Response(si.PN, si.id, si.count, si.CPSet)_message_to_its_parent_node;
```

Procedure 1. TDC algorithm

Terminologies:

si.PN: Parent Node of *si*;

si.CS: Child nodes set of *si*;

si.id: Node id of *si*;

si.count: Number of descendant nodes of *si* (including itself);

sj.CBegin: The starting position of descendant nodes of node *sj* in the package of node *si*;

sj.CCount: The number of descendant nodes of child node *sj* (including *sj*);

si.CPSet: The set of pair of child and parent node, which will be used by sink to construct the routing map.

Our TDC algorithm is reflected in the improved TAG response procedure, as shown in procedure 1. When parent node *si* receives a TAG_**Response** message from its child node *sj*, parent node *si* will extract the id and the number of descendant nodes of child node *sj*. Parent node *si* will also merge the pair of child and parent node { *sj.id, si. id*} to its set of pair of child and parent node *si.CPSet*. Parent node *si* will save the id of child node *sj* and the number of descendant nodes of *sj*, as well as cumulate the number of descendant nodes of *si* (Lines 7-12). When the timer t expires, parent node *si* will compute the ordinal position of child node *sj* in the package of parent node *si*. The smaller the child node id is, the more the probability is arranged in the front of the package of node *si*. The parent node *si* will select an optimal node as its parent node and send back the **TAG_Response** message (Lines 13-26). In the TAG routing tree formation process, the set of pair of child and parent node will send to the sink ultimately (Line 26). So the sink will know the whole network topology map. Finally, the sink will restore the compressed data according to the topology map.

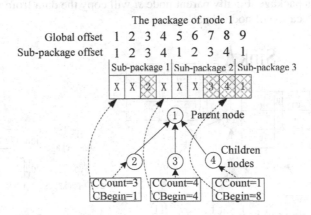

Fig. 4. The order of descendant nodes

Next, we will show an example of computing the ordinal position of descendant nodes in the package of parent node *si*. As shown in Figure 4, the global offset arranges in ascending order starting from 1, and the sub-package offset arranges from 1 to 4 and repeated. Parent node 1 has 3 child nodes, which are node 2, 3 and 4 (*s1.CS* = {2, 3, 4}). Parent node 1 has 9 descendant nodes (including itself), which means *s1*.count=*s2*.CCount+*s3*.CCount+*s4*.CCount+1=3+4+1+1=9. Child node 2 has 3 descendant nodes (including node 2) ($s2$.CCount=3). The starting position of descendant nodes of child node 2 in the package of parent node 1 is 1 ($s2$.CBegin=1; Global offset=1). Based on the **Procedure 1**, the ordinal position of child node 2 itself in the package of parent node 1 is 3 (Global offset = 3), which is the last position in all descendant nodes of node 2. As shown in Figure 4, the shadow lines parts represent the ordinal positions of child node 2, 3, 4 and parent node 1 themselves in the package of parent node 1.

Steady phase:
1: **if** si receiving_**Data_Package**_message_from_sj
2:    **if** sj.id ∈ si.CS
3:        **for** (sj.CBegin to (sj.CBegin +sj.CCount))
4:            compute_the_corresponding_subpackage_number_and_subpackage_offset( );
5:            copy_data_of_sj_to_package_of_si( );
6:        **end-for**
7:    **end-if**
8: **end-if**

Procedure 2. Data Package transmission procedure

Next, we will describe the Data Package message transmission in the steady phase. As shown in procedure 2, when parent node *si* receives a **Data Package** message from its child node *sj*, parent node *si* will extract the data package of child node *sj*, and obtain the sub-package number and the sub-package offset of each data. Then parent node *si* will compute the corresponding sub-package number and sub-package offset in its own package. Finally parent node *si* will copy the data from node *sj* to the corresponding location of node *si* package.

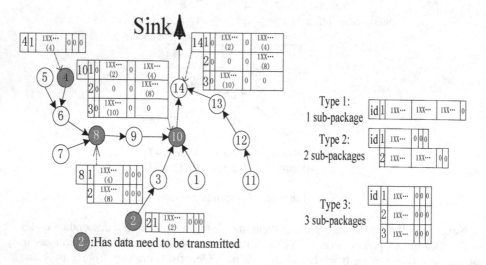

Fig. 5. Parts of nodes send their data to the sink

Fig. 6. 3 types of the package

Now, let us to describe the process of computing sub-package number and sub-package offset. We set the global offset to be o which starting from 1. We set the sub-package number to be p which starting from 1. We set the sub-package length to be *length*. We set the sub-package offset to be q which starting from 1. As shown in Figure 4, there are 4 data in a sub-package. So the *length* is 4. The global offset of child node 3 itself in the package of parent node 1 is 7 (Global offset $o = 7$). The sub-package number of child node 3 itself in the package of parent node 1 is 2 ($p=2$). The sub-package offset of child node 3 itself in the package of parent node 1 is 3 ($q=3$).

Assume a node *sm* is one of descendant nodes of node *sj*, and node *sj* is one of child nodes of node *si*. The sub-package number and the sub-package offset of descendant node *sm* in the package of child node *sj* are p and q respectively. The global offset o_{jm} of descendant node *sm* in the package of child node *sj* will be $o_{jm}=(p-1)*$ *length*$+q$. The global offset o_{im} of descendant node *sm* in the package of parent node *si* will be $o_{im} =sj.CBegin -1+o_{jm}=sj.CBegin -1+(p-1)*$ *length*$+q$. The sub-package number p_{im} of descendant node *sm* in the package of parent node *si* will be *ceil*$(o_{im}$ / *length*$)$. The function *ceil(A)* returns a minimum integer which is greater than or equal to *A*. For example: *ceil(5/ 4)=2*. The sub-package offset q_{im} of descendant node *sm* in the package of parent node *si* will be o_{im} *mod length; if(*o_{im} *mod length* $== 0$) *return length*. For example: 5 *mod* 4 =1. Finally, the data of descendant node *sm* will be copied to the location which is the p_{im}*th* sub-package and the q_{im}*th* sub-package offset of the package of parent node *si*.

Figure 5 shows an example of parts of data transmitted, where only node 2, 4, 8 and 10 need to send their data to the sink.

4.3   Performance of the TDC Algorithm

Now, we will analyze the performance of the TDC Algorithm.

If node *si* has 3 data needed to send to the sink, there are 3 types of the package. The parameters are shown in the table 1. As shown in the Figure 6, type 1 has 1 sub-package, and the bits want to be transmitted is: $6*8+(4+(1+4*8)+(1+4*8)+(1+4*8)+1)=152$; type 2 has 2 sub-packages, and the bits want to be transmitted is: $6*8+(4+(1+4*8)+1+1+1)+(4+(1+4*8)+(1+4*8)+1+1)=160$; type 3 has 3 sub-packages, and the bits want to be transmitted is: 168. If we use the Non-compression Algorithm, it means that each node will send common information and sensory attributes to the sink. The common information occupies 6 bytes and the sensory attribute occupies 4 bytes. The data transmitted using non-compression algorithm is: $(6+4)*8*3=240$ bits, which is greater than using TDC algorithm.

As shown in the Table 1, there is the performance of the TDC Algorithm and Non-compression Algorithm. If node *si* has *n* data needed to send to its parent node, and each data occupies a sub-package, such as the type 3 in Figure 6, there will be *n* sub-packages in the package of node *si*, which has the maximum number of sub-packages. The bits want to be transmitted is: $48+(4+33+1+1+1)*n=48+40*n$. If every sub-package is full, such as the type 1 in Figure 6, there will be *n/4 (if n mod 4 == 0) or ceil (n/4) (if n mod 4 != 0)* sub-packages in the package of node *si*, which is the minimum number of sub-packages. If *n mod 4 == 0*, the bits want to be transmitted is: $48+ (n/4)*136$, *if n mod 4 != 0*, the bits want to be transmitted is: $48+((ceil (n/4))-1)*136+(n mod 4)*32+8$. The bits want to be transmitted using the Non-compression Algorithm is $80*n$.

Table 1. The performance of the TDC Algorithm and Non-compression Algorithm

Parameters:	Common information includes node id (2 bytes) and sequence number or epoch (4 bytes). So common information occupied 6 bytes. Sub-package number occupied 4 bits. There is only one sensory attribute value which occupied 4 bytes. Data transmitted: 1XX... (1 bit+4Bytes); no data transmitted: 0 (1 bit).		
Number of data need to be transmitted	Number of sub-packages	TDC Algorithm (bits)	Non-compression Algorithm (bits)
1	1	88	80
2	1	120	160
	2	128	160
3	1	152	240
	2	160	240
	3	168	240
4	1	184	320
	2	192	320
	3	200	320
	4	208	320
n	minimum number of sub-packages: if n mod 4 == 0: $n/4$ else $ceil\ (n/4)$	if n mod 4 == 0: 48+$(n/4)$*136; if n mod 4 != 0: 48+$((ceil\ (n/4))-1)$*136+$(n$ mod 4)*32+8	80*n
	maximum number of sub-packages: n	48+40*n	80*n

If our TDC Algorithm is effective compare to the Non-compression Algorithm, the maximum bits transmitted should be less than the ones using Non-compression Algorithm, which means: $48+40*n< 80*n$. It can be derived that when $n>(48/40=1.2)$, which means that when n is greater than or equal to 2, our TDC Algorithm is better than Non-compression Algorithm.

If n is large enough, in the worst case, each data occupies a sub-package, the compression ratio is: $1-(48+40*n)/(80*n) \approx 1-(40*n)/(80*n)= 50\%$. In the best case, every sub-package is full (exclude the last sub-package), the compression ratio is: $1-(48+(n/4)*136)/(80*n) \approx 1-(34*n)/(80*n)=57.5\%$.

5   Simulation Results

To analyze the performance of our algorithm, we conduct experiments using omnetpp-4.1. We use two kinds of data sets. One is a real dataset collected by the Intel Berkeley Research Lab [14]. We use 1000 records from the real data set.

Another data set is a synthetic data set. We randomly deployed 300 homogeneous sensor nodes in the 400*400 m^2 rectangular region and sink is located at the center.

Our scheme is based on the TAG [7] routing algorithm, and assume communication links are error-free as well as MAC layer is ideal case. To compute the packets transmitted, we define a sampling period as a round.

The transmission range of sink node is usually greater than the transmission range of ordinary sensor nodes, so we assume that the transmission range of sink node can cover most regions of the monitoring networks.

The size of a message packet is 30 bytes. Other parameters are as shown in the table 1.

As mentioned above, energy consumption is a critical issue for wireless sensor networks and radio transmission is the most dominate source of energy consumption. Thus, we use the following two metrics to measure the performance in wireless sensor networks.

The Average Number of Packets Sent: the average number of packets sent by each node each round.

The Network Lifetime: the network lifetime refers to the number of rounds from the first round to the rounds of a node first failure. From the perspective of network availability, the network lifetime is a more useful metric than the average number of packets sent. The more the number of packets sent by a node, the more energy consumed, the faster the node died, the shorter the network lifetime is. So we use **the maximum number of packets sent** to measure **the network lifetime.**

The maximum number of packets sent means that a node which sends the largest number of packets per round to the sink among all nodes in wireless sensor networks.

Comparison of these two metrics, **the maximum number of packets sent** is more important than **the average number of packets sent**.

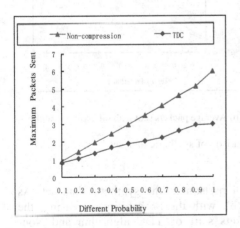 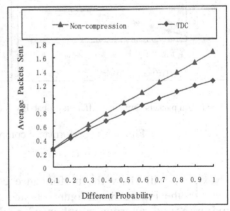

(a) Maximum packets sent with different probability   (b) Average packets sent with different probability

Fig. 7. The performance comparison of real data set

To simulate parts of nodes sending their data to the sink, we set a parameter which is **probability**. If **probability** is 0.1, only 10 percent of the nodes will send their data to the sink.

First, we will introduce the performance comparison in the real data set. As shown in the Figure 7 subfigure (a) and (b), with the probability increasing, the **maximum and average number of packets sent** of TDC algorithm and Non-compression algorithm increases too, and the performance of TDC algorithm is better than the Non-compression algorithm.

As described in section 4, the compression ratio fluctuates in the range *50%-57.5%*. Experimental results show that the experimental compression ratio is a little smaller than the theoretical. Why? The definition of package is different with the definition of message packet. As shown in Figure 2, if the package can be encapsulated into a message packet, it will be sent in one message packet. Otherwise, it may split into multiple message packets. When the probability equals to 1, the maximum number of packets sent of TDC algorithm will reduce to 50% than Non-compression algorithm, which means the network lifetime will extend 50% than Non-compression algorithm.

In summary, TDC algorithm is better than Non-compression algorithm, even the network size is small and where deployed few nodes.

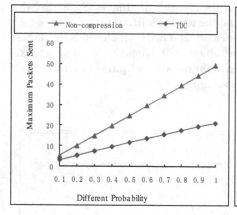 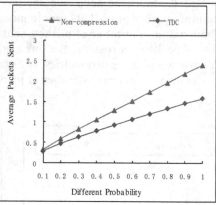

(a) Maximum packets sent with different probability   (b) Average packets sent with different probability

Fig. 8. The performance comparison of synthetic data set

Next, we will show the performance comparison in the synthetic data set. As shown in the Figure 8 subfigure (a) and (b), with the probability increasing, the **maximum and average number of packets sent** of TDC algorithm and Non-compression algorithm increases too, and the performance of TDC algorithm is better than the Non-compression algorithm.

6   Conclusions and Future Work

In this paper, we proposed a Topology-based Data Compression algorithm (TDC) in wireless sensor networks. The goal of the scheme is to save the energy consumption and extend the network lifetime. Our experimental result shows that the proposed TDC scheme can reduce the number of packets transmitted as well as extend the network lifetime.

In the future, we plan to extend the proposed Data Compression Algorithm to other aggregate functions, such as *join*, average, and sum.

Acknowledgements. This research was supported by the National Natural Science Foundation of China (61070056, 61033010), the HGJ Important National Science & Technology Specific Projects of China (2010ZX01042-001-002-002) and the Doctoral Fund of Ministry of Education of China.

References

1. Institute of Electrical and Electronics Engineers Inc. Ten emerging technologies that will change your world. IEEE Engineering Management Review 32, 20–30 (2004)
2. Zhang, H., Wu, Z., Li, D., Chen, H.: A Sampling-Based Algorithm for Approximating Maximum Average Value Region in Wireless Sensor Network. In: 2010 39th International Conference on Parallel Processing Workshops, ICPPW, pp. 17–23 (2010)
3. Rein, S., Reisslein, M.: Performance evaluation of the fractional wavelet filter: A low-memory image wavelet transform for multimedia sensor networks. Ad Hoc Networks 9(4), 482–496 (2011)
4. Xie, Z.-J., Wang, L., Chen, H.: Algorithm of Voronoi Tessellation Based Data Compression over Sensor Networks. Ruan Jian Xue Bao/Journal of Software 20(4), 1014–1022 (2009) (Language: Chinese)
5. Reinhardt, A., Christin, D., Hollick, M., Schmitt, J., Mogre, P.S., Steinmetz, R.: Trimming the Tree: Tailoring Adaptive Huffman Coding to Wireless Sensor Networks. In: Silva, J.S., Krishnamachari, B., Boavida, F. (eds.) EWSN 2010. LNCS, vol. 5970, pp. 33–48. Springer, Heidelberg (2010)
6. Yeo, M., Seong, D., Cho, Y., Yoo, J.: Huffman Coding Algorithm for Compression of Sensor Data in Wireless Sensor Networks. In: ICHIT 2009 - International Conference on Convergence and Hybrid Information Technology, pp. 296–301 (2009)
7. Madden, S., Franklin, M.J., Hellerstein, J., Hong, W.: TAG: a Tiny Aggregation Service for Ad-Hoc Sensor Networks. In: Proc. Usenix Fifth Symp. Operating Systems Design and Implementation (OSDI 2002), pp. 131–146 (December 2002)
8. Wang, J., Cho, J., Lee, S., Chen, K.-C., Lee, Y.-K.: Hop-based Energy Aware Routing Algorithm   forWireless   Sensor   Networks.   IEICE   Transactions   on Communications E93B(2), 305–316 (2010)
9. Heinzelman, W., Chandrakasan, A., Balakrishnan, H.: Energy efficient communication protocols for wireless microsensor networks. In: Proceedings of the 33rd Hawaiian International Conference on Systems Science (January 2000)
10. Li, C., Ye, M., Chen, G., Wu, J.: An Energy-Efficient Unequal Clustering Mechanism for Wireless Sensor Networks. In: IEEE International Conference on Mobile Adhoc and Sensor Systems Conference, November 7-10, pp. 1–8. IEEE Press, Washington (2005)

11. Younis, O., Fahmy, S.: HEED: a hybrid, energy-efficient, distributed clustering approach for ad hoc sensor networks. IEEE Transactions on Mobile Computing 3(4), 660–669 (2004)
12. Mo, S., Chen, H.: Competition-based Clustering and Energy-saving Data Gathering in Wireless Sensor Networks. In: 1st IET International Conference on Wireless Sensor Network, IET-WSN 2010 (2010)
13. Kim, D.-Y., Cho, J., Jeong, B.-S.: Practical Data Transmission in Cluster-Based Sensor Networks. KSII Transactions on Internet and Information Systems 4(3) (June 2010)
14. http://db.csail.mit.edu/labdata/labdata.html

A Residual Energy-Based Fairness Scheduling MAC Protocol for Wireless Sensor Networks[*]

Long Tan

School of Computer Science and Technology,
Heilongjiang University, 150080 Harbin, China
Tanlong01@163.com

Abstract. The Residual Energy-Based Fairness Scheduling MAC Protocol presented in this paper is aimed to solve the problems of how to ensure a balance energy consumption of WSN, how to reduce the collisions from interfering nodes in wireless sensor networks by introducing the fluid fair model into the wireless sensor network. We provide a conception of residual energy, and distributed energy fair scheduling MAC protocol for nodes to maintain a neighbor node's residual-energy queue so as to preserve the long-term balance energy consuming for every wireless sensor node and avoid the emerging of the network Island. The effectiveness of the protocol is validated through the simulation test to ensure the longest lifetime of WSN.

Keywords: wireless sensor network, Fair scheduling, residual energy.

1 Introduction

Wireless Sensor Network (WSN) has gained widespread attention in recent years. They can be used for testing, sensing, collecting and processing information of monitored objects and transferring the processed information to users[1]. The network has a wide range applications including health, military, and security.

Since Wireless Sensor Network works in a shared-medium, medium access control (MAC) is an important technique that enables the successful operation of the network. One fundamental task of the MAC protocol is how to avoid collisions among a large number of neighbor nodes which communicate each other at the same time in a share-medium. There are many MAC protocols that have been developed for wireless voice and data communication networks like IEEE802.11[2].

However, sensor nodes are constrained in energy supply and bandwidth[3]. Since WSN node is usually very cheap, charging or recharging the battery is not usually feasible and necessary. Therefore, how to maximize the node's battery lifetime is very important in WSN. Many research works have been finished to solve energy problem in [4][5][6][7][8].These works are mainly focused on MAC layer. Of course, we must know challenges necessitate energy-awareness at all layers of networking protocol

---

[*] Supported by the Science Technology Research Project of Heilongjiang educational office under Grant No.11551349.

H. Gao et al. (Eds.): WAIM 2012, LNCS 7418, pp. 35–46, 2012.

stack. For example, Network layer protocol, energy aware routing protocol [9], Directed Diffusion Protocol and so on., are all designed for reducing the energy consumption of nodes. Many solutions to the problem of idle listening have been proposed utilizing the technique of duty cycling [10][11].

In this paper, the fluid fair model of the traditional wireless channel is introduced into the wireless sensor network. By utilizing a variation fair scheduling algorithm of Start time Fair Queuing (STFQ) [12]based on the S-MAC protocol, we propose a fairness scheduling model based on the residual energy to realize a balanced consumption of node energy in WSN and eventually to ensure the longest lifetime of WSN. Specifically, we make the following work: In the event-driven WSN, the number of nodes responding to an event and the event-related neighbor nodes maintain neighbor node's residual-energy queue. When one node and all its neighboring nodes sense a event, according to their residual energy hierarchy, the nodes that have more residual energy have the priority to contending the channel and transmit the information. The priority is based on the distributed Coordination Function in the IEEE802.11 stand that the higher energy node has shorter back off time when conflicts occur and thus get the priority to transmit the report. In addition, the residual energy of nodes is introduced into the mechanism of data transmission, period sleep and collision handling in order to ensure a balanced energy consumption, prolong the lifetime of WSN and prevent the network island.

Meanwhile, in order to ensure the redundancy and reliability, in the sensor network, N sensor nodes are distributed in the transmitting station, When N neighboring nodes sense an event at the same time , not all nodes need to transmit report of it . Only R of N [11]need to report and N-R nodes come into the sleep state and give up transmission to save the energy so that the energy consumption is balanced. Finally, the protocol of this paper is introduced on the basis of S-MAC and the efficiency is tested through experiments.

The rest of the paper is organized as follows. Section Two introduces related work on wireless fair scheduling in Wireless Packet Switch Network and sensor networks' contention-based MAC protocol, their requirements and our ongoing research project. The protocol that this paper proposed is detailed in Section Three. Finally, simulations and results are presented in Section Four, and concluding in Section Five.

2   Related Work

The focus of this paper is to combine the traditional Fair Scheduling in Wireless Packet Switch Network with the MAC protocol of WSN in order to establish a fairness scheduling model based on residual energy to achieve the fair schedule of node energy in WSN.

2.1   Fair Scheduling in Wireless Packet Switch Network

The Fair Scheduling in Wireless Packet Switch Network aims to solve the conflicts of achieving the fairness. As we know, the locality of wireless transmission implies that collisions, and hence contention for the shared medium, are location dependent. When a single logical wireless channel is shared among multiple contending flows, due to the location-dependent contention and multi-hop nature of network, any two flows that are not interfering with each other can potentially transmit data packets over the physical channel simultaneously which leads to the conflicts and invalid service. So how to assign a minimum channel allocation to each flow proportional to its weight, and maximize the aggregate channel utilization is the goal of the fair scheduling fluid model[13].

In the fluid model, the granularity of channel sharing is a bit, and each flow f is assigned a weight. Intuitively, allocation of link bandwidth is fair if equal bandwidth is allocated in every time interval to all the flows. This concept generalizes to weighted fairness in which the bandwidth must be allocated in proportion to the weights associated with the flows.

Formally, if R_f is the weight of flow f and $W_f(t_1, t_2)$ is the aggregate service (in bits) received by it in the interval [t1,t2] , then an allocation is fair if, for all intervals [t1,t2] in which both flows f and m are backlogged Clearly, this is an idealized definition of fairness as it assumes that flows can be served in infinitesimally divisible units.

$$\frac{W_f(t_1,t_2)}{R_f} = \frac{W_m(t_1,t_2)}{R_m} \tag{1}$$

The earliest known fair scheduling algorithm is Weighted Fair Queuing (WFQ)[14]. WFQ was designed to emulate a hypothetical bit-by-bit weighted round-robin server in which the number of bits of a flow served in a round is proportional to the weight of the flow. Since packets cannot be serviced a bit at a time, WFQ emulates bit-by-bit round-robin by scheduling packets in the increasing order of their departure times in the hypothetical server. To compute this departure order, WFQ associates two tags—a start tag and a finish tag—with every packet of a flow. Specifically, if P_i^k and L_i^k denote the kth packet of flow R_i and its length, respectively, and if $V(A(P_i^k))$ denotes the arrival time of packet at the server, then start tag $S(P_i^k)$ and finish tag $F(P_i^k)$ of packet are defined as:

$$S(P_i^k) = MAX\{ V(A(P_i^k)), F(P_i^k) \} \tag{2}$$

$$F(P_i^k) = S(P_i^k) + (L_i^k)/R_i \tag{3}$$

Where V(t) is defined as

$$\frac{dV(t)}{dt} = C(t)/\sum_{i \in B(t)} R_i \tag{4}$$

C(t) is the capacity of the server and B(t) is the set of backlogged flows at time. Wireless Fair Queuing then schedules packets in the increasing order of their finish tags or start tags to ensure the fairness of the flow.

2.2   MAC Protocol in WSN

Since this paper focuses on a distributed protocol, we'd like to introduce some typical distributed MAC protocols for WSN.

The MAC protocol for WSN is often divided into Contention-based MAC and contention-free MAC. The contention-based MAC protocol is also known as random access protocol. Typically, some nodes communicate with each other in a shared-medium, and only one node can use the channel to communicate with others. Colliding nodes back off for a random duration and try to compete the channel again. Distributed coordination function (DCF) is an example of the contention-based protocol in the standardized IEEE 802.11.

The contention-free MAC is based on reservation and scheduling, such as TDMA-based protocols. Some examples of the kind of protocol are Time Division Multiple Access (TDMA); Frequency Division Multiple Access (FDMA); Code Division Multiple Access (CDMA).

S-MAC protocol is based on IEEE802.11 MAC protocol. It is a MAC protocol aiming at reducing energy consumption and support self-configuration. S-MAC protocol assumes that in normal circumstances, data transmission is little in WSN. The nodes coordinate to communication and the network can tolerate some additional message latency.

The idea of T-MAC[16] protocol is to reduce the idle listening time by transmitting all messages in burst of variable length m and sleeping between bursts. There are two solutions for the early sleeping problems: future request-to-send, FRTS and full buffer priority. But they are not very effective.

Sift MAC[11] protocol use a fixed-sized contention window and a carefully-chose, non-uniform probability distribution of transmitting in each slot within the window. Sift only sends the first R of N reports without collisions. But Sift doesn't consider the node's residual energy, that is, some low energy nodes may run out. The paper presents a new way of fair scheduling based on the residual energy.

In above-mentioned MAC layer protocols, in order to reduce the energy consumption, more is considered to lessen the idle listening period, to use period listening and sleeping, to avoid overhearing and to reduce bit numbers of MAC address of the communication data packet. But none of them has considered the issue of residual energy of nodes .Our idea is that the handling of residual energy is important to the lifetime of WSN. If the energy of some node is nearly exhausted , It can't be chosen as working node and needs to be replaced by its neighbor node which has more residual energy in case that the network island would appear.

At the same time, we believe that to solve the energy balancing problem at the lower layer has a more performance than higher-layer sensor network protocol. So, how to balance the energy consumption among the network neighbor nodes at MAC layer, and how to reduce the channel competition collision, ensure the effective communication and save energy are the problems that this paper tries to tackle.

The current MAC protocol use traditional binary exponential back off algorithm to tackle the collision problem in the course of communication. But the algorithm is designed for the fair distribution of shared channel. This paper has a revision of the back off mechanism of collision and introduces the idea of nodes residual energy, which can achieve the fair scheduling of all nodes.

3   Residual Energy-Based Fairness Scheduling MAC Protocol (REBFS MAC Protocol)

The main idea of this protocol is when N neighboring nodes sense an event at the same time, not all nodes need to transmit report of it. Only R of N need to report and the rest can give up transmission. When the first R nodes of N are chosen to transmit potential reports, they are chosen according to the residual energy. Fair scheduling MAC Protocol based on the model of the fluid fair schedule utilized to achieve the balanced energy consumption of each node in WSN.

3.1   Preliminary Conditions and Assumption

In the event-driven WSN, a single event-ID is distributed for every event. The node has an event table which includes event-ID, the number of nodes responding to an event and the event-related neighbor nodes.

Every node needs a neighboring node information table (see table 1) including neighboring Nodes ID, Residual Energy and the Hops of nodes. One-hop indicates the neighboring hop and two-hop indicates that the node is the next node of its neighboring node which can be picked up by the transmission of RTS-CTS packet. The condition of two-hop is taken consideration in our paper.

Table 1. Neighboring Node Information Table

Neighboring Node-ID	Residual Energy	Hops
A	55	1
B	30	2
C	80	2
D	10	1

The proposed protocol is derived from the distributed Coordination Function in the IEEE802.11 stand, which adopts RTS-CTS-DATA-ACK communication mechanism. Such mechanism can solve the problem of Hidden-station and overhearing. The following two conditions are necessary. The RTS packet needs to carry Residual Energy, Event-ID and residual Time of channel usage. The CTS packet needs to carry Node-ID, Residual Energy and residual Time of channel usage.

The proposed protocol is based on S-MAC. The nodes are put into periodic listen and sleep. Because of the residual time field of RTS-CTS-DATA-ACK packet, the neighboring nodes of source node and target node sense these packets, record the residual time and get into sleep state. The sleep time is equal to the residual time. When the sleep time is up, the nodes re-listen the channel. In order to reduce energy consumption, the nodes must be in sleep of low-level energy consumption. Meanwhile, the idea of residual energy is introduced which can sense collisions of multi-nodes in one event and balance the energy consumption of networks.

Next, the MAC protocol Design is discussed in Section 3.2

3.2   REBFS-MAC Protocol Design

3.2.1   Energy and Neighbor Information Transmission Mechanism

The transmission of the node's residual energy information among neighboring nodes is very important for the realization of the proposed protocol.

In this paper, the node's residual energy is picked up by RTS-CTS-DATA-ACK packets. For example, when node X acquires the channel, it'll pass RTS packet which picks up the node's residual energy, and X's neighboring nodes can sense and receive this information. When the target node Y has received the RTS packet passed by X, it will pass CTS packet which also carries Y's residual energy. Then Y's neighboring nodes receive the information of Y's residual energy. Every node having received the residual energy information will instantly update its local neighboring nodes information chart. It will modify the residual energy information of the existed node and append new information if it doesn't exist in the chart.

As we know, the periodic SYNC packet based on S-MAC Protocol also carries the information of the node's residual energy to ensure that the user can renew his information chart of the neighboring nodes residual energy. And the transmission of the node's residual energy information lays a foundation for the carry-out of the following mechanism.

3.2.2   Node's Resident Energy Scheduling Model

To every sensor node, a tag value E_i is attached which indicates residual energy. The energy consumption is shown in Neighboring Node Information Table from which we can get the set $B(t)$ which is the set of all neighboring nodes at time t, when contending the share channel.

In the average value of energy $E_{ave} = (\sum_{i \in B(t)} E_i)/n$, n is the number of nodes whose hop is 1 in Neighboring Node Information Table. When N neighboring nodes sense an event at the same time, not all nodes need to transmit report of it. When the energy of node E_i is R+1 in the energy queue, only R of N need to report and the rest can give up transmission. We can set the value of R is the numbers of node whose energy is bigger than E_{ave}. So E_i need not respond to the event, and only R can finish transmitting the report. Every node responding to the event maintains two variables, namely, the start tag and the finish tag. The former refers that the node has already got the channel service, while the latter means that new event occurs and new service is applied for. This means the finish tag includes the start tag.

To node E_i, the fluid fair model of the traditional wireless channel is introduced into the wireless sensor network. By utilizing a variation fair scheduling algorithm of Start Time Fair Queuing (STFQ) [11] based on the S-MAC protocol, we propose a fairness scheduling model based on the residual energy to realize a balanced consumption of node energy in WSN and eventually to ensure the longest lifetime of WSN. The specific process is as follows.

1. The initialization of the nodes. Every node has a quantifying number of its residual energy and exchanges information by S-MAC protocol. Every node maintains a neighboring node information table including neighboring Nodes ID, Residual Energy and the Hops of nodes.

2. When node I starts responding to event k, the start tag and the finish tag have to be computed. The start tag of node i is 0 at the initial stage.
3. The formula of the start tag and the finish tag is as follows:

$$S(Event_i^k) = MAX\{V(A(Event_i^k)), S(Event_i^{k-1}) + (L_i^{k-1}) * E_x/E_i)\} \qquad (5)$$

$$F(Event_i^k) = S(Event_i^{k-1}) + (L_i^{k-1}) * E_x/E_i)\} \qquad (6)$$

$S(Event_i^k)$ is the service which node i got when event k occurred.

$Event_i^k$ is event k to which node i is responding.

$V(t)$ is the virtual time to indicate the service got at time t.

$A(Event_i^k)$ is the arrival time of event k.

L_i^k is the length of data transmitted by event k of node i.

E_x is the energy consumed by transmitting one data.

$\frac{dV(t)}{dt} = C(t)/\sum_{i \in B(t)} E_i$; $C(t)$ is the channel capacity at time t.

4. When an event occurs at time t, the neighboring nodes in Neighboring Node Information Table begin contending the share wireless channel. The node with the least value of finish tag has the priority of the channel by choosing the shortest back-off time slot and send the RTS packet; if the node with the least value of finish tag is not available, the node with the least value of start tag is chosen to have the shortest back-off time slot and transmit the RTS packet.
5. To the node achieving the channel, the start tag equals to the finish tag and the residual energy after the communication is presented in the formula $E_i = E_i - E$ in which E is the consuming energy.
6. If the node fails to get the channel, it indicates that some neighboring node has successfully got the channel. It will update the neighboring node in the Information Table according to the energy information carried by the RTS packet and get into the sleep state based on the NAV information of the RTS packet.

3.2.3 Contention Window and Collision Handling

1. Contention Window

We assume that N neighboring nodes in one area sense an event simultaneously. But only R nodes need to pass the data(R/N) because of the redundant nodes. To ensure a balanced consumption of network energy, R nodes which have more residual energy will pass the data, and the rest nodes give up. This is the so-called R/N issue which has been discussed in Sift protocol. But here we propose a different solution.

Our solution is that we assume when one node and all it's neighboring nodes sense an event, according to their residual energy hierarchy, the nodes that have more residual energy have the priority to get the contending channel and transmit the information. Meanwhile, we choose different back off intervals for nodes with different residual energy when they sense the channel. The nodes that have more residual energy and less service are prior to transmit the information, whereas the nodes with less residual energy and more service have longer back off intervals. Hence a balanced energy consumption of network nodes is ensured.

Based on the above assumption, we can see the fair scheduling is achieved by combining the back-off interval with the service tags of the nodes. So we have a new formula of the start tag and the finish tag for E_i:

$$F\left(Event_i^k\right) = S\left(Event_i^{k-1}\right) + Scaling\_factor * \left(L_i^{k-1}\right) * \frac{E_x}{E_i} \tag{7}$$

In the formula, scaling_factor[15] is a proportion parameter, and it allows for a certain tag value of. Meanwhile, backoff interval is presented as B_i of nodes E_i. Different B_i is calculated for different E_i. High E_i value should have low B_i and low E_i should have high B_i.

$$B_i = \lfloor Scaling_{factor} * L_i^k * E_x/E_i \rfloor \tag{8}$$

For each node, It has a E_i and a back-off interval B_i. Because of the reciprocal relation between E_i and B_i, We define a Qi with:

$$Q_i = Ei/\textstyle\sum_{j\in B(t)} E_j \tag{9}$$

$Q_i = 1 - P_i$, $Q_i \in (0.1)$; then we set the value of B_i: $B_i = \lfloor B_i \times Q_i \rfloor$ .Despite that Bi may be chosen differently by different nodes in the first place and value Qi of residual energy may be different, Bi which is computed according to above formula can be same. In order to solve the problem, random number X_i is generated in [0.9,1.1]. Then the B_i, is set: $B_i = \lfloor B_i \times Q_i \times X_i \rfloor$ Hence, transmitting collision is decreased.

2. Collision Handling

Transmitting collision is unavoidable. For example when a node moves to the neighboring area of another node's wireless signal, it'll pass the information and cause the data collision due to the failure to sensing the signal.

Traditional IEEE802.1 uses binary exponential back off algorithm to tackle the collision problem. The mechanism is as the following:

If a collision occurs, then the following procedure is used. Let node i be one of the nodes whose transmission has collided with some other node(s). Node i chooses a new back-off interval as follows: (1) Increment Collision-Counter by 1. (2) Choose new B_i uniformly distributed in $\left[1, 2^{CollisionCounter-1} \times CW\right]$, where Collision Window is a constant parameter.

From above we can see, when collision occurs, contention window will double the size so that communication efficiency is decreased. So, for the second step, we take the same process:

$B_i = \lfloor B_i \times Q_i \times X_i \rfloor$ and the definition of B_i, Q_i and X_i is the same with above. The decrease of B i can increase the throughout of the system.

3.2.4   Message Passing

When an event occurs in WSN, N neighboring nodes sense it simultaneously, but we are interested in the first R of N potential reports because of the redundant nodes. The node will take the information down in its event information chart when it senses an event. At the same time, in order to solve R/N problem, it need to know which neighboring nodes also sense the event and prepare to transmit.

In other words, the RTS packet is passed by the node carries Event-ID. Once the node senses the Event-ID carried by the RTS packet, the number of the nodes responding to the event is added 1 to its counting of the same Event-ID. And the node will go through the periodic listen and sleep. Whenever the node senses the available channel and prepare to transmit the information, it'll check the counting machine first. If the result exceeds R, The node will give up the transmission, and delete the event information. It can also transmit by RTS-CTS-DATA-ACK packets and delete the information from the information chart when the result exceeds R.

The value of R is decided by the distribution density of the nodes, or by the numbers of a node's neighboring nodes.

3.2.5   Further Discussion

In the model of fair scheduling based on the residual energy, every node has its back-off interval according to the residual energy. Suppose node A and B are neighboring nodes within one hop, and B and C are neighboring nodes within one hop which means node A and C have two-hop distance. If we only put the residual energy of the neighboring nodes within one hop into consideration, there will be conflict between node A and C near B within two hops when both of them have the priority to their own shared channel. This is the issue of the hidden station. The conflicting nodes will back off and enlarge the contending window which eventually causes the unbalance of the energy. This problem will be tackled in this section.

The fair scheduling of all nodes energy within two hops of the node i is the solution of this problem. The preliminary condition is that when a node is sending CTS packet, it requires that CTS packet carry the energy of its own node and the energy of the node which has sent the RTS packet. At the initializing stage, the node i first acquires its neighboring node by exchanging the SYN, then in the course of communication, by utilizing the shakehand mechanism of RTS-CTS-DATA-ACK, the node i can acquire the node in two hops. We use set $B(i, t)$ to indicate the set of all nodes with two hops of node i and fair scheduling is applied within this set.

For each node of the set $B(i, t)$ it has a E_i and a back-off interval B_i. Because of the reciprocal relation between E_i and B_i, We define a Q_i with the following formula:

$$Q_i = E_i / \sum_{j \in B(i,t)} E_j \qquad (10)$$

In the formula, we set the value of B_i: $B_i = \lfloor R_i \times Q_i \times X_i \rfloor$ and the definition of B_i, Q_i and X_i is the same with 3.2.3. In this way, the collision of every node within two hops will be decreased and according to the characteristics of the wireless sensor network, the communication conflicts in the global network will also be decreased largely.

The Residual Energy-Based Fairness Scheduling Algorithm is describe as follows:

Residual Energy-Based Fairness Scheduling Algorithm

1. Start_tag$_i$: = 0;   /* start_tag of node i is set 0; */

2. E$_{ave}$: = $(\sum_{i \in B(t)} E_i)/n$;

3. R: = N$_{numbers\ of\ node}$ /*The value of R is set the numbers of node whose energy is bigger than E$_{ave}$*/

4. /* When node I starts responding to event k   */

5. if Sort_location(E$_i$)>=R+1 then

6.   Node i give up this communication and not update its start stag.

7. Elseif E$_i$ < E$_j$ and j $\in$ B(i, t) then

8.   Q$_i$: = Ei/$\sum_{j \in B(i,t)}$ E$_j$

9. Else   Q$_i$: = Ei/$\sum_{j \in B(t)}$ E$_j$

10. B$_i$: = $\lfloor B_i \times Q_i \times X_i \rfloor$

11./* The node i choose B$_i$ backoff intervals. */

12. IF (backoff time =0) then

13.   Send a RTS packet with own residual energy.

14. Elseif receive a RTS packet then

15.   Update or append the energy of the sender in neighbor nodes tables and node i will sleep with NAV times of RTS packet.

16. Elseif receive a CTS packet and E$_{Recevier}$ not in B(t) and E$_{Recevier}$ ≠ E$_i$ then

17.   The sender is inserted in neighbor nodes tables and his hops is set 2 hops. The energy of the sender and receiver are updated or inserted in neighbor nodes tables.

18. ElseIf receive a CTS packet and E$_{Recevier}$ = E$_i$ then

    Send data packet.

19. If receive a ACK packet and E$_{Recevier}$ = E$_i$ then

    Start_tag$_i$ = Finish_tag$_i$; E$_i$: = E$_i$ − E ;/* E is the consuming energy*/

---

4   Performance Evaluation

Given the difficulty in performing actual measurements in wireless networking, we first evaluate our system through simulation. We have created a simple simulator capable of creating an arbitrary multi-hop network topology of a group of networked sensors. Each program process represents a networked sensor, and a master process is responsible for synchronizing them to perform bit time simulation. Since our main focus is media access control. The simulator doesn't simulate the actual hardware operating in the Tiny OS environment. However, it preserves the event driven semantics.

We assume the max value of the original energy is 100 and the min value of the original energy is 80. The value of energy consumption is 1 for transmitting a RTS/CTS/ACK packet and the value of energy consumption is 5 for transmitting a DATA packet.

From Figure 1, we can see REBFS-MAC shows good features on saving the network nodes energy. REBFS-MAC introduces the node's residual energy based on S-MAC protocol and choose different back off intervals for nodes of different residual energy so that collisions are decreased and energy is saved.

Figure 1 demonstrates through simulation the balanced the node's energy by REBFS-MAC, S-MAC, Sift and IEEE802.11.

Ten neighbor nodes are checked after the node transmits the information. Figure 2 shows the energy consumption of the neighbor nodes. REBFS-MAC has better performance on balancing neighbor nodes' energy than others.

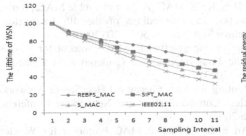
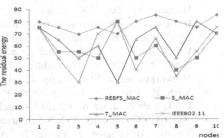

Fig. 1. The lifetime of WSN in different protocols

Fig. 2. The different balance of the node's energy in different protocols

5   Conclusion

The fluid fair model of the traditional wireless channel is introduced into the wireless sensor network. By utilizing a variation fair scheduling algorithm of Start time Fair Queuing (STFQ) based on the S-MAC protocol, we propose a fairness scheduling model based on the residual energy to realize a balanced consumption of node energy in WSN and eventually to ensure the longest lifetime of WSN. In this paper, the fluid fair model of the traditional wireless channel is introduced into the wireless sensor network and residual energy is introduced into the mechanism of data transmission, sleeping and collision handling to enhanced the communication efficiency, ensure the balanced consumption of the node's energy, and prevent the phenomenon of network island. The mechanism proposed in this paper is based on S-MAC protocol and is verified by the simulation test.

The simulations showed that the distributed handling mechanism utilized in this protocol adapts the moving nodes better, which is expected to be improved in our future work.

References

1. Li, J.-Z., Li, J.-B., Shi, S.-F.: Concepts, Issues and Advance of Sensor Networks and Data Management of Sensor. Journal of Software 14(10), 1717–1725 (2003)
2. LAN MAN Standards Committee of IEEE Computer Society, Wireless LAN medium control (MAC) and physical layer (PHY) specification, IEEE Std 802.11-1999 edition. IEEE, New York (1999)
3. Akyildiz, I.F., Su, W., Sankarasubramaniam, Y., Cayirci, E.: Wireless sensor network: A survey. Computer Networks 38, 393–422 (2002)
4. Yahya, B., Ben-Othman, J.: Towards a classification of energy aware MAC protocols for wireless sensor networks. Journal of Wireless Communication and Mobile Computing 4, 1572–1607 (2009)
5. Yahya, B., Ben-Othman, J.: An Energy Efficient Hybrid Medium Access Control Scheme for Wireless Sensor Networks with Quality of Service Guarantees. In: The Proceedings of GLOBECOM 2008, New Orleans, LA, USA, November 30-December 4, pp. 123–127 (2008)
6. Rhee, I., Warrier, A., Aia, M., Min, J., Sichitiu, M.L.: Z-MAC: A hybrid MAC for wireless sensor networks. IEEE/ACM Transactions on Networking 16, 511–524 (2008)
7. Buettner, M., Yee, G.V., Anderson, E., Han, R.: X-MAC: A short preamble MAC protocol for duty-cycled wireless sensor networks. In: Proceedings of the 4th International Conference on Embedded Networked Sensor Systems, pp. 307–320 (2006)
8. van Dam, T., Langendoen, K.: An adaptive energy-efficient MAC protocol for wireless sensor networks. In: Proceedings of the First International Conference on Embedded Networked Sensor Systems, pp. 171–179 (2003)
9. Shah, R., Rabaey, J.: Energy Aware Routing for Low Energy Ad Hoc Sensor Networks. In: The Proceedings of the IEEE Wireless Communications and Networking Conference (WCNC), Orlando, pp. 350–355 (March 2002)
10. Ye, W., Heidemann, J., Estrin, D.: An Energy-Efficient MAC Protocoal for Wireless Sensor Networks. In: Proc. 21st Int'l Annual Joint Conf. IEEE Computer and Communication Societies, InfoCOMM 2002 (2002)
11. Jamieson, K., Balakrishnan, H., Tay, Y.C.: Sift: A MAC protocol for wireless sensor networks. In: Proc. 1st Int'l Conf. on Embedded Network Sensor Systems (WenSys), Los Angeles, CA (2003)
12. Goyal, P., Vin, H.M., Chen, H.: Start-time fair queueing: A scheduling algorithm for integrated service access. In: ACM SIGCOMM 1996 (1996)
13. Demers, A., Keshav, S., Shenker, S.: Analysis and simulation of a fair queueing algorithm. In: Proc. ACM SIGCOMM, pp. 1–12 (September 1989)
14. Demers, A., Keshav, S., Shenker, S.: Analysis and simulation of a fair queueing algorithm. In: ACM SIGCOMM 1989 (1989)
15. Vaidya, N., Bahl, P.: Distributed Fair Scheduling in a Wireless LAN. IEEE Transactions on Mobile Computing 4(6), 616–629 (2005)
16. Van Dam, T., Langendoen, K.: An adaptive energy-efficient MAC protocol for wireless sensor network. In: Proc. 1st Int'l Conf. on Embedded Network Sensor System (November 2003)

Topology-Aided Geographic Routing Protocol for Wireless Sensor Networks

Guilin Li[1], Longjiang Guo[2], Jian Zhang[1], and Minghong Liao[1,*]

[1] School of Software, Xiamen University, Xiamen 361005, China
[2] School of Computer Science and Technology, Heilongjiang University, Harbin 150001, China
{liguilin.cn,longjiangguo,jianzhang9102}@gmail.com,
liao@xmu.edu.cn

Abstract. The problem of the traditional geographic routing protocols for the wireless sensor networks is that they can only adopt one kind of strategy, such as the right-hand rule, to bypass a hole, which is not always a proper choice. In this paper, we propose a topology-aided geographic routing protocol, which utilizes the topology of the network to guide a packet bypassing a hole in a shorter path than that of the traditional protocol. Experimental results show that our topology-aided geographic routing protocol can reduce the length of the path from the source to the destination.

Keywords: Routing Protocol, Geographic, Topology-Aided, Sensor Networks.

1   Introduction

Geographic routing protocol is one of the most important types of routing protocol for the wireless sensor networks [1,2,3]. The traditional geographic routing protocol routes a packet from the source to the destination only based on the local information of a node, which is not always optimal. In this paper, we propose a topology-aided routing protocol. With the help of the topology, our protocol can select a shorter path from the source to the destination than that selected by the traditional geographic routing protocol.

For example, Fig.1 shows a sensor network distributed in a rectangle area with a hole in it. The source A wants to send a packet to the destination B. The traditional geographic routing protocol first sends the packet greedily from A to B until the packet reaches an intermediate node C that can't find a neighbor whose distance to the destination is nearer than that of itself. The traditional geographic routing protocol then transmits the packet around the hole according to some strategy such as the right-hand rule [1]. In Fig.1, the package is transmitted along the left side of the hole. Obviously, it is not a good choice. As the traditional geographic routing protocol routes a packet purely based on the local geographic information, it can only adopt a certain strategy (such as the right-hand rule) to bypass a hole. In our protocol, when a packet can't be transmitted greedily, the intermediate node calculates an optimal path from itself to the destination according to the topology of the network. Based on the

---

* Corresponding author.

H. Gao et al. (Eds.): WAIM 2012, LNCS 7418, pp. 47–57, 2012.
© Springer-Verlag Berlin Heidelberg 2012

path just calculated, the intermediate node can choose a proper direction for the packet to bypass the hole. In this example, node *C* will choose the right direction for the packet to bypass the hole.

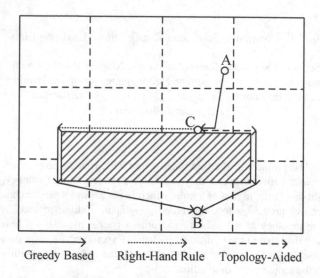

Greedy Based   Right-Hand Rule   Topology-Aided

Fig. 1. Problem of the Traditional Geographic Routing Protocols

In this paper, we first propose a grid based protocol to detect the topology of the network, which represents the topology of the network as a weighted graph. Based on the topology detected, a **T**opology-**A**ided **G**eographic **R**outing (**TAGR**) protocol is proposed to route the packet in a short path. Before giving details of the TAGR protocol, we propose some assumptions used in this paper. We assume that a homogeneous sensor network is distributed in a rectangular area with some holes in the area. According to [4], a hole is an empty area enclosed by a series of connected sensor nodes. Holes of considerable size (a percentage of the network diameter) break the isotropy of the network and may block the direct path of two nodes. We assume isotropy among the rest of the network excluding the holes. Each node knows its coordinate which can be estimated by the localization algorithms [5,6] and the area of the rectangle where the sensor network is distributed.

The general idea of the paper is as follows. The rectangle area is divided into a lot of grids, each of which has an ID. Each grid has 8 neighbor grids around it. The neighbor grids are numbered from 0 to 7 clockwise beginning at the grid to the left of the center grid. By using the topology detection protocol, each grid tests its connectivity with all its neighbor grids and forms a bit vector. One bit in the vector represents the connectivity of the center grid with its corresponding neighbor grid. After calculating the bit vector for each grid, the grid transmits the vector to other girds in the network. After receiving the bit vector from all the other grids, the topology of the network can be constructed as a weighted graph. When the packet is transmitted to an intermediate node at the edge of a hole, the node can guide a packet to bypass the hole in a short path with the help of the topology information.

The contribution of this paper is: First, we propose a topology-aided geographic routing protocol, which can select a shorter path for a packet than that of the traditional geographic routing protocols. Second, we verify the performance of our routing protocol by extensive experiments.

The rest of this paper is organized as follows. In section 2, we give some related works. We briefly introduce a grid based topology detection algorithm used by our routing protocol in section 3. The Topology-Aided Geographic Routing (TAGR) protocol is proposed in section 4. Section 5 evaluates the performance of the TAGR protocol by extensive experiments. Finally, we draw a conclusion in section 6.

2   Related Works

Routing is one of the most important and fundamental problems for wireless sensor networks. A lot of routing protocols have been proposed to fit the particular requirement of wireless sensor networks [7], which can be classified into several classes, such as the hierarchy based protocol, the geographic routing protocol, the virtual coordinate based geographic routing protocols, the data centric routing protocols and the Qos guaranteed protocols.

The hierarchy based protocol [8] divides sensors in a sensor network into two types, the cluster heads and the ordinary nodes. The ordinary nodes can just communicate with its cluster head. Communication with other nodes in the same cluster or in another cluster relies on the cluster heads. In the geographic routing protocols, every node needs to know its physical coordinate. The basic idea of geographic routing protocol is that, when transmitting a packet, a node selects the neighbor that is nearest to the destination as the next hop. The virtual coordinate based protocols [9,10] works as the same as the geographic routing protocol, except that they do not need nodes know their physical coordinates. Each node is assigned some kind of virtual coordinate. Unlike the geographic routing in which the source node searches for a route to the destination node according some kind of coordinate, in data centric routing the source searches for a particular data stored on an unknown subset of nodes. Hence the routing problem is actually a query problem. The routing algorithm must search for the route to a node with the desired data. Then the data can be transmitted along the discovered route. Direct diffusion [11] floods a query around the network so that a path to transmit events from the source node to sink can be constructed. TAG [12] constructs a tree in the sensor network to answer the aggregate query. Such kind of query processing algorithm is suitable for continuous query for a certain type of data. The data can also be distributed in some particular nodes in the network according to the feature of the data [13]. Queries for the data can be efficiently found in the structure according to their features. Such kind of method is suitable for ad hoc query processing. For example, data are stored in different sensors selected by some kind of hash function [14] according to their types. A query only needs to visit the hashed location to acquire data of a given type. DIM [15] embeds a k-d tree like index in the network, which can fulfill the range query easily. Finally, the Qos guaranteed protocols include the fault tolerant, the real time and long lifetime routing protocols etc. Fault tolerant protocols [16,17] guarantee the reliability by transmitting a packet multiple times or through different neighbors. Real-time routing protocols [18] select the neighbor to the destination with the minimum transmission latency as the next hop. Long lifetime routing protocol [19] tries to evenly consume the energy of different nodes to optimize the lifetime of the network.

3   The Grid-Based Topology Detection Protocol

The grid-based topology detection protocol is composed of two parts, which are the connectivity detection protocol for grids and the topology detection protocol for the network. The topology of the network detected by the protocol is represented as a weighed graph.

3.1   Connectivity Detection Protocol for Grids

Suppose a sensor network is distributed within a rectangle area whose height and width are h and w. The rectangle area is divided into a lot of identical grids whose height and width are h_g and w_g respectively. A sensor node s_i, whose communication range is r, knows its coordinate (x_i, y_i).

Nodes in a grid can be classified into three types, which are the border nodes, the boundary nodes and the inner nodes. The nodes that can communicate with nodes in other grids are called the border nodes. A node p is a boundary node if there exists a location q outside p's transmission range so that none of the 1-hop neighbors of p is closer to q than p itself. The other nodes in a grid are called inner nodes. Note a node can be both a border node and a boundary node. At this time, the node is considered as a border node.

A node can detect its type by using just the coordinates of its 1-hop neighbors. A node can determine whether it is a border node by checking the coordinates of its neighbors. [20] proposed a local algorithm named BOUNDHOLE to make a node detect itself a boundary node by just using 1-hop neighbor information. If a node is neither a border node nor a boundary node, it is an inner node.

If there is a communication path between two neighbor grids, we say that the two grids are connective. The Connectivity Detection Protocol is used by each grid to detect its connectivity to its 8 neighbor grids, which works as follows.

Firstly, each node in the network calculates the grid it lies in by using (x_i, y_i), h, w, h_g and w_g. Then each node detects its neighbors in the network. According to the coordinates of the neighbors, a node can decide whether it can communicate with nodes in other grids. If so, the node is a border node and it sets the corresponding bit in the bit vector to 1. Recall that each grid has 8 neighbor grids, which are numbered according to their relative position to the center grid. After constructing the bit vector, the border node broadcasts a message containing the partial connectivity information to the other nodes in the grid. A pure boundary node broadcasts a bit vector with all its bit set to 0, which means the boundary node does not connect to any neighbor grids. If a boundary node is also a border node, the bit vector is set as a border node. Only when an inner node receives bit vectors from all its neighbor nodes, whose distance to the center of the grid is farther than that of itself, it aggregates the connectivity information it received by doing an OR operation to the bit vectors and broadcasts the aggregated bit vector out. A node with no neighbor nearer to the center of a grid than that of itself is called a cluster head candidate. There may be multiple cluster head candidates in a grid. After the candidates collect the bit vectors from neighbors, they broadcast the aggregated bit vector to all nodes in the grid together with their coordinates and IDs. If there is only one candidate nearest to the center of a grid among all candidates, it is selected as the cluster head. Otherwise, the candidate with the smallest ID is selected as the cluster head. After the execution of the protocol, all nodes in a grid master the connectivity of the grid and the ID of the cluster head.

3.2   Topology Detection Protocol for the Network

There are holes in the sensor network, which can divide a grid into several parts. The connectivity detection protocol creates a cluster head for each part of a grid, which holds the complete connectivity information of the part in a grid.

The topology detection protocol spreads the connectivity information of each cluster head throughout the network. A cluster head broadcasts a message throughout the network, which is composed of three parts. The first part is the ID of the cluster head. The second part is the coordinate of the cluster head. The third part is the bit vector calculated by the cluster head. Based on the messages received from the cluster heads, a sensor node can construct the topology for the whole network, which is represented as a weighted graph. Nodes in the graph are the cluster heads. Edges are the connectivity between two cluster heads. The weight for each edge is the distance between two cluster heads of the edge. In this way, our topology-aided geographic routing protocol does not consume more energy of the cluster head than that of the ordinary node.

4   Topology-Aided Geographic Routing Protocol

In this section, we propose a topology-aided geographic routing protocol, which utilizes the topology of the network to select a short path for a packet when bypassing a hole. Before giving details of the protocol, we introduce the next hop selection algorithm used by a node to transmit a packet around a hole.

4.1   The Next Hop Selection Algorithm

To apply the geographic routing protocol, the boundary nodes in the network need to be planarized [1] as follows. An edge (u,v) exists between boundary node u and v if the distance between every other boundary node w, and whichever of u and v is farther from w. In equational form:

$$\forall w \neq u, v : d(u, v) \leq \max[d(u, w), d(v, w)] \tag{1}$$

We say that two boundary nodes are neighbors only when they satisfy the Eq.(1). Suppose a source node A transmits a packet to a destination node B in Fig.2. The packet is transmitted greedily in the same way as the traditional geographic routing protocol until it reaches a boundary node C, where the packet can't be transmitted in greedy mode any more. Node C needs to determine a transmission strategy for the packet to bypass the hole. It has two choices, the left-hand rule or the right-hand rule.

An intermediate node (node C in our example) adopts two steps to determine the transmission strategy for a packet to bypass a hole. First, it selects the next hop from all its neighbor boundary nodes. Second, it determines the proper transmission strategy based on the relationship among the intermediate node (node C), the previous hop of the intermediate node (node C') and the selected next hop (node R).

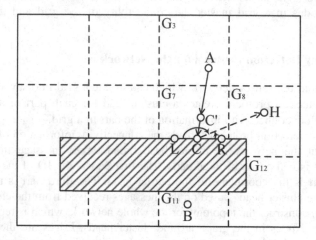

Fig. 2. The TAGR protocol

In Fig.2, the intermediate node C first decides the next hop for the packet. As node C knows the topology of the network represented by a weighted graph, it can calculate the shortest path for the packet to bypass a hole. Node C lies in grid G_7 and the destination node B lies in grid G_{11}. The nearest path from G_7 to G_{11} is $G_7 \rightarrow G_8 \rightarrow G_{12} \rightarrow G_{11}$, which means node C needs to transmit the packet to the direction towards G_8. Suppose the node H be the cluster head of G_8. The transmission direction can be determined by the vector $\overrightarrow{CH}$, called direction vector. The node C constructs vectors from itself to each of its neighbor boundary nodes ($\overrightarrow{CL}, \overrightarrow{CR}$ in Fig.4) and compares the angles formed by the vectors and the direction vector ($\angle HCL, \angle HCR$ in Fig.2). The vector with the minimum angle to $\overrightarrow{CH}$ is selected and the next hop is the neighbor boundary node of the selected vector, so node R is selected as the next hop in Fig.2.

The dot product between two vectors can be used to compare the size of different angles. According to the definition of dot product of two vectors, the angle $\angle HCL$ between $\overrightarrow{CL}$ and $\overrightarrow{CH}$ can be calculated by Eq.(2). As the value of cosine is a decreasing function from 0 to π, node C can find a smaller angle by selecting a larger cosine value. The next hop selection algorithm is given in algorithm 1. The input of the algorithm 1 is the direction vector and its output is the next hop.

$$\cos(\angle HCL) = \frac{\overrightarrow{CL} \cdot \overrightarrow{CH}}{|\overrightarrow{CL}| \, |\overrightarrow{CH}|} \tag{2}$$

```
Algorithm 1:
S1: Construct vectors to each neighbor boundary node
S2: Calculate the cosine values of the angles between
    each vector and the direction vector according to
    Eq.(2)
S3: Select the neighbor boundary node with the maximum
    cosine value as the next hop
```

After selecting the next hop, the intermediate node C needs to calculate the transmission strategy for the packet based on the next hop. In Fig.4, the packet reaches node C from node C', which forms a vector $\overline{CC'}$. To transmit the packet to the next hop node R, node C needs to use the left-hand rule, which turns the vector $\overline{CC'}$ clockwise around node C.

The cross product between two vectors can be used by the intermediate node to determine the transmission strategy for a packet. According to the definition of the cross product, if the sign of cross product result for two vectors is positive, the left-hand rule is selected. Otherwise, the right-hand rule is used. The transmission strategy selection algorithm is given in algorithm 2. The input of the algorithm 2 is two vectors originating from the current node and pointing to the previous hop and the next hop respectively, such as $\overline{CC'}$ and $\overline{CR}$ in Fig.2. The output of algorithm 2 is the transmission strategy.

```
Algorithm 2:
S1: Calculate the cross product of the two vectors
S2: if(the result is positive)
S3:    set the transmission strategy clockwise
S4: else
S5:    set the transmission strategy counterclockwise
```

4.2   The Topology-Aided Geographic Routing Protocol

The topology-aided geographic routing protocol works as follows. A packet in our protocol can be in one of two exclusive modes: the greedy mode or the perimeter mode. When a source node A wants to transmit a packet to a destination node B, it checks whether it has a neighbor whose distance is nearer to the destination than itself. If so, the packet is set to the greedy mode and the source node transmits the packet to the neighbor nearest to destination.

When an intermediate node receives a packet in the greedy mode, it repeats the above procedure until it finds that there is no neighbor whose distance to the destination is nearer than itself. In this case, the packet has reached a boundary node in the network. The current node changes the packet into the perimeter mode and does the following calculation:

```
S1: Save the distance between the current node and the
    destination in the packet
S2: Determine the destination cluster according to the
    coordinate of the destination
S3: Calculate the shortest path from the cluster head of
    the current node to the cluster head of the
    destination based on the weighted graph G(N,E)
S4: Calculate the direction vector originating from the
    current node to the cluster head of the next hop Grid
S5: Selects the next hop according to the algorithm 1
S6: Determine the transmission strategy for the packet
    according to the algorithm 2
S7: Store the strategy in a bit of the packet, where 0
    represents clockwise and 1 represents
    counterclockwise
S8: Transmit the packet to the selected next hop
```

In step 2, the current node determines the cluster that the destination belongs to. If we assume that each grid contains only one cluster, it is easy for the current node to calculate the cluster head of the destination according to the coordinate of the destination. We leave the case of multiple clusters in a grid as the future work. When an intermediate node receives a packet in the perimeter mode, the node does the following operations:

```
S1: Check the value of the bit representing
    transmission strategy
S2: If the value is 0, the node selects the next hop
    according to the left-hand rule.
S3: Otherwise, the node selects the next hop according to
    the right-hand rule.
S4: Check whether the selected next hop is nearer to the
    destination than the distance saved in the received
    packet
S5: If so, the node changes the packet into the greedy
    mode
S6: Transmit the packet to the selected next hop
```

The left-hand rule in step 2 means the current node selects its last neighbor boundary node, except the previous hop, met by turning the vector formed from the current intermediate node to its previous hop clockwise. The right-hand rule turns the vector counterclockwise.

5   Experiments

Two experiments were done to test the performance of our protocol. In the first experiment, we compare the differences of routing path selected by our protocol and the traditional geographic routing protocol in different network condition. In the second experiments, we compare the average length of path between our protocol and the traditional protocol. The setting of our experiments is as follows. 4379 nodes are uniformly distributed in a rectangle area of 600*600. The rectangle area is divided into 64 grids. The transmission range of a node is set to 20.

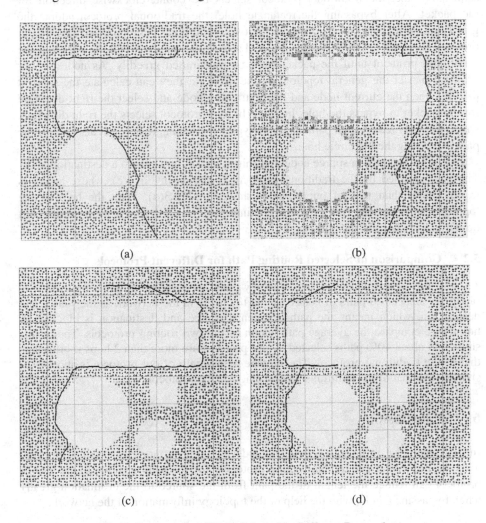

(a)   (b)

(c)   (d)

Fig. 3. Comparison of Paths Selected by Different Protocols

5.1   Comparison of Selected Routing Path for Different Protocols

In this experiment, we compare the different paths from a source to a destination selected by our protocol and the traditional geographic protocol. There are four holes in the network.

In Fig.3 (a) and (b), a source node transmits a packet downwards to a destination node. The routing path of the traditional geographic protocol is shown in Fig.3 (a). In the first phase, the packet is greedily transmitted until it meets the boundary of the rectangle hole. Then the traditional protocol adopts the perimeter mode. Based on the right-hand rule, the traditional protocol selects the counterclockwise direction for the packet. After bypassing the rectangle hole, the traditional protocol adopts the greedy mode for the packet to bypass the two circle holes. The path selected by our protocol is shown in Fig.3 (b). Our protocol selects the same path for the packet as the traditional protocol until the packet can't be transmitted greedily. As nodes in our protocol master the general topology of the network, they can calculate the shortest path between the current node and the destination node and select the right direction for the packet to bypass the hole. In Fig.3 (b), our protocol selects the right direction, which not only reduces the path length to bypass the rectangle hole but also avoids the packet meeting the big circle hole.

In Fig.3(c) and (d), a source node transmits a packet upwards to a destination node. The path selected by the traditional protocol is shown in Fig.3(c), whose length is much longer than that of our protocol shown in Fig.3(d). With the help of the topology information, our protocol continues to transmit the packet in the left direction and shortens the path length.

5.2   Comparison of Selected Routing Path for Different Protocols

To test the performance of the topology-aided geographic routing protocol, we randomly select 100 pairs of nodes around the holes in the network as the source and destination, which means the path between a source and destination is blocked by holes. We compare the average path length of the two routing protocols. The results show that about 40% paths selected by our protocol are more than 2 hops shorter than that of the traditional protocol. The average path length of our protocol is about 14% shorter than that of the traditional geographic routing protocol.

6   Conclusion

In this paper, we propose a topology-aided geographic routing protocol. Compared with the traditional geographic routing protocols for the sensor network, the TAGR protocol can reduce the length of the path from the source to the destination greatly when bypassing a hole, with the help of the topology information of the network.

Acknowledgements. This work was supported by a grant from the National Natural Science Foundation of China (No.61100032), Basal Research Fund of Xiamen University (No.2010121072) and the Natural Science Foundation of Fujian Province (No.2010J01342).

References

1. Karp, B., Kung, H.T.: GPSR: greedy perimeter stateless routing for wireless networks. In: Proceedings of ACM/IEEE MobiCom (2000)
2. Kuhn, F., Wattenhofer, R., Zhang, Y., Zollinger, A.: Geometric ad-hoc routing: of theory and practice. In: Proceedings of the 22nd Annual Symposium on Principles of Distributed Computing, pp. 63–72. ACM Press (2003)
3. Lee, S., Bhattacharjee, B., Banerjee, S.: Efficient Geographic Routing in Multihop Wireless Networks. In: Proceedings of MobiHoc. ACM Press (2005)
4. Wang, Y., Gao, J., Mitchell, J.S.B.: Boundary Recognition in Sensor Networks by Topological Methods. In: Proceedings of ACM MobiCom (2006)
5. Ji, X., Zha, H.: Sensor positioning in wireless ad hoc networks using multidimensional scaling. In: Proceedings of IEEE INFOCOM (2004)
6. Tran, D.A., Nguyen, T.: Localization in Wireless Sensor Networks Based on Support Vector Machines. IEEE Transactions on Parallel and Distributed Systems 19(7) (2008)
7. Akyildiz, I.F., Su, W., Sankarasubramaniam, Y., Cayirci, E.: A Survey on Sensor Networks. IEEE Communications Magazine 40(8) (2002)
8. Heinzelman, W.: Application-Specific Protocol Architectures for Wireless Networks. Ph.D. thesis, Massachusetts Institute of Technology (2000)
9. Bruck, J., Gao, J., Jiang, A.A.: MAP: Medial Axis Based Geometric Routing in Sensor Network. In: Proceedings of ACM MobiCom (2005)
10. Fang, Q., Gao, J., Guibas, L.J., de Silva, V., Zhang, L.: GLIDER: Gradient Landmark-Based Distributed Routing for Sensor Networks. In: Proceedings of IEEE INFOCOM (2005)
11. Intanagonwiwat, C., Govindan, R., Estrin, D.: Directed Diffusion: A Scalable and Robust Communication Paradigm for Sensor Networks. In: MobiCom 2000 (2000)
12. Madden, S., Franklin, M.J., Hellerstein, J.M., Hong, W.: TAG: a tiny AGgregation service for ad-hoc sensor networks. In: Proceedings of OSDI (2002)
13. Shenker, S., Ratnasamy, S., Karp, B., Govindan, R., Estrin, D.: Data-centric storage in sensornets. ACM SIGCOMM Computer Communication Review 33(1) (2003)
14. Ratnasamy, R., Karp, B., Yin, L., Yu, F., Estrin, D., Govindan, R., Shenker, S.: GHT: A geographic hash table for data centric stroage. In: Proceedings of WSNA (2002)
15. Li, X., Kim, Y.J., Govindan, R., Hong, W.: Multi-dimensional range queries in sensor networks. In: Proceedings of ACM SenSys 2003 (2003)
16. Zhang, W., Xue, G.L., Misra, S.: Fault-Tolerant Relay Node Placement in Wireless Sensor Networks: Problems and Algorithms. In: INFOCOM 2007, Anchorage, AL (2007)
17. Luo, X., Dong, M., Huang, Y.: On Distributed Fault-Tolerant Detection in Wireless Sensor Networks. IEEE Transactions on Computers 55(1), 58–70 (2006)
18. Kim, J., Ravindran, B.: Opportunistic Real-Time Routing in Multi-Hop Wireless Sensor Networks. In: SAC (2009)
19. Liu, H., Wan, P.J., Jia, X.: Maximal Lifetime Scheduling for Sensor Surveillance Systems with K Sensors to 1 Target. IEEE Transactions on Parallel and Distributed Systems 17(12) (2006)
20. Fang, Q., Gao, J., Guibas, L.J.: Locating and Bypassing Routing Holes in Sensor Networks. In: Proceedings of IEEE INFOCOM 2004 (2004)

Polaris: A Fingerprint-Based Localization System over Wireless Networks

Nan Zhang and Jianhua Feng

Department of Computer Science and Technology,
Tsinghua University, Beijing 100084, China
n-zhang10@mails.tsinghua.edu.cn,
fengjh@tsinghua.edu.cn

Abstract. As the foundation of location-based services, accurate localization has attracted considerable attention. A typical WI-FI localization system employs a fingerprint-based method, which constructs a fingerprint database and returns user's location based on similar fingerprints. Existing systems cannot accurately locate users in a metropolitan-scale because of the requirement of large fingerprint data sets, complicated deployment, and the inefficient search algorithm. To address these problems, we develop a localization system called POLARIS. By the contribution of users, we construct a large fingerprint database. We introduce an effective localization model based on novel similarity measures of fingerprints. For fast localization, we devise an efficient algorithm for matching similar fingerprints, and develop a cluster-based representative fingerprint selection method to improve the performance. We conduct extensive experiments on real data sets, and the experimental results show that our method is accurate and efficient, significantly outperforming state-of-the-art methods.

1 Introduction

Getting user's accurate location is the foundation of location-based service. Existing localization systems[1,2,3,4] return user's location by using GPS, GSM, WI-FI, etc. Using GPS [4] to find the location has its own limitation because of the requirement of an unobstructed line of sight to four or more GPS satellites. Some other methods use GSM [5,6,7] cellular tower, which has the drawback of a lower accuracy. Using WI-FI [8,9,1] to locate the user has been more and more popular because it can provide a more stable service than GPS and achieve a much higher accuracy than GSM localization systems. A fingerprint-based WI-FI localization method[2,10] first maps the existing fingerprints (a list of signals with unique identifer and signal strength) with locations and generates user's location based on the similar fingerprints. In this paper, we present a fingerprint-based localization system called POLARIS, solving three challenges: accuracy, efficiency and fingerprint database validity.

H. Gao et al. (Eds.): WAIM 2012, LNCS 7418, pp. 58–70, 2012.

(1) The accuracy of a fingerprint-based localization system mainly depends on the number of fingerprints. The fingerprints should cover all the regions and more fingerprints bring a higher accuracy. Existing methods either requires careful deployment of access points[2] or a professional crew with specialized and expensive equipment to perform wardriving[3]. In our system, we use fingerprints contributed by anonymous users and give them credits. In this way, POLARIS captures a large quantity of fingerprints and always keeps the database fresh.

(2) In order to achieve a higher accuracy, the system needs massive fingerprints; however, the similar fingerprint candidates for a single query can be several thousand; therefore, if the raw fingerprints are not preprocessed and indexed properly, the search efficiency would be low. We introduce an effective localization model and devise a novel cluster-based indexing method to select representative fingerprints, decreasing the number of candidates. Our search algorithm can effectively find the user's location and the pruning technique improves the performance greatly.

(3) For those changing conditions, i.e., existing WI-FI access points being moved or removed, the system has a mechanism to capture changes immediately and refresh the fingerprint data sets. POLARIS maintains the coarse locations of access points and when a fingerprint comes, it first inspects the validity of the index. We also give an expired time for fingerprints and index them incrementally. To summarize, we make the following contributions.

- We design and implement an innovative metropolitan-scale localization system POLARIS, which has been commonly used and widely accepted.
- We develop a localization model based on novel similarity measures of fingerprints and design an efficient algorithm to accurately locate the user.
- We devise a cluster-based incremental representative fingerprint selection schema to improve the performance, reducing the storage space and decreasing the search time significantly.
- We have conducted an extensive set of experiments and results show that POLARIS is more accurate than state-of-the-art methods and the indexing mechanism performs well.

The rest of this paper is organized as follows. Section 2 formulates the localization problem and gives an overview of the system. The POLARIS model is presented in Section 3. We introduce the detail of search algorithm and improved representative selection algorithm in Section 4. Experimental results are provided in Section 5. We review the related work in Section 6 and make a conclusion in Section 7.

2 Polaris Overview

Problem Formulation. Formally, our paper considers a set of fingerprint data $\mathcal{F}$. Each fingerprint $f \in \mathcal{F}$ contains a set of signal sources $\mathcal{S}$ and a location l, i.e., $f = \langle \mathcal{S}, l \rangle$. The similarity value between f_i and f_j is denoted by $\text{SIM}(f_i, f_j)$, which will be discussed in Section 3. Specifically, each $s \in \mathcal{S}$ has a unique

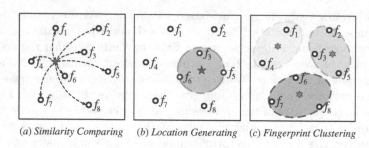

(a) Similarity Comparing   (b) Location Generating   (c) Fingerprint Clustering

Fig. 1. POLARIS Framework

identifier p and a value θ, representing the received signal strength, i.e., $s = \langle p, \theta \rangle$. The query from a user is a fingerprint without location, denoted by f_x. Giving a threshold τ, our problem is to find all the similar fingerprints $f_i \in \mathcal{F}$ such that $\text{SIM}(f_i, f_x) > \tau$ and uses f_i to generate the user's final location.

Polaris Framework. POLARIS framework contains a search part to find the accurate location and a representative selection part to cluster the fingerprints.

Our localization system first uses the collected fingerprints to construct a signal index in order to map each fingerprint to its location. When a user queries, system finds the fingerprints that have higher similarity values with the query and generates the location using those fingerprints. To reduce the index size and increase the search speed, our representative selection framework clusters the similar fingerprints and generates representative ones.

As an example, Figure 1(a) shows a signal map containing eight fingerprints $f_1, f_2, ..., f_8$. The map is established before the search process, using the data contributed by users. When a query f_x, the star, comes, it compares with $f_1, f_2, ..., f_8$ to find the fingerprints that have the higher similarity values with f_x, for example, f_3, f_5, f_6 in Figure 1(b). The next step is to generate the location using these fingerprints. For the similar fingerprints, we use a cluster-based representative selection method to merge them into one fingerprint. In Figure 1(c), the eight fingerprints are clustered into three regions and generate three new representative fingerprints based on our similarity function.

System Architecture. The POLARIS system includes three components:

1) Collector: This component collects the data submitted by anonymous users when they are able to access their locations. For example, users can submit data when they are on their way to supermarkets or to a park. We give them credits to encourage them to upload data; therefore, we can automatically collect data without carrying the burden of mapping the whole world by ourselves. The quantity and quality of the data can also be guaranteed.

2) Selector: Since users can submit millions of fingerprints each day, this component is crucial to our system. For those noise and repeated data, this component first cleans them to reduce the data size. It then clusters the cleaned data

to find representative fingerprints in order to further reduce the index size and decrease the search time, achieving the metropolitan-scale localization, based on the similarity function we design. Finally, it builds up a signal index for the search step.

3) Searcher: Using the signal index and a similarity function, this component finds the user's location. When a query comes, system generates the fingerprint candidates from the index and calculates the similarity value of each candidate with the query. The result location is generated from those fingerprints that have higher similarity values.

3   Polaris Model

We present the POLARIS models: a similarity model for measuring the similarity of two fingerprints and a localization model for generating the final location.

Similarity Model. The similarity function is used to represent how similar two fingerprints are. It groups the signals into levels and then calculates the similarity value of each signal pair to generate the final value of two fingerprints.

The signal strength θ, measured by dBm (decibel relative to one milliwatt), has a range from -60 dBm to -99 dBm. Values that are greater than -60 or less than -99 are considered as noise points caused by equipment mistakes or signal disturbances. $P = 10^{\theta/10}$ shows the relation between absolute power P and dBm. For a 3 dBm increase, the power is roughly doubled.

$$\theta' = 9 - \left\lceil \frac{|\theta + 59|}{5} \right\rceil \tag{1}$$

In real experiments, we find out that although in a fixed location, the θ from an access point may change for 4 dBm to 6 dBm from time to time caused by the disturbances of the other signals. To minimize fluctuations in the received signals, we group the signals into eight levels from 1 to 8 using Equation 1. For example, a signal with $\theta = -72$ is grouped into level 6. This level mechanism can describe the similarity value of two signals properly and precisely, taking both the number differences and the signal powers into consideration.

$$\omega(\theta_1, \theta_2) = 1 - |2^{\theta'_1} - 2^{\theta'_2}|/2^8 \tag{2}$$

Since we have already grouped the θ into different levels, we can use Equation 2 to calculate the similarity value, $\omega(\theta_1, \theta_2)$, of two signals. $|2^{\theta'_1} - 2^{\theta'_2}|$ represents the difference of two signals, measuring the distance between them. For ease to compare, we normalize the difference into $[0, 1]$ by dividing 2^8 and the normalized value is then subtracted by 1 to form the final similarity value. Here are some examples:

$$\omega(-73, -80) = 1 - |2^6 - 2^4|/2^8 = 0.813 \quad \omega(-83, -90) = 1 - |2^4 - 2^2|/2^8 = 0.953$$

For two fingerprints, the number of same signal pairs takes a large weight in the function. In Table 1(a), f_1 and f_x have four same signals, s_1, s_3, s_4, s_6 while

Table 1. Fingerprint & Index

(a) A Set of Fingerprints & A Query f_x

f_1		f_2		f_3		f_4		f_x	
s	θ	s	θ	s	θ	s	θ	s	θ
s_1	5	s_1	5	s_1	6	s_1	5	s_1	4
s_2	2	s_3	1	s_2	2	s_3	6	s_3	7
s_3	4	s_4	3	s_5	6	s_4	3	s_4	3
s_4	1	s_6	7	s_6	7	s_8	1	s_6	5
s_6	6	s_7	2	s_7	3	s_9	4	s_8	2
		s_9	4	s_8	2			s_9	4
				s_9	2				

(b) A Set of Inverted Index

$\mathcal{I}_{s_1}$	$\mathcal{I}_{s_2}$	$\mathcal{I}_{s_3}$	$\mathcal{I}_{s_4}$	$\mathcal{I}_{s_5}$	$\mathcal{I}_{s_6}$	$\mathcal{I}_{s_7}$	$\mathcal{I}_{s_8}$	$\mathcal{I}_{s_9}$
f_1	f_1	f_1	f_1	f_3	f_1	f_2	f_3	f_2
f_2	f_3	f_2	f_2		f_2	f_3	f_4	f_3
f_3		f_4	f_4		f_3			f_4
f_4								

f_4 and f_x have five same signals, s_1, s_3, s_4, s_8, s_9. Therefore, the similarity value $\text{SIM}(f_4, f_x)$ should be greater than $\text{SIM}(f_1, f_x)$. Equation 3 shows the similarity function of f_i and f_x; the similarity value is the sum of ω plus n, the number of same signals.

$$\text{SIM}(f_i, f_x) = \sum \omega + n \tag{3}$$

Suppose the example in the Table 1(a), $\text{SIM}(f_1, f_x)$. The two fingerprints have four same signals and the signal identifiers are s_1, s_3, s_4, s_6. $\text{SIM}(f_1, f_x) = \omega_1 + \omega_3 + \omega_4 + \omega_6 + 4 = 0.9375 + 0.5625 + 0.9766 + 0.875 + 4 = 7.3516$.

Localization Model. Based on the similarity model, we first find several fingerprints that have higher values. Using the locations of those fingerprints, system generates the final location for user.

The number of fingerprints used to generate the location is an important factor. We sort the values first. If the value differences are greater than a threshold, we stop getting fingerprints and use those to generate the location; however, when the differences are small, this process may add more fingerprints and the result location would be less accurate. Therefore, we use an alternative way, which is to use a constant number of fingerprints.

Giving a number α, we use the top α fingerprints to generate the final location. A straightforward method is to use the most similar fingerprint as the result. This method is called NN method. Another way is to use α fingerprints, which is called αNN. For αNN, we use the locations of the α fingerprints to generate a mini circle and the center of the circle is the user's location; however, this would bring errors since the mini circle can be large and the center may be influenced by less important fingerprints. An alternative way is to calculate the average location. In the real experiments, we find that using the average value as the final result has a better performance.

Suppose the example in Table 1(a). We first calculate $\text{SIM}(f_1, f_x) = 7.3516$, $\text{SIM}(f_2, f_x) = 9.0703$, $\text{SIM}(f_3, f_x) = 7.6406$ and $\text{SIM}(f_4, f_x) = 9.6797$ and then sort them as f_4, f_2, f_3, f_1. For a $\alpha = 1$, the location of f_x is the location of f_4. For $\alpha = 2$, the location is generated by the locations of f_4 and f_2.

Algorithm 1: Basic Indexing Algorithm

Input: $\mathcal{F}$: A collection of fingerprints
Output: $\mathcal{I}$: A collection of inverted indexes
1 **begin**
2   Initiate a *set*;
3   **for** f *in* $\mathcal{F}$ **do**
4   **for** s *in* $f.\mathcal{S}$ **do**
5           $s.\theta = 9 - |s.\theta + 59|/5$;
6           $\mathcal{I}_s$.add(f); set.add(s);

Fig. 2. Basic Indexing Algorithm

4 Polaris Algorithm

In this section, we introduce the detail of POLARIS algorithm. In Section 4.1, we describe the POLARIS index. The search algorithm and the pruning technique will be presented in Section 4.2. To further improve the search efficiency, we devise a cluster-based representative fingerprint selection algorithm in Section 4.3 and an incremental improvement of it in Section 4.4.

4.1 Indexing

Scanning each f, POLARIS builds an inverted index for each signal s, denoted by $\mathcal{I}_s$. $\mathcal{I}_s$ includes a list of f that contains s. Table 1(b) shows an inverted index example, using the raw data f_1, f_2, f_3, f_4 in Table 1(a). For example, for s_3, it appears in f_1, f_2 and f_4; therefore, the inverted index $\mathcal{I}_{s_3}$ contains those fingerprints. We give the pseudo-code of our indexing algorithm in Figure 2.

4.2 Search Algorithm

Finding Candidates. POLARIS finds out all the fingerprints f_i that has intersections with f_x ($\mathcal{S}_i \cap \mathcal{S}_x \neq \varnothing$) and puts the result in a set to avoid duplication. The finding process can be easily achieved by visiting all the $\mathcal{I}_s$ where $s \in \mathcal{S}_x$.

Sorting Candidates. For all the fingerprint candidates, POLARIS calculates the similarity value using Equation 3. All the values are maintained in a max-heap. The fingerprint algorithm terminates and returns several fingerprints that have greater similarity values.

Generating Location. The algorithm gets α top fingerprints that have larger similarity value, using the model mentioned in Section 3. After calculating the average location, it gives the location back to the user.

Algorithm 2: Search Algorithm

    Input: f_x: A query fingerprint; $\mathcal{I}$: A collection of inverted indexes
    Output: l: The user's location
1 **begin**
2   Initiate a *set*, a *heap*, and a *list*;
3   **for** s *in* $f_x.S$ **do**
4   *set*.add($\mathcal{I}$.get(s));

5     $list =$ SORTFINGERPRINTS(*set*);
6   **for** f *in list* **do**
7   **if** $2 * f.n > heap.least$ **then**
8            $w =$ SIM(f, f_x);
9   *heap*.add(w);

10   **else**
11   break;

12     $l =$ GENERATELOCATION(*heap*);
13   **return** l;

Fig. 3. Search Algorithm

Pruning Technique. From Equation 3, we see the similarity value is no more than twice the number of same signals, $\text{SIM}(f_i, f_x) = \sum \omega + n \le 2 * n$. Therefore, we use $2 * n$ as an upper bound and the algorithm can be terminated without calculating all the similarity values. Since we want to find the top α fingerprints, we first sort the fingerprint candidates using the value of n. During the query, we maintain α fingerprints in the heap. For the current fingerprint that we need to calculate the similarity value, if $2 * n$ is less than the smallest similarity value in the heap, it means from that fingerprint, all the remaining ones cannot be greater than the least in the heap so they cannot be added. Therefore, the search process can be terminated and the fingerprints in the heap are the ones we need.

For example, in Table 1(a), we set $\alpha = 2$. We first sort the fingerprints based on n, that is f_2, f_4, f_1, f_3. Since $\text{SIM}(f_2, f_x) = 9.0703$, we put f_2 in the heap. After calculating $\text{SIM}(f_4, f_x) = 9.6797$, the fingerprints in the heap are f_4, f_2. For the next fingerprint f_1, $2 * n = 8$, which is less than the smallest value $\text{SIM}(f_2, f_x)$. Therefore, the algorithm is terminated and the average location of f_4, f_2 is given to the user.

We give the pseudo-code of our search algorithm in Figure 3. POLARIS first initiates a set to maintain the fingerprint candidates and a max-heap for sorting the candidates with their similarity values (line 2). Then it finds all the fingerprint candidates that need to be compared with f_x (line 3), sorts them based on the number of same signals (line 5), and puts the sorted results in a list. The algorithm calculates the similarity value using Equation 3 and puts the result into the heap. It terminates based on the pruning mechanism (line 6 to 11). The final location is calculated using the model we mentioned in Section 3 (line 12).

4.3   Cluster-Based Representative Selection

For our metropolitan-scale localization system, the fingerprints collected each day can be more than a million. Maintaining all the fingerprints in the index lowers the search efficiency and gives burden to the server. Therefore, we present a cluster-based method to select representative fingerprints, indexing those fingerprints to get a better efficiency.

$$\text{SIM}'(f_i, f_j) = \left(\sum \omega\right)/n \tag{4}$$

For a single fingerprint f_i, the system first visits all the fingerprints that have already indexed, comparing with them using Equation 4. Equation 4 is the sum of similarity values of each similar signal $\sum \omega$ divide the number of same signals n. It has a slight difference with the similarity function we have defined in Equation 3. Using SIM', we can modify all the similarity values into $[0,1]$ in order to control the selection rate. Suppose the example in the Table 1(a), $\text{SIM}'(f_1, f_4)$. $\text{SIM}'(f_1, f_4) = (\omega_1 + \omega_3 + \omega_4)/3 = (1 + 0.8125 + 0.9766)/3 = 0.9297$.

After calculating all the similarity values, we find the fingerprint f_k with the greatest SIM' and compare it with a threshold β. If the value is greater than β, the two fingerprints f_i and f_k are similar to each other and can be merged together. We modify the signals in f_k and the location of f_k. However, if the value is less than or equal to β, it means that no fingerprints are similar to f_i so we index it and insert it into the fingerprint list as a new one. For example, in Table 1(a), $\text{SIM}'(f_2, f_4) = 0.9395$. If β is 0.95, then the two fingerprints cannot be merged. If β is 0.93, then they can be merged together.

We should also take the validity of the fingerprint into consideration since the access points are frequently modified. The basic method is to rebuild the index for every constant period t, for example, five days. During the t period, POLARIS accumulates the data submitted by the users and then selects the representative fingerprints. We give the pseudo code in Figure 4.

Algorithm 3: Cluster-Based Representative Selection Algorithm

---

    Input: $\mathcal{F}$: Fingerprint list
    Output: $\mathcal{I}$: The inverted index
1 **begin**
2   Initiate a *set*;
3   **for** f_i *in* $\mathcal{F}$ **do**
4          $w' = \text{FINDTHETOPSIMILARVALUE}(f_i)$;
5   **if** $w' > \beta$ **then**
6              $\text{MERGEFINGERPRINT}(f_i, f_j)$;
7   **else**
8              $\text{MODIFYINDEX}(f_i)$; $set.add(f_i)$;

Fig. 4. Cluster-Based Representative Selection Algorithm

4.4   Incremental Representative Selection

Using the basic method has some drawbacks. First and foremost, during the t period, some existing access points may be already relocated or removed; therefore, the data cannot be trusted. What is more, when we rebuild the index, the service may stop for a few moment, making it inconvenient to use. Therefore, our goal is to generate the representative fingerprints using an incremental algorithm.

The difference between basic and incremental method is that we rebuild each signal's index instead of rebuilding the whole index. For every signal, we generate a coarse location to approximately represent the location of that access point. When a fingerprint comes, POLARIS compares the coarse location of each signal in the fingerprint with the location of that fingerprint. If the location is far away from the coarse location for γ, for example, 3 kilometers, it means that access point may have been moved. That signal should be eliminated from all the fingerprints and the index should also be rebuild.

We give the pseudo-code in Figure 5. For each signal in f, it tests validity of the data and modifies the index if the access point has been moved (line 4). It then compares with the existing fingerprints to calculate the similarity value (line 7). If the top value in the heap is greater than β, then the two fingerprints are similar and can be merged together (line 10). Otherwise, it adds f into the set and indexes it as a new one (line 12).

5   Experiment

We have implemented the POLARIS system and conducted a set of experiments on real data sets. We used the data contributed by users in three different regions: a community, a commercial district and a high technology industrial area. We then collected data in these regions and randomly chose 50% of the data to be the test cases. Table 2 shows the detailed information of the data sets.

Fingerprint Evaluation. We evaluated the effects of using different α to generate the final location in the last step of search algorithm (Figure 3, line 12). We set the value of α from 1 to 10, running our algorithm on three data sets. Figure 6(a) shows the result.

We see that the average error of the final result increases with α from less than 5 meters to 15 meters. The reason is that when we choose more fingerprints to generate the final location, the less similar ones may be included; therefore, the result error increases. From the figure, we can see, when we set the α value to 1, that is, choosing the most similar fingerprint to be the result, our algorithm has a median accuracy less than 5 meters, which is better than the other localization systems.

Method Comparison. We evaluated our fingerprint algorithm and representative selection algorithm. Comparing with the state-of-the-art Google Localization System[1] and SOSO Localization System, our system has a higher accuracy.

---

[1] http://code.google.com/apis/gears/geolocation_network_protocol.html

---

Algorithm 4: Incremental Representative Selection Algorithm

Input: f: A Fingerprint, set: Indexed fingerprints, $\mathcal{L}$: The coarse locations
Output: $\mathcal{I}$: The inverted index

```
1  begin
2  │  for sᵢ in f.S do
3  │  │  if DISTANCE(L.sᵢ,f.l) > γ then
4  │  │  └  EXPIRE(sᵢ);
5  │  Initiate a heap;
6  │  for fₖ in set do
7  │  └  heap.add(SIM'(f, fₖ));
8  │  w' = heap.pop();
9  │  if w' > β then
10 │  └  MERGEFINGERPRINT(f, fₖ);
11 │  else
12 │  └  MODIFYINDEX(f); set.add(f);
```

Line 2: for s_i in $f.S$ do
Line 3: if DISTANCE$(\mathcal{L}.s_i,f.l) > \gamma$ then
Line 4: EXPIRE(s_i);
Line 5: Initiate a $heap$;
Line 6: for f_k in set do
Line 7: $heap$.add(SIM$'(f, f_k)$);
Line 8: $w' = heap$.pop();
Line 9: if $w' > \beta$ then
Line 10: MERGEFINGERPRINT(f, f_k);
Line 11: else
Line 12: MODIFYINDEX(f); set.add(f);

Fig. 5. Incremental Representative Selection Algorithm

Table 2. Data Sets

Name	Type	Fingerprints	Test Cases
Data #1	Residential District	21582	10791
Data #2	Commercial District	26890	13445
Data #3	High-Tech District	31863	15932

Figure 6(b) shows that the accuracy of POLARIS is at least 10 times better than Google and 15 times better than SOSO. After compressing the data with a threshold $\beta = 0.98$, the accuracy is at least 4 times better than Google and 6 times better than SOSO. The reason is that they use a triangulation method, which needs to estimate the location of access points and uses the estimated locations to generate the result. Therefore, the performance is worse than our method.

Index Evaluation. We evaluated the effectiveness and efficiency of our representative selection algorithm. The four aspects using to evaluate the algorithm are the number of fingerprints indexed, the number of candidates for each search, the search time and the result accuracy. We set β from 0.99 to 0.90 and set $\alpha = 1$, running the test on three data sets. Figure 7 shows the result.

We see that the size of the index, the fingerprint candidates that need to compare with f_x and the search time decrease greatly with β. Since β is the threshold we use to verify whether the two fingerprints are similar or not, a decreasing in β compresses more fingerprints into one; therefore, we reduce a great many of fingerprints and the index size decreases. As a result, when a query

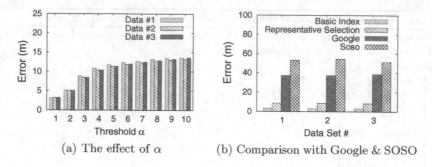

(a) The effect of α (b) Comparison with Google & SOSO

Fig. 6. Accuracy Comparison

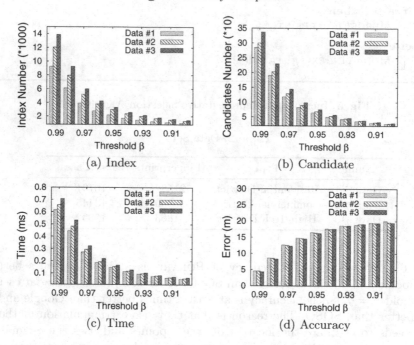

(a) Index   (b) Candidate

(c) Time   (d) Accuracy

Fig. 7. Effectiveness of Representative Selection

comes, we need to compare with less candidates and the search time decreases. However, since the index is small, we cannot map all the locations to their accurate fingerprints; therefore, the median error increases a little. Choosing the best β has great influence to POLARIS. For example, when we set $\beta = 0.96$, the result accuracy is about 15 meters while the other three factors decrease greatly.

The experimental results of POLARIS show that our system has a higher accuracy and precision than state-of-the-art systems and the incremental representative selection method can compress the fingerprint data effectively and efficiently.

6   Related Work

Many location systems[1] have been proposed for the past two decades. Some systems require special equipments in order to locate the user. The Active Badge system[11] relies on infrared technology, presenting the location of the badge that emits the infrared; however, the transmission range of infrared is the limitation. The Bat system[12] and Cricket system[13] use ultrasonic technology for getting a higher accuracy. Receivers get the ultrasonic pulses, which can be used to calculate the distance. The systems require a large infrastructure and difficult to deploy. TRIP[14] and SurroundSense[15] use image and adjacent contexts to help locate. With the help of the equipments, those systems can have a better accuracy; however, it is impossible for them to use in a large area. Some other systems[16,5,6,7] use GSM cellular tower; however, the lower accuracy cannot reach the standard of an accurate location. Place Lab[3] uses particle filter method to locate the user in the open area, which has the accuracy about 20 to 30 meters. [17] presents a case study of applying particle filters to location estimation; however, those methods require more computation and memory. RADAR[2,10] first proposes the fingerprint method using the previously collected data. It provides an accurate localization in building but the massive deployment make it unable to use in the large area. [18,19,20] use similar method like RADAR but taking into consideration of different kinds of sources. However, they cannot be used in the open area for the complicated deployment and inefficient algorithm.

7   Conclusion

In this paper, we present and implement a localization system, called POLARIS. Using a motivate mechanism, our system accumulates the data contributed by anonymous users, making it easy to deploy. We design a novel localization model to measure the similarity of two fingerprints and generate the user's location. Our search algorithm has a higher accuracy between 5m to 15m. To improve the performance, we present a novel incremental representative fingerprint selection schema to compress the massive raw data in order to support the metropolitan-scale localization. The result from experiments and real-world usage show that our system has a better performance than the existing state-of-the-art systems in both the accuracy and capability of dealing with the massive data.

References

1. Hightower, J., Borriello, G.: Location systems for ubiquitous computing. Computer 34(8), 57–66 (2001)
2. Bahl, P., Padmanabhan, V.N.: Radar: An in-building rf-based user location and tracking system. In: INFOCOM, pp. 775–784 (2000)

3. LaMarca, A., Chawathe, Y., Consolvo, S., Hightower, J., Smith, I., Scott, J., Sohn, T., Howard, J., Hughes, J., Potter, F., Tabert, J., Powledge, P., Borriello, G., Schilit, B.N.: Place Lab: Device Positioning Using Radio Beacons in the Wild. In: Gellersen, H.-W., Want, R., Schmidt, A. (eds.) PERVASIVE 2005. LNCS, vol. 3468, pp. 116–133. Springer, Heidelberg (2005)
4. Enge, P., Misra, P.: Special issue on global positioning system. Proceedings of the IEEE 87(1), 3–15 (1999)
5. Chen, M.Y., Sohn, T., Chmelev, D., Haehnel, D., Hightower, J., Hughes, J., LaMarca, A., Potter, F., Smith, I., Varshavsky, A.: Practical Metropolitan-Scale Positioning for GSM Phones. In: Dourish, P., Friday, A. (eds.) UbiComp 2006. LNCS, vol. 4206, pp. 225–242. Springer, Heidelberg (2006)
6. Otsason, V., Varshavsky, A., LaMarca, A., de Lara, E.: Accurate GSM Indoor Localization. In: Beigl, M., Intille, S.S., Rekimoto, J., Tokuda, H. (eds.) UbiComp 2005. LNCS, vol. 3660, pp. 141–158. Springer, Heidelberg (2005)
7. Benikovsky, J., Brida, P., Machaj, J.: Localization in Real GSM Network with Fingerprinting Utilization. In: Chatzimisios, P., Verikoukis, C., Santamaría, I., Laddomada, M., Hoffmann, O. (eds.) MOBILIGHT 2010. LNICST, vol. 45, pp. 699–709. Springer, Heidelberg (2010)
8. Cheng, Y.C., Chawathe, Y., LaMarca, A., Krumm, J.: Accuracy characterization for metropolitan-scale wi-fi localization. In: MobiSys, pp. 233–245 (2005)
9. Haeberlen, A., Flannery, E., Ladd, A.M., Rudys, A., Wallach, D.S., Kavraki, L.E.: Practical robust localization over large-scale 802.11 wireless networks. In: MOBICOM, pp. 70–84 (2004)
10. Bahl, P., Balachandran, A., Padmanabhan, V.: Enhancements to the radar user location and tracking system. Microsoft Research Technical Report (2000)
11. Want, R., Hopper, A., Falcao, V., Gibbons, J.: The active badge location system. ACM Trans. Inf. Syst. 10(1), 91–102 (1992)
12. Harter, A., Hopper, A., Steggles, P., Ward, A., Webster, P.: The anatomy of a context-aware application. In: MOBICOM, pp. 59–68 (1999)
13. Priyantha, N.B., Chakraborty, A., Balakrishnan, H.: The cricket location-support system. In: MOBICOM, pp. 32–43 (2000)
14. de Ipiña, D.L., Mendonça, P.R.S., Hopper, A.: Trip: A low-cost vision-based location system for ubiquitous computing. PUC 6(3), 206–219 (2002)
15. Azizyan, M., Constandache, I., Choudhury, R.R.: Surroundsense: mobile phone localization via ambience fingerprinting. In: MOBICOM, pp. 261–272 (2009)
16. Laitinen, H., Lahteenmaki, J., Nordstrom, T.: Database correlation method for gsm location. In: Vehicular Technology Conference, vol. 4, pp. 2504–2508 (2001)
17. Hightower, J., Borriello, G.: Particle Filters for Location Estimation in Ubiquitous Computing: A Case Study. In: Davies, N., Mynatt, E.D., Siio, I. (eds.) UbiComp 2004. LNCS, vol. 3205, pp. 88–106. Springer, Heidelberg (2004)
18. Pandya, D., Jain, R., Lupu: Indoor location estimation using multiple wireless technologies. In: PIMRC, pp. 2208–2212 (2003)
19. Smailagic, A., Kogan, D.: Location sensing and privacy in a context-aware computing environment. IEEE Wireless Communications 9(5), 10–17 (2002)
20. Papapostolou, A., Chaouchi, H.: WIFE: Wireless Indoor Positioning Based on Fingerprint Evaluation. In: Fratta, L., Schulzrinne, H., Takahashi, Y., Spaniol, O. (eds.) NETWORKING 2009. LNCS, vol. 5550, pp. 234–247. Springer, Heidelberg (2009)

A High-Performance Algorithm
for Frequent Itemset Mining

Jun-Feng Qu and Mengchi Liu

State Key Lab. of Software Engineering, School of Computer,
Wuhan University, Wuhan 430072, China
cocoqjf@gmail.com, mengchi@sklse.org

Abstract. Frequent itemsets, also called frequent patterns, are important information about databases, and mining efficiently frequent itemsets is a core problem in data mining area. Pattern growth approaches,
such as the classic FP-Growth algorithm and the efficient FPgrowth*
algorithm, can solve the problem. The approaches mine frequent itemsets by constructing recursively conditional databases that are usually
represented by prefix-trees. The three major costs of such approaches are
prefix-tree traversal, support counting, and prefix-tree construction. This
paper presents a novel pattern growth algorithm called BFP-growth in
which the three costs are greatly reduced. We compare the costs among
BFP-growth, FP-Growth, and FPgrowth*, and illuminate that the costs
of BFP-growth are the least. Experimental data show that BFP-growth
outperforms not only FP-Growth and FPgrowth* but also several famous
algorithms including dEclat and LCM, ones of the fastest algorithms, for
various databases.

Keywords: Algorithm, data mining, frequent itemsets.

1 Introduction

Frequent itemsets derived from databases have been extensively used in association rule mining [1], clustering [2], classification [3], and so on. Therefore, mining
efficiently frequent itemsets is very important in data mining area.

1.1 Problem Definition

Let $I = \{i_1, i_2, i_3, \ldots, i_n\}$ be a set of n distinct items. An itemset X is a subset
of I, i.e., $X \subseteq I$, and X is called a k-itemset if $|X| = k$. DB is a transaction
database, where each transaction T is also a subset of I, i.e., $T \subseteq I$. We say
that transaction T *satisfies* itemset X if $X \subseteq T$. Let $S(DB)$ be the number
of transactions in DB and $C(X)$ the number of transactions satisfying X, and
then $C(X)/S(DB)$ is the *support* of X. Given a user specified minimum support
threshold ξ, an itemset X is called *frequent* if $(C(X)/S(DB)) >= \xi$. The frequent itemset mining problem [4] is to enumerate all the frequent itemsets with
their supports given a DB and a ξ. For a database with n items, 2^n itemsets
must be checked, and thus the problem is intractable.

H. Gao et al. (Eds.): WAIM 2012, LNCS 7418, pp. 71–82, 2012.
© Springer-Verlag Berlin Heidelberg 2012

1.2   Previous Solutions

Apriori [5] is one of the most well-known algorithms for frequent itemset mining, and it uses the downward closure property of itemset support: any superset of an infrequent itemset is infrequent and any subset of a frequent itemset is frequent. After scanning a database, Apriori find out all the frequent 1-itemsets from which it generates the candidate 2-itemsets. Afterwards, Apriori scans iteratively the database to find out all the frequent k-itemsets ($k >= 2$) from which it generates the candidate $(k + 1)$-itemsets. The qualification as a candidate $(k + 1)$-itemset is that all of its subsets containing k items, namely $k+1$ k-itemsets, are frequent. The Apriori-based approaches such as [6], [7] are called candidate generation-and-test approaches.

Pattern growth approaches such as FP-Growth [8] adopt the divide-and-conquer strategy to mine frequent itemsets. They first identify all the frequent items in a database, and subsequently divide the database into the disjoint conditional databases according to the frequent items. After that, each conditional database is recursively processed. For a frequent k-itemset, each frequent item in its conditional database can be appended to the k-itemset, which makes it grow into a frequent $(k+1)$-itemset. Many pattern growth approaches employ prefix-trees to represent (conditional) databases. Prefix-trees are highly compressed on which both database scan and support counting can be performed fast.

There are a number of other mining approaches. Using a vertical database layout, the Eclat algorithm [9] links each item up with a set of transaction identifiers and then intersects the sets to mine frequent itemsets. The TM algorithm [10] is a variant of Eclat, and the dEclat algorithm [11] incorporating the "diffset" technique significantly improves Eclat's performance. The FIUT algorithm [12] mines frequent itemsets by gradually decomposing a length k transaction into k length $(k - 1)$ transactions. Tiling [13] makes the best of CPU cache to speed previous algorithms up; CFP-growth [14] consumes less memory than other algorithms; LCM [15] integrates multiple optimization strategies and achieves good performance.

1.3   Contribution

The two major costs for a mining algorithm are database scan (or prefix-tree traversal) and support counting. For the algorithms constructing conditional databases, the construction cost of conditional databases is also nontrivial.

For a large database and/or a small minimum support, the mining task usually becomes very intractable because numerous transactions need to be scanned and a very large number of itemsets must be checked. In this case, a high-performance algorithm is indispensable. To obtain better performance, a common method is to reduce the costs of a previous algorithm as much as possible. For example, the FPgrowth* algorithm [16] significantly improves FP-Growth's performance by reducing half the traversal cost. The difficulty of the method is that the decrease

in a cost can lead to the increase in another cost. Although FPgrowth* outperforms FP-Growth, it is usually neglected that the counting cost of FPgrowth*
is more than that of FP-Growth.

The paper presents a novel algorithm, called BFP-growth, for frequent itemset
mining. BFP-growth employs prefix-trees to represent (conditional) databases
as most pattern growth algorithms do. We compare the traversal, counting, and
construction costs among BFP-growth, FP-Growth, and FPgrowth* in details,
and demonstrate that these costs in BFP-growth are greatly reduced. We conduct extensive experiments in which several famous algorithms are tested besides
the three algorithms aforementioned. Experimental data show that BFP-growth
achieves significant performance improvement over previous works. The rest of
the paper is arranged as follows. Section 2 looks back on the classic pattern
growth approach. Section 3 presents the BFP-growth algorithm. The three costs
of BFP-growth, FP-Growth and FPgrowth* are analyzed in Section 4. Section
5 gives experimental data. The paper ends in the conclusion of Section 6.

2   Pattern Growth Approach

The classic pattern growth approach, FP-Growth [8], employs extended prefix-
trees called FP-trees to represent (conditional) databases. FP-Growth first identifies all the frequent items by a scan over a (conditional) database. After that, it
constructs an (conditional) FP-tree by processing each transaction as follows: (1)
pick out the frequent items from the transaction; (2) sort the items in frequency-
descending order to generate a branch; (3) insert the branch into the FP-tree.
Fig. 1(a) and (b) show a transaction database and the corresponding FP-tree.
A prefix-tree's node contains two fields: an *item* and a *count*, and an FP-tree's
node holds two extra pointers: a *parent-link* pointing to its parent node and a
node-link pointing to another node containing the same item. There is a header
table for each FP-tree, in which an entry registers a frequent item, its support,
and the head of the list that links all the nodes containing the item.

FP-Growth processes all the items in a header table one by one. For item i,
the paths from all the nodes containing i to the root constitute the conditional

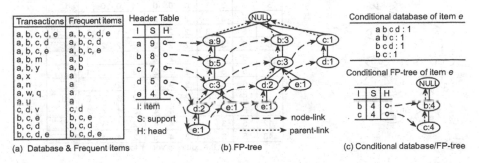

(a) Database & Frequent items   (b) FP-tree   (c) Conditional database/FP-tree

Fig. 1. Database, FP-tree, and conditional database/FP-tree (minimum support=30%)

database of i. FP-Growth traverses the paths along both node-links and parent-links to count for the items in the paths. After identifying the frequent items in the conditional database, FP-Growth traverses the paths again to construct the conditional FP-tree of i. Fig. 1(c) shows the conditional database and conditional FP-tree of item e.

FPgrowth* is an efficient variant of FP-Growth, and it counts for the items in an FP-tree when constructing the FP-tree. In this way, FPgrowth* reduces half the traversal cost of FP-Growth and thereby gains significant performance improvement, although it increases the counting cost. FPgrowth* [16] is the fastest algorithm in IEEE ICDM Workshop on frequent itemset mining implementations (FIMI'03). To obtain better performance, the questions are: (1) Can we further reduce the traversal cost? (2) Why does FPgrowth* increase the counting cost? Can that be avoided? (3) Can the FP-Growth-based methods mine frequent itemsets using plain prefix-trees (e.g., Fig. 2) rather than extended prefix-trees?

3   BFP-Growth Algorithm

BFP-growth mines frequent itemsets by constructing recursively conditional prefix-trees, as most pattern growth approaches do. Different from previous approaches, for a (conditional) prefix-tree, BFP-growth first builds the counting vectors for all the items in the tree, and subsequently constructs simultaneously all the conditional prefix-trees of next level.

3.1   Building Counting Vectors

Given a prefix-tree, the paths from all the nodes containing item i to the root constitute the conditional database of i. To construct the conditional prefix-tree

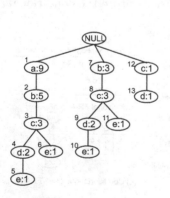

Fig. 2. Prefix-tree

Node's number	CV$_b$	CV$_c$		CV$_d$			CV$_e$				stack
	a	a	b	a	b	c	a	b	c	d	
1	0	0	0	0	0	0	0	0	0	0	
2	5	0	0	0	0	0	0	0	0	0	a
3	5	3	3	0	0	0	0	0	0	0	ab
4	5	3	3	2	2	2	0	0	0	0	abc
5	5	3	3	2	2	2	1	1	1	1	abcd
6*	5	3	3	2	2	2	2	2	2	1	abc
7	5	3	3	2	2	2	2	2	2	1	
8	5	3	6	2	2	2	2	2	2	1	b
9	5	3	6	2	4	4	2	2	2	1	bc
10	5	3	6	2	4	4	2	3	3	2	bcd
11	5	3	6	2	4	4	2	4	4	2	bc
12	5	3	6	2	4	4	2	4	4	2	
13	5	3	6	2	4	5	2	4	4	2	c

CV$_i$: the counting vector for item i

Fig. 3. Building counting vectors

of i, all the frequent items in its conditional database should be first identified. For the purpose, BFP-growth will build the counting vector for i denoted as CV_i. Each item in the conditional database of i corresponds to a component of CV_i. For example, the counting vector for item d of the prefix-tree in Fig. 2 contains the three components corresponding to items a, b, and c in the conditional database of d. After initializing the counting vectors for all the items of a prefix-tree, BFP-growth starts to count for the items in all the conditional databases.

Using a work stack, BFP-growth continually updates the counting vectors for a prefix-tree in the process of traversing the prefix-tree in depth-first way. The stack stores the items in the path from the parent node of the current node to the root. Fig. 3 shows how the counting vectors for the prefix-tree in Fig. 2 are updated when BFP-growth processes each node numbered at its upper left corner according to the depth-first order. For example, when BFP-growth arrives at the node numbered 6, the path from the node to the root is a part of item e's conditional database. Therefore, items a, b, and c in the path stored in the stack are counted and the corresponding components in CV_e are increased by 1 (1 is the *count* of the node numbered 6 and indicates one occurrence of a, b, and c in e's conditional database). In this way, BFP-growth builds the counting vectors for all the items in a prefix-tree by one traversal of the prefix-tree.

3.2   Constructing Conditional Prefix-Trees

After building the counting vectors for all the items in a prefix-tree, BFP-growth can identify the frequent items in any conditional database. Subsequently, BFP-growth will traverse the prefix-tree again to construct simultaneously all the conditional prefix-trees of next level.

When processing the node containing item i, BFP-growth picks first out the frequent items from the items in the path from the parent node of the node to the root according to CV_i. These frequent items are sorted in frequency-descending

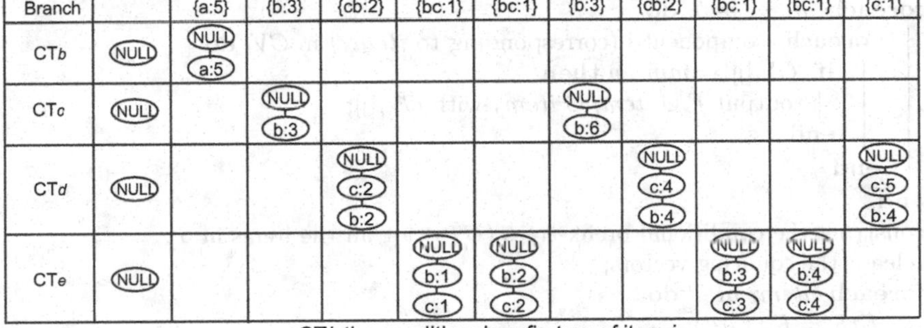

Node's number	1	2	3	4*	5	6	8	9	10	11	13
Stack		a	ab	abc	abcd	abc	b	bc	bcd	bc	c
Branch		{a:5}	{b:3}	{cb:2}	{bc:1}	{bc:1}	{b:3}	{cb:2}	{bc:1}	{bc:1}	{c:1}
CTb	NULL	NULL a:5									
CTc	NULL		NULL b:3					NULL b:6			
CTd	NULL			NULL c:2 b:2				NULL c:4 b:4			NULL c:5 b:4
CTe	NULL				NULL b:1 c:1	NULL b:2 c:2			NULL b:3 c:3	NULL b:4 c:4	

CTi: the conditional prefix-tree of item i

Fig. 4. Constructing conditional prefix-trees

order and subsequently inserted into CT_i (the Conditional prefix-Tree of item i). Fig. 4 demonstrates BFP-growth's construction procedure for the prefix-tree in Fig. 2. Only when a conditional prefix-tree is updated is it depicted in the figure. For example, when BFP-growth arrives at the node numbered 4, there are items a, b, and c stored in the stack. According to the counting results in CV_d, only b and c are frequent (the minimum support is 30%). They are sorted in frequency-descending order, and a branch $\{cb : 2\}$ is generated (2 is the *count* of the node numbered 4 and indicates two occurrences of transaction cb in the conditional database of item d.). Afterwards, the branch is inserted into CT_d.

3.3   Pseudo-code of BFP-Growth

Algorithm 1 shows the pseudo-code of BFP-growth.

BFP-growth first traverses prefix-tree T (the second parameter) to build the counting vectors for all the items in T (line 1). Component j in counting vector CV_i denoted by $CV_i[j]$ stores the support of $item_j$ in the conditional database of $item_i$. If $CV_i[j]$ exceeds minimum support threshold $minsup$ (the third parameter), $item_j$ is frequent in the conditional database of $item_i$. Then, the two items and prefix itemset F (the first parameter) constitute a new frequent itemset, and the itemset with its support $CV_i[j]$ is outputted (lines 2-8). From another perspective, the counting vectors actually store the supports of all the 2-itemsets of T. After outputting the frequent itemsets, BFP-growth constructs simultaneously all the conditional prefix-trees as stated in Section 3.2 (line 9). The counting vectors are released before the algorithm enters the recursions of next level (line 10). At last, for each conditional prefix-tree CT_i, BFP-growth generates its prefix itemset ExF (line 12) and processes it recursively (line 13).

Algorithm 1. *BFP-growth*

Input: F is a frequent itemset, initially empty;
$\qquad\quad$ T is the conditional prefix-tree of F;
$\qquad\quad$ $minsup$ is the minimum support threshold.
Output: all the frequent itemsets with F as prefix.
1 build the counting vectors (CVs) for all the items in T;
2 **foreach** $item_i$ in T **do**
3 $\quad$ **foreach** component j (corresponding to $item_j$) in CV_i **do**
4 $\quad\quad$ **if** $CV_i[j]>=minsup$ **then**
5 $\quad\quad\quad$ output $F \cup item_i \cup item_j$ with $CV_i[j]$;
6 $\quad\quad$ **end**
7 $\quad$ **end**
8 **end**
9 construct the conditional prefix-trees (CTs) for all the items in T;
10 release the counting vectors;
11 **foreach** $item_i$ in T **do**
12 $\quad$ $ExF = F \cup item_i$;
13 $\quad$ BFP-growth(ExF, CT_i, $minsup$);
14 **end**

Given database DB and minimum support $minsup$, after prefix-tree T is constructed from DB and all the frequent 1-itemsets are outputted, BFP-growth($\varnothing$, T, $minsup$) can generate all the frequent k-itemsets ($k >= 2$).

4   Time Analysis

Most pattern growth algorithms derive from FP-Growth [8], in which FPgrowth* [16] is very efficient. Prefix-tree traversal, support counting, and prefix-tree construction are the major costs for these algorithms. The section compares the costs among FP-Growth, FPgrowth*, and BFP-growth.

4.1   Less Traversal Cost

Prefix-tree traversal is always necessary for both support counting and conditional prefix-tree construction, and the traversal cost takes a very large proportion in the whole cost for a mining task. FPgrowth* gains significant performance improvement over FP-Growth by reducing half the traversal cost.

A prefix-tree/FP-tree T with n items has 2^n nodes in the worst case. If the root is at level 0, the number of nodes at level i is combination number C_n^i. For a node at level i, FP-Growth counts for the items in the path from the node to the root, and thereby i nodes are accessed. Let

$$f(m) = C_n^m + C_n^{m+1} + \cdots + C_n^n \quad (1 \leq m \leq n)$$

and n is an even number for convenience of computation. Then, the number of accessed nodes in FP-Growth's counting phase for T is:

$$
\begin{aligned}
\sum_{i=1}^{n} iC_n^i &= 1C_n^1 + 2C_n^2 + 3C_n^3 + \cdots + nC_n^n \\
&= f(1) + f(2) + f(3) + \cdots + f(n) \\
&= (f(1)+f(n)) + (f(2)+f(n-1)) + \cdots + (f(\frac{n}{2})+f(\frac{n}{2}+1)) \\
&= 2^n + 2^n + \cdots + 2^n \\
&= \frac{n}{2} \times 2^n
\end{aligned}
$$

The same number of nodes are accessed in FP-Growth's construction phase. Hence, the total number of nodes accessed by FP-Growth for T is:

$$Accessed\_node\_number(FP\text{--}Growth) = n \times 2^n \tag{1}$$

FPgrowth* merges the counting procedure into the construction procedure (see Section 4.2), and thus the accessed nodes in FPgrowth* for T are half those in FP-Growth, namely:

$$Accessed\_node\_number(FPgrowth^*) = \frac{n}{2} \times 2^n \tag{2}$$

BFP-growth traverses a whole prefix-tree in the counting phase and does it again in the construction phase, and hence the total number of nodes accessed by BFP-growth for T is:

$$Accessed\_node\_number(BFP-growth) = 2 \times 2^n \qquad (3)$$

On the one hand, there are usually many items in a database, and namely n is very large (see Fig. 6); On the other hand, there are a large number of prefix-trees generated during a mining process [17]. Therefore, the traversal cost of BFP-growth is far less than that of FP-Growth and that of FPgrowth*.

4.2   Should the FP-Array Technique Be Incorporated?

FPgrowth* counts for the items in the conditional databases of all the items in an FP-tree when constructing the FP-tree (the FP-array technique) and thereby reduces its traversal cost. Although the technique can also be applied to BFP-growth, we find out that the FP-array technique leads to the increase in counting cost. The following example explains this point.

Suppose the minimum support is 15% (rather than 30%), and then items a, b, c, and d are all frequent in the conditional database of item e for the prefix-tree in Fig. 2 according to the counting results in CV_e in Fig. 3. FPgrowth* constructs CT_e and simultaneously counts for the items in all the conditional databases of next level, which is demonstrated in Fig. 5(a) (It is the FP-tree that FPgrowth* constructs, but both parent-links and node-links don't relate to the analysis here.). FPgrowth* performs 13 times of counting labeled as shaded blocks when constructing CT_e. The counting procedure of BFP-growth performed after CT_e has been constructed is demonstrated in Fig. 5(b), and there are only 8 times of counting labeled with asterisks.

The fundamental reason why the times of counting in BFP-growth are fewer than those in FPgrowth* is that BFP-growth counts for the items in a compressed database (namely a prefix-tree) but FPgrowth* counts in an uncompressed database. Especially for a dense database, there are a relatively large number of transactions and a corresponding relatively highly-compressed prefix-tree, and thus counting on the prefix-tree is more efficient than counting on the

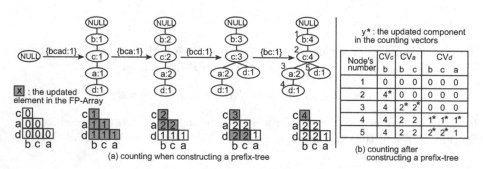

Fig. 5. Counting comparison between FPgrowth* and BFP-growth

transactions. In this case, the FP-array technique cannot significantly speed the algorithm up because the increase in the counting cost counteracts to a large extent the decrease in the traversal cost. It is also the reason why FPgrowth* gives up the FP-array technique when confronted with dense databases [16]. We observed by preparatory experiments that BFP-growth's performance was not significantly improved and was even deteriorated a little in some databases when the counting procedure is merged into the construction procedure. Therefore, BFP-growth doesn't incorporate the technique.

4.3   Plain Prefix-Trees

Another advantage of BFP-growth is that it employs plain prefix-trees, but FP-Growth and FPgrowth* employ extended prefix-trees, namely FP-trees. The following lemma holds for BFP-growth and FP-Growth/FPgrowth*.

Lemma 1. *Given a database and a minimum support, there is a one-to-one correspondence between the FP-trees constructed by FP-Growth/FPgrowth* and the prefix-trees constructed by BFP-growth.*

Proof. (1) BFP-growth constructs the initial prefix-tree from the database as FP-Growth/FPgrowth* constructs the initial FP-tree (see Section 2). Hence, the initial prefix-tree is the same as the initial FP-tree, except that the latter holds a parent-link and a node-link for each node. (2) Without regard to both parent-links and node-links, suppose that FP-tree FPT is the same as prefix-tree PT. For item i in FPT, FP-Growth/FPgrowth* takes the paths from the nodes containing i to the root as the conditional database of i, and BFP-growth does so for item i in PT. Therefore, the conditional FP-tree of i constructed by FP-Growth/FPgrowth* is the same as the conditional prefix-tree of i constructed by BFP-growth. (3) FP-Growth/FPgrowth* processes all the items in FPT (one by one), and BFP-growth processes (simultaneously) all the items in PT as well. The Lemma can be deduced from (1), (2), and (3).   □

Because of extra overheads for building parent-links and node-links for FP-trees, we can conclude from Lemma 1 that the construction cost of BFP-growth is less than that of FP-Growth/FPgrowth* for a mining task.

5   Experiments

Our experiments include the algorithms: BFP-growth, FP-Growth, FPgrowth* (the fastest algorithm on FIMI'03), AFOPT [18], dEclat [11], [19], and LCM [15] (the fastest on FIMI'04). We implemented BFP-growth. To avoid implementation bias, the implementation of FP-Growth was downloaded from [20], and the implementations of the other algorithms downloaded from [21]. All of the codes were written in C/C++, used the same libraries, and were compiled using gcc (version 4.3.2). Fig. 6 shows the statistical information about the experimental databases from [21]. The experiments were performed on a 2.83GHz

Database	Size(bytes)	#Trans	#Items*	AvgTransLen	MaxTransLen
accidents	35509823	340183	468	33	51
chess	342294	3196	75	37	37
connect	9255309	67557	129	43	43
kosarak	32029467	990002	41270	8.099	2498
T40I10D100K	15478113	100000	942	39.605	77
webdocs	1481890176	1692082	5267656	177.229673	71472

Fig. 6. Statistical information about experimental databases

PC (Intel Core2 Q9500) with 4×10^9 bytes memory, running on a Debian (Linux 2.6.26) OS. Running time contains input time, CPU time, and output (directed to "/dev/null") time.

The experimental results are depicted in Fig. 7 (we did not plot when an implementation terminated abnormally.). For almost all the databases and minimum supports, BFP-growth performs the best. For example, in Fig. 7(b), the running times of the algorithms are respectively: BFP-growth (10.975 seconds), FP-growth* (143.173s), FP-Growth (400.934s), AFOPT (139.097s), dEclat (23.875s), LCM (19.153s) when the minimum support is 25% for real dense database *chess*. For synthetic sparse database $T40I10D100K$ in Fig. 7(e), their running times are respectively: BFP-growth (45.777 seconds), FPgrowth* (168.714s), FP-Growth (2890.735s), AFOPT (186.062s), dEclat (763.604s), LCM (114.614s) when the minimum support is 0.08%.

Fig. 8 shows BFP-growth's performance improvement over the previous pattern growth algorithms, in which the execution speed of FP-Growth is normalized as 1. For dense databases, for example, in Fig. 8(b) and (c), FPgrowth* has a speedup of less than 5-fold compared with FP-Growth, and BFP-growth

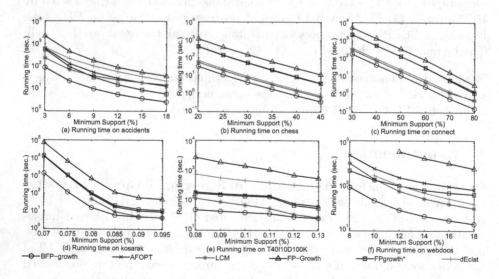

Fig. 7. Performance comparison

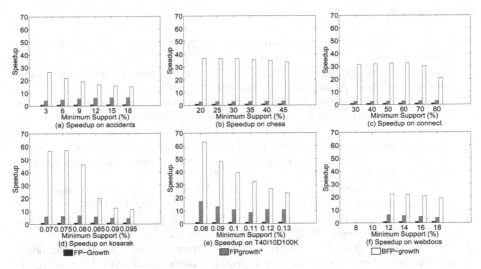

Fig. 8. Performance improvement

has a speedup of about 30-fold. A small-size and highly-compressed prefix-tree is usually constructed from a dense database, which means the relatively small traversal cost and the relatively large counting cost in the mining task. Based on FP-Growth, FPgrowth* reduces half the traversal cost but increases the counting cost, whereas BFP-growth reduces the more traversal cost without increasing the counting cost. Therefore BFP-growth has a larger speedup than FPgrowth*. For sparse databases, the corresponding bushy prefix-trees mean the relatively large traversal cost and the relatively small counting cost. In this case, for example, in Fig. 8(d) and (e), the speedup of FPgrowth* is more than 5-fold, whereas BFP-growth is even over 50 times faster than FP-Growth for low minimum supports. Compared with FPgrowth*, BFP-growth accesses fewer nodes, constructs simpler trees, and does not increase times of counting, thereby gaining more performance improvement.

6 Conclusion

In this paper, we proposed the BFP-growth algorithm for frequent itemset mining. The advantages of BFP-growth over the previous pattern growth algorithms are as follows. (1) For any prefix-tree generated during a mining process, BFP-growth traverses it only twice and thus dramatically reduces the traversal cost. (2) The counting cost of BFP-growth is less than that of FPgrowth*, one of the fastest algorithms. (3) BFP-growth employs plain prefix-trees to represent databases, and therefore the construction cost of BFP-growth is less than that of FP-Growth/FPgrowth* representing databases by extended prefix-trees. Extensive experimental data show that BFP-growth outperforms several famous algorithms including FPgrowth*, dEclat, and LCM, ones of the fastest algorithms, for various databases.

References

1. Ceglar, A., Roddick, J.F.: Association mining. ACM Comput. Surv. 38(2), 1–42 (2006)
2. Wang, H., Wang, W., Yang, J., Yu, P.S.: Clustering by pattern similarity in large data sets. In: Proc. ACM SIGMOD, pp. 394–405 (2002)
3. Cheng, H., Yan, X., Han, J., Yu, P.S.: Direct discriminative pattern mining for effective classification. In: Proc. ICDE, pp. 169–178 (2008)
4. Agrawal, R., Imieliński, T., Swami, A.: Mining association rules between sets of items in large databases. In: Proc. ACM SIGMOD, pp. 207–216 (1993)
5. Agrawal, R., Srikant, R.: Fast algorithms for mining association rules in large databases. In: Proc. VLDB, pp. 487–499 (1994)
6. Savasere, A., Omiecinski, E., Navathe, S.B.: An efficient algorithm for mining association rules in large databases. In: Proc. VLDB, pp. 432–444 (1995)
7. Bastide, Y., Taouil, R., Pasquier, N., Gerd, S., Lakhal, L.: Mining frequent patterns with counting inference. SIGKDD Explor. Newsl. 2(2), 66–75 (2000)
8. Han, J., Pei, J., Yin, Y., Mao, R.: Mining frequent patterns without candidate generation: A frequent-pattern tree approach*. Data Min. Knowl. Disc. 8(1), 53–87 (2004)
9. Zaki, M.J.: Scalable algorithms for association mining. IEEE Trans. Knowl. Data Eng. 12(3), 372–390 (2000)
10. Song, M., Rajasekaran, S.: A transaction mapping algorithm for frequent itemsets mining. IEEE Trans. Knowl. Data Eng. 18(4), 472–481 (2006)
11. Zaki, M.J., Gouda, K.: Fast vertical mining using diffsets. In: Proc. ACM SIGKDD, pp. 326–335 (2003)
12. Tsay, Y.J., Hsu, T.J., Yu, J.R.: Fiut: A new method for mining frequent itemsets. Inf. Sci. 179(11), 1724–1737 (2009)
13. Ghoting, A., Buehrer, G., Parthasarathy, S., Kim, D., Nguyen, A., Chen, Y.K., Dubey, P.: Cache-conscious frequent pattern mining on modern and emerging processors. The VLDB Journal 16(1), 77–96 (2007)
14. Schlegel, B., Gemulla, R., Lehner, W.: Memory-efficient frequent-itemset mining. In: Proc. EDBT, pp. 461–472 (2011)
15. Uno, T., Kiyomi, M., Arimura, H.: Lcm ver. 2: Efficient mining algorithms for frequent/closed/maximal itemsets. In: Proc. IEEE ICDM Workshop FIMI (2004)
16. Grahne, G., Zhu, J.: Fast algorithms for frequent itemset mining using fp-trees. IEEE Trans. Knowl. Data Eng. 17(10), 1347–1362 (2005)
17. Liu, G., Lu, H., Yu, J.X., Wang, W., Xiao, X.: Afopt: An efficient implementation of pattern growth approach. In: Proc. IEEE ICDM Workshop FIMI (2003)
18. Liu, G., Lu, H., Lou, W., Xu, Y., Yu, J.X.: Efficient mining of frequent patterns using ascending frequency ordered prefix-tree. Data Min. Knowl. Disc. 9(3), 249–274 (2004)
19. Schmidt-thieme, L.: Algorithmic features of eclat. In: Proc. IEEE ICDM Workshop FIMI (2004)
20. FP-Growth Implementation, http://adrem.ua.ac.be/~goethals/software/
21. Frequent Itemset Mining Implementations Repository, http://fimi.ua.ac.be/

Mining Link Patterns in Linked Data

Xiang Zhang[1], Cuifang Zhao[1], Peng Wang[1], and Fengbo Zhou[2]

[1] School of Computer Science and Engineering, Southeast Univesity, Nanjing, China
{x.zhang,cuifangzhao,pwang}@seu.edu.cn
[2] Focus Technology Co., Ltd., Nanjing, China
zhoufengbo@made-in-china.com

Abstract. As the explosive growth of online linked data, an emerging problem is what and how we can learn from these data. An important knowledge we can obtain is the link patterns among objects, which are helpful for characterizing, analyzing and understanding of linked data. In this paper, we present a novel approach of mining link patterns. A Typed Object Graph is proposed as the data model, and a gSpan-based algorithm is proposed for pattern mining. A type determination policy is introduced in cases of multi-types and a data clustering algorithm is proposed to improve scalability. Time performance and mining results are discussed by experiments.

Keywords: linked data, frequent link pattern, semantic web.

1 Introduction

As the rapid growth of semantic web in this decade, there is an exponential growth in the scale of online linked data. As indicated by W3C wiki of Linked Datasets[1], about 70 publishers open their linked data to public and lots of them contain a scale of billion triples. There is still enormous amount of linked data produced by social communities, companies, and even on the desktop of end-users. All these efforts are now producing an unprecedented huge web of data, bringing the semantic web from vision to practice. We have faced the problem of lack of semantic data, but now, the problem is what and how we can learn from these data.

One important thing to learn is the link patterns. Link patterns are consensus practices characterizing how different types of objects are typically interlinked. For example, in certain linked data, a *Researcher* may *focuses* on a *ResearchArea*, and *publishes* some *Papers* in *Proceedings* of a *Conference*. Besides, he *knows* some *Researchers* in the same *ResearchArea*. Link patterns describe widely-used relations among objects, which may or may not be defined explicitly in ontologies.

Link patterns are critical in several topics of research. First, they can be used to find useful semantic associations between objects by indicating what kind of object links are typical and thus significant [1]; Second, when accessing distributed RDF stores relying on different ontologies within different domains, link patterns are helpful in evaluating the potential contribution of each store to the processing of RDF queries,

---

[1] http://www.w3.org/wiki/DataSetRDFDumps

H. Gao et al. (Eds.): WAIM 2012, LNCS 7418, pp. 83–94, 2012.
© Springer-Verlag Berlin Heidelberg 2012

because link patterns characterizes the content of RDF repositories [2]; finally, in producing concise and comprehensible summaries for human understanding of linked data, link patterns can play a role of indicating the most important part of the link data from the point of view of data providers[3].

Although usage mining in semantic web gains a lot of interest in these years [4, 5], mining link patterns is still a topic not well-discussed. This is because the lack of a formal definition of link pattern in linked data, and the complex graph structure of link patterns also limits the scalability and efficiency of mining. Our contributions in this paper are: first, we formally define link pattern by a notion of Typed Object Graph; second, we propose a pattern mining algorithm based on gSpan [6], a clustering algorithm is also proposed to improve the mining scalability and efficiency.

The rest of the paper is organized as following: Typed Object Graph and Link Pattern are defined in Section 2. We explain the policy of type determination in building Typed Object Graph in Section 3. A data clustering algorithm and a gSpan-based pattern mining approach are proposed in section 4. Experiments on time performance and mining results are discussed in Section 5.

2   Preliminary Concepts

Link patterns are frequent and typical styles of how different types of object are interlinked. In order to clearly define the notion of link patterns, we have to introduce a new data model, which embody object types as well as object links in the model. In this section, we name our new data as Typed Object Graph, and then we define Link Pattern based a notion of RDF2Pattern Homomorphism.

2.1   Typed Object Graph

Link patterns cannot be directly mined from RDF graphs. Object types are core elements in link patterns. In RDF graphs, object types are implicit and can only be determined by reasoning according to RDF semantics [7]. A novel graph model is needed for pattern mining, in which object types should be explicitly embodied.

Definition 1 (Link Quintuple): Given an RDF document d, and an Triple $t = \langle s, p, o \rangle$ in d, an Link Quintuple $q = \langle s, p, o, type(s, d), type(o, d) \rangle$ is an extension of t, where s and o must represent object nodes, p must be object property, $type(s,d)$ and $type(o,d)$ represent the *rdf:type* of s and o defined in d respectively.

Definition 2 (Typed Object Graph): A Typed Object Graph g comprises a set of Link Quintuple: $\{q_1, q_2, \dots, q_n\}$, which is extended from an RDF Graph g' comprising a set of RDF Triple: $\{t_1, t_2, \dots, t_n\}$. And $\forall q_i \in g$ and $t_i \in g'$, q_i is extended from t_i.

For simplicity, multiple types are not allowed for an object in a link quintuple. We will introduce a type determine policy in Section 3.2. Shown in Figure 1, a fragment of RDF graph is extracted from a linked data of Semantic Web Dog Food[2], and the derived Typed Object Graph is shown aside (object URIs are omitted for simplicity).

---

[2] http://data.semanticweb.org/

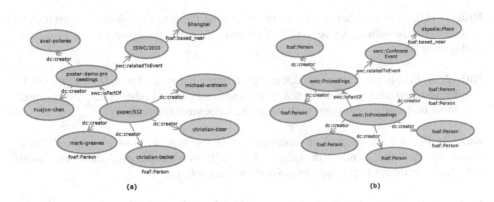

Fig. 1. (a) A fragment of an RDF Graph;   (b) A derived Typed Object Graph

2.2   Link Pattern

Definition 3 (RDF2Pattern Homomorphism): Given a subgraph g of a Typed Object Graph derived from RDF document d and a graph p, An RDF2Pattern Homomorphism is an injective function $f: V(g) \rightarrow V(p)$, such that (1) $\forall u \in V(g), type(u, d) = f(u) \in V(p)$, and (2) $\forall (u, v) \in E(g), (f(u), f(v)) \in E(p)$, in which $E(g)$ is the edge set of g.

Definition 4 (Link Pattern): Given an RDF document dataset $D = \{d_1, d_2, ... d_n\}$, a Link Pattern p is a graph, satisfying that:

1) $p = \langle V, E \rangle$ is a directed graph, in which each vertex is an user-defined class and each edge is an object property.
2) There exists RDF2Pattern homomorphism in D, and $support(p) \geq min\text{-}sup$. $support(p)$ is the number of RDF documents, in which p is an RDF2Pattern homomorphism of derived Typed Object Graphs; $min\text{-}sup$ is a predefined minimum support threshold.

3   Type Determination

Before mining link patterns, triples in linked data should be cleaned in advance in order to remove annotations, schema-level definitions, only object links are preserved. There is another issue we have to consider. In the definition of Typed Object Graph, an object must have a single type for the computational complexity of pattern mining. However practically, an object in linked data often be defined to multi-types. For example, in Semantic Web Dog Food, an accepted paper can be defined as an *swc:Paper* and meanwhile as a *swrc:InProceedings*. It should be eliminated before building Typed Object Graph, since multi-types will bring extra complexity in pattern mining, especially to subgraph homomorphism detection. Given $D = \{d_1, d_2, ..., d_n\}$ is the document set, and an object o defined in document d_i with multi-types $\{t_1, t_2, ..., t_m\}$, a set of heuristic rules are used to determine the single type of o:

Rule 1: If there is no other document in D defining the type of object o, $type(o, d_i)$ will be assigned with t_k, where $1 \le k \le m$ and type t_k has the largest set of instances in document d_i.

Rule 2: If there is more than one document defining the type of o in D, $type(o, d_i)$ will be assigned with t_k, where $1 \le k \le m$, and comparing to other types, o is defined to t_k most frequently, or saying, with the highest documents frequency.

Rule 3: For a triple $\langle s, p, o \rangle$ in document d, if the type of s or o is not defined explicitly, and the domain or range of p is defined explicitly, then we have: $type(s, d) = domain(p)$ and $type(o, d) = range(p)$.

In Rule 1, the single type of an object is assigned to a local dominated type; while in Rule 2, the single type is assigned to a global dominated type. If the dominance of each type is not obvious, the type determination becomes random.

There is still a case we have to consider, in which the type of an object may be not defined explicitly. In this case, Rule 3 will be applicable, and the domain and range definitions of an ObjectProperty are utilized to determine an objects' type.

Here we only try to determine the type of objects by exploring the domain and range of ObjectProperty. More powerful reasoning ability could be utilized for type determination, such as described in [8]. Here we use a rather lightweight reasoning to make the pattern mining scalable.

4   Mining Link Patterns

Among some frequent pattern mining algorithms, we select gSpan algorithm for mining link patterns in linked data. It is efficient in mining frequent graph patterns in massive data. Before pattern mining, we first cluster related Typed Object Graphs into groups for the sake of scalability. And then pattern mining is performed on each group separately in a divide-and-conquer manner.

4.1   Graph Clustering

Given a whole set of a linked data, the occurrences of a link pattern are often localized in some RDF documents of the same topic, while not in documents of other topics. For example, in Falcons dataset[3], we can find link patterns linking persons and organizations from some FOAF documents, or patterns linking genes markers and chromosomes from other bio2rdf documents (originated from bio2rdf[4]). If we can cluster RDF documents according to the link patterns being concerned, we can perform pattern mining in a divide-and-conquer way.

There have been several works on RDF clustering or classification, such as the RDF metadata clustering [9], instance clustering [10], clustering for snippets generation [11], and ontology classification [12] [13]. Here we introduce a clustering

---

[3] http://ws.nju.edu.cn/falcons
[4] http://bio2rdf.org/

approach for Typed Object Graphs. The intuition is: if two Typed Object Graphs share a common link pattern, they are affinitive and should be clustered together.

Definition 5 (Direct Connections between Typed Object Graphs): Given two Typed Object Graphs: $g = \{q_1, q_2, ..., q_n\}$ derived from RDF document d, and $g' = \{q_1', q_2', ..., q_m'\}$ derived from RDF document d'. g and g' are connected, denoted as $connected(g, g')$, iff $\exists q_i = \langle s_i, p_i, o_i, type(s_i, d), type(o_i, d) \rangle \in g$ and $q_j' = \langle s_j', p_j', o_j', type(s_j', d'), type(o_j', d') \rangle \in g'$, satisfying (1) $p_i = p_j'$; (2) $type(s_i, d) = type(s_j', d')$; (3) $type(o_i, d) = type(o_j', d')$.

In Definition 5, two Typed Object Graphs are directly connected if and only if they share at least one common link pattern. Given a set of Typed Object Graph $G = \{g_1, g_2, ... g_n\}$, it can be divided into a set of disjoint connected clusters through a testing of connectivity. It is easy to prove that we can use each cluster as a standalone dataset for pattern mining (although the support is still calculated globally), because different clusters don't share common link patterns. An example from Falcons dataset is shown in Figure 2, in which five fragments of Typed Object Graphs are clustered into two groups.

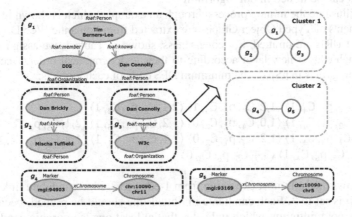

Fig. 2. Four Typed Object Graphs in Falcons are clustered into groups

4.2   gSpan-Based Pattern Mining

Our pattern mining approach follows the idea of pattern-growth-based frequent graph pattern mining algorithms, in which gSpan is a typical and efficient one. Basically, a candidate frequent pattern is produced by extending a mined frequent pattern by adding a new edge.

The kernel ideas in gSpan are the *minimum DFS code* and the *rightmost extension.* The *minimum DFS code* is introduced to canonically identify a pattern by a DFS traverse path; the *rightmost extension* is used producing candidates based on mined patterns. Both ideas can reduce the generation of duplicated candidates. In gSpan, the minimal frequent patterns are firstly discovered (with 0-edges), and gSpan is called recursively to make a rightmost extension on mined patterns so that their frequent descendants (with 1-edges, 2edges and so on) are found until their support is lower than *min-sup* or its DFS code is not minimum any more. All mined patterns comprise a *lexicographic search tree*. More details of gSpan and its expansion can be found in [14].

The original gSpan algorithm is designed for undirected and simple graphs. However in our scenario, Typed Object Graphs are directed and non-simple graphs. Self-loops (edges join vertexes to themselves) and multiple edges (multiple edges connecting two same vertexes) should be taken into consideration. According to suggestions proposed in CloseGraph [15], we modify gSpan algorithm, especially the DFS coding, to make it adaptable to Typed Object Graphs.

Given an link quintuple $q = \langle s, p, o, type(s,d), type(o,d) \rangle$ in a Typed Object Graph, it is represented in our implementation by a 6-tuple: $\langle i, j, l(i), l(i,j), l(j), \kappa \rangle$, in which i and j are DFS subscripts of vertex s and o; $l(i)$ and $l(j)$ are labeling functions of i and j, which equals to $type(s,d)$ and $type(o,d)$ respectively; $l(i,j)$ is the edge label, which equals to p; and κ represents the direction of the edge between i and j, in which $\kappa = 0$ represents $i \rightarrow j$, and $\kappa = 1$ represents $i \leftarrow j$.

There are two parameters to control the mining results in our implementation. One is *min-sup*, which indicates the minimum threshold of supports that a discovered link pattern can be seen as "frequent"; the other is *max-edge*, which limits the size of link patterns that can be mined in our algorithm.

We will illustrate the mining process through an example, which is shown in Figure 3. Three fragments of Typed Object Graphs are extracted from Semantic Web Dog Food, in which object URIs are omitted. For conciseness, shown in Table 1, we assign with each node and each edge a new label according to their lexicographical order. According to the DFS coding rules in gSpan, the minimum DFS code of each graph is:

g_1: $\langle 0,1,C_2,p_1,C_4,1 \rangle \langle 1,2,C_4,p_4,C_5,1 \rangle \langle 2,1,C_5,p_5,C_4,1 \rangle$

g_2: $\langle 0,1,C_2,p_1,C_4,1 \rangle \langle 1,0,C_4,p_2,C_2,0 \rangle \langle 1,2,C_4,p_4,C_5,1 \rangle \langle 2,1,C_5,p_5,C_4,1 \rangle$

g_3: $\langle 0,1,C_1,p_3,C_4,1 \rangle \langle 1,2,C_4,p_1,C_2,0 \rangle \langle 2,1,C_2,p_2,C_4,1 \rangle \langle 1,3,C_4,p_4,C_5,1 \rangle$
$\langle 3,1,C_5,p_5,C_4,1 \rangle \langle 3,4,C_5,p_6,C_3,0 \rangle$

Figure 3 shows the rightmost extension from 1-edge to 3-edge candidate link patterns. The *min-sup* is set to 3. Each candidate in dotted-line should be pruned, because their DFS code is not minimum, which indicates that at least one isomorphic candidate has already been mined. Other candidates with $min\_sup \geq 3$ will be qualified as link patterns in the final result.

Table 1. Relabeling Nodes and Edges

Node URI	Label	Edge URI	Label
foaf:Document	C_1	dc:creator	p_1
foaf:Person	C_2	foaf:maker	p_2
swrc:ConferenceEvent	C_3	ical:url	p_3
swrc:InProceedings	C_4	swc:hasPart	p_4
swrc:Proceedings	C_5	swc:isPartOf	p_5
		swc:relatedToEvent	p_6

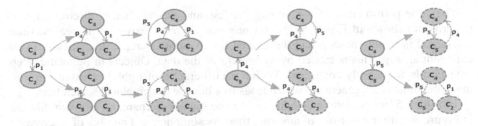

Fig. 3. Rightmost extension from 1-edge to 3-edge candidates

5   Experiments

We evaluate our approach on two sets of linked data. The first is Semantic Web Dog Food, which has a highly unified topic, and the other is a random selected subset of Falcons dataset, which is a collection of online linked data with various topics. In our evaluation, we mainly discuss the time performance and the number of discovered link patterns under the impact of various *min-sup* and *max-edge*. Our algorithm is implemented in C++ and experiments are performed on a 3GHZ Intel Core2 Duo PC with 4G main memory, running on Windows 7.

5.1   Experiment on Semantic Web Dog Food

Semantic Web Dog Food is a well-known and widely-used linked data for scholars. In the dataset, detailed information is provided on accepted papers, people who attended, and other things that have to do with the main conferences and workshops in the area of Semantic Web research.

This linked data is small in size and it describes limited types of objects. The main feature of it is that objects in this dataset are densely linked, and RDF documents in this dataset are highly unified in style: each describes a same topic using a same link pattern. It is very typical for those centrally generated, domain-specific linked data, such as each topic of dbpedia, GO annotations, etc.

Table 2 shows the statistics of Semantic Web Dog Food. Since the dataset has a highly unified topic, all derived Typed Object Graphs are connected, thus they are clustered into a single group.

Table 2. Statistics of Semantic Web Dog Food

RDF Document	Triple	Object
47	166,083	16,281
Quintuple	**Typed Object Graph**	**Cluster**
54,540	47	1

The time performance of the mining process and the number of discovered link patterns are shown in Figure 5. We fix *min-sup* to 3. As *max-edge* increases, time consumed in mining process and discovered link patterns increase dramatically in an exponential way. This is caused by the nature of the data. Objects in Semantic Web Dog Foods are densely connected. Within each iteration of rightmost extension, too many candidates are generated, which leads to a huge lexicographical search tree.

In Figure 5, we perform an evaluation on the impact of changing *min-sup*. Shown in Figure 8, with the increase of *min-sup*, time consumption and number of discovered link patterns both keep declining. This result indicates that a higher threshold of supports will lead to a loss of frequent patterns, but meanwhile it improves the time efficiency of mining.

Besides, experiments shows that link patterns can be huge. Figure 6 shows a very large link patterns discovered in this dataset with *support* = 5 . This pattern describes a typical scenario when two researchers are co-authors of two papers, and the papers are both included in proceedings of a conference and are both accessible via an online document.

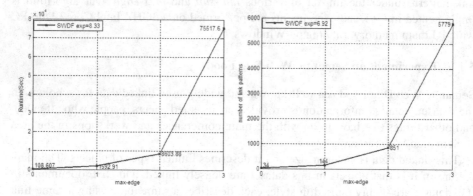

Fig. 4. Time Performance and number of link patterns with various max-edge (SWDF)

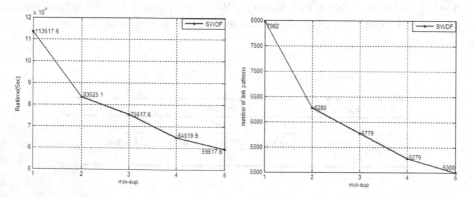

Fig. 5. Time Performance and number of link patterns with various min-sup (SWDF)

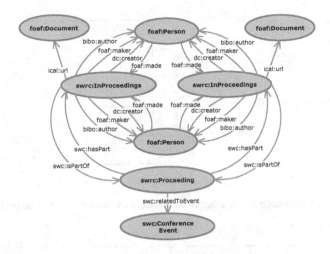

Fig. 6. A huge pattern discovered in Semantic Web Dog Food with 23 edges

5.2   Experiment on Falcons Subset

Falcons is a semantic web search engine. It collects online linked data, and provides searching service of objects, ontologies and RDF documents. The full dataset of Falcons contains over 2 billion triples. We randomly select a subset of Falcons, and the statistics in shown in Table 3.

The selected subset is diverse in topics. The 394 derived Typed Object Graphs are clustered into 8 groups. This dataset is similar to those domain-independent or crawled linked data, such as Billion Triple Challenge Dataset, etc.

Figure 7 shows the time performance and number of discovered link patterns with various *max-edge* and a fixed *min-sup* = 3:

Table 3. Statistics of Falcons

RDF Document	Triple	Object
394	481,213	87,135
Quintuple	**Typed Object Graph**	**Cluster**
96,103	394	8

We can see for this dataset, the increasing rate of both time consumption and discovered link patterns are smaller than the ones of Semantic Web Dog Food. In Falcons, objects are not linked so densely, and link patterns are not shared frequently among documents. The nature of sparsely-linked objects results in a relatively smaller lexicographical search tree, and fewer candidate patterns. The nature also leads to a comparatively more remarkable decrease rate when increasing *min-sup*, as shown in Figure 8.

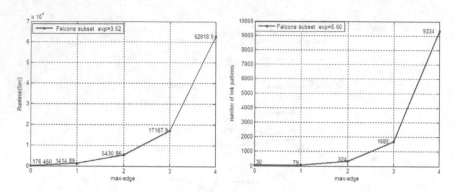

Fig. 7. Time Performance and number of link patterns with various max-edge (Falcons)

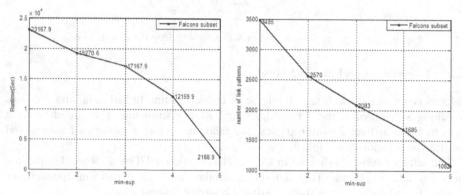

Fig. 8. Time Performance and number of link patterns with various min-sup (Falcons)

6   Related Works

Finding frequent patterns has been studied for years and has become a focused theme in data mining research. Algorithms of frequent graph pattern mining can be roughly classified into three categories: greedy algorithms, Inductive-Logic-Programming-based algorithms, and graph-theory-based algorithm. The last category can be further classified into Apriori-based [16] and Pattern-Growth Approach. The Apriori-based approach has to use the breadth-first search strategy because of its level-wise candidate generation, and thus consumes too much memory and produces too many candidate patterns. gSpan and CloseGraph [15] are two typical Pattern-Growth algorithms, both employing the depth-first search strategy.

Basse et al. has proposed in [2] a DFS-based approach for extracting frequent patterns in triple store, which is much closed to our approach. They improve the canonical representation of RDF graphs based on DFS code proposed by gSpan, and then provide a join operator to significantly reduce the number of frequent graph patterns generated from the analysis of the content of triple stores. The differences between our work and theirs lie in that: their motivation of pattern mining is to describe the schematic characteristic of the content of RDF bases, to benefit queries involving distributed RDF bases maintained in different servers, while our motivation is to extract typical link patterns for a general purpose of retrieving, understanding and creating linked data; besides, our definitions of data model, as well as the type determination policy, data clustering approach, are fully novel.

7   Conclusion and Future Works

With the explosive growth of online linked data, usage mining on linked data is becoming crucial. In this paper, we present an approach for mining link patterns that describe how different types of objects are frequently interlinked. A Typed Object Graph is defined as the data model, and a gSpan-based algorithm is proposed to discover link patterns. In our algorithm, a policy of type determination is introduced to handle cases of multi-types and an algorithm of data clustering is proposed to improve the scalability. Experiments are performed on two dataset. Time performance and mining results are fully discussed with various parameters.

In our future work, we will explore and compare different pattern mining approaches, to improve the time efficiency of mining. And we will also study the approach of reducing candidate patterns by means of semantic filtering, which will fully utilize the semantics in linked data.

Acknowledgements. The work is supported by the NSFC under Grant 61003055, 61003165, and by NSF of Jiangsu Province under Grant BK2009136, BK2011335. We would like to thank Huaping Chen for his valuable suggestions and their work on related experiments.

References

1. Sheth, A., Aleman-Meza, B., Arpinar, B., et al.: Semantic Association Identification and Knowledge Discovery for National Security Applications. Journal of Database Management 16(1), 33–53 (2005)
2. Basse, A., Gandon, F., Mirbel, I., et al.: DFS-based Frequent Graph Pattern Extraction to Characterize the Content of RDF Triple Stores. In: Proceedings of the WebSci1 2010: Extending the Frontiers of Society Online (2010)
3. Thor, A., Anderson, P., Raschid, L., Navlakha, S., Saha, B., Khuller, S., Zhang, X.-N.: Link Prediction for Annotation Graphs Using Graph Summarization. In: Aroyo, L., Welty, C., Alani, H., Taylor, J., Bernstein, A., Kagal, L., Noy, N., Blomqvist, E. (eds.) ISWC 2011, Part I. LNCS, vol. 7031, pp. 714–729. Springer, Heidelberg (2011)
4. Dai, H., Mobasher, B.: Integrating Semantic Knowledge with Web Usage Mining for Personalization. In: Web Mining: Applications and Techniques, pp. 273–306 (2004)
5. Xu, X., Cong, G., Ooi, B.C., et al.: Semantic Mining and Analysis of Gene Expression Data. In: Proceedings of the 30th International Conference on Very Large Data Bases, pp. 1261–1264 (2004)
6. Yan, X., Han, J.W.: gSpan: Graph-based Substructure Pattern Mining. In: Proceedings of the 2002 IEEE International Conference on Data Mining, pp. 721–724 (2002)
7. Hayes, P.: RDF Semantics. W3C Recommendation (February 10, 2004),
   http://www.w3.org/TR/rdf-mt/

8. Cheng, G., Qu, Y.: Integrating Lightweight Reasoning into Class-Based Query Refinement for Object Search. In: Domingue, J., Anutariya, C. (eds.) ASWC 2008. LNCS, vol. 5367, pp. 449–463. Springer, Heidelberg (2008)

9. Maedche, A., Zacharias, V.: Clustering Ontology-Based Metadata in the Semantic Web. In: Elomaa, T., Mannila, H., Toivonen, H. (eds.) PKDD 2002. LNCS (LNAI), vol. 2431, pp. 348–360. Springer, Heidelberg (2002)

10. Grimnes, G.A., Edwards, P., Preece, A.D.: Instance Based Clustering of Semantic Web Resources. In: Bechhofer, S., Hauswirth, M., Hoffmann, J., Koubarakis, M. (eds.) ESWC 2008. LNCS, vol. 5021, pp. 303–317. Springer, Heidelberg (2008)

11. Penin, T., Wang, H., Tran, T., Yu, Y.: Snippet Generation for Semantic Web Search Engines. In: Domingue, J., Anutariya, C. (eds.) ASWC 2008. LNCS, vol. 5367, pp. 493–507. Springer, Heidelberg (2008)

12. Patel, C., Supekar, K., Lee, Y., Park, E.K.: OntoKhoj: A Semantic Web Portal for Ontology Searching, Ranking and Classification. In: Proceedings of 5th ACM International Workshop on Web Information and Data Management, pp. 58–61 (2003)

13. Seidenberg, J., Rector, A.: Web Ontology Segmentation: Analysis, Classification and Use. In: Proceedings of 15th International Word Wide Web Conference, pp. 13–22 (2006)

14. Han, J.W., Kamber, M.: Data Mining Concepts and Techniques, 2nd edn. Elsevier Inc. (2006)

15. Yan, X., Han, J.W.: CloseGraph: Mining Closed Frequent Graph Patterns. In: Proceedings of the 9th ACM SIGKDD Internal Conference on Knowledge Discovery and Data Mining, pp. 285–295 (2003)

16. Inokuchi, A., Washio, T., Motoda, H.: An Apriori-Based Algorithm for Mining Frequent Substructures from Graph Data. In: Zighed, D.A., Komorowski, J., Żytkow, J.M. (eds.) PKDD 2000. LNCS (LNAI), vol. 1910, pp. 13–23. Springer, Heidelberg (2000)

Detecting Positive Opinion Leader Group from Forum[*]

Kaisong Song[1], Daling Wang[1,2], Shi Feng[1,2], Dong Wang[1], and Ge Yu[1,2]

[1] School of Information Science and Engineering, Northeastern University
[2] Key Laboratory of Medical Image Computing, Northeastern University,
Ministry of Education, Shenyang 110819, P.R. China
songkaisongabc@126.com,
{wangdaling,fengshi}@ise.neu.edu.cn

Abstract. Forum has long been the main way of communication, and more and more users publish their opinions by it. The most influential users or opinion leaders will contribute to the formation of information, especially the positive influential users who can guide public opinions and make positive influence. Positive Opinion Leader Group (*POLG*) represents a group of users, each of who expresses the similar content and same sentiment orientation with their followers to a great extent, who are regarded as the most influential men during the information dissemination process. However, most existing researches pay less attention to the implicit relationship, heterogeneous structure and positive influence. In this paper, we focus on modeling multi-themes user network of forum with explicit and implicit links for this purpose. In detail, we put forward a data structure **L**ongest **S**equence **P**hrase **T**ree (LSP-Tree) for representing comments on forum, measuring the similarity between comments based on LSP-Tree to obtain implicit links, and further detecting positive opinion leader group. Experiments using dataset from Tianya forum show that our method can detect positive opinion leaders group effectively and efficiently.

Keywords: sentiment analysis, opinion mining, positive opinion leader group, multi-themes user network.

1   Introduction

In recent years, forums have been the main way of communication and lots of rich reviews contribute to the formation of viewpoints. Detecting opinion leaders is important in promoting products and guiding public opinions. Although researchers have done some pioneering work, there are still limitations. (1) Most existing work neglects the implicit links such as similarity opinion. (2) Many researchers introduce classical models, but neglect the special structure. (3) As for the implicit relationships, there are no effective ways to find them. (4) Sentiment orientation is often neglected.

---

[*] Project supported by the State Key Development Program for Basic Research of China (Grant No. 2011CB302200-G), National Natural Science Foundation of China (Grant No. 60973019, 61100026), and the Fundamental Research Funds for the Central Universities(N100704001).

H. Gao et al. (Eds.): WAIM 2012, LNCS 7418, pp. 95–101, 2012.

In this paper, we focus on modeling multi-themes user network of forum with explicit and implicit links. In detail, we put forward Longest Sequence Phrase Tree (LSP-Tree) structure for representing comments in forum, measuring the similarity between comments to obtain implicit links and further detecting positive opinion leader group. Experiments show that our methods are effective and efficient.

The rest of the paper is organized as follows. Section 2 introduces the related work. Section 3 gives the problem definition. Section 4 presents various links detection concerning with multi-themes user network. Section 5 describes the process of detecting positive opinion leader group. Section 6 shows our experiment results. Section 7 concludes the research and gives directions for future studies.

2   Related Work

In detecting opinion leader, some researchers have done much related work. Budak [2] defined four important roles and introduced their functions. Goyal [9] proposed a method based on time window for detecting opinion leader from community. Xiao [13] proposed a LeaderRank algorithm to identify the opinion leaders in BBS [12]. Freimut [6, 7] identified opinion leaders and analyzed opinion evolvement by social network analysis. Zhou [15] introduced the concept of opinion networks and proposed OpinionRank algorithm to rank the nodes in an opinion network. Zhai [14] proposed interest-field based algorithms to identify opinion leaders in BBS. Feng [5] proposed a framework to identify opinion leaders for a marketing product. We [11] proposed an approach detecting opinion leader based on single theme from Sina news community.

3   Problem Description

Different from web pages, the structure of forum has different properties, such as topic, theme and comment. Let $C=\{c_0, c_1, ..., c_n\}$ be a comment set, and c_i $(0 \leq i \leq n)$ be an item of comment. We can obtain the sentiment orientation O_i $(0 \leq i \leq n)$ for every $c_i \in C$ by sentiment analysis, and the value of O_i is defined as P, N, and M corresponding to positive (support), negative (oppose), and neutral sentiment. Moreover, we give the following definitions.

Definition 1 (explicit link and implicit link). For c_i and c_j ($c_i \in C$, $c_j \in C$, $0 \leq i,j \leq n$), suppose c_i is published earlier than c_j. If c_j is a reply of c_i, c_j explicitly links to c_i. If c_j isn't, but has semantic similarity with c_i, c_j is regarded as having an implicit link to c_i.

Definition 2 (positive link and negative link). If c_j has the same sentiment orientation with c_i, the link (explicit or implicit) is called as "positive link", otherwise as "negative link".

Based on definition 1 and 2, comment set C can be transferred into a multi-themes comment network. Then the comment network can be mapped into a user network.

Definition 3 (multi-themes user network). The network can be represented with a graph $UNG(V, E)$. Where E is edge set including all links of definition 1 and 2, and V is vertex set including all users corresponding every comment.

Definition 4 (positive opinion leader group). For user set $U=\{u_0, u_1, ..., u_m\}$, i.e. the V in $UNG(V, E)$, any $u_i \in U$ has an authority score $u_i.score$. We rank these users by their scores. Without loss of generality, suppose $u_1.score > u_2.score > ... > u_m.score$, we select top-$k$ users as positive opinion leader group $POLG=\{u_1, u_2, ..., u_k\}$.

Based on above definitions, for detecting $POLG$, we do the following work in this paper. (1) Detect explicit and implicit links between comments. (2) Map various links to ones between users. (3) Model multi-themes user network $UNG(V, E)$ based on the links. (4) Detect $POLG$ from $UNG(V, E)$.

4   Detection of Links between Comments

As mentioned above, the links between comments include explicit and implicit links, and every link may be positive or negative. In this section, we obtain sentiment analysis in Section 4.1, explicit and implicit links in Section 4.2.

4.1   Sentiment Analysis for Positive and Negative Links

Here we also use the same method proposed in [11] for calculating SO (orientation of each comment). The method accumulates sentiment value, and transfers orientation by negative words number. SO is 1 (positive), 0 (neutral) and -1 (negative).

4.2   Explicit and Implicit Links Detection

As for explicit links, we detect them according to Definition 1, however implicit links detection is complex. In order to compare comment similarity, we propose a structure called Longest Sequence Phrase (LSP) to represent a sentence (have been partitioned into words). All LSPs compose a LSP-Tree to represent comment, and we measure similarity by comparing LSP-Trees.

There are several classical phrase structures such as V+(D), V+(D)+O, S+V+(D), S+V+O+(D). We use ICTCLAS [8] to split sentences into words and extract structures of each sentence, and then construct LSP-Trees.

In LSP-Tree, the format of each node is "$w: ct$". Where w is word corresponding S/V/D/O, and ct is the count of w in the comment. Moreover, we assign different values to the words with different pos (part of speech) for different expression ability. Verb/noun=0.7, adjective=0.5, and pronoun/adverb=0.2. For every path from root node to leaf node, the words of the path construct a weighted LSP. The Formula (1) below is used for computing LSP weight value LV.

$$LV = \sum_i w_i.weight \times ct_i \quad w_i.weight \in \{0.7, 0.5, 0.2\} \tag{1}$$

Suppose LSP-Tree1 and LSP-Tree2 are two LSP-Trees of c_1 and c_2, c_2 is replier. We firstly find the common path cp_1 and cp_2, and then calculate common path value CV similar to LV in Formula (1) according to LSP-Tree1. CV is the value of common path corresponding to cp. LV is the biggest value of c_1's path containing cp. After detecting all common paths, we finally calculate the value of $\sum CV / \sum LV$. The calculation is based on LSP-Tree1, and LSP-Tree2 only contributes some common paths. Finally, the similarity $Sim(c_1, c_2)$ between c_1 and c_2 is calculated with Formula (2), k is any common path and m is the number of common paths.

$$Sim(c_1, c_2) = \frac{\sum_{k=1}^{m} CV_k | \quad \text{for common paths of } c_1 \text{ and } c_2}{\sum_{k=1}^{m} LV_k | \quad \text{for LSPs in } c_1} \tag{2}$$

In addition, we set threshold $P_{threshold}$ for CV and $C_{threshold}$ for Sim. $P_{threshold}$ avoids short common path, and $C_{threshold}$ excludes small Sim. Both are given proper values.

5   Positive Opinion Leader Group Detection

In Section 4, we have prepared related links. In this section, we use these links to model multi-themes user network, and then detect $POLG$.

User-Item Table containing user-value pairs is a simplified form of multi-themes user network $UNG(V, E)$. Our purpose modeling $UNG(V, E)$ is detecting $POLG$, it only concerns with the number of links, so modeling $UNG(V, E)$ maintains User-Item Table. Based on the idea, table will be updated while scanning the dataset. The weight of link between comments is mapped into corresponding users' in the process below.

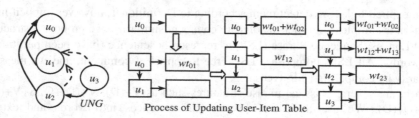

Fig. 1. An Example of Maintaining User-Item Table

In Fig.1, suppose u_j published comment c_j aiming at u_i's comment c_i (explicit or implicit), when processing u_j, a weight wt_{ij} will be added to u_i's score. According to content of Section 4.1 to 4.2, we give wt_{ij} as Formula (3).

$$wt_{ij} = \begin{cases} 1 & c_j \text{ explicitly links to } c_i \text{ and their } SO \text{ is the same} \\ -1 & c_j \text{ explicitly links to } c_i \text{ and their } SO \text{ is different} \\ Sim(c_i, c_j) \big/ n_j & c_j \text{ implicitly links to } c_i \text{ and their } SO \text{ is the same} \\ -Sim(c_i, c_j) \big/ n_j & c_j \text{ implicitly links to } c_i \text{ and their } SO \text{ is different} \end{cases} \tag{3}$$

In Formula (3), SO is sentiment orientation, $Sim(c_i, c_j)$ is from Formula (2), and n_j is the number of comments linked implicitly by c_j before c_i (including c_i). Suppose c_j implicitly links to c_k ($1 \leq k \leq n_j$), then c_j has the probability $1/n_j$ to be influenced by c_k, so the suppose is appropriate. In following, we give the algorithm as Algorithm 1.

Algorithm 1: Building Multi-Themes User Network;
Input: multi-themes comment set C, $P_{threshold}$, $C_{threshold}$; //see Section 4.2
Output: User-Item Table UT;
Method:
   1) transfer C into XML file X;
   2) for every theme
   3)   {create User-item for u_0;
   4)   for every user u_i ($i>0$)
   5)   {for $j=0$ to i-1
   6)   {analyze u_i's comment c_i aiming to u_j's comment c_j;
   7)   if (c_i implicitly links to c_j and $Sim(c_i, c_j) \geq C_{threshold}$ and $CV \geq P_{threshold}$)
   8)   or (c_i explicitly links to c_j)
   9)   {calculate wt_{ji} with Formula (6);
   10)   modify User-Item of u_j with wt_{ji};}}}}

As mentioned above, detecting $POLG$ only concerns with the number of links, so User-Item Table can replace multi-themes user network. For user u, score is authority score of u. We rank users according to their score, and top-k users compose $POLG$.

6   Experiments

Here we download dataset from "天涯杂谈" (http://www.tianya.cn/bbs/index.shtml). By multi-themes user network model, we can detect top-k users as $POLG$. The results are showed in Fig.2. Parameter $k=10$, $P_{threshold} = 0.5$ and $C_{threshold} = 0.65$.

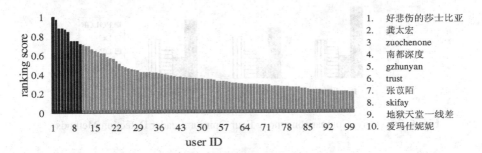

Fig. 2. Positive Opinion Leader Group from "天涯杂谈"

We have designed experiments for comparison. In Fig.3, There are more implicit links influencing the ranking result vastly. Fig.4 shows the distribution of each leader's positive, negative and neutral comments, and they usually have more positive and neutral comments. In Fig.5, POPLGR is positive opinion leader group rank proposed in this paper, DC is Degree Centrality [10], DP is Degree Prestige [10] and ACSC is accumulated scores. A standard POLG considering activity, influence range and ACSC is selected. From Fig.5, our method has better precision, recall and F-score.

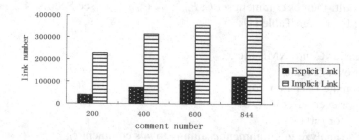

Fig. 3. Distribution of Explicit and Implicit Links

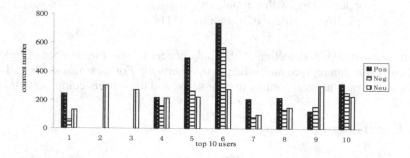

Fig. 4. Distribution of Top k users' Positive, Negative and Neutral comments

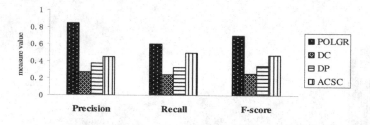

Fig. 5. Effectivity Comparison among Four Opinion Leader Detection Methods

7   Conclusion

In this paper, we focus on obtaining explicit, implicit links and sentiment orientation of comments. According to the links, we model multi-themes user network and detect positive opinion leader group effectively.

Although this paper proposed some useful methods for detecting *OPLDG*, there are some limitations. The phrase structure sometimes can't be extracted exactly. A convictive *OPLDG* for comparison has no standard. So they will be our future work.

References

1. Brin, S., Page, L.: The Anatomy of a Large-Scale Hypertextual Web Search Engine. Computer Networks (CN) 30(1-7), 107–117 (1998)
2. Budak, C., Agrawal, D., Abbadi, A.: Where the Blogs Tip: Connectors, Mavens, Salesmen and Translators of the Blogosphere,
   `http://snap.stanford.edu/soma2010/papers/soma2010_15.pdf`
3. Dong, Z., Dong, Q.: HowNet (2003),
   `http://www.keenage.com/html/e_index.html`
4. Ester, M., Kriegel, H., Sander, J., Xu, X.: A Density-Based Algorithm for Discovering Clusters in Large Spatial Databases with Noise. In: KDD 1996, pp. 226–231 (1996)
5. Feng, L., Timon, C.: Who is talking? An ontology-based opinion leader identification framework for word-of-mouth marketing in online social blogs. Decision Support Systems (DSS) 51(1), 190–197 (2011)
6. Freimut, B., Carolin, K.: Detecting opinion leaders and trends in online social networks. In: CIKM-SWSM 2009, pp. 65–68 (2009)
7. Freimut, B., Carolin, K.: Detecting Opinion Leaders and Trends in Online Communities. In: ICDS 2010, pp. 124–129 (2010)
8. Golaxy. ICTCLAS, `http://www.ictclas.org`
9. Goyal, A., Bonchi, F., Lakshmanan, L.: Discovering leaders from community actions. In: CIKM 2008, pp. 499–508 (2008)
10. Liu, B.: Web Data Mining: Exploring, Hyperlinks, Contents, and Usage Data. Springer, Heidelberg (2007)
11. Song, K., Wang, D., Feng, S., Yu, G.: Detecting Opinion Leader Dynamically in Chinese News Comments. In: Wang, L., Jiang, J., Lu, J., Hong, L., Liu, B. (eds.) WAIM 2011. LNCS, vol. 7142, pp. 197–209. Springer, Heidelberg (2012)
12. Xiao, Y., Xia, L.: Understanding opinion leaders in bulletin board systems: Structures and algorithms. In: LCN 2010, pp. 1062–1067 (2010)
13. Yu, X., Wei, X., Lin, X.: Algorithms of BBS Opinion Leader Mining Based on Sentiment Analysis. In: Wang, F.L., Gong, Z., Luo, X., Lei, J. (eds.) WISM 2010. LNCS, vol. 6318, pp. 360–369. Springer, Heidelberg (2010)
14. Zhai, Z., Xu, H., Jia, P.: Identifying Opinion Leaders in BBS. In: Web Intelligence/IAT Workshops 2008, pp. 398–401 (2008)
15. Zhou, H., Zeng, D., Zhang, C.: Finding leaders from opinion networks. In: ISI 2009, pp. 266–268 (2009)

D'MART: A Tool for Building and Populating Data Warehouse Model from Existing Reports and Tables

Sumit Negi[1], Manish A. Bhide[2], Vishal S. Batra[1],
Mukesh K. Mohania[1], and Sunil Bajpai[3]

[1] IBM Research, New Delhi, India
{sumitneg,vibatra1,mkmukesh}@in.ibm.com
[2] IBM Software Group, Hyderabad, India
abmanish@in.ibm.com
[3] Center for Railway Information System, Indian Railways, Delhi, India
bajpai.sunil@cris.org.in

Abstract. As companies grow (organically or inorganically), Data Administration (i.e. Stage 5 of Nolans IT growth model) becomes the next logical step in their IT evolution. Designing a Data Warehouse model, especially in the presence of legacy systems, is a challenging task. A lot of time and effort is consumed in understanding the existing data requirements, performing Dimensional and Fact modeling etc. This problem is further exacerbated if enterprise outsource their IT needs to external vendors. In such a situation no individual has a complete and in-depth view of the existing data setup. For such settings, a tool that can assist in building a data warehouse model from existing data models such that there is minimal impact to the business can be of immense value. In this paper we present the D'MART tool which addresses this problem. D'MART analyzes the existing data model of the enterprise and proposes alternatives for building the new data warehouse model. D'MART models the problem of identifying Fact/Dimension attributes of a warehouse model as a graph cut on a Dependency Analysis Graph (DAG). The DAG is built using the existing data models and the BI Report generation (SQL) scripts. The D'MART tool also uses the DAG for generation of ETL scripts that can be used to populate the newly proposed data warehouse from data present in the existing schemas. D'MART was developed and validated as part of an engagement with Indian Railways which operates one of the largest rail networks in the world.

1   Introduction

Many organization today use IT to support decision making and planning. A precursor for achieving this is the need to integrate disparate IT systems and the data that flows through them. Data Warehousing has been a widely accepted approach for doing this. However, many firms, especially those with extensive legacy systems find themselves at a great disadvantage. Having started late, these firms have to play catch up with new companies that are way ahead in their adoption and use of Decision Support Systems (DSS). Tools/Processes that can accelerate the data warehouse or data mart development in such an environment are of significant value to such companies. Implementing

H. Gao et al. (Eds.): WAIM 2012, LNCS 7418, pp. 102–113, 2012.

a data warehouse solution is a challenging and time consuming process. The various challenges in building a data warehouse include:

- Most warehousing projects begin by understanding how data is currently managed and consumed in the organization. Even though most of this knowledge exists in the IT systems that are in place, there is a limited capability to extract this information from these systems. For instance the existing data models and target reports have to be manually scanned to create a "data-inventory". This can be a very time consuming process.
- The development of the ETL scripts (which populate the existing tables in the organization) is typically outsourced to external consultants/vendors. In order to build the ETL scripts, the data modelers first need to understand the large and complex data environment which can be a very cumbersome task.
- The above problem is exacerbated by the fact that the data vocabulary could be inconsistent across departments thereby leading to errors in the (newly) generated ETL workflows.

One option that has been used by companies is to use a domain dependent warehouse data model and move all the data from different Lines of Business (LOB) to this new data model. However, using such a pre-defined data model does not avoid any of the problems mentioned above. Furthermore, a primary requirement of customers is to ensure that all BI Reports that were running on existing data model be re-engineered to run on the new data model. Considering these challenges and requirements what is required is an automated approach that utilizes the information present in the existing data models and BI Reports to recommend an "optimal" and "sufficient" data warehouse model and ETL scripts.

The D'MART tool presented in this paper analyzes the existing data model and BI Report generation scripts. It uses this information to do the following:

1. D'MART proposes a new data warehouse schema (Fact and Dimensions) such that the proposed model reuses (to the maximum possible extent) parts of the original schema. The tool also identifies shared dimensions and possible hierarchies in the dimensions.
2. D'MART identifies common attributes across the merging data models that have different names but similar content. These data elements attributes are candidates for merging in the new data warehouse. This avoids any duplicate/redundant data in the warehouse.
3. D'MART ensures that the data characteristics of the newly proposed data warehouse adheres to various design principles such as dimensions being in second normal form and fact tables being in third normal form.
4. D'MART generates a skeleton of the ETL scripts for populating the data from the base tables directly to the new data warehouse.

DMARTs is designed to reuse the existing data model as far as possible. This ensures that the amount of effort needed to understand, build and migrate to the new data warehouse/mart is kept to a minimum.

2   Related Work

In this section we describe the different Data Warehouse development approaches and the uniqueness of the D'MART approach. Existing Data Warehouse development processes can be broadly categorized into three basic groups:

1. *Data-Driven Methodology*: This approach promotes the idea that data warehouse environments are data driven, in comparison to classical systems, which have a requirement driven development life cycle. As per this approach business requirements are the last thing to be considered in the decision support development life cycle. These are understood after the data warehouse has been populated with data and results of queries have been analyzed by users. [5] propose a semi-automated methodology to build a dimensional data warehouse model from the pre-existing E/R schemes that represent operational databases. However, like other data-driven approaches their methodology does not consider end user requirements into the design process.

2. *Goal Driven Methodology*: (Business Model driven data warehousing methodology) [2]: This approach is based on the SOM (Semantic Object Model) process modeling technique. The first stage of the development cycle determines goals and services the company provides to its customers. Then the business process is analyzed by applying the SOM interaction schema. In the final step sequences of transactions are transformed into sequences of existing dependencies that refer to information systems. In our opinion this highly complex approach works well only when business processes are designed throughout the company and are combined with business goals.

3. *User-Driven Methodology* [6]: This approach is based on the BI needs. Business users define goals and gather, priorities as well as define business questions supporting these goals. Afterwards the business questions are prioritized and the most important business questions are defined in terms of data elements, including the definition of hierarchies, dimensions.

The User-Driven and Data-driven methodologies are two ends of a spectrum with their fair share of pros and cons. Our approach adopts a middle ground. The D'MART tool simultaneously analyzes the schema of the operational databases (bottom up) and the reporting requirements i.e. BI Reports (top down) to recommend a data warehouse model.

Paper Organization: We present an overview of the usage scenario of D'MART in the Indian Railways context in Section 3. The details of the D'MART tool are presented in Section 4. The experimental evaluation of D'MART is presented in Section 5 and Section 6 concludes the paper.

3   Indian Railways Scenario

Indian Railways is the state-owned railway company of India, which owns and operates the largest and busiest rail networks in the world, transporting 20 million passengers and more than 2 million tons of freight daily. CRIS (Center for Railway Information

System), which is an umbrella organization of the Indian Railways, caters to all IT needs of the Indian Railways. CRIS is entrusted with the design and development of IT applications that serve different LOB (Lines of Business) within the railways such as freight, passenger services etc.

The IT application landscape at CRIS is as follows. Each application has its own operational data store and an off-line data summary store. Periodically, data is moved from the applications operational data store to its off-line data summary store. This is done using a set of complex stored procedures that read data from the operational data source perform the required summarization and copy the data to the off-line data summary store. Reports are generated from the off-line summary database. As each of these applications was developed at different points in time, each application uses a separate and isolated data model and vocabulary that is unique to itself. The LOBs (line of business) are a logical representation of the morphology of the enterprise, and therefore of its data. The D'MART tool was conceived to accelerate the design and development of the data mart schema for different lines of business and to merge these marts into a single, integrated data warehouse. While the tool has this integration capability, in this paper we focus on the features of D'MART which were used in CRIS to build a specific data mart. CRIS is currently in the process of building a data warehouse, and the efficacy of the D'MART tool in assisting the process would be clearer as this exercise is taken to its conclusion.

4   D'MART Approach

The D'MART tool recommends a schema for the data warehouse/mart by analyzing the existing data model and the scripts used to generate the BI reports. The process of identifying the best possible data warehouse/mart schema consists of the following steps:

1. Fact Identification
2. Dimension Identification
3. Push Down Analysis
4. Data Analysis
5. Redundancy Analysis.

The first two steps are responsible for finding the set of tables which will form the fact and dimension tables in the proposed data warehouse. Another task done by the D'MART tool is that of generating the ETL scripts for populating the new data warehouse using the existing base table. This task is accomplished by the *Push Down Analysis* step. The *Data Analysis* step ensures that the selected dimensions and facts adhere to the standard design principles. The final task is that of *Redundancy Analysis* which tries to merge similar attributes and tables from multiple base tables. This task might look similar to the problem of schema matching ([9] [7] [3]), however it has some subtle but significant differences which are gainfully used by D'MART to improve its performance. We now explain each of these steps in the following sections.

4.1 Fact Identification

The fact identification step finds the set of tables which will form the fact tables of the proposed data mart. This process of identifying the fact table consists of *Fact Attribute Identification* followed by *Affinity Analysis*.

Fact Attribute Identification. In the first step the tool scans the BI report generation SQLs to identify set of attributes on which aggregate operation (such as sum, min, max, average, etc.) is defined. In addition to these attributes, the tool also identifies those attributes which are referred directly in the reports. These attributes can be of two types namely *direct projection attribute* and *indirect projection attribute*. The first type of attribute (i.e. *direct projection attribute*) is that which is present in the outermost "select" clause of the report generation SQL, whereas the second type of attribute (i.e. *indirect projection attribute*) is the one which is used in the inner query, but is projected out unchanged (possibly after being renamed) and used in the report. In order to understand the use of an indirect projection attribute, consider a *Delay Report* that displays the list of delayed trains along with the delay (in minutes) during their last run. This report shows only those trains which were delayed more than 80% of the times in the last one month. Notice that this report will find the difference between the scheduled arrival time and actual arrival time for each train in the last one month and will do a count to identify whether the train was delayed more than 80% of the times. It will then report the difference between the scheduled arrival time and actual arrival time during its last run. Thus these two attributes will be indirect projection attributes. They will also have an aggregate operation defined on it (count), but it will not be used in the report. The SQL query used to generate this report is given below:

```
select Name, DATEDIFF(minute, SCHEDULED_ARR, ACTUAL_ARR) as Delay from
(select Name, SCHEDULED_ARR, ACTUAL_ARR from TRAIN_STATUS ii1 where
SCHEDULED_ARR >= (select max( SCHEDULED_ARR) from TRAIN_STATUS ii2 where
ii1.Name = ii2.Name) and DATEDIFF(minute, ii1.SCHEDULED_ARR,
ii1.ACTUAL_ARR) < 0 and ii1.Name in (select Name from
(select Name, count(*) as AA from TRAIN_STATUS where
t1.late = 1 group by Name) t12 where t12.AA > (select 0.8*count(*)
from TR where t2.Name = t12.Name group by t2.Name)))
```

In our work with Indian Railways, we found that there were very few direct projection attributes and a large number of indirect projection attributes. Finding the indirect projection attribute is a very challenging but important task. In the above query, the two attributes SCHEDULED_ARR and ACTUAL_ARR are used at multiple places. However, we are only interested in those attributes which are projected out. In order to find the right attributes, D'MART uses a graph based representation system to address this challenge.

D'MART represents the report generation SQL in the form of a *Dependency Analysis Graph*. This graph represents the transformation that each attribute undergoes before it is eventually used in the report. At the lowest level of the graph are the (source) attributes of the various tables. At the highest level of the graph are the attributes which are present in the report. There are multiple paths from the source to the output during which the data undergoes various transformations such as sum, min, count, case statement, etc. We categorize each operation as either being an *aggregate operation* or a *cardinality preserving* operation. The aggregate operation generates one output for multiple rows in the input, where as the cardinality preserving operation generates one row in output for each input row. We are interested in identifying both types of attributes as they will be part of the fact table. In addition to this, the aggregate attribute are used for Dimension identification (details in next section).

Building the Dependency Analysis Graph becomes very tricky when an attribute is renamed, merged and reused in a different form. The Dependency Analysis Graph of the *Delay Report* described earlier is shown in Figure 1. In this figure, notice that at first glance the output attribute ii1.Name appears to be a non aggregate attribute. However, a closer analysis shows that it has a path to an aggregate operator via T2.Name, AGG, T2. In order to find such paths, D'MART uses the following rules for traversing the Dependency Analysis Graph:

- Rule 1: A path starting from the output attribute must not traverse through another output attribute.
- Rule 2: The path should always start from an output attribute and should terminate in a node representing a table.
- Rule 3: If for a given output node, there exists at least one path that goes via an aggregate operator, then the node is deemed as an aggregate node.

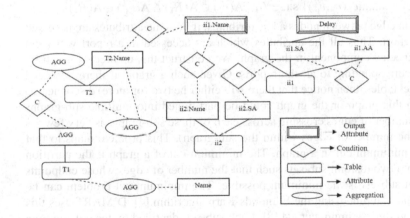

Fig. 1. Dependency Analysis Graph

D'MART uses a Breadth First Search based algorithm to start searching from each of the output nodes. For each of these output nodes, it finds all paths to the base tables such that they adhere to Rule 1 above. Once these paths are found, all those output attributes which do not have any path traversing through an aggregation operation are categorized as either *direct* or *indirect projection* attributes. The rest of the attributes are those which either belong to the dimension tables or are used in the "where" clause of the report generation SQL. Once the *direct* and *indirect projection* attributes have been found the next step that of *Affinity Analysis* is performed.

Affinity Analysis. Many-a-time, multiple fact tables are required either because these fact tables contain unrelated data (E.g., invoices and sales) or for reasons of efficiency. For example, multiple fact tables are often used to hold various levels of aggregated (summary data). The Affinity Analysis step checks whether there is a need to have multiple fact tables. Typically when multiple fact tables are used to store unrelated data no (or very few) reports access data from these different fact tables. D'MART uses this idea to identify the grouping of attributes identified in Section 4.1. We map this problem to that of finding a minimum cut in a graph as follows.

Let A be the set of attributes identified in Section 4.1 i.e. $A=\{A_1,A_2,.....A_n\}$. Let R be the set of reports which access these attributes, $R=\{R_1,R_2,....R_k\}$. Let $A(R_i)$ represent the set of attributes accessed by report R_i then $A(R_i)=\{A_{i1},A_{i2},.....,A_{im} \mid A_{ik} \in A\}$. Inversely, let $R(A_i)$ denote the set of reports, in which attribute A_i is used. We map this set to an undirected graph $G=\{V,E\}$ where V is the set of vertices in the graph and E is the set of edges. $V = \{v_1,v_2,......,v_n \mid A(v_i) \in A\}$. Thus in this graph, we have one vertex for each attribute in the schema. The function $A(v_i)$ above takes as input a vertex v_i and gives the attribute of A which is represented by the input vertex. Notice that the function $A(v_i)$ is overloaded and it can take as input either a report or an attribute. If the input is a report then it gives as output the set of attributes accessed by that report whereas if the input is a vertex, then it outputs the attribute which is represented by that vertex. The set of edges E in the graph G is defined a follows

$$E=\{e_1(v_1',v_1''), ...,e_p(v_p',v_p'')\} \text{ s.t. } \exists\, R_i \mid A(v_i') \in A(R_i) \wedge A(v_i'') \in A(R_i)$$

Thus, there is an edge between nodes a.k.a attributes, if both the attributes are accessed by the same report. Thus all the attributes which are accessed in a report will form a strongly connected component in the graph. We construct this undirected graph by adding edges corresponding to all the reports. Given such a graph, if there is a need for multiple fact tables, then notice that there will either be two (or more) disconnected components in this graph or the graph could be partitioned into multiple sub-graphs such that the number of edges crossing across each of these sub-graphs is very few (as compared to the number of edges within the sub-graph). This problem maps to that of finding the minimum cut in a graph. The minimum-cut of a graph is the partition of the graph into two disjoint sub-sets such that the number of edges whose endpoints are in different subsets is the minimum possible. The minimum-cut problem can be solved in polynomial time using the Edmonds-Karp algorithm [4]. D'MART uses this algorithm to find the minimum cut [1] [8]. Each sub-set identified by the cut can map to an independent fact table. In some cases where no natural cut exists, the algorithm finds a cut whose cut size (i.e., the number of edges whose ends points are in different sub-sets) is very large (as compared to the number of edges in the smaller sub-set). In that case D'MART does not suggest the use of multiple fact tables.

Another scenario where D'MART can suggest the use of multiple fact tables is when multiple reports aggregate data from the fact table at different levels of aggregation. For example, if 50% of the reports are reporting results on a daily basis where as the rest of the reports are reporting results on a monthly basis. In such cases D'MART suggests the use of two fact tables, one aggregating data on a daily basis, whereas the other aggregating data on a monthly basis. Due to space constraints we skip the details of this approach.

4.2 Dimension Identification

A dimension in a data warehouse is responsible for categorizing the data into non-overlapping regions. In other words, a dimension captures (in a loose sense) the distinct values of some attributes present in the fact table. Hence, attributes of the dimension table are typically used as a "group by" column in the BI Report generation SQL. We use this fact to find the set of attributes that can be part of the dimension table. The procedure of finding the dimension table is divided into two parts namely *Candidate Set Generation* and *Hierarchy Generation*.

Candidate Set Generation. In the first part, D'MART identifies the set of all attributes which are used in a "group by" clause of a report generation SQL. Notice that finding these attributes is a non-trivial task as the report generation SQLs are fairly complex and large. D'MART uses the Dependency Analysis Graph to find the set of attributes on which "group by" is defined. These attributes could be anywhere within the SQL such as a nested query, sub-query, etc. The set of attributes identified by the above procedure form what we call as the *"Candidate Attribute Set"*. The set of tables which have at least one of the attributes from the *"Candidate Attribute Set"* form the *"Candidate Table Set"*. The *"Candidate Table Set"* is the set of tables which can potentially form a dimension in our new data warehouse schema. Once the candidate table set has been identified, we need to identify whether we need to generate a star schema or a snow-flake schema using these candidate dimension attributes. This is done in the *Hierarchy Generation* step.

Hierarchy Generation. In order to identify whether we need to use a star schema or a snowflake schema, we essentially need to find whether any of the candidate dimension table can be represented as a hierarchy of multiple tables or if a single table representation suffices. In case we can split a dimension table into multiple tables then we need to use the snowflake schema, else we use the star schema. As we explain next, there are two steps for identifying the presence (or absence) of a hierarchy in the dimension table. When the data warehouse has a hierarchical dimension, the reports which use these dimensions would involve multiple joins across all the dimensions in the hierarchy. We use this information to decide between using a star schema or a snow flake schema as follows. Notice that if a set of attributes are used together in the "group by" clause, we would exploit this fact to suggest a hierarchical dimension to improve the efficiency of report generation. For example, consider the following report generation SQL:

```
Select T3.city_id,
count(T1.units_sold) from T1, T2 where T1.location_id =
T2.location_id and T2.city_id = T3.city_id group by
T2.location_id, T3.city_id
```

In the above query the attributes T2.location_id and T3.city_id appear together in a "group by" clause and have a "join" between them. This suggests that T2 and T3 form a hierarchy. In cases where the tables are incorrectly designed, we could have a case where the *city* and *location* information is present in a single table. Even in those cases, we are able to suggest the use of dimension hierarchy.

We use the fact that attributes appearing together in a "group by" clause could suggest the use of a hierarchy of dimensions. We first identify the set of mutually exclusive super-sets of the candidate attribute set which are used together in the "group by" clause of the various report generation SQLs. We explain this task with the following example: Let the *candidate attribute* set be: {A, B, C, D, E, F}. Let the attributes {A, B, C}, {B}, {A, B}, {D, E, F} and {D} be each used together in the same "group by" clause of a report generation SQL, i.e., {A, B, C} is used in one "group by" clause of a report generation SQL where as {A, B} is used in another "group by" clause of (possibly) another report generation SQL. What we are interested in are those set of attributes which are used together in the same "group by" clause. The mutually exclusive super-set, for the above example, will be {A, B, C} and {D, E, F}. The key property of this set is that any member (attribute) of one super set is never used with a member of another super set, i.e., A is never used together with say, D in the same "group by" clause of a report generation SQL. This property helps us to identify the set of attributes which will be part of the same dimension (or dimension hierarchy).

Given the mutually exclusive super-set, for each super-set we form the set of tables whose attributes are part of the super-set. As the existing schema in the enterprise may not well defined, we could end up with a case where the same table could be part of multiple super set. For example, we could have the following super set for the above scenario, {T1, T2}, {T1, T3, T4} (mapping to {A, B, C} and {D, E, F}). A common reason for this is that the table (T1) is not in second normal form, i.e., the table T1 has some amount of redundant data. If we remove this redundant data, then we can possibly avoid the overlap across the two super sets. In order to do this we convert the table T1 into second normal form which leads to a split of the table into multiple tables. We then reconstitute the super set of tables and check if there is any overlap across super set. In case the overlap still exists, then the same procedure is repeated. This is done for a fixed number of times (currently set at 3). If the overlap problem is still not solved, then we report this to the administrator. However, this is rarely required in practice. During our work for Indian railways, D'MART could automatically remove the overlap in all the cases.

Once the overlap has been removed, we identify the dimension for each of the super set. If the set of tables in the super set already have a primary key-foreign key relationship amongst them, then we use it to form a snowflake schema. In case there is no relationship, we check if each of these tables is in second normal form. If yes, then each of these tables form a separate dimension as part of a star schema. If a table is found as not being in the second normal form, then we convert it to second normal form and repeat the same procedure again.

4.3 Push Down Analysis

The Push Down Analysis phase tries to suggest the right granularity for the fact table. As described earlier, D'MART scans the report generation SQLs and identifies whether all the reports are using a common aggregation before generating the reports. In such a case, D'MART suggests changes to the granularity of the fact table. In order to do so, it suggests the aggregation to be pushed to the ETL scripts used to populate the fact table. It extracts the aggregation operator from the report generation SQL and suggests the same to be used in the ETL scripts. D'MART also suggests changes in the report generation SQL due to the changes in the fact table. This immensely helps the administrator to quickly generate the necessary ETL scripts for populating the newly defined data warehouse from the base tables.

4.4 Redundancy Analysis

A key aspect of the bottom-up approach is the identification of *conformed dimensions*. A *conformed dimension* is a set of data attributes that have been physically implemented in multiple database tables using the same structure, attributes, domain values, definitions and concepts in each implementation. Thus, conformed dimension define the possible integration "points" between the data marts from different LOBs. The final phase of D'MART, Redundancy Analysis phase, is responsible for finding candidates for conformed dimension. At first glance, this looks very similar to schema matching. However, the key advantage that we have in our setting is that we also have access to the SQL queries which are used to populate the data in the original data model. We make use of these queries to identify the commonalities in the schema across departments.

D'MART creates a *data model tree* for each attribute of the original schema. The data model tree tries to capture the origins of the attribute, i.e., from where is the data populated in this attribute, what kinds of transformations are applied to the data before it is populated in the attribute, etc. The data model tree is created by analyzing the SQL scripts which populate the attribute. D'MART scans the scripts and converts it into a graph (Data model tree) similar to that used in Section 4.1. The only difference is that the graph in Section 4.1 was used to analyze how reports are generated from the existing schema, whereas the data model tree is used to analyze how existing summary tables are populated from the source tables. Further, there is only one graph model per report, where as there is one data model tree per attribute of the original schema.

Once the data model tree has been generated, we find similar trees by comparing the structure and source of the trees. If two trees are similar, then D'MART suggests them as candidates for conformed dimension to the administrator. Once these candidates have been generated, D'MART can then use existing schema matching tools to do further redundancy analysis. Thus using the five steps described in this section, D'MART finds the new schema for the enterprise. A byproduct of this is the suggestion for writing ETL scripts that can be used to populate data in a new data warehouse. The next section presents an overview of the experimental evaluation of the D'MART tool.

5 Experimental Evaluation

The D'MART tool was applied to multiple Lines of Business at CRIS. In this section we talk about our experience in applying D'MART to a specific system namely the PAMS systems (Punctuality Analysis and Monitoring System). The PAMS system is used to generate reports on punctuality of different trains running in different zones and division within the Indian Railway network. We were provided with the operational database schema, BI Reports and corresponding stored procedure details. For instance for the PAMS system we received 27 reports and 65 stored procedures. The following SQL snippet has be taken from the *Train Status* Report, which is one of the 27 reports of the PAMS systems. This report displays the current status (e.g. *canceled,rescheduled* etc) of a particular train running on a given date in a particular region or division.

```
SELECT TRAIN_GAUGE, PAM_TYPE, TRAIN_STATUS, COUNT(TRAIN_STATUS) AS
DELAY_COUNT FROM( SELECT DISTINCT A.TRAIN_NUMBER, A.TRAIN_START_DATE,
DECODE(A.TRAIN_TYPE, 'EXP', 'MEX', A.TRAIN_TYPE) AS TRAIN_TYPE, CASE
WHEN  EXCEPTION_TYPE = 1 THEN 'CANCELLED' WHEN EXCEPTION_TYPE = 2
THEN 'SHORT DESTIN' WHEN EXCEPTION_TYPE = 3 THEN 'DIVERTED' WHEN
EXCEPTION_TYPE = 4 THEN 'MAJOR UNUSUAL' WHEN EXCEPTION_TYPE = 5
THEN 'RESCHEDULED' END AS TRAIN_STATUC, A.TRAIN_GAUGE, B.PAM_TYPE,
A.STATION_CODE FROM COISPHASE1.ST_SCHEDULED_TRAINS_DIVISION A,
COISPHASE2.MAT_TRAIN_TYPE B WHERE A.SCHEDULED_DIVISION="||LOC_CODE||"
AND TRUNC(A.SCHEDULED_DATE)>=TO_DATE("||FROM_DATE||",'DD-MM-YYYY')
AND TRUNC(A.SCHEDULED_DATE)<=TO_DATE("||TILL_DATE||",'DD-MM-YYYY')
AND A.EXCEPTION_TYPE IN(1,2,3,5) AND A.TRAIN_TYPE) GROUP BY
TRAIN_GAUGE,PAN_TYPE, TRAIN_STATUS ORDER BY TRAIN_GAUGE,
PAM_TYPE, TRAIN_STATUS
```

For illustration the extracted *Dependency Analysis Graph* is shown in Figure 2. The recommended Fact and Dimension attributes for this SQL are: *Fact Attributes* = {ST_SCHEDULE_TRAIN_DIV.EXCEPTION_TYPE} , *Dimension Attributes* ={TRAIN_GAUGE, PAM_TYPE, TRAIN_STATUS}.

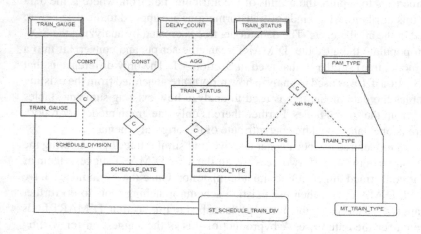

Fig. 2. Dependency Analysis Graph

D'MART performed similar processing on the other SQLs from the PAMS system. For the PAMS system D'MART recommended 1 Fact and 4 Dimension tables. The Fact table has a set of 18 attributes. Two of the four dimensions have a hierarchy which is one level deep. To validate the correctness and usefulness of the recommended schema a user-study was conducted. Data administrators were shown the old and new schema and asked to rate it on certain parameters. These parameters included factors such as completeness, correctness (*Are all modeled aspects correct with respect to the requirements and terms of the domain?*), consistency (*Are all modeled aspects free of contradictions?*), coverage, level of detail and minimality (*Is the schema modeled compactly and without redundancies?*). The schema recommended by the D'MART tool was rated highly on all these parameters.

6 Conclusion

In this paper we presented the D'MART tool that helps companies accelerate the development of a data warehouse/data mart from existing data models. The key advantage of D'MART is that it proposes a new data warehouse schema with minimal changes to the existing setup. Such an approach is extremely useful when companies need a phased approach for building the warehouse. D'MART works by analyzing the existing data model and BI Reports of the enterprise. D'MART models the problem of identifying Fact/Dimension attributes of a warehouse model as a graph cut problem on a *Dependency Analysis Graph* DAG. The DAG is built using the existing data models and the BI Report generation SQL scripts. The D'MART tools also uses a variant of the DAG for generation of ETL scripts that can be used to populate the newly proposed data warehouse from data present in the existing schemas. D'MART was developed and validated as part of an engagement with Indian Railways which operates one of the largest and busiest rail networks in the world.

Acknowledgement. We would like to acknowledge and express our heartfelt gratitude to Mr Vikram Chopra, without whose support this work would not have been possible, and also thank the CRIS team led by Ms Priya Srivastava for generously contributing their effort and knowledge.

References

1. Arora, S., Rao, S., Vazirani, U.: Expander flows, geometric embeddings and graph partitioning (2009)
2. Chowdhary, P., Mihaila, G., Lei, H.: Model driven data warehousing for business performance management (2006)
3. Doan, A., Domingos, P., Halevy, A.: Reconciling schemas of disparate data sources: a machine-learning approach (2001)
4. Edmonds, J., Karp, R.: Theoretical improvements in algorithmic efficiency for network flow problems (1972)
5. Golfarelli, M., Maio, D., Rizzi, S.: Conceptual design of data warehouses from e/r schemes (1998)
6. Kimball, R.: The Data Warehouse Toolkit: Practical Techniques For Building Dimensional Data Warehouse. John Wiley & Sons (1996)
7. Rahm, E., Bernstein, P.: A survey of approaches to automatic schema matching (2001)
8. Cormen, T.H., Leiserson, C.E., Rivest, R.L., Stein, C.: Introduction to Algorithms. MIT Press and McGraw-Hill (2009)
9. Westerman, P.: Data Warehousing using the Wal-Mart Model. Morgan Kaufmann (2001)

Continuous Skyline Queries with Integrity Assurance in Outsourced Spatial Databases

Xin Lin[1,2], Jianliang Xu[2], and Junzhong Gu[1]

[1] Department of Computer Science, East China Normal University
{xlin,jzgu}@ecnu.edu.cn
[2] Department of Computer Science, Hong Kong Baptist University
xujl@comp.hkbu.edu.hk

Abstract. Integrity assurance is an important problem for query processing in outsourced spatial databases, where the location-based service (LBS) provides query services to the clients on behalf of the data owner. If the LBS server is not trustworthy, it may return incorrect or incomplete query results intentionally or unintentionally. Therefore, to ensure the query integrity, the data owner needs to build additional authenticated data structures so that the clients can authenticate the soundness and completeness of query results. In this paper, we study the integrity assurance problem for continuous location-based skyline queries. We propose three novel techniques based on MR-Sky-tree, i.e., using *valid scope*, *visible region*, and *incremental VO* to reduce the computation and communication cost. Experimental results show that our proposed techniques achieve shorter computation time and lower communication cost than the existing approach.

1 Introduction

In outsourced spatial databases, the LBS provides query services to the clients on behalf of data owners (e.g., government land survey department or non-profit organizations) for better service quality and lower cost. However, such an outsourcing model brings great challenges to query integrity. Since the LBS server is not the real owner of data, it may return incorrect or incomplete query results intentionally or unintentionally. Thus, there is a need for the clients to authenticate the *soundness* and *completeness* of query results, where *soundness* means that the original data is not modified by the LBS server and *completeness* means that no valid result is missing. This problem is known as *authenticated query processing* in the literature [3, 6, 7, 10–13, 17, 18].

A typical framework of authenticated query processing is shown in Fig. 1. The data owner (DO) builds an authenticated data structure (ADS) for the spatial dataset before outsourcing it to the LBS. To support efficient query processing, the ADS is often a tree-like index structure, whose root is signed by the DO using her private key. In addition to the spatial dataset, the DO transfers the ADS and its root signature to the LBS server. Upon receiving a query from the client, the LBS server returns the query results, the root signature, and a verification object (VO) that is constructed from the ADS. The correctness of the query results can be verified by the client using the VO, the root signature, and the DO's public key.

H. Gao et al. (Eds.): WAIM 2012, LNCS 7418, pp. 114–126, 2012.
© Springer-Verlag Berlin Heidelberg 2012

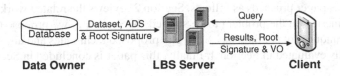

Fig. 1. Authenticated Query Processing

Authenticated processing of location-based queries is a hot topic in the database area [16, 18, 19]. Nevertheless, most of the existing studies focused on spatial-only queries, including range queries [3,16], nearest-neighbor (NN) queries [19], and shortest-path queries [18]. These queries are not sufficient to support LBS applications that need to consider both spatial and non-spatial attributes of queried objects. A typical scenario is finding nearby car parks with cheap parking fees, where the distance is a spatial attribute and the parking fee is a non-spatial attribute. To deal with this kind of multi-criterion search, location-based skyline queries (LSQs) have received considerable attention in LBS research (e.g., [4, 9, 15, 20]). In general, the dynamic nature of spatial attributes makes LSQs unique and challenging, which implies that the skyline results would differ with respect to different query locations. Taking the above car-park-finding scenario for example, the distance from a client to a car park varies with the location of the client. Such dynamic nature of query results signifies the necessity of authenticating LSQ results.

In a previous work [8], we have proposed an efficient index, called MR-Sky-tree, for authenticating one-shot LSQs. However, in LBS applications, users may sometimes prefer continuous queries; e.g., a driver may issue a continuous LSQ like "finding nearby car parks with cheap park fees" as he drives on the street. The authentication problem for continuous LSQs is more challenging since it is not efficient to repeat one-shot query authentication whenever the client changes location. In this paper, we propose three novel techniques, i.e., using *valid scope*, *visible region*, and *incremental VO* to reduce both the computation and communication cost. Specifically, the *valid scope* defines an area in which the query results do not change. The *visible region* is a super set of valid scopes in which the client is able to compute the new results locally without contacting the server. Lastly, if the query point moves out of the visible region, the server can return only a part of VO, called *incremental VO*, to the client so as to save the communication cost.

On the whole, our contributions made in this paper can be summarized as follows:

- We propose a valid scope computation and authentication algorithm for continuous LSQs, which can save both the computation and communication cost on the client.
- We propose a concept of visible region, by which the LSQ results can be computed locally on the client without contacting the server.
- We propose the construction and merge algorithms for incremental VO, which avoids sending the complete VO and reduces the communication cost when the query point moves out of the visible region.
- We conduct extensive experiments to evaluate the performance of the proposed techniques and algorithms. The results show that our proposed techniques perform efficiently under various system settings.

The rest of this paper proceeds as follows. Section 2 reviews the related work and gives some preliminaries on the problem to be studied. In Section 3, we propose three optimization techniques for continuous LSQs. The proposed techniques and algorithms are experimentally evaluated in Section 4. Finally, this paper is concluded in Section 5.

2   Preliminaries and Related Work

In this section, we give the formal definition of location-based skyline queries (LSQs) and describe our previous work MR-Sky-tree [8], which can be used to authenticate one-shot LSQs. We consider a set of data objects O. Each object $o \in O$ is associated with one spatial location attribute (denoted by $o.x$ and $o.y$) and several non-spatial attributes (e.g., parking fee and service quality, denoted by $o.A_i$ for the i-th non-spatial attribute). In this paper, we employ the Euclidean distance metric to measure the spatial proximity.

Definition 1. *(Dominance) Given two objects o and o', if $o'.A_i$ is not worse than $o.A_i$ for any non-spatial attribute A_i, then we say o' non-spatially dominates o, and o' is a non-spatial dominator of o. Formally, it is denoted as $o' \triangleleft o$. The set of o's non-spatial dominator objects is denoted as $Dom(o)$. Given a query point q, if (1) o' non-spatially dominates o, and (2) o' is no farther away from q than o (i.e., o' also spatially dominates o), then we say o' dominates o w. r. t. the query point q. Formally, it is denoted as $o' \triangleleft_q o$.*

Definition 2. *(Location-based Skyline Query (LSQ)) Given an object set O, the location-based skyline of a query point q is a subset of O, $LSQ(O, q)$, in which each object is not dominated by any other object in O w. r. t. q.*

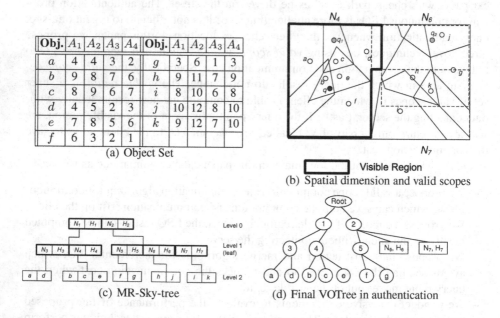

(a) Object Set

(b) Spatial dimension and valid scopes

(c) MR-Sky-tree

(d) Final VOTree in authentication

Fig. 2. LSQ Example and MR-Sky-Tree

By definition, LSQs have an intuitive characteristic, i.e., for a given query point q, if an object o is closer than all its non-spatial dominators, o must be a member of $LSQ(O, q)$. With this characteristic, we define a concept *skyline scope*, which can be used to accelerate LSQ processing and authentication.

Definition 3. *(Skyline Scope) For any object $o \in O$, its skyline scope in a 2D plane P is denoted as $SS(o) = \{q \mid q \in P \wedge o \in LSQ(q, O)\}$, where $o \in LSQ(q, O)$ means $\forall m \in Dom(o)$, o is closer to q than m, i.e., $dist_S(o, q) < dist_S(m, q)$.*

The skyline scope of an object o is essentially the Voronoi cell of o in the object set $\{o\} \cup Dom(o)$. If $Dom(o)$ is empty, o's skyline scope is the entire space. In general, the Voronoi cell can be computed using a divide-and-conquer algorithm, with a time complexity of $O(|O|log|O|)$ [2].

With the skyline scopes for all objects, a location-based skyline can be found by searching the objects whose skyline scopes cover the query point q. For example, Fig. 2(a) shows the non-spatial attributes ($A_1 - A_4$) of objects a - k, where a smaller non-spatial value is preferred. The spatial location and skyline scope of each object are shown in Fig. 2(b). Since the query point q is covered by the skyline scopes of a, b, d, e, f and g (the skyline scopes of a, d, f and g are the entire scope), the $LSQ(O, q)$ results are these six objects.

To support query authentication, the skyline scopes of all objects are inserted into an MR-tree [16] as data entries. The MR-tree is a combination of MH-tree [5] and R*-tree [1]. Each leaf node in the MR-tree is identical to that of R*-tree, which stores pointers pointing to actual data objects. The digest of a leaf node is obtained by hashing the concatenation of the binary representations of all objects in the node. Each internal node contains a number of entries in the form of (ptr_i, MBR_i, H_i),[1] where ptr_i is the pointer pointing to the i-th child, MBR_i and H_i are the minimum bounding rectangle and the digest of the i-th child, respectively. The digest of an internal node summarizes the MBRs and digests of all children nodes. The use of digests makes possible the pruning of index nodes in the VO while being able to verify the correctness of query results. We call this MR-tree indexing skyline scopes as MR-Sky-tree. Fig. 2(c) shows the MR-Sky-tree for our running example, where N_1, N_2, N_3 and N_5 are the entire space.

Given $LSQ(O, q)$, the VO of this query is represented by a subtree of the MR-Sky-tree index (termed as *VOTree*). In detail, the server checks, from the root and downwards, whether each child of an MR-Sky-tree node covers the query point q. If it does not, the child is pruned, while its MBR and digest are inserted into the *VOTree*. Otherwise, the node is unfolded and the children of the node are checked recursively by repeating the above procedure. Fig. 2(d) shows the final *VOTree* for our running example. Since N_6 and N_7 does not cover q, their digests and MBRs are inserted into *VOTree* instead of their children.

To verify the query results, the client checks the following three facts: 1) the skyline scopes of all objects in the result set should cover the query point q; 2) no MBRs of the pruned nodes and no skyline scopes of the non-result objects cover q; 3) the root

---

[1] In the actual implementation, all the digests of a node can be stored in a separate page and pointed by the node. In this way, the original R*-tree structure is not modified.

signature matches the digest computed from *VOTree*. The fact 3 ensures the soundness of the results, and facts 1-3 ensure the completeness of the results.

3 Authentication for Continuous LSQs

Algorithm 1. Valid Scope Computation and VO Construction

INPUT: Skyline set S, root of MR-Sky-Tree $mrRoot$
OUTPUT: Valid scope VS, VOTree *votree*

1: $VSVR \leftarrow$ The whole space
2: **for** each member o in the S **do**
3:     $VSVR \leftarrow VSVR \cap SS(o)$
4: $VS \leftarrow VSVR$
5: initialize the root of *voTree* with *mr-Root* (excluding the digest)
6: insert *mrRoot* into a queue Q
7: **while** Q is not empty **do**
8:     get the top element e from Q
9:     insert e's children into *voTree*
10:    **if** e is an index node **do**
11:        **for** each child c of e **do**
12:            **if** c intersects with $VSVR$ **do**
13:                insert c to Q
14:        **else**
15:                prune c and keep its digest in the parent node
16:    **else** // e is a leaf node
17:        **for** each child c of e **do**
18:            **if** c intersects with VS and $c \notin S$ **do**
19:                remove the c part from VS

Algorithm 2. Incremental VO Construction

INPUT: Previous query point q, current query point q', root of MR-Sky-Tree $mrRoot$
OUTPUT: Incremental VOTree *votree*, skyline set S

1: initialize the root of *voTree* with *mr-Root* (excluding the digest)
2: insert *mrRoot* into a queue Q
3: **while** Q is not empty **do**
4:     get the top element e from Q
5:     insert e's children into *voTree*
6:     **if** e is an index node **do**
7:         **for** each child c of e **do**
8:             **if** c covers q' **do**
9:                 insert c to Q
10:        **else**
11:                prune c and replace its corresponding node with label "Hit" in the *votree*.
12:    **else** // e is a leaf node
13:        **if** e covers q **do**
14:            replace e's corresponding node with label "Hit" in the *votree*.
15:        **else**
16:            **for** each child c of e **do**
17:                **if** c covers q' **do**
18:                    $S \leftarrow S \cup \{c\}$

To authenticate continuous LSQs such as "finding nearby car parks with cheap parking fees," a naive method is to repeat the query processing and VO construction processes shown in Section 2, whenever the client changes location. Obviously, this method is not efficient on both computation and communication cost. In this section, we propose three techniques to improve the performance. First, we compute the *valid scope* of query point q, in which the LSQ results remain the same as LSQ(q, O). By the valid scope, the clients need not send location updates to the server or re-compute the query results by itself. Second, we introduce a concept of *visible region* for each VO. If the query point is located in the visible region, the new LSQ results can be computed by the client locally, without contacting the server. Lastly, when the client moves out of the visible region, we propose an *incremental VO* technique so that only a small part of VO needs to be sent back to the client.

3.1   Valid Scope and Its Authentication

As mentioned above, the valid scope of q is the area that shares the same LSQ results with q. This means if the client moves from q to any point in the valid scope, no skyline object leaves the skyline set and no non-skyline object enters the skyline set. Hence, the valid scope can be computed by:

$$VS(q) = \bigcap_{o \in LSQ(q,O)} SS(o) - \bigcup_{u \notin LSQ(q,O)} SS(u). \qquad (1)$$

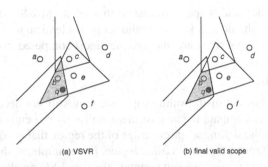

(a) VSVR   (b) final valid scope

Fig. 3. An Illustration of Valid Scope

Take Fig. 3 for example. Firstly, we compute the intersection area of the skyline scopes of all skyline objects. Since the skyline set in our running example is $\{a, b, d, e, f, g\}$, the intersection area of their skyline scopes is the shaded area in Fig. 3(a). Secondly, the part which intersects with the skyline scope of any non-skyline object is removed from the valid scope. In Fig. 3(a), since c is not a skyline object, the part which intersects with c's skyline scope should be removed. As a result, the final valid scope is the shaded area in Fig. 3(b).

After computing the valid scope, we construct the VOTree, with which the client can verify the soundness of the valid scope. We define the intersection area of the skyline scopes of all skyline objects as the *valid scope verification region* ($VSVR$) (just like the shaded area in Fig. 3(a)). The idea is that, if a node in the MR-Sky-tree does not intersect with $VSVR$, the node need not be unfolded and only its digest and MBR are inserted into the VOTree. The pseudo-code of the valid scope computation and VO construction algorithm is summarized in Algorithm 1.

The following theorem proves the correctness of Algorithm 1.

Theorem 1. *The VOTree constructed by Algorithm 1 can be used to verify the correctness of valid scope VS.*

Proof. *The client can verify VS by the following two facts: 1) all query points inside VS share the same LSQ results with LSQ(q, O); 2) any query point p outside VS satisfies LSQ(p, O) ≠ LSQ(q, O).*

For fact 1, since VS is contained in $VSVR$ and all nodes intersects with $VSVR$ are unfolded and stored in the VOTree, the client has a full view about the skyline scopes that intersect with VS. As such, the client can verify the LSQ results for any point inside VS.

For fact 2, if p is outside VS but inside $VSVR$, all the nodes that cover p would be unfolded according to Algorithm 1. Hence, the client can verify $LSQ(p, O)$ by the covering skyline scopes. Otherwise, if p is outside $VSVR$, we show $LSQ(p, O) \neq LSQ(q, O)$ by contradiction. Suppose there exists such a point p that satisfies $LSQ(p, O) = LSQ(q, O)$. Since p is outside $VSVR$, p must be outside the skyline scope of some member(s) in $LSQ(q, Q)$. Hence, such member(s) will not be in the result set of $LSQ(p, O)$ according to the definition of skyline scope, which leads to $LSQ(p, O) \neq LSQ(q, O)$. □

The verification of the valid scope is similar to that of the MR-Sky-tree method (see Section 2). The client should check: 1) the valid scope is identical to the area computed by Equation (1); 2) the root signature matches the digest computed from the VO.

3.2   Visible Region

When the client moves out of the valid scope, we may still not need to issue a new query. It is possible to compute the new results locally. By the current VO, the client has a full view about the skyline scope coverage of the region that is not covered by any pruned node. We call such a region as *visible region*. For example, as shown in Fig. 2(b), if the client moves form q to q_1, we can compute the new LSQ results as $\{a, c, d, f, g\}$ locally, since no pruned object in the current VO covers q_1.

Formally, the visible region can be obtained as follows:

$$VR = \mathcal{P} - \bigcup_{n \in PN} MBR(n), \tag{2}$$

where $\mathcal{P}$ denotes the entire space and PN denotes the index nodes which were pruned in the VO. Take Fig. 2(b) for example. Since the nodes N_6 and N_7 were pruned in the VO construction, the client has no idea about the skyline scope distribution in the region that is cover by these two nodes. Thus, the visible region is $\mathcal{P} - (N_6 \bigcup N_7)$.

3.3   Incremental VO

When the client moves out of the visible region, we have to issue a new query to update the results. A simple method for the server is to reconstruct a new VO and return it to the client. However, this method is inefficient because it does not reuse the previous VO to reduce the communication cost. In this section, we propose an incremental VO technique to address this problem.

The main idea of incremental VO is to reuse the VOTree of the last query, since it may overlap with the VOTree of the new query. Thus, we modify the VO construction algorithm in the MR-Sky-tree method by the following changes (Algorithm 2): 1) if a node does not cover the new query point, its related information (including MBR and digest) need not be stored in the incremental VO since the client can get them from the previous VO (we use a label "Hit" to represent such nodes; see Line 11); 2) if a leaf node covers the last query point, it also need not be returned to the client (Line 14), since all its members have been available in the previous VO.

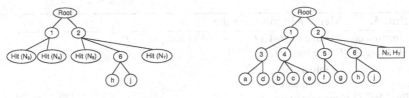

Fig. 4. An Incremental VO Example   **Fig. 5.** Merged VO

Revisit our running example. If the query point move from q to q_2, the incremental VO is shown in Fig. 4. The node N_7 does not cover the new query point q_2, it is replaced by label "Hit." Since the leaf nodes N_3, N_4 and N_5 cover the last query point q, they are also represented by label "Hit." Only the node N_6 covers q_2 and does not cover q. Hence, it is unfolded and all its children skyline scopes are inserted into the incremental VO. While the storage overhead for label "Hit" is negligible, the main communication cost here is incurred by the unfolded leaves. Fortunately, the number of such leaves is usually small for continuous queries.

After the client receives the incremental VO, it will merge the incremental VO with the previous VO. The merge algorithm traverses the incremental VO and the previous VO in parallel to construct the merged VO. The type of the incremental VO node determines what is inserted into the merged VO. We use two heaps $\mathcal{INC\_Q}$ and $\mathcal{PRE\_Q}$ to temporarily store the nodes of the incremental VOTree and the previous VOTree. During the merge process, if the incremental VO node is an index node, we just insert the corresponding node of the previous VO into the merged VO. And then their children are inserted into $\mathcal{INC\_Q}$ and $\mathcal{PRE\_Q}$ for further processing (Line 6-9). If the incremental VO node is a leave labeled "Hit," the whole subtree of the corresponding node in the previous VO is inserted the merged VO without any change (Line 10-11). Otherwise, the incremental VO node must be an unfolded leaf and is inserted into the merged VO instead of the previous one (Line 12-13).

4 Experiments

4.1 Experiment Setup

In this section, we evaluate the performance of our proposed techniques and algorithms through experiments. The spatial object set used in the experiments contains 2,249,727 objects representing the centroids of the street segments in California [14]. All testing datasets draw objects randomly from this set. The data space is normalized to a 100,000 Unit × 100,000 Unit square, where 1 Unit represents about 1 m. The non-spatial attribute values of these objects are synthesized with a uniform distribution in the interval [0, 100,000]. The page size is 4K bytes and the size of each object is 320 bytes. The hashing function we use for the MR-Sky-tree is SHA-512, and the size of each digest is 64 bytes. Table 1 summarizes the default settings and value ranges of various system parameters.

We measure the performance of continuous LSQ authentication with three metrics: communication cost, server computation time and client verification time. We compare the effects of the three optimization techniques proposed in in Section 3 with the naive

---

Algorithm 3. VO Merge Algorithm on the Client

INPUT: Incremental VO tree *increVO*, previous VO tree *preVO*
OUTPUT: Merged VO tree *merVO*

```
 1:  insert increVO.root into a queue INC_Q
 2:  insert preVO.root into a queue PRE_Q
 3:  while INC_Q is not empty do
 4:      get the top element inc_e from INC_Q
 5:      get the top element pre_e from PRE_Q
 6:      if inc_e is an index node do
 7:          insert pre_e into the merVO
 8:          insert the children of inc_e into INC_Q
 9:          insert the children of pre_e into PRE_Q
10:      else if inc_e is a label "Hit" do
11:          insert subtree of pre_e into the merVO
12:      else if inc_e is an unfolded leaf
13:          insert subtree of inc_e into the merVO
```

Table 1. Parameter Settings

Parameter	Default	Value Range
Dataset cardinality	100K	[10K, 1,000K]
# non-spatial attributes	2	2, 4, 6 ,8
Non-spatial attribute values	[0, 100,000]	
Digest size	64 Bytes	
Object size	320 Bytes	

method, which repeatedly updates the query results and the corresponding VO when the client changes location. We assume that the client moves with an average speed of $6\ m/s$ and the location sampling period is 100 seconds.

We conducted the experiments on a workstation (Intel Xeon E5440 2.83GHz CPU) running on Ubuntu Linux Operating System. The simulation codes were written in Java (JDK 1.6). Each measurement is the average result over 100 sampling periods.

Communication Cost. We first measure the size of the VO transferred between the server and clients as the metric of communication cost. We compare the performance of four schemes: 1) the naive method without any optimization (denoted as *Naive*); 2) the MR-Sky-tree based method optimized with the valid scope technique (denoted as *VS*); 3) the MR-Sky-tree based method using both the valid scope and visible region techniques (denoted as *VS+VR*); 4) the MR-Sky-tree based method optimized with all proposed techniques (denoted as *All*). The evaluation results are shown in Figs. 6(a) through 6(c). We can observe that the communication cost of *VS* is close to that of the naive method, while *VS+VR* and *All* achieve a much lower cost. This is because the area of the valid scope is generally very small and the adjacent query points are seldom located in the same valid scope. If we promote the location update frequency by reducing the sampling period (see the left part of Fig. 6(c)), this increases the chance

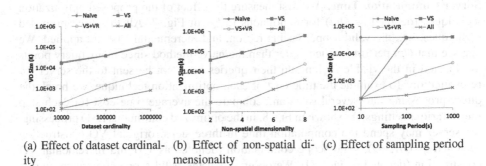

(a) Effect of dataset cardinal- (b) Effect of non-spatial di- (c) Effect of sampling period
ity   mensionality

Fig. 6. Average VO Size

that the adjacent query points are located in the same valid scope, which demonstrates the advantage of *VS* over the naive method. The visible region is much larger than the valid scope since many nodes in the MR-Sky-tree will be unfolded as long as they cover the query point. As a result, *VS+VR* gains much performance improvement over *VS* in most cases tested. Finally, employing the incremental VO technique further reduces the VO size by a factor of up to 10.

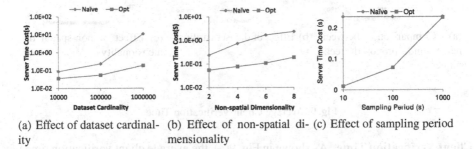

(a) Effect of dataset cardinal- (b) Effect of non-spatial di- (c) Effect of sampling period
ity   mensionality

Fig. 7. Average Sever Computation Cost

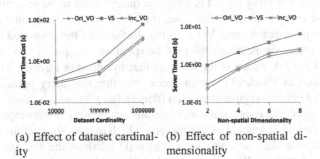

(a) Effect of dataset cardinal- (b) Effect of non-spatial di-
ity   mensionality

Fig. 8. Step-Wise Server Computation Time

Server Computation Time. We first measure the effect of the proposed optimization techniques (denoted as *Opt*) for continuous LSQs in Fig. 7. All the three proposed techniques, i.e., the valid scope, visible region, and incremental VO, are applied. We can see that *Opt* performs much better than the naive method since many query points are located in the visible region and their queries need not be sent to the server for re-evaluation. To examine the time cost of each optimization technique, we break the query processing into several steps and compare the average time cost for each step under various settings. As shown in Fig. 8, in the optimized continuous LSQ processing, the server may spend the computation time in three steps: original VO construction (denoted as *Ori_VO*), valid scope computation (denoted as *VS*) and incremental VO construction (denoted as *inc_VO*). We observe that the valid scope construction costs the most time since it needs to combine the skyline scopes of all skyline objects and then remove the part of non-skyline objects. For all three steps, the computation time increases as increasing dataset cardinality and non-spatial dimensionality. This is as expected due to the increased size of the skyline result set.

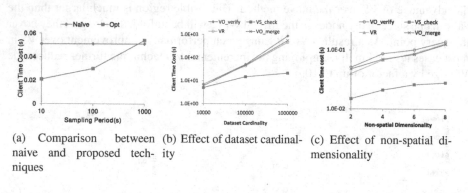

(a) Comparison between naive and proposed techniques

(b) Effect of dataset cardinality

(c) Effect of non-spatial dimensionality

Fig. 9. Average Client Verification Time

Client Verification Time. As shown in Fig. 9(a), the average client verification time of the proposed optimization techniques is much better than the naive method for short location sampling periods, due to the same reason explained in the previous experiments. When the location sampling period is long, the query needs to be re-evaluated by the server at almost every period. In this case, the client would incur some unnecessary cost in valid scope checking and VO merging. In fact, for the proposed optimization techniques, the client verification time may be spent on four parts: 1) VO verification (denoted as *VO_verify*), which is the same as that in one-shot queries; 2) valid scope checking (denoted as *VS_check*), which checks whether the query point is located in the valid scope; 3) visible region utilization (denoted as *VR*), which checks weather the visible region covers the query point and use the visible region to compute new LSQ results; 4) VO merging (denoted as *VO_merge*), which merges the incremental VO and the previous VO to make a new VO. Figs. 9(b) and 9(c) show the average cost of each step under various settings. We can observe that the valid scope checking uses the least time, while the costs of the other three parts are similar.

5 Conclusions and Future Work

In this paper, we have studied the problem of authenticating continuous location-based skyline queries with integrity assurance. We have developed three optimization techniques based on the MR-Sky-tree method, i.e., using valid scope, visible region, and incremental VO to reduce both the computation and communication cost. The experiment results show that our proposed techniques perform better than the existing method in terms of various performance metrics.

As for future work, we will extend the authentication problem to road network environments, where the query distance is defined by network distance. As the skyline scope does not work for network distance, new query authentication algorithms need to be developed.

Acknowledgments. This work is supported by GRF Grant HKBU210811, NSFC Grant 60903169, and the Hong Kong Scholars Program. Lin Xin's research is also supported by the Fundamental Research Funds for the Central Universities.

References

1. Beckmann, N., Kriegel, H.-P., Schneider, R., Seeger, B.: The R*-tree: An Efficient and Robust Access Method for Points and Rectangles. In: SIGMOD (1990)
2. Berg, M., Cheong, O., Kreveld, M.: Computational Geometry: Algorithms and Applications, 3rd edn., ch. 7 (2008) ISBN: 978-3-540-77973-5
3. Hu, H., Xu, J., Chen, Q., Yang, Z.: Authenticating Location-based Services without Compromising Location Privacy. In: SIGMOD (2012)
4. Huang, Z., Lu, H., Ooi, B.C., Tong, K.H.: Continuous Skyline Queries for Moving Objects. IEEE Trans. on Knowledge and Data Engineering (TKDE) 18(12), 1645–1658 (2006)
5. Merkle, R.C.: A Certified Digital Signature. In: Brassard, G. (ed.) CRYPTO 1989. LNCS, vol. 435, pp. 218–238. Springer, Heidelberg (1990)
6. Li, F., Hadjieleftheriou, M., Kollios, G., Reyzin, L.: Dynamic Authenticated Index Structures for Outsourced Databases. In: Proc. SIGMOD (2006)
7. Li, F., Yi, K., Hadjieleftheriou, M., Kollios, G.: Proof-Infused Streams: Enabling Authentication of Sliding Window Queries On Streams. In: VLDB (2007)
8. Lin, X., Xu, J., Hu, H.: Authentication of Location-based Skyline Queries. In: CIKM (2011)
9. Lin, X., Xu, J., Hu, H.: Range-based Skyline Queries in Mobile Environments. TKDE (in press, 2012)
10. Pang, H., Jain, A., Ramamritham, K., Tan, K.-L.: Verifying Completeness of Relational Query Results in Data Publishing. In: Proc. SIGMOD (2005)
11. Pang, H., Mouratidis, K.: Authenticating the Query Results of Text Search Engines. PVLDB (2008)
12. Papadopoulos, S., Yang, Y., Bakiras, S., Papadias, D.: Continuous Spatial Authentication. In: Mamoulis, N., Seidl, T., Pedersen, T.B., Torp, K., Assent, I. (eds.) SSTD 2009. LNCS, vol. 5644, pp. 62–79. Springer, Heidelberg (2009)
13. Papadopoulos, S., Yang, Y., Papadias, D.: CADS: Continuous Authentication on Data Streams. In: VLDB, pp. 135–146 (2007)
14. R-tree Portal, http://www.rtreeportal.org/
15. Sharifzadeh, M., Shahabi, C.: The Spatial Skyline Queries. In: Proc. VLDB (2006)

16. Yang, Y., Papadopoulos, S., Papadias, D., Kollios, G.: Authenticated Indexing for Outsourced Spatial Databases. VLDB Journal 18(3), 631–648 (2009)
17. Yang, Y., Papadias, D., Papadopoulos, S., Kalnis, P.: Authenticated Join Processing in Outsourced Databases. In: SIGMOD (2009)
18. Yiu, M.L., Lin, Y., Mouratidis, K.: Efficient Verification of Shortest Path Search via Authenticated Hints. In: ICDE, pp. 237–248 (2010)
19. Yiu, M.L., Lo, E., Yung, D.: Authentication of Moving kNN Queries. In: ICDE (2011)
20. Zheng, B., Lee, C.K., Lee, W.-C.: Location-Dependent Skyline Query. In: Proc. Int'l. Conf. Mobile Data Management (2008)

Assessing Quality Values of Wikipedia Articles Using Implicit Positive and Negative Ratings

Yu Suzuki

Nagoya University, Furo, Chikusa, Nagoya, Aichi, Japan
suzuki@db.itc.nagoya-u.ac.jp

Abstract. In this paper, we propose a method to identify high-quality Wikipedia articles by mutually evaluating editors and text using implicit positive and negative ratings. One of major approaches for assessing Wikipedia articles is a text survival ratio based approach. However, the problem of this approach is that many low quality articles are misjudged as high quality, because of two issues. This is because, every editor does not always read the whole articles. Therefore, if there is a low quality text at the bottom of a long article, and the text have not seen by the other editors, then the text survives beyond many edits, and the survival ratio of the text is high. To solve this problem, we use a section or a paragraph as a unit of remaining instead of a whole page. This means that if an editor edits an article, the system treats that the editor gives positive ratings to the section or the paragraph that the editor edits. This is because, we believe that if editors edit articles, the editors may not read the whole page, but the editors should read the whole sections or paragraphs, and delete low-quality texts. From experimental evaluation, we confirmed that the proposed method could improve the accuracy of quality values for articles.

Keywords: Wikipedia, Edit History, Quality, Reputation.

1 Introduction

Wikipedia[1] is one of the most successful and well-known User Generated Content (UGC) websites. It has more and fresher information than existing paper-based encyclopedias, because any user can edit any article. Many experts submit texts, and texts submitted by them should be informative for all who read it. Therefore, as well as being very large, Wikipedia is also very important. However, a dramatic increase in the number of editors causes an increase in the number of low-quality articles. Kittur et al. [6] showed that about 78.6% of 147,360 articles had not reached "start" status[2]. Therefore, automatic or semi-automatic systems should be developed to identify which part of article is high-quality and which is not.

In this paper, we propose a method to identify high quality texts using edit history. Here we define the word "quality" as the degrees of excellence. The definition of quality has many aspects such as credibility, expertise, and correctness.

---

[1] http://www.wikipedia.org/
[2] http://en.wikipedia.org/wiki/Template:Grading

H. Gao et al. (Eds.): WAIM 2012, LNCS 7418, pp. 127–138, 2012.
© Springer-Verlag Berlin Heidelberg 2012

Therefore, measuring excellence is difficult task. To solve this problem, we measure the number of editors who consider the article excellent, which is one of the important aspects of quality. When many editors consider excellent for an article, the quality of this article is high, but when a small number of editors consider the article excellent, the quality is low. In the latter case, even if only a small number of readers read the article, and these readers consider the article excellent, we decide that the quality of the article is low. This is because, there is a small number of evidence to decide whether the quality of the article is high or low.

If editors find low quality texts, the editors generally reject and delete them. Adler et al. [2] investigate that the recall for bad-quality short-lived text is 79%. This means that if a text survives beyond multiple edits by the other editors, the text should be high-quality. Therefore, using the *survival ratio* of texts, the system calculates the quality value of a text.

Example 1: Let us consider a motivating example. One editor e_a writes a part $p(e_a)$ of an article. Then, another editor e_b edits another part of this article, but keeps $p(e_a)$ intact. In this case, we assume e_b remains $p(e_a)$ as it is because s/he judged $p(e_a)$ to be high-quality. Next, another editor e_c deletes $p(e_a)$. We assume that e_c judged $p(e_a)$ to be low-quality, hence s/he deleted the text. As a result, the paragraph $p(e_a)$ is confirmed by e_b, but not confirmed by e_c. If e_c had not delete $p(e_a)$, the quality in this case would have been higher than that in the former case, because the paragraph $p(e_a)$ is trusted by one editor in the former case whereas it was trusted by two editors in the latter case. In this case, the *survival ratio* of $p(e_a)$ is 1 when e_b edits, and 0 when e_c edits. Therefore, the overall survival ratio of $p(e_a)$ is 0.5.

In this method, when a text survives beyond multiple edits, the text is judged as high quality. However, the problem is that every editor does not always read the whole articles, then if there is low-quality text on long articles, the text is treated as high-quality. In this paper, to solve this problem, we introduce to use section and paragraph as a unit instead of the whole page. This means that if an editor edits an article, the system treats that the editor gives positive ratings to the section or the paragraph which the editor edits. This is because, we believe that if editors edit articles, the editors may not read whole pages but should read whole sections or paragraphs, and delete low-quality texts.

2   Related Work

Adler et al. [1–3], Hu et al. [5], Wilkinson et al. [10], and Suzuki et al. [8] proposed a method for calculating quality values from edit histories. These methods are based on survival ratios of texts and is similar to the basic idea described at section 3.2. In these methods, edit distance is used for measuring difference between old and new versions. In this case, the impact for text quality by deletion and that by remaining are the same. However, these impacts of these two operations should be different, because the editors can delete a text only once whereas they can remain a text many times. Therefore, if we treat deletion and

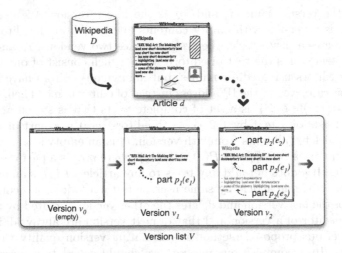

Fig. 1. Notations of this paper

remaining texts equivalently, we should separately calculate the impacts from these operations, and then integrate after normalizing.

Moreover, these methods proposed in the past do not use negative ratings. Therefore, when an editor writes many texts, and these texts are deleted immediately, the editor of the text do not imposed a penalty, then the quality of the editor do not decrease. We believe that the quality value of this editor should be low, then we use both positive ratings and negative ratings.

3   Proposed Method

Our goal is to assess quality value of articles by mutually evaluating quality values of texts and editors. The process is as follows:

1. The system extracts edit histories of articles
   from dumped data, and identify editors of texts.
2. It calculates positive and negative ratings for editors from edit histories.
3. It calculates editor quality values by combining positive and negative ratings.
4. It calculates positive and negative ratings using editor quality values.
5. It calculates editor quality values using modified positive and negative ratings.
6. Repeat 4. and 5. until editor quality value converges.
7. Calculate version quality values using editor quality values.

3.1   Modeling

In this section, we define notations that are used throughout this paper as shown in Fig. 1. On Wikipedia, every article has a version list $V = \{v_i | i = 0, 1, \cdots, N\}$

where i is the version number, and v_N is the latest version. We denote that if $i = 0$, v_0 is a version with empty contents and no editor. If editor e in all editors E creates a new article, the system makes two versions, v_0 and v_1, and then the system stores the text of editor e in v_1 which consist of one part $p(e)$. We identify editors using editor names, but anonymous editors have no editor name. In this case, we use the IP address instead of editor name. Then, we define version $v_i = \{p(e)|e \in E\}$ as a set of complete parts that is stored at i-th edit and that consists of a text by $1, 2, \cdots, i$-th editors. $p(e)$ is a part of article by editor e. If e deletes all texts from i-th version, v_i is an empty set.

Editor e creates a set of parts $P(e) = \{p(e)\}$ where $p(e)$ is a part created by e for all articles. If e does not add any texts to any articles, $P(e)$ is an empty set. When editors edit one article by same user more than twice consecutively, the system keeps the last version and deletes the other versions created by the users. That is, the editor of a version and that of next version are always different.

The aim of our proposed method is calculating version quality value $T(v_i)$ of version v_i. To accomplish our mission, we should calculate converged parts quality value $\tau_K(d, e)$ of parts on article d by editor e, and converted editor quality value $u'_K(e)$ of editor e. $\tau_0(d, e)$ is an initial parts quality value, and $U_0(e)$ is an initial editor quality value. In step 6. at section 3.3, we repeat calculating k-th parts quality value $\tau_k(d, e)$ and k-th editor quality value $U_k(e)$ until converge. $k = 1, 2, \cdots, K$ is the number of repetitions of steps 4 and 5. $\tau_K(d, e)$ and $u'_K(e)$ is the converged value of $\tau_k(d, e)$ and $u'_k(e)$.

3.2   Key Idea

We show how to calculate quality values of articles using an example of edit history in Fig. 3. Using this example, we explain how to calculate quality values of parts $p(e_1)$ that are added by editor e_1 in version v_1. First, we identify the texts that are added in version v_1. In this example, the editor e_0 writes all texts of v_0, and the editor e_1 adds the texts "Ueshima" and "Prime Minister" as $p(e_1)$ to version v_2. At this time, we suppose that e_1 gives positive and negative ratings to e_0. e_1 remains 21 letters written by e_0 ("Yoshihiko", "is", "the", "of Japan"), then e_1 gives 21 letters of positive ratings to e_0. e_1 deletes 13 letters written by e_0 ("Noda", "president"), then e_1 gives 13 letters of negative ratings to e_0.

However, the problem of this method is that if e_1 edits small edits, e_1 writes only a small number of letters and remain all texts, the system decides that e_1 permits to remain almost all texts. We believe that all editors do not always read whole articles. Therefore, when there are many editors who do not read whole articles, the accuracy of quality values should decrease. To solve this problem, we use section and paragraph as a reading unit.

3.3   Quality Value

In this section, we describe how to calculate positive and negative ratings. Fig. 4 shows an example to explain how to calculate positive and negative ratings. First, we define a unit of article, section, and paragraph. A unit of whole article

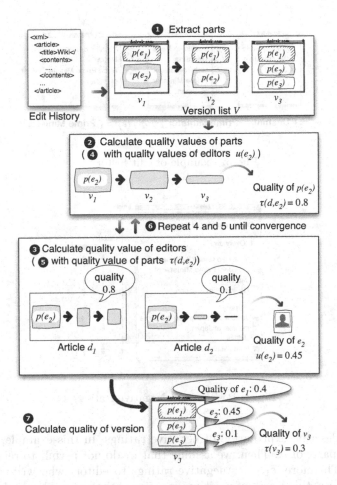

Fig. 2. Overview of our proposed method

is defined as texts in the whole article. A unit of section is defined as the texts divided by symbols which indicate separation of sections. In this example, A and B belong to the same section, C and D belong to different section. A unit of paragraph is defined as texts divided by special, not linguistic characters, such as HTML tags and line break code. In this example, A and B belong to different paragraph because A and B is divided by line break.

We describe intuitive explanation of positive ratings. In this example, editor e_1 edits a part of article A. When we use whole article as a unit, we assume that e_1 permits to remain parts A, B, C, and D. Therefore, e_1 gives positive ratings to editors who write When we use section as a unit, we assume that e_1 permits to remain parts A and B. In this case, we suppose that e_1 do not read C and D because e_1 do not edit. When we use paragraph as a unit, we assume that e_1 permits to remain only A. In this case, we suppose that e_1 do not read B, C, and D.

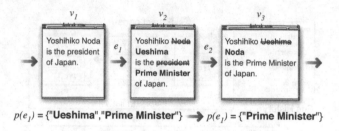

$p(e_1) = \{$"Ueshima","Prime Minister"$\} \longrightarrow p(e_1) = \{$"Prime Minister"$\}$

Fig. 3. Example of edit history

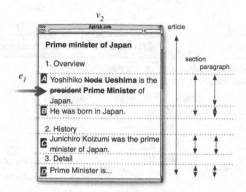

Fig. 4. Positive and Negative Ratings

We also describe explanation of negative ratings. In this example, editor e_1 deletes two parts of A. Then, we assume that e_1 do not permit to remain these two parts. Therefore, e_1 gives negative ratings to editors who write these two parts. However, the degree of positive/negative ratings of e_1 should depends on the quality value of e_1. This means that if e_1 is a high-quality editor, the positive and negative ratings by e_1 should be large. To solve this problem, we mutually calculate quality values of parts and editors.

Positive Ratings. Next, we calculate the quality values of parts using the quality values of editors.

Whole Article as a Unit. We calculate the positive ratings using whole article $\tau_k^{pw}(d, e)$ as follows:

$$\tau_k^{pw}(d, e) = \alpha \cdot \tau_0^{pw}(d, e) + (1 - \alpha) \sum_{e' \in E} |\delta(e')| u_{k-1}(e') \qquad (1)$$

where E is a set of editors who delete $p(e)$, $\delta(e')$ is the letters in $p(e)$ deleted by e', $|\delta(e')|$ is the number of letters in $\delta(e')$, and α $(0 < \alpha \leq 1)$ is the parameter to control the effect of editor quality value. $u_{k-1}(e')$ is a quality value of e' which is calculated by the method using whole article as a unit.

The first part of expression of this equation means the part quality values, which is defined $\tau_0^{pw}(d, e) = \sum_{p(e) \in \bar{P}_w} (\log |p(e)| + 1)$. When we calculate quality value of parts, we use a log scale instead of the raw number of letters, because we face a problem when the editor adds long texts. If an editor adds $10,000$ letters to the article, and the texts survive only one edit, this text quality value is $10,000$, which is the same quality value as a 100-letter texts that survives beyond 100 edits. We think that the latter text is higher quality than the former text. Therefore, we count the number of characters using a log scale.

The second part of expression means the number of deleted letters with quality values of editors who delete them. If an editor who has a low-quality value deletes a part $p(e)$, then the value of the second expression is high, then the value of $\tau_k^{pw}(d, e)$ is almost the same as $\tau_{k-1}^{pw}(d, e)$. Therefore, the editor quality value does not affect the part quality value. In this case, if the editor who deletes the part has a high-quality value, the second expression has a high value. Thus, the value of $\tau_k^{pw}(d, e)$ decreases more than $\tau_{k-1}^{pw}(d, e)$.

Section as a Unit. We calculate the positive ratings of parts using section $\tau_k^{ps}(d, e)$ by equation (1). E is a set of editors who delete $p(e) \in \bar{P}_s$, and $u_{k-1}(e')$ is a quality value of e' which is calculated by the method using section as a unit.

Paragraph as a Unit. We calculate the positive ratings of part using paragraph $\tau_k^{pa}(d, e)$ by equation (1). E is a set of editors who delete $p(e) \in \bar{P}_a$, and $u_{k-1}(e')$ is a quality value of e' which is calculated by the method using section as a unit.

Negative Ratings. We calculate the negative ratings $\tau_k^n(d, e)$ as follows:

$$\tau_k^n(d, e) = \alpha \cdot \tau_{k-1}^n(d, e) + (1 - \alpha) \sum_{e' \in E^n} |\delta(e')| u_{k-1}^{n'}(e') \tag{2}$$

where E^n is a set of editors who delete $p(e) \in N$, and $u_{k-1}^{b'}(e')$ is a quality value of e' which is calculated by the method using negative ratings. We set $\tau_0^n(d, e) = \sum_{p(e) \in N} (\log |p(e)| + 1)$.

Quality Values of Parts. We define the quality values $\tau_k(d, e)$ of part of an article d by editor e using positive and negative ratings as follows:

$$\tau_k(d, e) = \tau_k^p(d, e) - \tau_k^n(d, e) \tag{3}$$

where $\tau_k^p(d, e)$ is one of three text quality values $\tau_k^{pw}(d, e)$, $\tau_k^{ps}(d, e)$, and $\tau_k^{pa}(d, e)$.

Quality Values of Editors Using Adjusted Quality Values of Parts. Using part quality values $\tau_k(d, e)$, we define the editor quality values $u_k(e)$ of e as follows:

$$u_k(e) = \frac{1}{|D(e)|} \cdot \sum_{d \in D(e)} \tau_k(d, e) \tag{4}$$

We normalize $u_k(e)$ to range between 0 and 1 as follows:

$$u'_k(e) = \frac{u_k(e) - \min_{e' \in E} u_k(e')}{\max_{e' \in E} u_k(e') - \min_{e' \in E} u_k(e')} \tag{5}$$

We repeat the processes until the values of $\tau_k(d, e)$ and $u_k(e)$ converge.

Quality Values of Versions. Using the converged value of $u'_K(e)$, we define the version quality value $T(v)$ of version v as follows:

$$T(v) = \frac{1}{|v|} \cdot \sum_{e \in P(e)} u'_K(e) \cdot |p(e)| \tag{6}$$

where $|v|$ is the number of letters in v, $|p(e)|$ is the number of letters in $p(e)$, and $u'_K(e)$ is the editor quality value of e. This function means that the version quality value is the weighted averaging ratio of part quality values, and the weight is the number of letters in the parts.

4 Experiments

To determine the accuracy of the quality values calculated by our proposed system, we did experimental evaluation. In this evaluation, we tried to confirm that when we use editor quality values to calculate text quality values, the article quality values are accurate.

In this experiment, we compared 7 systems. 3 systems used only positive ratings; *page* is the system using article based, *sec* is the system using section based, and *par* is the system using paragraph based positive ratings. 1 system, *delete*, using only negative ratings. 3 systems used both positive and negative ratings; *del+page* used both article based positive ratings and negative ratings, *del+sec* used both section based positive ratings and negative ratings, and *del+par* used both paragraph based positive ratings and negative ratings.

We compared these systems using average precision ratio, which is an averaging value of precision ratios at each recall level [4]. We compared the answer set with the list of articles in ascending order of their quality values. If articles in the answer set are ranked higher, we will be able to confirm that the system calculates accurate quality values. The key in this evaluation is the appropriateness of answer sets. In current information system retrieval evaluation, observers create answer sets by judging relevance of articles. However, judging the quality of articles is difficult, so we cannot confirm the appropriateness of quality judgments of articles. Therefore, we put featured and good articles selected by Wikipedia users in the answer set.

We set α to 0.7 used at equations from (1) to (2). Before these experiments, we set α from 0.1 to 0.9 in 0.1 increments and calculate averaging precision ratio as preliminary experiment. In this result, when we set 0.7, we got the highest averaging precision ratio of our proposed system.

4.1   Data Sets

In this experiment, we used the Japanese version of Wikipedia edit history dumped on Jan. 4, 2012, which can be downloaded at the Wikipedia dump download site[3]. We randomly select 1,000 articles which contain 192,227 versions. The number of editors is 65,909 including not registered editors who are identified by IP addresses, and bots which are listed[4]. When we select articles, we referred to Wikipedia statistics[5] to decide which articles we select. We do not select the articles that do not contains at least one link to Wikipedia articles. We also do not select the articles for specific purposes, such as redirect pages, notes, and rules of Wikipedia.

In this experiment, we set the answer set of "featured" and "good" articles as a correct answer set. Featured and good articles are selected by the votes of Wikipedia users (mainly readers). These articles are evaluated by "Featured article criteria"[6] and peer reviewed by many active users. If vandals nominate low-quality articles for featured or good articles, the nomination is rejected by administrators. Therefore, we believe that these articles are high-quality, so we could use featured and good articles as high-quality articles for the test set. The number of featured and good articles are 72 and 611 respectively. In our selected articles, the number of featured and good articles are 2 and 5 respectively. We decide these 7 articles as answer set.

4.2   Experimental Results and Discussions

Fig. 5 shows an average precision ratio per each repeated count. The meaning of each line is described at the top of section 4. From this graph, we unveil that *del+par*, the system which use paragraph based positive ratings and negative ratings calculates article quality values more accurately than the other methods. When we compare the results of *par* and *delete*, the order of articles dramatically changes, different high-quality articles have high quality values. Therefore, when we combine these positive and negative ratings, we can calculate more accurate quality values.

However, the repeats of calculation of positive/negative ratings and editor quality values are not effective in this experiment. From Fig. 5, when repeated count increases, average precision ratio increases at most 0.02%, almost 0%. The reason of this is that 61% of all editors edit only one article more than once. Therefore, when we construct bipartite graph of texts and editors, the graph is very sparse. When editors edit small number of articles, small number of the other editors review the articles, then the editors get small number of positive/negative ratings. To be concrete, second expressions of equation (1) almost equals to 0, then if we repeated to calculate quality values of parts, these

---

[3] http://dumps.wikimedia.org/jawiki/20120104/
[4] http://ja.wikipedia.org/wiki/WP:BOTST
[5] Wikipedia: What is an article?:
     http://en.wikipedia.org/wiki/Wikipedia:What_is_an_article
[6] http://en.wikipedia.org/wiki/Wikipedia:Featured_article_criteria

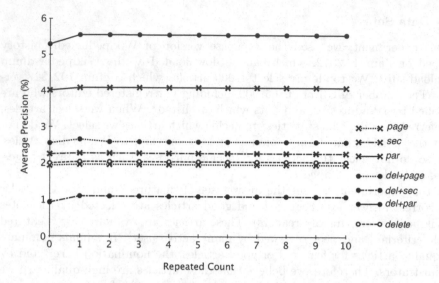

Fig. 5. Average precision ratio vs. repeated count

quality values do not change. Therefore, repeats of calculating editor quality values are not effective. This is because, in this experiment, we selected only 1,000 articles. If we calculate all articles at the Wikipedia, averaging precision should increase by repeating calculation of editor quality values.

5   Conclusion

Wikipedia is the most popular and highest quality encyclopedia to be created by many editors. The information on Wikipedia keeps expanding, but its quality is not proportional to its quantity. In this paper, we propose a method to identify high-quality articles mutually evaluating editors and text to improve the accuracy of quality values of versions.

In our method, we introduce a combination of a peer-review based quality value calculation method and a link analysis method, which is based on quality values of editors and texts. Not all editors are honest, and many vandals attack Wikipedia by deleting high-quality texts. Using our proposed method, quality values of editors affect that of texts. Therefore, if vandals delete high-quality texts, they do not affect the survival ratio of the texts, so the quality values of versions which are attacked by vandals do not decrease. As a result, we can calculate accurate quality values of parts, editors, and versions without the activities of vandals.

Moreover, in this paper, we tackled two problems. One problem is that every editors do not always read whole articles, then if there is low-quality text on long articles, the text is treated as high-quality. To solve this problem, we used section and paragraph as a unit instead of whole page. This means that if an editor edits an article, the system treats that the editor gives positive ratings to the section or the paragraph which the editor edits. This is because, we believe that if editors edit articles, the editors should read whole sections or paragraphs, and delete low-quality texts. Another problem is that if there is a low-quality editor, the editor writes texts many times, and the texts are deleted frequently, the editor is treated as high-quality. This is because, survival ratio based methods do not use features of negative ratings. To solve this problem, we used features of negative ratings. From experimental evaluation, we confirm that our proposed system can calculate accurate quality values if we use paragraph as a unit for positive ratings and we also use negative ratings.

Quality of information is becoming increasingly important in information retrieval research field. An information retrieval system retrieves the documents that are relevant to the user's query, but the system is not concerned about whether the documents are high-quality or not. However, if the retrieved documents are low-quality, they should not be retrieved even if they are relevant. Therefore, as Toms et al. [9] already mentioned, when we combine an information retrieval systems with our proposed high-quality article retrieval system, we will develop an information retrieval system more accurate than current information retrieval systems.

Finally, we describe three open problems:

Vagueness of Quality Value: In this paper, we calculated quality values for editors described at section 3.3. However, this editor quality value is not always distinct because the frequency of editing is different for each editor. We suppose that if an editor rarely edits articles, the editor may just happen to obtain a high quality value, but vagueness of the editor quality value should be high. Therefore, we should develop a method to calculate vagueness of editor quality values that does not depend on editor quality value.

Use of Natural Language Processing Techniques: In our proposed method, we do not analyze linguistic structures; we only count the number of letters in texts. A strong point of our proposed system is that it can adapt to different language versions of Wikipedia articles. However, a weak point is that it cannot use important features that come from linguistic features. In our experiment, we found that high-quality articles are always written in formal language. Moreover, Sabel et al. [7] says that text analysis is useful for calculating quality values. For example, if an editor changes "A is a B." to "A is not a B." the number of letters changes by only three, but the meanings of these sentences are completely different. Therefore, we should analyze texts using natural language analysis techniques for calculating survival ratio of texts.

Scalability: We implemented our system on a single PC with two CPUs and a 16 GB memory. As a result, this system took about 20 minutes to analyze $1,000$ articles on the Japanese version of Wikipedia. Therefore, to analyze the whole Wikipedia, we will need more and more calculation costs. If we use multiple cluster PCs and map/reduce frameworks such as Hadoop, we will reduce calculation time of article analysis. We will therefore work on developing systems that are more scalable.

Acknowledgment. This work was partly supported by Japan Society for the Promotion of Science, Grants-in-Aid for Scientific Research (23700113).

References

1. Adler, B.T., Chatterjee, K., de Alfaro, L., Faella, M., Pye, I., Raman, V.: Assigning Trust to Wikipedia Content. In: Proceedings of the International Symposium on Wikis (WikiSym 2008). ACM (2008)
2. Adler, B., de Alfaro, L.: A content-driven reputation system for the Wikipedia. In: Proceedings of the 16th International Conference on World Wide Web (WWW 2007), pp. 261–270 (2007)
3. Adler, B., Chatterjee, K., de Alfaro, L., Faella, M., Pye, I., Raman, V.: Measuting Author Contributions to the Wikipedia. In: Proceedings of the 2008 International Symposium on Wikis (WikiSym 2008) (2008)
4. Baeza-Yates, R., Ribeiro-Neto, B.: Modern Information Retrieval: the concepts and technology behind search. Addison-Wesley (2011)
5. Hu, M., Lim, E., Sun, A., Lauw, H.W., Vuong, B.: Measuring Article Quality in Wikipedia: Models and Evaluation. In: Proceedings of ACM International Conference on Information and Knowledge Management (CIKM 2007), pp. 243–252 (2007)
6. Kittur, A., Kraut, R.E.: Harnessing the wisdom of crowds in wikipedia: quality through coordination. In: Proceedings of the 2008 ACM Conference on Computer Supported Cooperative Work, CSCW 2008, pp. 37–46. ACM, New York (2008)
7. Sabel, M.: Structuring Wiki revision history. In: Proceedings of the 2007 International Symposium on Wikis (WikiSym 2007), pp. 125–130. ACM (2007)
8. Suzuki, Y., Yoshikawa, M.: Qualityrank: Assessing quality of wikipedia articles by mutually evaluating editors and text. In: Proceedings of the 23rd ACM Conference on Hypertext and Social Media (HT 2012). ACM Press (to appear, 2012)
9. Toms, E.G., Mackenzie, T., Jordan, C., Hall, S.: wikiSearch: enabling interactivity in search. In: Proceedings of the 32nd Annual International ACM SIGIR Conference on Research and Development in Information Retrieval (SIGIR 2009), p. 843 (2009)
10. Wilkinson, D.M., Huberman, B.A.: Cooperation and quality in wikipedia. In: Proceedings of the 2007 International Symposium on Wikis (WikiSym 2007), pp. 157–164. ACM (2007)

Form-Based Instant Search and Query Autocompletion on Relational Data

Hao Wu and Lizhu Zhou

Department of Computer Science and Technology
Tsinghua University, Beijing 100084, China
haowu06@mails.tsinghua.edu.cn, dcszlz@tsinghua.edu.cn

Abstract. Finding information in relational data is a key task of data management in the age of information explosion. However, state-of-the-art search paradigms in relational data, such as *Structured Query Language* (SQL), *keyword search*, and *query-by-example* (QBE) have their own limitations. In this paper we address the issue of enhancing the power of QBE-like search interface, i.e. form-style interface, by incorporating instant search and query autocompletion, which are useful features for traditional keyword search but have not been paid enough attention specifically for form-style interfaces by database researchers. We devise effective index and efficient algorithms that can support instant search and query autocompletion using distinct attribute string values simultaneously. In addition, we also consider the issue of query autocompletion for those attributes in which each distinct attribute string value has few occurrences in the table, and devise the algorithm that provides complete words as completion results instead. We conducted extensive experiments on real-world datasets, and results show that our methods outperform baseline methods in terms of both efficiency and scalability.

1 Introduction

Finding information in structured data, specifically relational data, is a key task of data management in the age of information explosion. In early days, people could only use *Structured Query Language* (SQL) to find information in a relational database. SQL provides to users sufficient abilities to express their query intents, but it is hard to use and master even by experienced users. In recent years, because of its ease of use, keyword search on relational data has attracted more and more interests from database researchers. Although keyword search is a easy way for users to find information they need, sometimes its limited expression power cannot meet users' search requirements. For example, if we want to find all papers whose titles contain the word "database" with CompleteSearch[1] [1], which is a search engine on the Computer Science Bibliography (DBLP)[2] dataset, and pose the keyword "database" as the query to the system,

---

[1] http://dblp.mpi-inf.mpg.de/dblp
[2] http://www.informatik.uni-trier.de/~ley/db

H. Gao et al. (Eds.): WAIM 2012, LNCS 7418, pp. 139–151, 2012.
© Springer-Verlag Berlin Heidelberg 2012

then some of the top-ranked results are irrelevant because in each of them the word is contained in conference name instead of paper's title. Another example is that, if we want to find all of Wei Wang's publications, and simply input "`wei wang`" as the query to CompelteSearch, none of the top 20 results are relevant because the system interprets "`wei`" and "`wang`" as two names.

A trade-off between the ease of use and expression power is *Query-By-Example* (QBE), in which users input their queries by filling in forms with keywords. By specifying the query condition to different input boxes, the user can express her query intent. However, since a form-style interface is more complicated than a single input box that is used for keyword search, its usage is limited to those "advanced search" systems in real applications, such as eBay Advanced Search[3] and PubMed Advanced Search[4], etc.

In this paper we investigate the problem of enhancing the ease of use of form-style interface by introducing the *instant search* feature to it. With this feature users can get results instantly right after each keystroke when they are filling in the form. The feature of instant search has been proved to be useful in improving the user experience of traditional keyword search, but it has never been considered to be applied in form-style interfaces before. In addition, to further improve the user experience, we investigate the problem of *query autocompletion* for form-style interfaces, with which users can get lists of suggested attribute values or keywords that can help them extend their current queries to get the results they need more quickly when they are typing in the form. The contributions of this paper are summarized as follows.

- We designed a trie-based index and an efficient incremental algorithm that support simultaneous instant search and query autocompletion for querying relational tables. We also considered the techniques to improve the performance of the algorithm by avoiding unnecessary computations.
- We addressed the problem of query autocompletion for *textual attributes*, in which each distinct attribute value has very few occurrences in the data table, using keywords as completion results. We proposed an efficient algorithm, namely `HistScan`, that avoid brute-force scan operations by traversing the trie structure in a best-first manner instead of depth-first manner with the support of converting inverted lists and result lists into histograms.
- We conducted extensive experiments to evaluate our algorithms on real datasets. Results showed that our algorithms can achieve interactive speed for instant search and real time query autocompletion, and can greatly outperform baseline algorithms in terms of both efficiency and scalability.

Besides, we also built a real system that can provide a fully-functional form-style instant search and autocompletion support on the DBLP dataset for computer science researchers. The system is called Seaform[5], which stands for *Search-as-you-type on forms*. The system has been online since Oct. 2009, and a brief

---

[3] `http://shop.ebay.com/ebayadvsearch`
[4] `http://www.ncbi.nlm.nih.gov/pubmed/advanced`
[5] `http://dbease.cs.tsinghua.edu.cn/seaform/`

introduction of it can be found in [16]. All of our experiments in this paper were based on the real query log of this system.

The rest of the paper is organized as follows. In Section 2 we formally define the problems we address. We propose our basic index structures and algorithms in Section 3, and introduce the techniques of query autocompletion for textual attributes in Section 4. Experimental results are presented in Section 5. We give a review of related work in Section 6 and conclude this paper in Section 7.

2   Problem Formulation

In this paper, we use a single relational table as our underlying data. We denote the set of tuples of the table we are going to search by $T = \{t_1, t_2, \ldots, t_M\}$. The attribute set of the table is denoted by $A = \{a_1, a_2, \ldots, a_N\}$. The value of the attribute a_n of the tuple t_m is denoted by $t_{m,n}$, where $m \in [1, M]$ and $n \in [1, N]$. For example, Figure 1(a) illustrates a sample relational table.

ID	Title	Conf.	Author
1	xml database	VLDB	albert
2	xml database	SIGMOD	bob
3	xml search	VLDB	albert
4	xml security	VLDB	alice
5	xquery optimization	SIGMOD	charlie

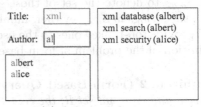

(a) A sample dataset.   (b) A form-style query.

Fig. 1. A sample relational table (left) and a typical form-style query (right). The second figure also illustrates the result list of instant search in the right-side box, as well as the result list of query autocompletion in the lower-left box.

A form-style keyword search interface consists of several input boxes (see Figure 1(b)). Each of these input boxes corresponds to an attribute of the table. When a user wants to search for something with this interface, she fills in the input boxes with query keywords, and a form-style query is then posed to the system. Without loss of generality, in this paper we only consider the case in which each attribute has exactly one corresponding input box on the form, and as a result we can avoid the distinguishing of attributes and input boxes in the rest of the paper. Formally, a form-style query is defined by $q = \{f_n | n \in [1, N]\}$, where f_n, called the n-th field of the query, denotes the non-empty query string in the input box that corresponds to the attribute a_n. For example, the query { Title:"xml",Author:"al" } has two fields: one is "xml" that belongs to the Title input box, the other is "al" that belongs to the Author input box.

Each field in a form-style query denotes the search condition to the corresponding attribute values of the tuples in the relational table. For instant keyword search, keywords in form-style queries are usually *prefix keywords*, i.e. each keyword is meant to be a prefix of a set of possible complete words appear in the data table. Formally, we use $p \sqsubseteq w$ to denote that a prefix keyword p is a prefix of a complete word w. If each prefix keyword in a query q has a matched complete word in the corresponding attribute in a tuple t_i, we say that q *matches* t_i. In other words, q matches t_i (denoted by $q \sqsubseteq t_i$) if and only if $\forall f_n \in q, \forall p \in q_n, \exists w \in t_{i,n}$, s.t. $p \sqsubseteq w$. With the help of above concepts, we can formally define the problem of form-based instant search as follows.

Definition 1 (Form-Based Instant Search). *Given a relational table T, form-based instant search answers the form-style query q with a set of tuples $R = \{t_i | t_i \in T \text{ and } q \sqsubseteq t_i\}$ as the result.*

To further improve the user experience, we can provide all possible distinct attribute string values of the currently-editing input box as query completions to the user when they are filling in the form. Assume that a user is inputting her query q in the n-th input box and the result of form-based instant search is R. We use S_n to denote the set of distinct string values of the attribute a_n, and use S_{n,q_n} to denote the set of those strings in S_n that are matched by q_n. We rank the strings in S_{n,q_n} by their number of occurrences, a.k.a. *frequencies*, in R, and provide top-ranked ones to the user as completions. To summary, the formal definition of the problem of form-based query autocompletion is as follows.

Definition 2 (Form-Based Query Autocompletion). *Given a form-style query q and its result set R, the completion list for the n-th attribute is the top-k ranked distinct string values in S_{n,q_n}, in which the score of each string value $s \in S_{n,q_n}$ is calculated by its number of occurrences (frequencies) in R.*

In the following section, we will introduce the detail of our index structures and algorithms that support form-based instant search and query autocompletion simultaneously. However, for some kind of attribute such as Title, returning distinct attribute string values to the user is useless for guiding her compose a better query because the number of occurrences of a specific attribute string value will be small. We call this kind of attribute the *textual* attribute (correspondingly, we call the other attributes the *categorical* attribute). We discuss the techniques of providing complete words as autocompletion results instead of attribute string values to the user in Section 4.

3   Basic Index Structures and Algorithms

3.1   Form-Based Prefix Keyword Search

According to the definition, the main difference between form-based keyword search and traditional keyword search is that a user's query conditions can be split into fields and specified to different attributes. As a result, each attribute of the data table should have its own index. Specifically, to support prefix-matching-based keyword search, which is the very basic requirement of instant search, we index all of the words appear in each attribute into a *trie* structure.

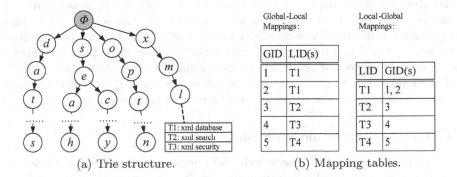

(a) Trie structure.   (b) Mapping tables.

Fig. 2. The index structures of the Title attribute for our sample dataset

A trie is a special tree structure in which a path from the root to a leaf corresponds to a word and all the words with common prefixes share the same path from the root. For example, Figure 2(a) shows the trie structure of the Title attribute of our sample dataset shown in Figure 1(a).

We can attach an inverted list of tuple ids to each leaf node of the trie to support simple keyword search by prefixes (see Figure 2(b)). For each field in the query, we first locate the trie node for each of the prefix keyword. We get the id list of each prefix keyword by calculating the union of all the inverted lists of the leaf nodes under the sub-trie of the previously located trie node. Finally, we intersect the id lists of all the prefix keywords of all the query fields to get the final id list of the result set, called *result list* for short.

However, the problem we should solve is not the simple prefix keyword search only. We should also consider the problem of: 1) query autocompletion; and 2) making prefix keyword search achieve interactive speed. To this end, in the following two sub-sections, we will introduce our adapted index and algorithms that support simultaneous query autocompletion and our cache-based techniques that can greatly improve the overall performance of our online processing by making all the online calculations incremental.

3.2   Simultaneous Query Autocompletion

The data structures that is used to support this new method are several mapping tables. First, for each attribute, we construct a so-called *local-global mapping table*. The l-th row of the mapping table stores the ids of all the tuple in the data table containing the l-th distinct string value (for each distinct string value we assign an id to it, called *local id*, as well). Second, we also construct a *global-local mapping table*, in which the g-th row contains the local ids of distinct string values that is contained in the g-th tuple in the data table (accordingly, the id of a tuple in the data table is called a *global id*). For example, Figure 2(b) shows the two mapping tables for the Title attribute.

To make the autocompletion more efficiently, we split the process of original prefix keyword search into three steps, and computes the search results and completion results simultaneously. Specifically, the first step of our new method is to find matched attribute string values according to each of the query fields, the second step is to find result set of tuples using these attribute string values, and the final step is to verify the attribute string values of the currently-editing attribute according to the result set to compose the completion list.

3.3   Incremental Online Processing

The key requirement of instant search and query autocompletion is efficiency. As a result, it is reasonable to utilize the previously-computed results and use the difference of the queries to compute new results. This is called *incremental online processing*. Specifically, we cache the previous query and its results. When a new query is submitted, we first check the cache to see whether the query can be answered from the cached results. If the new query can be obtained by extending the cached query with one or more letters, then we have a cache hit. We perform the non-incremental search algorithm described previously if there is no cache hit, or perform an incremental search based on the base query and base results if there is a cache hit. We use *global results* to denote the result id list of tuples in the data table, and use *local results* to denote the id list of distinct string values of an attribute. The incremental algorithm can be described as follows.

Step 1. Identify the difference between the cached query and the new query. We use a_n to denote the currently-editing attribute (i.e. input box), and use p to denote the newly appended prefix keyword.

Step 2. Calculate the local ids of a_n based on the query string in a_n. This is done by merging the id lists of all the leaf nodes on the sub-tree rooted at the node corresponding to w in the trie, and then intersecting the merged list with the local base results of a_n.

Step 3. Calculate the global results. This is done by first calculating the set of global ids corresponding to the local results of a_n calculated in Step 2 using the local-global mappings in the index, and then intersecting it with the cached global results.

Step 4. Calculate the local results of a_n. This step is called "synchronization". It is done by first calculating the set of local ids corresponding to the global results using the global-local mapping table in the index, and then intersecting it with the local base results of a_n.

Dual-List Trie Structures. In step 3 of the incremental algorithm, to obtain the global results, we map the local ids calculated in step 2 to lists of global ids, merge these lists, and then intersect the merged list with the global base results. If there are many local ids, the merge operation could be very time consuming. To address this problem, we can attach an inverted list of global ids to each of the corresponding trie leaf nodes. In this way, given a prefix keyword, we can identify the matched tuple in the data table without any mapping operation.

We call the adapted trie structures *dual-list tries*. With the help of dual-list tries, the overall search time can be reduced compared with that of using original tries, which are called *single-list tries*.

On-Demand Synchronization. In step 4 of the incremental algorithm we use a brute-force method to keep the local result list of the currently-editing attribute (a_n) up to date. However, if the user does not switch the input box to another one, it is unnecessary to perform synchronization for this input box. As a result, we do not perform synchronization for the query that has the same currently-editing input box with the cached query. On the other hand, if the user changes her focus to another input box, we must synchronize for (and only for) the corresponding attribute at once. We call this mechanism the *on-demand synchronization*. It requires one merge operation and one intersection operation. In contrast, the brute-force synchronization requires one merge operation and one intersection operation whenever the user types in a letter.

4   Autocompletion for Textual Attributes

As is discussed in Section 2, if a user is typing in her query in the input box of a textual attribute (for example, in the Title input box), showing a list of distinct attribute string values (i.e. complete title strings) to the user cannot help her much to extend the query. This is because each attribute string value has very few occurrences in the dataset. To address this issue, for textual attributes, we provide a list of ranked keywords instead of attribute string values as query-extending suggestions (a.k.a. query completions). Specifically, we calculate the scores of each possible keywords according to the current search result set and the current keyword prefix the user is inputting, and return top-k keywords as completions. The score of a keyword to suggest is defined by the number of tuples it appears (i.e. its frequency) in the result set of search. The rational of using this definition is straightforward: if the user uses the top-ranked keyword to extend her query, she will get the largest result set compared with using other keywords. This kind of query autocompletion is called *frequency-based keyword suggestion* for form-based instant search, and is defined as follows.

Definition 3 (Frequency-Based Keyword Suggestion). *Given the search result set R and the currently-input keyword prefix p in the input box of a textual attribute A_n, we return the top-k frequent keywords appear in the attribute A_n in the tuples of R. In addition, these keywords should also take p as their prefixes.*

4.1   ScanCount: Exact Method as Baseline

The most straightforward method can be derived directly from the definition of the problem. Conceptually, after we get the result set R, we scan each of the tuples in R, and count the occurrences of each word in attribute A_n. Meanwhile, we should also avoid those words that do not take p as their prefixes. The detail of this baseline algorithm is omitted due to space limitation.

4.2   HistScan: Histogram-Based Estimation

Although the `ScanCount` algorithm is straightforward to implement, its performance highly depends on the size of the result set: if the result set is large, the algorithm should scan through too many tuples, making the processing time too long for short queries (since short queries usually lead to large result sets). An alternative method is to scan all possible keywords instead of tuples in the result set, calculate their frequencies, and then return top-k most frequent ones as completions. Obviously, the performance of this method does not depend on the size of the result set, making it specially fit for short queries.

The key of this keyword-scan method is the calculation of the frequency values of keywords. With the help of the index we proposed in Section 3, we can get the frequency of a keyword by intersecting its inverted tuple id list with the id list of the result set (called the *result list* for short). However, this method requires $\mathcal{O}(|I| \cdot \log(|R|))$ or $\mathcal{O}(|R| \cdot \log(|I|))$ time where I denotes the inverted list of the keyword, making it unsatisfactory in terms of performance.

Our solution is to convert the id lists (the inverted list and the result list) into histograms, and estimate the frequency value of the keyword using these histograms in $\mathcal{O}(1)$ time. The histogram of an id list is defined as follows.

Definition 4 (Histogram of ID List). *Given an id list $L = \langle l_1, l_2, \ldots, l_{|L|} \rangle$, its corresponding histogram $H^{(L)}$ is a B-length array $\langle h_1, h_2, \ldots, h_B \rangle$, where $h_b \in [1, B]$ is the number of ids in L that are in $\left(\frac{M}{B} \cdot (b-1), \frac{M}{B} \cdot b \right]$. Here $B \in [1, M]$ is a user-defined number, and M is the number of tuples in the dataset.*

Using histograms, the frequency of a word w, whose inverted list is I, in the result list R can be estimated by

$$\texttt{freq}(w|R) \approx \sum_b \left(h_b^{(I)} \cdot h_b^{(R)} \cdot \frac{B}{M} \right) = \sum_b h_b^{(I)} \cdot h_b^{(R)} = H^{(I)} \cdot H^{(R)}, \quad (1)$$

i.e. it is the dot product of the two arrays. It can be proved that this estimated frequency can get more accurate when we use a larger B (proof omitted due to space limitation). However, a larger B value also leads to more computation and more storage space for histogram maintain. As a result, adjusting B it is a trade-off between accuracy and efficiency.

Now we can describe our new query autocompletion algorithm as follows. During the offline process of index building, we calculate the histogram for each of the words appear in attribute a_n. During the online process, we first calculate the histogram of the result list, and then estimate the frequencies of all the matched words using Equation (1). The matched words can be obtained by first identifying the node on the trie of a_n in the index corresponding to the currently-inputting keyword prefix in the query, and then traversing the sub-trie rooted at this node to find all of the leaf nodes. Finally, we sort the matched words in descending order of their estimated frequencies and return the top-k of them as query autocompletion results. This algorithm is called `HistScan`.

4.3   HistBFS: Scan-Free Histogram-Based Estimation

Both of the previous two algorithms have poor performance since they cannot avoid costly scan operations. In this sub-section we propose an algorithm that can avoid the scan on either of the result list or the matched words by applying a best-first traversal on the trie structure of the attribute a_n.

For each non-leaf node x on the trie, we estimate the maximum frequency among all of the leaf nodes (complete words) in the sub-trie rooted at x using the pre-calculated histogram for x and the histogram of R. We traverse the trie by enumerating nodes in descending order of the estimated maximum frequency values. Using this traversal strategy, we can avoid visiting unpromising sub-tries, as well as the scan of all the words or result tuples. To estimate the maximum frequency value for a node x without enumerating its corresponding leaf nodes, the histogram of x should be set as the *maximum-merging* of the histograms of its child nodes, which is defined as follows.

Definition 5 (Maximum-Merging of Histograms). *Given a set of B-length histograms* $\mathcal{H} = \{H^{(i)}\}$, *in which* $H^{(i)} = \langle h_1^{(i)}, h_2^{(i)}, \ldots, h_B^{(i)} \rangle$, *its maximum-merging is a histogram* $H^* = \langle h_1^*, h_2^*, \ldots, h_B^* \rangle$, *where* $h_b^* = \max_i h_b^{(i)}$.

It can be proved that, if we set the histogram of x to be the maximum-merging of all of its child nodes, the estimated frequency value is an upper bound of the actual maximum frequency value among all the leaf nodes of x, no matter what the result set R is. The detail of proof is omitted here.

The maximum-merging histograms for each node on the trie can be calculated in a bottom-up manner after we build the trie structure. During online processing, we use a priority-queue to maintain the traversal state. The detail of this algorithm, called `HistBFS`, is illustrated in Figure 3.

Algorithm 1: HistBFS(p, R)

---

    input : p The currently-inputting keyword prefix;
            R The ID list of the result set.
    output: C The list of keyword completions.

1 Locate the corresponding trie node x according to p;
2 Push $\langle x, f_x \rangle$ into an empty priority-queue $\mathcal{Q}$, where $f_x = H^{(x)} \cdot H^{(R)}$;
3 **while** $\mathcal{Q} \neq \emptyset$ *and* $|C| < k$ **do**
4   Pop the top element $\langle y, f_y \rangle$ in $\mathcal{Q}$;
5   **if** y *is a leaf node* **then** Add the word corresponding to y to C;
6   **else if** y *is a non-leaf node* **then**
7   Let Z be the set of y's child nodes;
8   **foreach** $z \in Z$ **do** Push $\langle z, f_z \rangle$ into $\mathcal{Q}$, where $f_z = H^{(z)} \cdot H^{(R)}$;

9 **return** C;

Fig. 3. The `HistBFS` algorithm

5   Experiments

In this section, we evaluate our proposed algorithms on a real-world dataset, DBLP, which contains 1.3 million computer-science publications. The dataset is delivered in XML format, as a result we converted it into a single relational table, in which each tuple contains the following attributes of a published paper: ID, Title, Journal Name, Authors, and Year. The whole table is approximately 400MB large. We implemented our algorithms in C++ and compiled our code using gcc with -O3 flag. All the experiments were done on a Ubuntu computer with Intel Xeon CPU 2.50GHz and 16GB of RAM.

5.1   Evaluation of Incremental Algorithms

We used a workload of 45,276 real queries collected from the query log of our Seaform system to evaluate our proposed algorithms. Figure 4 shows the comparison of average processing time per query of four algorithms: (1) SL-BF, which uses Single-List tries and Brute-Force synchronization, (2) SL-OD, which uses Single-List tries and On-Demand synchronization, (3) DL-BF, which uses Dual-List tries and Brute-Force synchronization, and (4) DL-OD, which uses Dual-List tries and On-Demand synchronization.

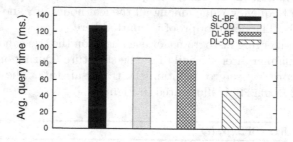

Fig. 4. Performances of the incremental algorithm

We can see that both the dual-list tries and on-demand synchronization can improve the performance speed. If we use these two together, the DL-OD algorithm can answer a query two times faster than using none of the optimizations, at an average speed of 50 milliseconds per query. Figure 5 shows the scalability of the DL-OD algorithm. The processing time and index size increase linearly as the dataset increases. The index size gets slightly larger if we use dual-list tries compared with using single-list tries (10% larger), while the algorithm becomes about 2 times faster.

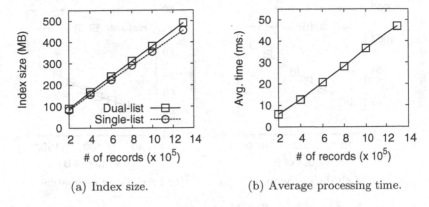

(a) Index size.   (b) Average processing time.

Fig. 5. The scalability of the DL-OD algorithm

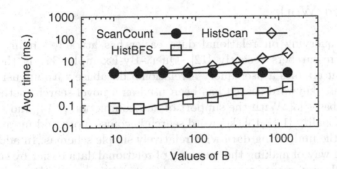

Fig. 6. Performances of the three query autocompletion algorithms

5.2 Evaluation of Query Autocompletion

In the DBLP dataset, the Title attribute is a textual attribute. As a result, we evaluated our three algorithms for query autocompletion specifically on this attribute. The query workload we used is a subset of the 45,276 queries: we only consider the queries whose Title fields are not empty. Figure 6 shows the performance of the three algorithms we proposed. As we can see, when the length of the histogram, B, varies, the performance of HistScan and HistBFS gets worse. However, HistBFS still greatly outperforms the other two.

Figure 7 illustrates the comparison of the size of the histograms and the accuracy. The precision values used in the accuracy computation is obtained by comparing the results of HistBFS and ScanCount (here ScanCount is used for ground-truth). We can see that, when B varies, the total size increases while the accuracy also increases. This result tells us that we can trade-off between the scalability and accuracy by choosing different B values.

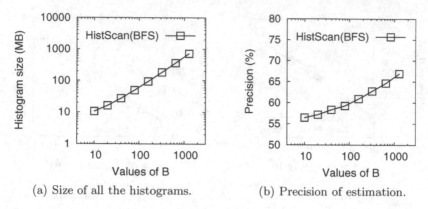

(a) Size of all the histograms.   (b) Precision of estimation.

Fig. 7. Scalability vs. accuracy

6   Related Work

Making the querying on relational data easier has attracted many interests of database researchers recently [12]. Query-By-Example [17] is the earliest paradigm that enables a user query a relational database without using SQLs. In recent years, keyword search has been used as a novel search method in relational databases [3]. With the support of autocompletion [7,14] and the 'type-ahead' functionality [1,10,13], keyword search becomes more and more powerful in searching the underlying data with relatively simple schemas. In addition, [6] takes another way of making the querying of relational data easier by suggesting complete SQL statements to users according to their keyword queries. These works are all based on single-input-box interfaces.

Query autocompletion on relational databases has also been researched for years. [2] provides users the ability to navigate a relational database in different facets. [1] and [5] also provide query autocompletion over the DBLP dataset. In addition, [11] and [15] enable users to navigate the underlying dataset by choosing one of the frequently occurred terms. There are also recent works on keyword search in form-style interfaces, in which [8] and [9] focus on form creation, and [4] focuses on finding the most possible interfaces for keyword search. Obviously, the goals of these works are different from ours.

7   Conclusions

In this paper we proposed new methods of instant search and query autocompletion that can greatly improve the ease of use of traditional QBE search paradigm. Experimental results show that our algorithms achieve scalability and accuracy, and also have high performance.

Acknowledgement. This work is supported by the National Natural Science Foundation of China (Grant No. 60083003).

References

1. Bast, H., Weber, I.: Type less, find more: fast autocompletion search with a succinct index. In: SIGIR, pp. 364–371 (2006)
2. Basu Roy, S., Wang, H., Das, G., Nambiar, U., Mohania, M.: Minimum-effort driven dynamic faceted search in structured databases. In: CIKM, pp. 13–22 (2008)
3. Chen, Y., Wang, W., Liu, Z., Lin, X.: Keyword search on structured and semi-structured data. In: SIGMOD Conference, pp. 1005–1010 (2009)
4. Chu, E., Baid, A., Chai, X., Doan, A., Naughton, J.F.: Combining keyword search and forms for ad hoc querying of databases. In: SIGMOD Conference, pp. 349–360 (2009)
5. Diederich, J., Balke, W.-T.: The Semantic GrowBag Algorithm: Automatically Deriving Categorization Systems. In: Kovács, L., Fuhr, N., Meghini, C. (eds.) ECDL 2007. LNCS, vol. 4675, pp. 1–13. Springer, Heidelberg (2007)
6. Fan, J., Li, G., Zhou, L.: Interactive sql query suggestion: Making databases user-friendly. In: ICDE, pp. 351–362 (2011)
7. Grabski, K., Scheffer, T.: Sentence completion. In: SIGIR, pp. 433–439 (2004)
8. Jayapandian, M., Jagadish, H.V.: Automated creation of a forms-based database query interface. PVLDB 1(1), 695–709 (2008)
9. Jayapandian, M., Jagadish, H.V.: Automating the design and construction of query forms. IEEE Trans. Knowl. Data Eng. 21(10), 1389–1402 (2009)
10. Ji, S., Li, G., Li, C., Feng, J.: Efficient interactive fuzzy keyword search. In: WWW, pp. 371–380 (2009)
11. Koutrika, G., Zadeh, Z.M., Garcia-Molina, H.: Data clouds: summarizing keyword search results over structured data. In: EDBT, pp. 391–402 (2009)
12. Li, G., Fan, J., Wu, H., Wang, J., Feng, J.: Dbease: Making databases user-friendly and easily accessible. In: CIDR, pp. 45–56 (2011)
13. Li, G., Ji, S., Li, C., Feng, J.: Efficient type-ahead search on relational data: A TASTIER approach. In: SIGMOD Conference, pp. 695–706 (2009)
14. Nandi, A., Jagadish, H.V.: Assisted querying using instant-response interfaces. In: SIGMOD Conference, pp. 1156–1158 (2007)
15. Tao, Y., Yu, J.X.: Finding frequent co-occurring terms in relational keyword search. In: EDBT, pp. 839–850 (2009)
16. Wu, H., Li, G., Li, C., Zhou, L.: Seaform: Search-as-you-type in forms. PVLDB 3(2), 1565–1568 (2010)
17. Zloof, M.M.: Query-by-example: the invocation and definition of tables and forms. In: VLDB, pp. 1–24 (1975)

Range Query Estimation
for Dirty Data Management System[*]

Yan Zhang, Long Yang, and Hongzhi Wang^{**}

Department of Computer Science and Technology
Harbin Institute of Technology
`zhangy@hit.edu.cn`, `{yanglonghit,whongzhi}@gmail.com`

Abstract. In recent years, data quality issues have attracted wide attention. Data quality is mainly caused by dirty data. Currently, many methods for dirty data management have been proposed, and one of them is entity-based relational database in which one tuple represents an entity. The traditional query optimizations having the ability to estimate the cost of execution of a query plan have not been suitable for the new entity-based model. Then new query optimizations need to be developed. In this paper, we propose new query selectivity estimation based on histogram, and focus on solving the overestimation which traditional methods lead to. We prove our approaches are unbiased. The experimental results on both real and synthetic data sets show that our approaches can give good estimates with low error.

Keywords: query estimation, histogram, dirty data, data quality.

1 Introduction

Data quality has been addressed in different areas, such as statistics, management science, and computer science [1]. Dirty data is the main reason to cause data quality. Many surveys reveal dirty data exists in most database systems. The consequences of dirty data may be severe. Having uncertain, duplicate or inconsistent dirty data leads to ineffective marketing, operational inefficiencies, and poor business decisions. For example, it is reported [2] that dirty data in retail databases alone costs US consumers $2.5 billion a year. Therefore, several techniques have been developed to process dirty data to reduce the harm of dirty data.

Existing work on processing dirty data can be divided into two broad categories. The first category is data cleaning [3], which is to detect and remove errors and inconsistencies from data to improve data quality. However, data cleaning cannot clean the dirty data exhaustively and excessive data cleaning may lead to the loss of information. Besides this, existing data cleaning techniques are generally time-consuming. Therefore, some researchers propose algorithms in the other category, to

---

[*] This paper was partially supported by NGFR 973 grant 2012CB316200 and NSFC grant 61003046, 6111113089. Doctoral Fund of Ministry of Education of China (No. 20102302120054).
^{**} Corresponding author.

H. Gao et al. (Eds.): WAIM 2012, LNCS 7418, pp. 152–164, 2012.

perform queries on dirty data directly and obtain query results with clean degree from the dirty data [4-6].

Several models for dirty data management without data cleaning have been proposed [7-9]. But most of these models only consider the uncertainty in values of the attributes and the quality degree of the data without the consideration of the entities in real world and their relationships. In this paper, we focus on entity-based relational database model in which one tuple represents an entity. This model can better reflect the real world entities and their relationships.

In applications, the different representations of the same real-word entities often lead to inconsistent data, uncertain data or duplicate data, especially when multiple data sources need to be integrated [10-11]. In the entity-based relational database, for the duplicate data referring to the same real-world entity, we combine these data, and for inconsistent data (or uncertain data), we endow each of them a value (we call it as quality degree) which reflects its quality. *Example 1* shows this process.

Table 1. A Dirty Data Fragment

ID	Name	City	Zipcode	Phn	Reprsnt
1	Wal-Mart	Beijing	90015	80103389	Sham
2	Carrefour	Harbin	20016	80374832	Morgan
3	Wal-Mart	BJ	90015	010-80103389	Sham
4	Walmart	Harbin	20040	70937485	Sham
5	Carrefour	Beijing	90015	83950321	Morgan
6	Mal-Mart	Beijing	90015	80103389	Sham

Example 1: Consider a fragment of the dirty data shown in Table 1. We can easily identify that tuples 1, 3 and 6 refer to the same entity in the real world even though their representations are different. By preforming entity resolution and combining these three tuples, we can get one entity tuple. In this process, we don't remove any data, which implies that the value of one attribute in a tuple may be uncertain, and it may contain multiple values. We endow possible each attribute value with a quality degree in accordance with their proportion, as shown in Table 2. In tuples 1, 3 and 6, the value "Wal-Mart" appears twice, so the quality degree is $2/3 \approx 0.67$. Similarly, other quality degrees can be given. Then we get an entity tuple as shown in Table 2.

Table 2. An Entity Tuple

ID	Name	City	Zipcode	Phn	Reprsnt
1	(Wal-Mart, 0.67), (Mal-Mart, 0.33)	(Beijing, 0.67), (BJ, 0.33)	(90015, 1.0)	(80103389, 0.67), (010-80103389, 0.33)	(Sham, 1.0)

As *Example 1* shows, the entity-based relational database ingeniously processes the dirty data by entity resolution [12-13] and quality degree. In the implementation of this model, query optimization techniques are in demand. As the base of the query optimization, the estimation technique computing the size of the results of an operator is crucial. Even though over the past few decades, there has been a lot of work on query estimation for traditional relational database management systems. Most approaches for query estimation are based on histogram [14], which records data distributions. However, the traditional histograms are not suitable for entity-based relational database, and often lead to overestimation, especially for range queries. One reason is that query processing on the entity-based relational database need to consider the effect of the quality degrees of values, but the traditional histograms are only concerned about the attribute value without the consideration of the quality degree. The other reason is that traditional approaches based on histogram often lead to overestimation, especially for range queries. Because one attribute of a tuple may contain multiple values in the entity-based relational database, if all values are partitioned into different buckets, this tuple is counted for multiple times. Thus, the overestimation occurs.

Therefore, the traditional query estimation approaches based on histogram cannot be applied to our problem. Unfortunately, there is no work for the query estimation of the entity-based relational database. New query estimation approaches are in demand.

Our Contributions: In this paper, we propose new range query estimation methods suitable for entity-based relational database. As we know, this is the first paper considering such problem. These algorithms are demonstrated in details and the complexity of these algorithms is analyzed. We theoretically prove our algorithms are unbiased. Last, we experimentally validate the effectiveness of our algorithms and show that our methods are accurate.

The rest of this paper is organized as follows. Section 2 introduces entity-based relational database model and some related conceptions. Section 3 presents our range query estimation methods. We show our experimental results in Section 4. We conclude our paper and discuss the future work in Section 5.

2   Preliminaries

2.1   Entity-Based Relational Database Model

We firstly define the *Uncertain Attribute Value* in Definition 1. An uncertain attribute value not only contains possible values, but also contains the corresponding quality degrees. Then we give the definition of *Entity* in Definition 2. Entity is the basic unit of storage in the entity-based relational database system, containing a set of uncertain attribute values.

Definition 1 (Uncertain Attribute Value): *An uncertain attribute value is a set of pair* $A = \{(v, p) / v$ *is possible value of the attribute and p is the quality degree of the value v}.*

Definition 2 (Entity): *An entity is a pair* $E = (K, A)$, *where A is a set of uncertain attribute values and K is a set of keys that is to identify the entity uniformly (e.g., entity-ID).*

Table 2 can help to understand these two definitions. Since we introduce the quality degree dimension, we need to define a new conception to reflect whether a tuple satisfies a query, and we call this conception *Similarity*.

Definition 3 (Similarity): *For an uncertain values V in attribute a and an atom constraint C in form of $a @ v$ where $@$ is a predicate symbol (e.g. $>$, $<$ ...) and v is a constraint, the similarity between them is defined as follows:*

$$Sim(V@C) = \sum_{(v_i,p_i)\in V} sim(v_i@v)p_i ,\tag{1}$$

$$sim(v_i@ v) = \begin{cases} 1 & where\ v_i@v \\ 0 & otherwise \end{cases} .\tag{2}$$

For a selection query with a constraint $a < v$ (for the convenience, we use this form $a < v$ to represent a selection query in this paper), we consider that one tuple satisfies query with a similarity S, which can be calculated by Equation (1) and (2). We use an example to illustrate it.

With the support of the conceptions, some query operators are defined.

2.2   Operators

In this paper, we focus on the estimation of selection operation. Each query result satisfies the query with a similarity, since results with a low similarity are generally less interesting than higher similarity answers, we consider those results with a similarity less than a threshold τ (this parameter can be provided by user or the system sets a default value) as unsatisfied for a query. Therefore, the results of queries should be those answers that have a similarity exceeding a threshold τ. So a query given by $a <_\tau x$ can be defined as an operator as follows:

$$Sim(a < x) > \tau \Leftrightarrow \sum_{(v_i,p_i)\in V} sim(v_i < x)p_i > \tau .\tag{3}$$

The goal of this paper is to propose new query estimation techniques for the entity-based relational database. In next section, we will give our approaches in details.

3   Range Query Estimation

In this section, we describe our estimation methods in details. First, we give a preliminary query estimation method in Section 3.1, and this method can well estimate unbounded range query (e.g., $a >_\tau x$ or $a <_\tau x$) result size, but for general range queries (e.g., $x_1 < a < x_2$), it often leads to underestimation. Then, a more accurate range query estimation method is proposed in Section 3.2, and it can well solve the underestimation problem which the former method encounters.

3.1   Preliminary Range Query Estimation Method

As discussed in Section 1, existing query estimation methods are not suitable for range queries on entity-based relational database management system for two reasons that with the quality degree, the existing methods often lead to the overestimation. To solve these problems, we consider an unbounded range query Q by $a <_\tau x$ firstly, where a is an uncertain attribute value and τ is the similarity threshold. This query returns all tuples satisfying $Sim(a < x) > \tau$, which means that a satisfies the following relationship:

$$\sum_{(v_i,p_i)\in V} sim(v_i < x)p_i > \tau .\tag{4}$$

If all possible values of an uncertain value are sorted in database system, the relationship (3) is equivalent to calculate the cumulative distribution function $F_a(x)$, where $F_a(x) = \sum_{v_i < x} p_i$, and return the values satisfying $F_a(x) > \tau$.

Fig. 1 shows an example of the cumulative distribution functions (CDF) of several tuples on attribute A, whose corresponding values are shown in Table 3. In the figure, each stacked line represents one tuple. The meaning of every stacked line is like the cumulative distribution function of every uncertain value. For example, the point P on the stacked line represents that the value of attribute A of tuple 3 is smaller than 35 with similarity 0.6. Therefore, with such a figure containing all tuples, for a given query Q ($a <_\tau x_0$), the total number of tuples which satisfy query Q can be estimated directly. It is the number of stacked lines crossing the line segment l given by $x = x_0, \tau < y \le 1$.

Table 3. A Data Fragment

ID	A	B	......
1	((10, 0.1), (35, 0.3), (65, 0.5), (80, 0.1))	......	......
2	((20, 0.3), (50, 0.5), (80, 0.2))	......	......
3	((15, 0.6), (60, 0.2), (70, 0.2))	......	......

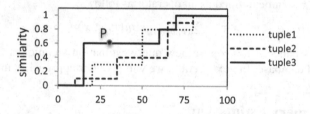

Fig. 1. Example for showing the histogram structure

Histogram Structure

Based on the above discussion, we define a basic two-dimensional histogram. The range of input values is partitioned into $n * m$ buckets where n and m are the lengths of each dimension. A histogram bucket $H(i, j)$ covers the area given by $(i * \delta_x, j * \delta_s, (i + 1) * \delta_x, (j + 1) * \delta_s)$, where δ_x and δ_s are the widths of histogram bucket along x and y axis. Each bucket has a value, which stores the height of this bucket that records the number of tuples whose stacked lines intersect this bucket.

Obviously, the errors of the estimations using this histogram are associated to the number of the stacked line inflection points in buckets and do not exceed them. If there is no inflection points in bucket H_i, the estimation for queries which are located in H_i must be accurate. Hence in order to make the estimation more accurate, we need to ensure that the number of the inflection points in each bucket H_i is small enough.

In our approach, the histogram is firstly partitioned into p equal-width buckets, and we set the number of the inflection points in each bucket should not exceed ε

($\varepsilon = M/p$, where M is the total number of inflection points, which equals the number of all possible attribute values). When a bucket contains more than ε inflection points, this bucket is partitioned into q equal-width buckets (generally, $q \ll p$, and q can be considered as a constant) and we set each new bucket containing ε/q inflection points. In the next process, for the buckets which do not meet the requisition, they are partitioned until that all buckets contain less than ε inflection points.

We now present the histogram construction algorithm. To facilitate the description of algorithm, we firstly summarize the main notations that will be used in our paper in Table 4. With these notations, *Algorithm 1* illustrates the detailed steps of histogram construction. Note that, we assume all possible values of an uncertain value are increasing in the database system. For each uncertain value, it is supposed that all possible values are $v_0, v_1 \dots v_{m-1}$ and the corresponding quality degrees are $p_0, p_1 \dots p_{m-1}$, the interval $[v_{min}, v_{max}]$ can be divided into $m+1$ intervals: $[v_{min}, v_0), [v_0, v_1) \dots [v_{m-1}, v_{max}]$. In each interval, we need to record value a in correct histogram buckets. For example, in $[v_{min}, v_0)$, value a should be recorded in buckets $H([v_{min}, v_0), 0)$, which represents $Sim(a < x)$ is 0, where $x \in [v_{min}, v_0)$. Similarly, value a should also be recorded in buckets: $H([v_0, v_1), p_0), H([v_1, v_2), p_0 + p_1) \dots H([v_{m-1}, v_{max}], 1)$. Meanwhile, the number of inflection points is stored in each bucket, and when the size of some bucket exceeds ε, it is partitioned into q equal-width buckets and the histogram is adjusted. *Algorithm 1* is the pseudo-code of this process, where symbol P_i represents the number of inflection points in bucket H_i.

Table 4. Main Notations

Notation	Meaning
τ	Similarity threshold
p, q	Initial granularity of partition and granularity of repartition
ε	Threshold of the number of inflection points in one bucket
l_i, r_i	Left boundary and right boundary of bucket H_i along x axis
δ_s, s	Width of buckets along y axis, where $s = \lceil 1/\delta_s \rceil$
a_i	Uncertain values of attribute A $(0 \leq i < N)$
v_i, p_i	Possible values and quality degrees of an uncertain attribute value a
v_{min}, v_{max}	Minimum and maximum among all possible values of an attribute A
$sim(x_1, x_2)$	Similarity of $x_1 < a < x_2$ where a is an uncertain value
C	The number of tuples satisfying query Q

Algorithm 1

---

1   Initialize $H \leftarrow \emptyset$
2   **for** each uncertain value a **do**
3   **for** each possible value v_k of an uncertain value **do**
4   **for** all buckets meeting $v_{k-1} < l_i < v_k$ **do**
5               $H(i, \lfloor F_a(l_i)/\delta_s \rfloor)$++
6   P_i++
7   **if** $P_i > \varepsilon$ **then**
8   **partition** and **adjust** this bucket
9   **for** all buckets meeting $r_i > max(v_k)$ **do**
10          $H(i, \lceil 1/\delta_s \rceil)$++

Theorem 1: *The time complexity of Algorithm 1 is* $O(p(N + s))$ *and the space complexity is* $O(ps)$.

Proof: This algorithm scans each tuple once and records each tuple in m appropriate histogram buckets, where m is the length of the histogram in x dimension. In the worst case, partition occurs per $\varepsilon(q - 1)/q$ tuples, and partition times does not exceed $pq/(q - 1)$ (i.e., $M/(\varepsilon(q - 1)/q)$). Each partition adds $q - 1$ buckets and adjusts s buckets along y axis, so $m < (q - 1)pq/(q - 1) + p = p(q + 1)$. Thus the time complexity is $O(p(q + 1)N) + O(qspq/(q - 1))$ and the space complexity is $O(p(q + 1)s)$. Thus the time complexity is $O(p(N + s))$ and the space complexity is $O(ps)$, because q can be considered as a constant.   □

Query Estimation Method

With the histogram structure, we can easily estimate query result size. Given a query $a <_\tau x_0$, query result size is estimated as the sum of the heights of the buckets where x_0 is located in and meet the similarity threshold. *Algorithm 2* shows this algorithm in details.

However, this algorithm is only applicable for the unbounded range queries in form of $a <_\tau x_0$. For anther unbounded range queries in the form of $a >_\tau x_0$, we need to perform an equivalent transformation to make *Algorithm 2* suitable for such form.

$$a >_\tau x_0 \Leftrightarrow Sim(a > x_0) > \tau \Leftrightarrow Sim(a < x_0) < 1 - \tau . \tag{4}$$

Such that *Algorithm 2* can also be used to estimate queries in the form as $a >_\tau x_0$, with a modification of the loop range in *line 4*, and the loop range should be modified to $[0, \lfloor (1 - \tau)/\delta_s \rfloor]$. Theorem 2 proves this estimation method is unbiased when p tends to infinity.

Algorithm 2
1   **if** $x_0 < v_{min}$ **then return** 0
2   **if** $x_0 > v_{max}$ **then return** N
3   **let** $C = 0$ **and** find H_i meeting $l_i < x_0 \leq r_i$
4   **for** j **from** $\lfloor \tau/\delta_s \rfloor$ **to** $\lceil 1/\delta_s \rceil$ **do**
5        $C = C + H(i, j)$
6   **return** C

Theorem 2: *The estimation method in Algorithm 2 is unbiased when p tends to infinity.*

Proof: To facilitate the proof, we assume $q = 2$, for other cases, the proof process is similar. As proved in Theorem 1, partition times does not exceed $2p$ (i.e., $pq/(q - 1)$), and each partition adds 1 (i.e., $q - 1$) buckets. Given a query $a <_\tau x_0$, we make the following assumptions. First, x_0 falls each bucket with equal probability. Second, n times partitions occur. Last, m buckets contain more than ε inflection points where $m \leq n$. With these assumptions, the total number of buckets along x axis is $p + n$. We have known the estimation error does not exceed the number of inflection points in bucket which x_0 is located in. Hence the expectation of estimation error is:

$$E(e) < \frac{p+n-m}{p+n}\varepsilon + \frac{m}{p+n}E'(e) = \frac{p+n-m}{p+n}\varepsilon + \frac{m}{p+n}\left(\varepsilon + \frac{M-m\varepsilon}{m}\right)$$
$$= \varepsilon + \frac{M-m\varepsilon}{p+n} < \varepsilon + \frac{M}{p} = 2\frac{M}{p}.$$

Therefore, when p tends to infinity, the expectation of estimation error tends to 0, and this approach is unbiased.   □

We have discussed the unbounded range queries. Consider the general range query Q $(x_1 < a < x_2)$, and that is $Sim(x_1 < a < x_2) > \tau$. The unbounded range queries can be considered as a special case of the general range query. To estimate the general range query result size, with the application of the techniques in this section, a naïve method is proposed. Firstly, the numbers of tuples that satisfy query Q1 ($a <_\tau x_1$) and query Q2 ($a <_\tau x_2$) are estimated by *Algorithm 2*, and they are denoted by $C1$ and $C2$ respectively. We can use $C2 - C1$ as the estimation of query Q $(x_1 < a < x_2)$. Clearly, if we do not consider the threshold, this method is correct. However, it often leads to underestimation with the consideration of the effect of the similarity threshold on query result sizes, and it is related to the width of query range and the threshold. We show the effect by experiments in Section 4. With the shortcoming of this naïve method, we propose more accurate query range estimation method in next section.

3.2   Accurate Range Query Estimation

In this section, we present an accurate range query estimation algorithm, and it can solve the underestimation problem discussed in Section 3.1. In order to adapt to general range queries, we add another dimension to the histogram proposed in Section 3.1. The meanings of two original dimensions do not change (the x axis and y axis respectively represent the end point of the query and the similarity), and the new additional dimension (z axis) represents the beginning of the query. Therefore, given a general range query $Q(x_1 < a < x_2)$, we can estimate the size of query result set by counting the number of stacked lines crossing the line segment l given by $x = x_2$, $\tau < y \le 1$ and $z = x_1$, similar to Fig. 1. That is equivalent to executing a query Q' ($a <_\tau x_2$) on the plane, where $z = x_1$.

We call such new histogram as improved histogram. In this histogram, every plane on z axis is a basic histogram proposed in Section 3.1, corresponding to the constraint $z \le x < v_{max}$ (clearly, it is not necessary to store the whole range). The width of a bucket on z axis is controlled by an input parameter δ_z(in general, δ_z can be equal to $(v_{max} - v_{min})/p$). The detailed algorithms for constructing this improved histogram and estimating the result size of a general range query are respectively presented in *Algorithm 3* and *Algorithm 4*. Compared with *Algorithm 1*, *Algorithm 3* only adds another layer of loops on z axis, but this improved histogram structure can give more accurate estimation than the basic histogram of Section 3.1. Theorem 3 gives the time and space complexity of the construction algorithm, and Theorem 4 proves this estimation algorithm using this improved histogram is also unbiased when p tends to infinity.

Algorithm 3
1   Initialize $H \leftarrow \emptyset$
2   **for** each uncertain value a **do**
3   **for** each possible value v_n of an uncertain value **do**
4   **for** k **from** 0 to $\lfloor (v_k - v_{\min})/\delta_z \rfloor$ **do**
5   **for** all buckets meeting $v_{n-1} < l_i < v_n$ **do**
6                    $H(k, i, sim(k * \delta_z, l_i)/\delta_s)$++
7                    $P_{k,i}$++
8   **if** $P_{k,i} > \varepsilon$ **then**
9   **partition** and **adjust** this bucket
10   **for** k **from** 0 to $\lfloor (v_k - v_{\min})/\delta_z \rfloor$**do**
11   **for** all buckets meeting $r_i > max\,(v_k)$ **do**
12                $H(k, i, sim(k * \delta_z, l_i)/\delta_s)$++

Theorem 3: *The time complexity of Algorithm 3 is* $O(p^2(N + s))$ *and the space complexity is* $O(p^2 s)$ *with the assumption:* $\delta_z = (v_{max} - v_{min})/p)$.

Proof: Compared with *Algorithm 1*, this algorithm only adds another dimension, and the length of this dimension is p. Therefore, similarly the analysis of the complexity of *Algorithm 1*, the time complexity of *Algorithm 3* is $O(p^2(N + s))$ and the space complexity is $O(p^2 s)$.   □

Theorem 4: *The estimation method in Algorithm 4 is unbiased when* p *tends to infinity.*

Proof: Compared with the basic histogram of Section 3.1, this improved histogram with more detailed information can get a more accurate estimation for general queries. Therefore, with the conclusion of Theorem 2, the estimation method using this improved histogram is also unbiased when p tends to infinity.   □

Algorithm 4
1   **if** $x_1 < v_{min}$ **then** let $x_1 = v_{min}$
2   **if** $x_2 > v_{max}$ **then** let $x_2 = v_{max}$
3   **let** $C = 0; k = \lfloor (x_1 - v_{min})/\delta_z \rfloor$ and find $H_{k,i}$ meeting $l_i < x_2 \leq r_i$
4   **for** j **from** $\lfloor \tau/\delta_s \rfloor$ to $\lceil 1/\delta_s \rceil$ **do**
5        $C = C + H(k, i, j)$
6   **return** C

4   Experimental Evaluation

In this section, we study the performance of our proposed algorithms experimentally. Our experiments are conducted on a 2.93 GHz Inter(R) Core(TM)2 Duo CPU with 2 GB main memory.

4.1   Data Sets

The data sets used for estimate query result size can be categorized into two main parts of synthetic data sets and real-world data sets. Table 5 summarizes some information about these data sets.

Synthetic Data Sets: We generate the synthetic data sets and each tuple has a Tuple ID, along with an uncertain value. The number of the possible values of an uncertain value is uniformly distributed between 1 and 5. The quality degree of each possible value is randomly generated from 0.01 to 1 and these quality degrees sum up to 1 for an uncertain value. To evaluate the robust of our approaches, we consider three synthetic data sets with different distributions: uniform distribution, normal distributions and zipfian distribution.

Real-World Data Sets: One of the most important applications of the histograms is for those cases in which the distribution of the data is unknown or cannot be simply modeled. Therefore, in order to validate our approaches over such kind of data, we consider the real-world data sets: eCommerce data. We respectively collect book information about *Computers & Technology* from eBay (http://www.ebay.com) and Amazon (http://www.amazon.com). After the processes for original data, we get the real data set with 10053 entities. In this data set, each tuple represents a book which contains four uncertain values: *title*, *author*, *press* and *price*. We perform our experiments by building the histograms on attribute *price*.

Table 5. Data sets used for the experimental results

Name	Distribution	Size	Parameter
Uniform	Uniform	$1m$	$min0, max1k$
Normal	Normal	$1m$	$\mu500, \sigma100$
Zipf	Zipfian	$1m$	$\alpha1.0$
Real	eCommerce	10K	-

4.2   Query Set and Error Metric

Without loss of generality, we ran every experiment on a variety of queries. All queries are in form of $\{x_1 < A < x_2: x_1, x_1 \in U\}$, where A is an attribute and U is its domain. We measure the error of estimation made by histograms on the above query set by using the average of the relative error: $\frac{1}{N}\sum_{q \in Q}\frac{|\hat{C}_q - C_q|}{C_q}$, where N is the cardinality of the query set, $\hat{C}_q$ and C_q are the actual and the estimated size of the query result set, respectively. $|\hat{C}_q - C_q|/\hat{C}_q$ represents the relative error of query q. In our experiments, we randomly generate 100 queries for each query set.

4.3   Experimental Results

Our experiments compare the two estimation algorithms proposed in Section 3. We denote them by $H1$ (in Section 3.1) and $H2$ (in Section 3.2). Without explicit explanation, the default value of the similarity threshold is 0.2 for all experiments; the default size of the real and synthetic data sets are 10,000 and 200,000 tuples; the default size of initial granularity p of partition and granularity q of repartition are 50 and 4; the default size of bucket width on similarity dimension δ_s is 0.1.

4.3.1   Effect of Data Distribution

For query estimation algorithms based on histogram, data distribution is an important factor affecting the accuracy of estimation. Fig. 2 shows the effect of data distribution to our estimation algorithms. It can be seen that for each data distribution, both two estimation algorithms can get a good estimation (relative errors are less than 40%) and algorithm $H2$ is always more accurate than algorithm $H1$.

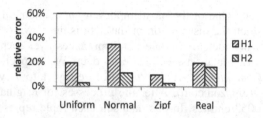

Fig. 2. Effect of data distribution

4.3.2   Effect of Data Set Size

Fig. 3 shows the effect of data set size. In this experiment, we vary the data set size by selecting the desired number of tuples from the synthetic data set T, and the data set sizes are 100K, 200K, 400K, 600K, 800K and 1000K (for eCommerce data, data set sizes are 1K, 2K, 4K, 6K, 8K and 10K). It is observed from Fig. 3 that the relative error is not sensitive to the dataset size for both two algorithms.

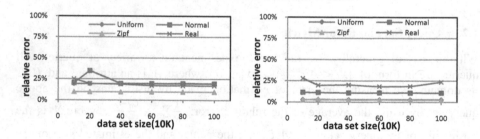

Fig. 3. Effect of data size on $H1$ and $H2$

4.3.3   Effect of Threshold

For the entity-based relational database, the threshold plays an important role in query processing. Fig. 4 shows the impact of the threshold with different thresholds from 0.1 to 0.9. For algorithm $H2$, no matter how the data is distributed, the change is not very significant. Therefore, algorithm $H2$ is relatively stable for different thresholds. However, for algorithm $H1$, the relative error decreases at first and then increases with the threshold. When the threshold is in $[0.4, 0.6]$, the relative errors is minimum. Hence the accuracy of algorithm $H1$ is related to threshold τ as mentioned in Section 3.1, and it can give accurate estimations when $\tau \in [0.4, 0.6]$.

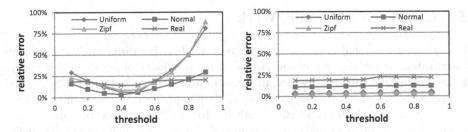

Fig. 4. Effect of threshold on $H1$ and $H2$

4.3.4   Effect of Granularity of Partition

Another important factor of our estimation algorithms is the granularity of partition. We show the effect of the initial granularity of partition p in Fig. 5, and set p at 10, 20, 40, 60, 80 and 100 respectively. We observe that there is a similar trend of the relative error both algorithm $H1$ and $H2$. As initial granularity of partition p increases, the relative error decreases. This phenomenon empirically verifies the conclusion that our approaches are unbiased when p tends to infinity.

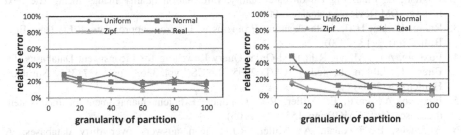

Fig. 5. Effect of granularity of partition on $H1$ and $H2$

4.3.5   Effect of Width of Query Range

In Section 3.1, we mentioned that the accuracy of the estimation using the basic histogram $H1$ is related to the width of query range and the threshold τ. Fig. 4 has showed us the effect of the threshold. In this experiment, we show the effect of the width of query range. Fig. 6 gives the experimental results. We can observe that with an increasing width of query range, the relative error has a decreasing trend for both $H1$ and $H2$, especially for $H1$, and the estimations using histogram $H2$ are more accurate. As a result, the wider query range is, the more accurate estimation is.

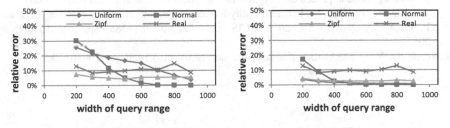

Fig. 6. Effect of width of query range on $H1$ and $H2$

5   Conclusion and Future Work

Entity-based relational database is a practical method for dirty data management. Intermediate result size estimation is crucial for the query optimization for entity-based relational database. However, traditional estimation methods cannot be applied to this problem directly. In this paper, we study this problem. To solve this problem, we propose two histogram-based methods for different form of queries and requirements. It is proven that they are unbiased. The experimental results validate the effectiveness of our algorithms, and they can indeed give good estimations for range queries. For future work, we plan to continue to study query optimization based on the cost of estimation, especially the estimation of join result size.

References

1. Batini, C., Scannapieco, M.: Data quality: concepts, methodologies and techniques. Springer (2006)
2. English, L.: Plain English on data quality: Information quality management: The next frontier. DM Review Magazine (2000)
3. Rahm, E., Do, H.H.: Data cleaning: Problems and current approaches. IEEE Data Eng. Bull. 23(4), 3–13 (2000)
4. Fuxman, A.D., Miller, R.J.: First-Order Query Rewriting for Inconsistent Databases. In: Eiter, T., Libkin, L. (eds.) ICDT 2005. LNCS, vol. 3363, pp. 337–351. Springer, Heidelberg (2005)
5. Fuxman, A., Fazli, E., Miller, R.J.: Conquer: Efficient management of inconsistent databases. In: SIGMOD, pp. 155–166 (2005)
6. Andritsos, P., Fuxman, A., Miller, R.J.: Clean answers over dirty databases: A probabilistic approach. In: ICDE, p. 30 (2006)
7. Boulos, J., Dalvi, N., Mandhani, B., Mathur, S., Re, C., Suciu, D.: MYSTIQ: a system for finding more answers by using probabilities. In: SIGMOD, pp. 891–893 (2005)
8. Widom, J.: Trio: a system for integrated management of data, accuracy, and lineage. In: CIDR, pp. 262–276 (2005)
9. Hassanzadeh, O., Miller, R.J.: Creating probabilistic databases from duplicated data. The VLDB Journal, 1141–1166 (2009)
10. Lenzerini, M.: Data integration: A theoretical perspective. In: PODS, pp. 233–246 (2002)
11. Dong, X.L., Halevy, A., Yu, C.: Data integration with uncertainty. The VLDB Journal, 469–500 (2009)
12. Benjelloun, O., Garcia-Molina, H., Menestrina, D., Whang, S.E., Su, Q., Widom, J.: Swoosh: a generic approach to entity resolution. The VLDB Journal, 255–276 (2008)
13. Li, Y., Wang, H., Gao, H.: Efficient Entity Resolution Based on Sequence Rules. In: Shen, G., Huang, X. (eds.) CSIE 2011. CCIS, vol. 152, pp. 381–388. Springer, Heidelberg (2011)
14. Ioannidis, Y.E.: The history of histograms (abridged). In: VLDB, pp. 19–30 (2003)

Top-k Most Incremental Location Selection
with Capacity Constraint

Yu Sun[1], Jin Huang[2], Yueguo Chen[3], Xiaoyong Du[1,3], and Rui Zhang[2]

[1] School of Information, Renmin University of China, Beijing, China
[2] University of Melbourne, Melbourne, Australia
[3] Key Laboratory of Data Engineering and Knowledge Engineering
(Renmin University of China), MOE, China
{yusun.aldrich,soone.enoos,chenyueguo}@gmail.com,
duyong@ruc.edu.cn, rui@csse.unimelb.edu.au

Abstract. Bichromatic reverse nearest neighbor (BRNN) based query uses the number of reverse nearest customers to model the influence of a facility location. The query has great potential for real life applications and receives considerable attentions from spatial database studies. In real world, facilities are inevitably constrained by designed capacities. When the needs of service increase, facilities in those booming areas may suffer from overloading. In this paper, we study a new kind of BRNN related query. It aims at finding most promising candidate locations to increase the overall service quality. To efficiently answer the query, we propose an $O(n \log n)$ algorithm using pruning techniques and spatial indices. To evaluate the efficiency of proposed algorithm, we conduct extensive experiments on both real and synthetic datasets. The results show our algorithm has superior performance over the basic solution.

Keywords: capacity constraint, location selection, BRNN, spatial database.

1 Introduction

A common problem many companies are facing is to establish new facilities in most suitable places. Take online shopping as an example, to improve the shipping service and minimize storage cost, online sales giants such as Amazon and 360Buy have built numerous warehouses throughout the country. Since each warehouse is constrained by its designed capacity (e.g. storage ability or the number of employees), it is desired to choose good locations for new warehouses. The mobile service is another example showing the important application of this problem.

The problem is based on bichromatic reverse nearest neighbor (BRNN) query [3]. Let W and R denote two sets of locations in the same space. Given a location $w \in W$, a BRNN query returns all locations $r \in R$ whose nearest neighbor is w. Those locations form the BRNN set of w, denoted by $B(w, W)$. In our problem, we denote the facility set as W and the customer set as R. In addition to the common BRNN query, the capacity constraint of facilities is considered. We use L_2 (Euclidean) distance metric and assume that: (1) a customer is served by the nearest facility and (2) service providers have the knowledge of customer distribution and can offer a serving order that

H. Gao et al. (Eds.): WAIM 2012, LNCS 7418, pp. 165–171, 2012.

determines the sequence of locations being served. Let $c(w)$ denote the corresponding capacity of a facility and $wt(r)$ denote the weight of a customer location r. A w is responsible to serve its reverse nearest neighbors. And a r that has w as its nearest facility is *within* w's capacity only if w provides service to some customers reside in r.

We use a running example for illustration throughout the paper. For presenting convenience, we set all customer locations the same weight 1, although the proposed techniques can be applied to any weights. We also adopt the serving order that makes w serve its customers r in an increasing order of $dist(r, w)$ for the same reason. Here we use $dist(r, w)$ to denote the distance between r and w. In Fig.1 and 2, circles, small and

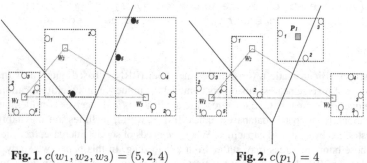

Fig. 1. $c(w_1, w_2, w_3) = (5, 2, 4)$ **Fig. 2.** $c(p_1) = 4$

big rectangles denote customer locations, facilities and BRNN sets respectively, and the number besides each r denotes the serving sequence. In Fig.1, w_2 serves 3 locations. Location 1 and 2 use up $c(w_2)$, thus 3 is out of its capacity. Here black circles denote poorly served ones. After adding p_1, as shown in Fig.2, three locations are well served by p_1. Since p_1 shares the workload of w_2, w_2 has the ability to satisfy previous black locations. The total weight of newly served parts is 3, so p_1 has an increment of 3. Our aim is finding top-k locations that have maximal increments from a candidate set P. We make the following contributions:

– We formulate the problem of top-k most incremental location selection query.
– We propose pruning techniques and an $O(n \log n)$ algorithm to reduce the complexity of query processing.
– We perform extensive experiments and the results confirm the effectiveness of the pruning techniques and the efficiency of the algorithm.

The rest of the paper is organized as follows. Section 2 reviews previous studies on related topics. Section 3 defines the problem and gives a baseline solution. Section 4 describes pruning techniques and the proposed algorithm. The empirical study is given in Section 5, followed by the conclusion in Section 6.

2   Related Work

The concept of *reverse nearest neighbor (RNN)* query [3] has been raised more than ten years. Korn et al.[3] first propose methods to solve RNN query. Paper [5] studies how to discover influence sets based on RNN in frequently updated databases and proposes an efficient algorithm to solve BRNN query. Since we focus on location selection, without proper modification, the proposed methods cannot be applied. Wong et al.[6] study the problem called MaxBRNN. It aims at finding an optimal region that maximizes the size of BRNN set. Zhou et al.[10] extend the problem to MaxBRkNN. Their problems are different from ours. First, their studies attempt to retrieve the optimal region with highest influence, while ours focuses on selecting top-k ones from a candidate set. Second, capacity constraint is not considered in their studies.

Paper [9] and [2] propose and solve the min-dist optimal location query, which minimizes the average distance from each customer location to its closest facility after adding a new one. Some other papers [8,1] describe another BRNN related query, called maximal influence query. The influence of a facility is defined as the total weight of its BRNN members. Though these problems are similar to ours, they fail to consider the capacity of each facility. The authors of [7] and [4] study the capacity constrained assignment problem. Paper [7] proposes an algorithm that assigns each $r_i \in R$ to a $w_j \in W$ without exceeding w_j's capacity. Additionally, paper [4] tries to minimize the assignment cost of $\sum_{(r_i, w_j) \in R \times W} dist(r_i, w_j)$. However, they are more suitable for profile-matching applications which provide service using existing facilities. Our problem considers both existing and to-be built ones.

3 Preliminary

3.1 Definitions

First, we give the definition of ε-*served location* to indicate the percentage of customers who are under service in a customer location.

Definition 1. *An ε-served location is a location r that has $\varepsilon \cdot wt(r)$ customers under service, noted as $\varepsilon(r), 0 \leq \varepsilon \leq 1$.*

Given the serving order, w serves its BRNN set members one by one, until some r uses up $c(w)$ or all r are fully served. Hence the ε factor of r can be calculated. Most r are 1 or 0-served. Only those use up $c(w)$, while still have unserved customers reside in are partly served.

Definition 2. *Given W and R, service quality noted as $sq(W)$ equals to $\sum_{r \in R} \varepsilon(r)wt(r)$.*

Service quality is the number of all customers who are under service. To increase $sq(W)$, new facilities should be built and locations are chosen from a candidate set. The concept of *candidate increment* indicates the increment of service quality when a new facility p is added. Let W' denote $W \cup \{p\}$ in the following of this paper.

Definition 3. *Given a p, its increment denoted by $inc(p)$ equals to $sq(W') - sq(W)$.*

The larger increment, the better a candidate location is. Thus we formulate the proposed query as follows: given a constant $k, 0 \leq k \leq |P|$, it finds a set $P' \subseteq P$, so that $|P'| = k$ and $\forall p_i \in P', p_j \in P \setminus P', inc(p_i) \geq inc(p_j)$. Answering the query means finding k most promising candidates that maximize the increase of service quality.

3.2 Basic Solution

The query definition gives a basic solution. It takes R, W, P as input and returns P', which also applies to other algorithms. It scans the candidate set, adds p to W, calculates the new service quality, gets $inc(p)$ then removes p and tries next candidate. Finally it picks out those have top k increments as results.

4 Algorithm

4.1 Notation and Properties

The challenge of the query are two folds. One is that the influence set is not only re-lated to customer locations, as addressed in [8,1], but also related to facilities. When a new facility is added, it becomes the new nearest facility of some customers and their old facilities are also affected. Here we give the exact definition the *influence set* of a candidate location.

Definition 4. *The influence set of a p denoted by $I(p)$ is $B(p, W') \cup \{r \in B(w, W)|w$ is old nearest facility of $r \in B(p, W')\}$.*

The other challenge is to efficiently deal with large amounts of input data. Hence we introduce the concept of *nearest facility circle* [3] (NFC) of r, which is a circle denoted by $nfc(r, w)$ that centers at r and has a radius of $dist(r, w)$. If p falls into $nfc(r, w)$, then p will become the new nearest facility of r, which is useful to efficiently get $I(p)$.

4.2 Pruning Techniques

To calculate $inc(p)$, it is unnecessary to assign every r to a facility again. As Fig.1 and 2 show, after adding p_1, same customers are assigned to w_1. We know that $w_1 \notin I(p_1)$. So we have the following theorem (proof can be found in the full paper).

Theorem 1. *Given a p, $\forall w \in W \setminus I(p)$, we have $B(w, W) = B(w, W')$.*

Hence, it is possible that we set aside those unaffected locations and focus on a candi-date's influence set. This idea is supported by the following theorem.

Theorem 2. *Given a p, $inc(p) = \sum_{r \in I(p)} (\varepsilon_{new}(r) - \varepsilon(r))wt(r)$, $\varepsilon_{new}(r)$ denotes the new ε factor of r after p is added.*

Maintaining the nn and rnn lists, we can easily calculate $\varepsilon_{new}(r)$ for every $r \in I(p)$, then get $inc(p)$ according to theorem 2, which effectively bounds the search space.

4.3 Spatial Index

We apply several spatial indices to further lower the time complexity. Two time-consuming parts are: (1) to find nearest facility for each $r \in R$ and (2) to get the influence set for each $p \in P$. For the first part, we introduce a spatial structure s_1 that supports NN query to index W. Whereas there is no direct tool to handle the second part. We use NFC and a spatial index s_2 that supports point enclosure query to achieve our goal. The first part has assigned each r to a w, so we can build $nfc(r, w)$ and insert it into s_2. When considering p, we form a p point enclosure query, search s_2 and process all NFCs returned by s_2 to get $I(p)$.

Theorem 3. *Given a structure s that indexes all $nfc(r,w), r \in R$, if we search s with a p-enclosure query, then $\{nfc(r,w)|\forall r \in B(p,W')\} = \{all\ NFCs\ returned\ by\ s\}$.*

According to theorem 3 and definition 4, we can get the influence set of p efficiently. The detailed steps of the proposed algorithm are given as follows. And a formal analysis shows the basic solution has a $O(n^3)$ time complexity. After applying pruning techniques, it reduces to $O(n^2)$. When we introduce the spatial indices, the proposed algorithm has a $O(nlogn)$ complexity.

Algorithm 1. Index

---

1. index each $w \in W$ with s_1
2. **for** each $r \in R$ **do**
3. $w \leftarrow$ NN query result of r on s_1, and assign r to w
4. calculate ε factor for all $r \in R$
5. build $nfc(r,w)$ for each $r \in R$ and insert it into s_2
6. **for** each $p \in P$ **do**
7.   search s_2 with a p point enclosure query
8.   **for** each $nfc(r,w) \in \{all\ NFCs\ returned\ by\ s_2\}$ **do**
9.   add r and $r' \in B(w,W')$ to $I(p)$
10.   **for** each $r \in I(p)$ **do**
11.   calculate $\varepsilon_{new}(r)$
12. $inc(p) \leftarrow inc(p) + (\varepsilon_{new}(r) - \varepsilon(r))wt(r)$
13. sort $p \in P$ according to $inc(p)$, and get first k candidates as P'

---

5 Empirical Study

5.1 Experimental Setup

We use C++ to implement all algorithms and conduct all experiments on a PC with an Intel(R) Core 2 Duo 1.7 GHz processor, 2 GB memory and running Window XP platform. We use both real-world and synthetic datasets. Two real-world datasets NE and NA are downloaded from http://www.rtreeportal.org/spatial.html. NE contains 123,593 points in north east of USA. NA contains 24,493 locations in North America. When using NE or NA, we uniformly sample from it to get R,W and P. To simulate real-life scenarios, we generate R, W with Zipfian distribution and P with uniform distribution. The serving sequence we adopt is an increasing order of the distances between customer locations and facilities. The weight of each r in both real and synthetic datasets is set to 1, whose results are similar to those set to any positive integer. The capacity of w and p is generated with uniform distribution ranging [1,40]. Cardinalities are set to $|R| : |W| : |P| = 20 : 2 : 1$ in all experiments. We adopt kd-tree as the spatial index for the nearest neighbor query and R* tree for the point enclosure query.

5.2 Comparisons

We compare the proposed algorithm with the basic solution. Since the basic solution is not scalable for large data, we conduct this set of experiments with relative small datasets. Fig.3 shows the proposed algorithm performs nearly 4 orders of magnitude faster than the basic solution when $|P|$ is 200. And the gap will be bigger with the growth of $|P|$. Only adopting the pruning techniques, the basic solution runs 10^2 times faster. Both real and synthetic datasets show similar results.

5.3 Scalability

Then we study the scalability of the proposed algorithm. Apart from two real-world datasets, we synthesize R, W and P with cardinality of 200K, 20K and 10K respectively. Each dataset is used from a quarter to the whole of it. To get the influence of capacity, we sample 3 subsets from them and vary the range of distribution from [1,20] to [1,60]. Both the execution time and index size are evaluate. Fig.4 and Fig.5 show both the cpu time and index size linearly increase with the growth of input data. Fig.6 shows the influence of different capacities is minor. All datasets support the above observations. So we conclude that the proposed algorithm has good scalability.

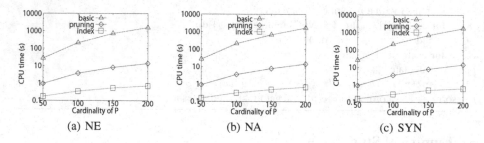

(a) NE   (b) NA   (c) SYN

Fig. 3. Execution time comparison on NE, NA and synthetic datasets

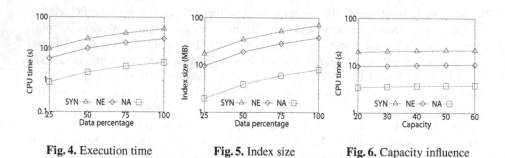

Fig. 4. Execution time   **Fig. 5.** Index size   **Fig. 6.** Capacity influence

6 Conclusion

In this paper, we formulate the top-k most incremental location selection query. Through analyzing the properties of the query, we propose pruning techniques and an $O(n \log n)$ algorithm to answer the query. The results of experiments confirm the effectiveness of the pruning techniques and the efficiency of the proposed algorithm.

Acknowledgements. This work is supported by NSFC under the grant No. 61003085 and HGJ PROJECT 2010ZX01042-002-002-03.

References

1. Huang, J., Wen, Z.: Top-k most influential locations selection. In: CIKM 2011 (2011)
2. Jianzhong, Q., Rui, Z.: The min-dist location selection query. In: ICDE 2012 (2012)
3. Korn, F.: Influence sets based on reverse nearest neighbor queries. SIGMOD Rec. (2000)
4. Leong Hou, U.: Capacity constrained assignment in spatial databases. In: SIGMOD 2008 (2008)
5. Stanoi, I.: Discovery of influence sets in frequently updated databases. In: VLDB 2001 (2001)
6. Wong, R.C.-W., Özsu, M.T.: Efficient method for maximizing bichromatic reverse nearest neighbor. Proc. VLDB Endow. (2009)
7. Wong, R.C.-W., Tao, Y.: On efficient spatial matching. In: VLDB 2007 (2007)
8. Xia, T., Zhang, D.: On computing top-t most influential spatial sites. In: VLDB 2005 (2005)
9. Zhang: Progressive computation of the min-dist optimal-location query. In: VLDB 2006 (2006)
10. Zhou, Z., Wu, W.: Maxfirst for maxbrknn. In: ICDE (2011)

An Approach of Text-Based and Image-Based Multi-modal Search for Online Shopping[*]

Renfei Li[1], Daling Wang[1,2], Yifei Zhang[1,2], Shi Feng[1,2], and Ge Yu[1,2]

[1] School of Information Science and Engineering, Northeastern University
[2] Key Laboratory of Medical Image Computing, Northeastern University,
Ministry of Education, Shenyang 110819, P.R. China
lirenfei@foxmail.com,
{wangdaling,zhangyifei,fengshi,yuge}@ise.neu.edu.cn

Abstract. Nowadays, more and more people prefer online shopping to physical store shopping for its convenience, cheapness and timesaving. Customers visit some commercial shopping websites, and select their favorite commodities by accessing links or retrieving by search box. However, in our real life, most online shopping websites provide a simple and single text retrieval method only, to some extent it's difficult for customers to submit query and retrieve satisfactory results. In this paper, a multi-modal search approach combining text-based and image-based search techniques is presented. Besides text search, a two-stage image search approach is proposed, which utilizes basic features consisting of color and textural features to filter mismatching images in first stage, and further uses SIFT features for accurate search in second stage. Moreover, a prototype system has been developed for multi-modal search on online shopping websites. By submitting some words, phrases, images or their combination, customers can search out what they want. The experiments compared with traditional algorithms based on single visual feature validate that our approach and multi-modal search prototype system are effective, and the retrieval results can satisfy customers' requirements well for online shopping.

Keywords: multi-modal search, online shopping, textual feature, visual feature.

1   Introduction

With the popularization and application of E-commerce extending, more and more people go shopping online rather than shopping out. Searching for product information and buying commodities online have become popular activities [1]. Empirical research shows that, nowadays, many individuals tend to start their shopping process with an information search on the Internet before they go to the store [2]. Many online shopping websites become popular consequently, such as Taobao, 360buy, Amazon. It is convenient to buy commodities on Internet, and customers can buy almost everything without going out, furthermore the commodities

---

[*] Project supported by the State Key Development Program for Basic Research of China (Grant No. 2011CB302200-G), National Natural Science Foundation of China (Grant No. 60973019, 61100026), and the Fundamental Research Funds for the Central Universities(N100704001).

H. Gao et al. (Eds.): WAIM 2012, LNCS 7418, pp. 172–184, 2012.

which customer have bought can even be delivered to them. Thus, E-shopping could lift the time and space constraints of shopping process and bring more flexibility [3].

As a commodity has different prices in different online stores, customers need to select an approving store for buying when they see a favorite commodity. Customers even actively seek commodities they want to buy. Most search engines of online shopping websites provide search based on text words currently. However, the search can not meet customers' requirement as: (1) Sometimes, text words are more difficult to express customers' requirements than images. For example, if a customer wants to buy a "sark with latticework", an image of the latticework is easier for expressing clearly than text description about it. (2) Text search is based on tags of commodities, but a commodity has different tags in different online stores. Thus search will return many irrelevant results or loss many relevant results.

Obviously, search based on images is required. Nowadays some search engines have provided image search, such as image Baidu, but it can't retrieve the commodities on shopping websites. We have noted that commodities of online shopping websites have a character that there are about five images and a paragraph of text to describe a commodity. The images can be divided into three kinds, i.e. "big image", "middle image", and "small image". Customers achieve a general impression of the commodity through "big image", and details from "small image". Based on the character, in this paper, we extract the visual features from these images without their background, and apply the features to a two-stage image search proposed, which utilizes basic features consisting of color and textural features to filter mismatching images in first stage, and further uses SIFT features for accurate search in second stage. Moreover, we develop a prototype system for multi-modal search including text-based and image-based search. Using the prototype, by submitting words, phrases, images or their combination, customers can retrieve what they want to buy. The experiments of comparing with traditional algorithms based on single visual feature show that our approach and multi-modal search prototype system are effective, and the retrieval results can satisfy customers' requirements well for online shopping.

The remainder of the paper is structured as follows. Section 2 introduces the related work. Section 3 gives problem description. Section 4 presents how to offline extract commodities' features. Section 5 describes online multi-modal search process. Section 6 shows our experiment results. Section 7 concludes our work and gives directions for future studies.

2   Related Work

In this paper, our purpose is to provide a multi-modal search prototype system about commodity for online shopping. In this field, researchers have done some pioneering work. Davis [4] proposed a multi-modal shopping assistant which provides users with a service capable of reducing the time spent on grocery shopping and the stress that occurs during this activity. It helps consumers to find commodities quickly at shopping mall in reality. Anil [5] proposed an algorithm to search commodities with trademark. Some companies have developed some image retrieval systems like QBIC of IBM [6, 7], Virage of company Virage [8], and MARS of Illinois University of United States [9]. All of them are great prototype systems for image retrieval.

However, these studies above have some obvious shortcomings. Firstly, few of them do research based on both visual features and textual features and apply them into online shopping websites. Secondly, although some of the works take multimodal into consideration, they do not mention the impact of other factors such as time. To overcome these shortcomings, we propose a multi-modal searching for online shopping websites in this paper. In our work, we will utilize HSV [10], GLCM [13], and SIFT [14] algorithm to extract color, textual, and SIFT features, respectively. About these algorithms, researches have done related work.

Color histogram is one of the most commonly used methods of image retrieval based on color feature, and HSV is an approximately-uniform color space: Hue, Saturation, and Value [10]. One of the reasons inhibiting these spaces from being widely used in image processing tasks is their noise-sensitivity due to the nonlinear transformations involved [11].

Texture is a visual feature which is produced by spatial distribution of tonal variations over relatively small areas [12], and a common technique in texture analysis is GLCM [13], which describes the frequency of one gray tone appearing in a specified spatial linear relationship with another gray tone, within the area under investigation [12].

SIFT descriptors are computed for normalized image patches with the code provided by Lowe [14]. The resulting descriptor is of dimension 128, and it is invariant to image rotation and scale and robust across a substantial range of affine distortion, addition of noise, and change in illumination [14].

Locality-sensitive hashing (LSH) was introduced as an approximate high-dimensional similarity search scheme with provably sublinear dependence on the data size [15, 16]. Instead of using tree-like space partitioning, the key idea is to hash the points using several hash functions so as to ensure that. For each function, the probability of collision is much higher for objects which are close to each other than for those which are far apart [17].

Based characteristics of HSV, GLCM, and SIFT algorithm above, in our work, we extract visual features using them, and search related image with the merged features roughly by LSH, then use SIFT features to search accurately.

3   Problem Description

For a shopping website, suppose $C=\{c_1, c_2, ..., c_n\}$ is a commodity data set (for short commodity without ambiguity) consisting of texts and images of all n commodities. For $c \in C$, $c=\{T\text{-Set}, I\text{-Set}\}$, where T-Set means c's text data such as tag, introduction, and description, and I-Set means c's image data such as big image, middle image, and small image. Moreover, we give more detailed descriptions as follows.

Let $TF=\{tf_1, tf_2, ..., tf_m\}$ be c's textual feature set, every tf_i $(i=1, 2, ..., m) \in TF$ be obtained from T-Set of c by natural language process techniques. Let $VF=\{VFC, VFT, VFS\}$ be the c's visual feature set and obtained from I-Set of c by image process techniques. Where $VFC=\{vfc_1, vfc_2, ..., vfc_{nc}\}$, $VFT=\{vft_1, vft_2, ..., vft_{nt}\}$, and $VFS=\{vfs_1, vfs_2, ..., vfs_{ns}\}$ represent the color feature set, textural feature set, and SIFT feature set, respectively, nc, nt, and ns are numbers of color features, textural features, and SIFT features, respectively.

For the shopping website, we will extract TF and VF from T-Set and I-Set of every commodity $c \in C$, and save TF, VFC, VFT, and VFS. On the other hand, for further search, we generate basic feature set $BF=f_1(VFC, VFT)$ and refined SIFT feature set $RF=f_2(VFS)$. Fig.1 gives an example of generating BF and RF.

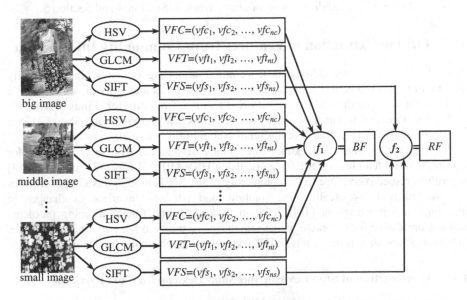

Fig. 1. Structure of VF Set and Generation of Basic Features and Refined Features

Moreover, for a customer's query CQ, let $CQ=\{T\text{-}CQ, I\text{-}CQ, \Re\}$, where $T\text{-}CQ$, $I\text{-}CQ$ be text set and image set submitted by the customer for query, respectively, $\Re$ be the relation of $T\text{-}CQ$ and $I\text{-}CQ$, and $\Re=\{$and, or$\}$. $\Re$ can be obtained from customers' feedback or other approaches.

For multi-modal search in shopping website, TF, $VF=\{VFC, VFT, VFS\}$, $BF=f_1(VFC, VFT)$ and $RF=f_2(VFS)$ need to be extracted and generated from $T\text{-}CQ$ and $I\text{-}CQ$ using the same approach with T-Set and I-Set above. Then besides search based textual feature set TF, we use two-stage strategy for image search.

Stage 1: For commodity set C, utilize $BF=f_1(VFC, VFT)$ to filter mismatching images (presenting commodities), i.e. compare BF from $I\text{-}CQ$ of customer's query CQ with one from I-Set of every commodity $c \in C$, and the result is $C' \subseteq C$.

Stage 2: Based on the results of Stage 1, use $RF=f_2(VFS)$ for image search accurately, i.e. in the commodities set C' after filtering in Stage 1, compare RF from $I\text{-}CQ$ of customer's query CQ with one from I-Set of every commodity $c \in C'$.

In this paper, we build a multi-modal search prototype system with above functions. It includes offline and online process.

The offline process includes: (1) Download data $C=\{c_1, c_2, ..., c_n\}$ from shopping websites; (2) Extract TF, VFC, VFT, and VFS, from every $c \in C$; (3) Generate $BF=f_1(VFC, VFT)$ and $RF=f_2(VFS)$; (4) Save BF into BF-Base and RF into RF-Base, respectively; (5) Update these sets regularly according to new commodities.

The online process includes: (1) Accept customers' query $\{T\text{-}CQ, I\text{-}CQ, \Re\}$ and obtain TF about $T\text{-}CQ$, BF and RF about $I\text{-}CQ$; (2) Extract textual features from $T\text{-}CQ$ and visual features from $I\text{-}CQ$, and generate BF and RF; (3) Execute text search based on *TF* for result *TR* (text results) and two-stage image search based on *BF* and *RF* for result *IR* (image results); (4) Return *R=TR℞IR* to customers.

We will give a detailed description of above work in Section 4 and Section 5.

4   Offline Extraction of Features from Commodity Information

After downloading commodity data from online shopping websites, the commodity data include two kinds of data, texts and images, i.e. *T*-Set and *I*-Set in Section 3. We apply different algorithms to extract different features from different kinds of data.

To obtain textual features, we build two lexicons, Stop Lexicon and Shop Lexicon to help us analyze the textual description of commodities more effectively.

To deal with images, we extract three kinds of features from the images, color feature, textural feature, and SIFT feature, using HSV [10], GLCM [13], and SIFT [14] algorithm, respectively. Due to the complexity of the SIFT, though this kind of features is invariant to image scaling and rotation, and partially invariant to changes in illumination and viewpoint [18], the high dimension and time-consuming problems make it unsuitable for complete image search, so we utilize other two kinds of features to filter first and then further utilize SIFT features to search smaller image set.

4.1   Construction of Stop Lexicon and Shop Lexicon for Text Partition

In our prototype system, we apply JE [19] to split *T*-Set into words. Because there are many onomasticon in shopping websites such as "抓绒"(fleece), "马海毛"(mohair), "清仓特价"(discount), JE can't split them correctly. For solving the problem, we build two lexicons, the Stop Lexicon and the Shop Lexicon.

(1) **Shop Lexicon.** For *T*-Set of a commodity, JE will split it to a string of words. However, JE often get too fine-grained words, so we build the Shop Lexicon to help JE to split appropriately. Our Shop Lexicon has 59 common words such as "豆豆鞋", "蝙蝠袖", and 10 onomasticons such as "颜色分类", "组合形式".

(2) **Stop Lexicon.** When we calculate the frequency to sort the result set, some words are unnecessary to the ranking, so we build the Stop Lexicon to filter the result. Our Stop Lexicon has 45 words, such as "打底衫", "春装", "2012".

With two lexicons, we apply Lucene [19] to create index of text and implement text search. Moreover, the two lexicons can be updated with new commodity data.

4.2   Generation of Basic Feature Set from Commodity Image

Empirically, in image search, if using single image features, the effectivity will be poor, but using SIFT features will be much time-consuming. To solve the problem, we propose a two-stage image search approach combining color, textural, and SIFT features, which utilizes basic features consisting of color and textural features to filter mismatching image in first stage, and further uses SIFT features for accurate search in second stage. For this purpose, we offline extract these features from commodity images, generate basic feature set *BF* and refined feature set *RF*.

We first use HSV algorithm to extract color features and GLCM algorithm to extract textural features, and use f_1 function to merge the two kinds of features for generating basic feature set *BF*. On shopping websites, the relation between commodity and images is one-to-many. Thus, we merge the color features and textural features extracted from images belonging to the same commodity and generate *BF* of the commodity.

It is notable that background of commodity image can make big noise to the image search, so we preprocess it firstly. In preprocessing, background interference is removed to get interest region, and merge the images features which describe one commodity. We describe the algorithm as Algorithm 1.

Algorithm 1: Basic Feature Generation
Input: *I*-Set;   // *I*-Set is image set belongs to one commodity;
Output: *BF*;   // *BF* is Basic Features of the commodity;
    1) call Algorithm 2 for getting interest region *I*-Set'$\subseteq$*I*-Set of each image;
    2) for every image *I*∈ *I*-Set'
    3)   {extract color features of *I* using HSV algorithm and get *VFC*;
    4)   extract textural features of *I* using GLCM algorithm and get *VFT*;
    5)   generate *BF*=f_1(*VFC*, *VFT*) and append *BF* into *BF*-Base;}

In the Algorithm, line 4) is for getting textural features using GLCM. In three kinds of images of a commodity, because small images describe details of the commodity, we extract textural features of small image as the commodity's image textural features. It is a vector of 5 dimensions. Because the value in this vector is very small, we amplify the value of the vector 100 times for smoothing it with color features. In this paper, we compare the area of interest region to judge which image is "small image". Next, line 5) is for getting basic feature set *BF* using f_1 function. In detail, for a commodity *c*, which has *m* images such as big images, middle images, and small images, we extract *m* interest regions I_1-Set', I_2-Set', ..., I_m-Set' from the *m* images, respectively. For any I_i-Set' (*i*=1, 2, ..., *m*), its color feature and textural feature set are represented as VFC_i, VFT_i, respectively. Here f_1 function is shown as Formula (1).

$$BF = f_1(VFC, VFT)$$

$$= \frac{\sum_{i=1}^{m} VFC_i}{m} \cup \left(VFT_j \bigg| \underset{j}{argmax}\, area(I_1 - Set', ..., I_j - Set', ..., I_m - Set') \right) \quad (1)$$

From Formula (1), f_1 includes average of color features of all I_i-Set' (*i*=1, 2, ..., *m*) and the textural feature of maximal area in all I_i-Set' (*i*=1, 2, ..., *m*) (small image has maximal area), *BF* is an union features finally. Here *VFC* is a 128 dimensions vector, *H* component is divided into 16 levels and *S* component is divided into 8 levels.

In line 1) of Algorithm 1, Algorithm 2 is called to get interest region. Algorithm 2 is described as follows.

Algorithm 2: Interest Region Getting
Input: I-Set; // I-Set is image set belongs to one commodity;
Output: I-Set'$\subseteq I$-Set; // I-Set' is interest region set of each images;
  1) for every image $I \in I$-Set
  2) {detect Canny operators CI of I;
  3)   compute CI' by $\; CI' = B \oplus I = \bigcup_{b \in B} CI_b \;$; //B is a structuring element;
  4)   detect Contours CCI' of I;
  5)   compute I' by $\; I' = B \ominus CCI' = \bigcap_{b \in B} CCI'_b$ and add I' to I-Set';}

In Algorithm 2, line 2) is for image edge extraction using Canny edge detection operator, because the operator is sensitive, many edge inside of the interest region can be detected. Hence, we need further computing. In line 3), $\oplus$ is dilation operation. Line 4) is to get the contours of the CI'. In line 5), $\ominus$ is erosion operation to eliminate the noise. At last, we get the produced image set I-Set' which has interest areas only.

4.3 Generation of Refined Feature Set from Commodity Image

Generating refined feature set is for second stage image search. We apply SIFT algorithm to extract SIFT feature set VFT from interest region of every commodity's images. According to SIFT algorithm, for an image, its VFT is a 128 dimensions vector. So a commodity with m images, its $RF = f_2(VFT_1, VFT_2, \ldots, VFT_m)$ is a matrix of $m \times 128$. Obviously, f_2 function is constructing a matrix with m vectors. Algorithm 3 gives the process of generating RF.

Algorithm3: Generation of Refined Feature Set
Input: I-Set'; //I-Set' is interest region set of all images;
Output: RF; //RF is refined feature set from I-Set';
  1) for every image $I \in I$-Set' extract SIFT feature set VFS;
  2) $RF = f_2$(all VFS) and save RF into RF-Base; //from vector to matrix;

After first stage image search, result set become a smaller set than initial image set. We can execute accurate search based on RF because of its invariant to image scaling and rotation. Because the search is executed in smaller set, the time cost is acceptable.

5   Multi-modal Online Search for Shopping

Our multi-modal search prototype system is a middleware between shopping websites and customers. In this prototype system, customer can submit phrases, sentences, or images for searching. We use multi-modal search algorithm to implement the process. Algorithm 4 gives the process of multi-modal search.

Algorithm 4: Multi-modal Online Search

Input: $CQ=\{T\text{-}CQ, I\text{-}CQ, \Re\}$; //$T$-$CQ$ and I-CQ are texts and images submitted by
a customer, $\Re$ is the relation of T-CQ and I-CQ;

Output: $R=TR\Re IR$; //TR and IR are returned results for text and image search;
  1) split T-CQ into $TF'=(tf_1, tf_2, \ldots, tf_m)$ with Shop Lexicon;
  2) delete stop words from TF with Stop Lexicon;
  3) execute text search based on $TF'=(tf_1, tf_2, \ldots, tf_m)$ for getting result TR;
  4) extract VFC, VFT, and VFS from I-CQ using HSV, GLCM, and SIFT method;
  5) generate BF using f_1 function, i.e. $BF=f_1(VFC, VFT)$;
  6) call Algorithm 5 based on BF for getting filtered image set C';
  7) generate RF using f_2 function, i.e. $RF=f_2(VFS)$;
  8) call Algorithm 6 based on RF and C' for getting exact result IR;
  9) if $\Re=$"or" return $R=TR\cup IR$;
  10) if $\Re=$"and" return $R=ISearch(TR)\cap TSearch(IR, n)$;

In Algorithm 4, line 1)~line 3) is for text search, where Shop Lexicon and Stop Lexicon have been introduced in Section 4.1. Next, line 4) is for extracting visual features from the image submitted by customers, the approach is the same with line 3) and line 4) in Algorithm 1, and line 1) in Algorithm 3. Line 5) is for generating basic features BF from color features and textural features using function f_1, and line 7) is for generating refined features RF from SIFT features using function f_2. Here f_1 and f_2 have been introduced in Section 4.2 and Section 4.3, respectively. Line 6) is for executing first stage image search, and line 8) is for executing second stage image search. In line 10), $ISearch(TR)$ is for searching matching commodity images in TR, and $TSearch(IR, n)$ is for searching the commodities by text which contains the top n of the word frequency from high to low of the TR. Algorithm 5 and Algorithm 6 show the two-stage search processes.

Algorithm 5: First Stage Image Search

Input: $BF=\{bf_1, bf_2, \ldots\}$; // BF is basic features from customers' query I-CQ;
Output: $C' \in C$; // C' is search result set with basic features
  1) for BF-Base of all commodities // BF-Base is generated in Algorithm 1;
  2) {transform every $bf \in BF$-Base into a binary vector c-B_H by

$$c\text{-}B_H=Unary(c)(bf_1)\ldots Unary(c)(bf_k); \qquad (2)$$

//$Unary$ is a function to transform an integer into a binary vector;
  3) transform every $bf \in BF$ into a binary vector q-B_H also using Formula (2);
       // c-B_H is from BF-Base and q-B_H is from BF
  4) compute c-B_H, q-B_H by one of a bunch of hash function g_i using Formula (3)

$$c\text{-}B_H'=g_i(\sigma, c\text{-}B_H) \qquad q\text{-}B_H'=g_i(\sigma, q\text{-}B_H) \qquad (3)$$

  5) compute the MD5 value I_i of c-B_H' and q-B_H' using Formula (4);}

$$Mc\text{-}B_H'=MD(c\text{-}B_H') \qquad Mq\text{-}B_H'=MD(q\text{-}B_H') \qquad (4)$$

  6) save all Mc-B_H' into c-bucket and get $hashtableNum$ buckets;
       // Mc-B_H' is the key value of the hashmap
  7) save the hashmap and hashfamily which include all hash function g_i;
  8) for $i=1$ to $hashtableNum$;
  9) find the same key c in ith-c-bucket with Mq-B_H' and add c to C';

In Algorithm 5, line 1) to line 7) is to analyze all *BF* in *BF*-Base for generating a hashmap and hashfamily and saving them. It will be executed only if the database has been updated. In line 4), the hash function g_i is to choose σ numbers of the vector randomly. The smaller σ is, the greater the ability of approximate searching is, but the false positive corresponding is bigger too. When σ equals to an appropriate size, the result will be best. Line 8) to line 9) is to search the similar data in every hash table.

In the first stage of image search, *BF* is high dimensional, we can't use traditional index technology like R-tree because of dimension curse. Hence, we choose LSH index technology to search based on *BF*. We put similar *BF* into one hash bucket by hash functions, which can assure that if *BF* is more similar, the probability that they are in the same bucket is more higher. Because of the uncertainty, *BF* are hashed in *hashtableNum* hash tables using related series of hash functions. When we query a commodity by *BF*, the hash value is calculated by related hash functions and the data in corresponding hash bucket is the candidate set which is the result of first stage.

Moreover, Algorithm 6 gives second stage image search process.

Algorithm 6: Second Stage Image Search
Input: *C'*, *RF*; // *C'* is search result set of first stage; *RF* is refined feature set
Output: *IR*; //*IR* is final image search result set
  1) find *RF'*⊆*RF*-Base;   // *RF'* is *RF* of *C'*
  2) find *IR* by comparing *RF* with every $rf \in RF'$;

The process is simple. In line 1), *RF'* is refined feature set of *C'* and obtained from *RF*-Base, i.e. refined SIFT features. Line 2) uses *RF* from *I-CQ* to search in *RF'*, and final result *IR* is image set found by refined SIFT features.

6   Experiment Results

In order to test the algorithm we proposed in this paper, we download the data from the hottest online website in China, Taobao (http://www.taobao.com/). We get about 9632 images in 309 stores, and searching data in the database. Some images are shown as Fig.2, and the category distribution of the data is shown in Fig.3.

Fig. 2. Image Data from Taobao

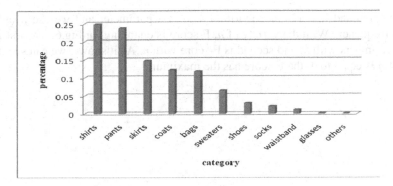

Fig. 3. Category Distribution of Data from Taobao

6.1   The Parameters in Search and Effectivity Comparison

In line 4) of Algorithm 5, σ is the unknown numbers of the vector in hash function g_i, with different σ, the F-score is changing. As Fig.4 shows, the three charts present precision, recall and F-score vary with σ respectively. We can see when the value of σ is 70, F-score has the maximum. In this experiment, the data is 200 commodities, and every value is the average of ten experiments.

In line 6) of Algorithm 5, *hashtableNum* is another parameter. As Fig.5 shows, the three charts present precision, recall and F-score varies with *hashtableNum*. When *hashtableNum* increases, F-score increases too. But when *hashtableNum* equals 10, the F-score is almost invariant, thus we choose 10 to be the value of *hashtableNum*.

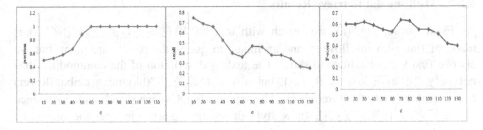

Fig. 4. The Variety of σ Impact

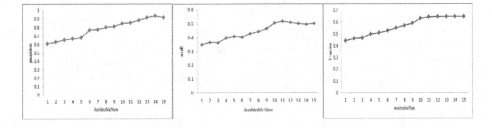

Fig. 5. The Variety of *hashtableNum* Impact

In line 11) and line 12) of Algorithm 4, n is a parameter and we use it as top-n highest frequency. With the variety of n, F-score is changing. In Fig.6, the first figure is result numbers with n, and second is F-score with n. As it shows, we can see, as the value of n is equal to 4, the F-score has the maximum.

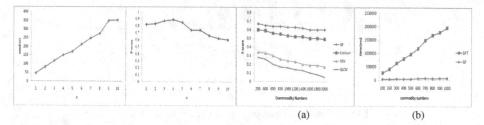

(a)   (b)

Fig. 6. The Variety of n Impact   **Fig. 7.** Different Kinds of Features

Fig.7(a) shows F-score decreasing with commodity number increasing. In Fig.7(a), *GF* is our method, means image features merged by HSV, GLCM and SIFT. Contour means using color features from interest area, HSV and GLCM represent using simple features only. As the curve shows, if we use single simple features, the accuracy will be very low, and even more, with the number of images increase, it decrease rapidly. After we choose interest area, the accuracy increases observably. The best result is *GF*, because we merge the SIFT features in it, the accuracy is high and stable. Fig.7(b) shows the comparison of time between SIFT and GF (using Base and SIFT features).

6.2 Multi-modal Retrieve Results

As Fig.8 shows, Fig.8(a), we search with text "水墨花朵花色的亚麻裙子"(skirt made of flax with ink flower), and an image to query, the results are many but we choose first 9 commodities to show. The textual description of the commodities has not only "水墨花朵"(ink flower), but also "木棉花开"(blooming ceiba flower) because of the sort of the frequency. The same as Fig.8(a), we search with a sentence "拉夏贝尔今年新款连衣裙"(new style dress) in Fig.8(b), and get the results as below. In both parts, the parameter $\Re$ is "and". The results in Fig.8(c) and Fig.8(d) are in the same condition except the parameter $\Re$ is "or".

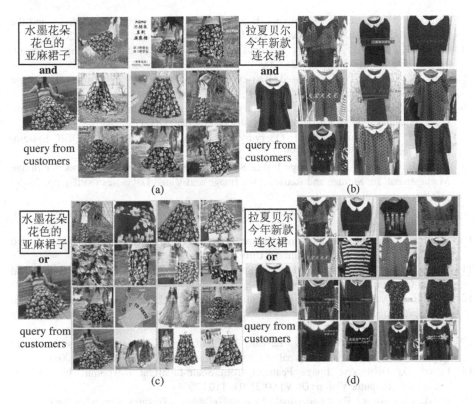

Fig. 8. Multi-modal Retrieved Results

7 Conclusions and Future Work

In this paper, we focus on searching commodities by both image and text from online shopping websites. For text search, we use Lucene to construct index. For image search, we propose two-stage search. The first stage is to filter obvious mismatching images by color and textural features using algorithm LSH, and the second stage is to match with SIFT features. For online search, customers may submit texts and (or) images for obtaining exact or extensive commodities' information they want to buy.

Although this paper puts forward many actual effective methods, there are still some to be improved such as applying and improving more image process methods, including more data modalities. All of them are our future research topics.

References

1. TNS Interactive. Global e-Commerce Report (2002),
   http://www.tnsofres.com/ger2002/ (accessed July 5, 2004)
2. Michael, R., Michelle, M.: Consumer acquisition of product information and subsequent purchase channel decisions. Volume Advances in Applied Microeconomics Issue, 231–255 (2002)

3. Couclelis, H.: Pizza over the Internet: E-commerce, the fragmentation of activity and the tyranny of the region. Entrepreneurship & Regional Development 16, 41–54 (2004)
4. Davis, Z., Hu, M., et al.: A Personal Handheld Multi-Modal Shopping Assistant. Networking and Service, 117–125 (2006)
5. Anil, K., Jain, A.: Shape-Based Retrieval: A Case Study With Trademark Image Databases. Pattern Recognition (PR) 31(9), 1369–1390 (1998)
6. Liu, Y., Zhang, D., Lu, G., Ma, W.: A survey of content-based image retrieval with high-level semantics. Pattern Recognition (PR) 40(1), 262–282 (2007)
7. Flickner, M., Sawhney, H., et al.: Query by Image and Video Content: The QBIC System. IEEE Computer 28(9), 23–32 (1995)
8. Bach, J., Fuller, C., et al.: Virage Image Search Engine: An Open Framework for Image Management. In: Storage and Retrieval for Image and Video Databases (SPIE), pp. 76–87 (1996)
9. Huang, T., Mehrotra, S., Ramchandran, K.: Multimedia Analysis and Retrieval System (MARS) Project. Data Processing Clinic (1996)
10. Smith, J.: Integrated spatial and feature image systems: Retrieval, compression and analysis, Ph.D. dissertation, Columbia Univ., NewYork (1997)
11. Song, K., Kittler, J., Petrou, M.: Defect detection in random color textures. Image and Vision Computing, 667–684 (1996)
12. Baraldi, A., Parmiggiani, F.: An investigation of the textural characteristics associated with gray level cooccurrence matrix statistical parameters. Geoscience and Remote Sensing Society 33(2), 193–304 (2002)
13. Malik, J., Belongie, S., Leung, T., Shi, J.: Contour and Texture Analysis for Image Segmentation. International Journal of Computer Vision (IJCV) 43(1), 7–27 (2001)
14. Lowe, D.: Distinctive Image Features from Scale-Invariant Keypoints. International Journal of Computer Vision (IJCV) 60(2), 91–110 (2004)
15. Indyk, P., Motwani, R.: Approximate Nearest Neighbors: Towards Removing the Curse of Dimensionality. In: STOC, pp. 604–613 (1998)
16. Gionis, A., Indyk, P., Motwani, R.: Similarity Search in High Dimensions via Hashing. In: VLDB, pp. 518–529 (1999)
17. Datar, M., Immorlica, N., Indyk, P., Mirrokni, V.: Locality-sensitive hashing scheme based on p-stable distributions. In: SoCG, pp. 253–262 (2004)
18. Skrypnyk, L.D.: Scene Modelling, Recognition and Tracking with Invariant Image Features. In: ISMAR, pp. 110–119 (2004)
19. Doug Cutting. Lucene, http://lucene.apache.org/

Categorizing Search Results
Using WordNet and Wikipedia

Reza Taghizadeh Hemayati[1], Weiyi Meng[1], and Clement Yu[2]

[1] Department of Computer Science, Binghamton University, Binghamton, NY 13902, USA
{hemayati,meng}@cs.binghamton.edu
[2] Department of Computer Science University of Illinois at Chicago, Chicago, IL 60607, USA
cyu@uic.edu

Abstract. Terms used in search queries often have multiple meanings and usages. Consequently, search results corresponding to different meanings or usages may be retrieved, making identifying relevant results inconvenient and time-consuming. In this paper, we study the problem of grouping the search results based on the different meanings and usages of a query. We build on a previous work that identifies and ranks possible categories of any user query based on the meanings and common usages of the terms and phrases within the query. We use these categories to group search results. In this paper, we study different methods, including several new methods, to assign search result record (SRRs) to the categories. Our SRR grouping framework supports a combination of categorization, clustering and query rewriting techniques. Our experimental results show that some of our grouping methods can achieve high accuracy.

Keywords: Search Result Clustering and Categorization, Search Engine.

1 Introduction

One common complaint about current search engines is that they return too many irrelevant results for users' queries. The reasons include (1) current search engines retrieve results mainly based on query words match, capturing only the main meanings or usages of query words, and (2) Internet users tend to submit very short queries which often do not provide enough context to determine the users' intentions. One way to tackle this problem is to group the search results into multiple categories such that all results in the same category correspond to the same meaning or usage of the query. This makes it much easier for users to identify useful results. Most current result clustering techniques are based on word-match similarity. Although a few techniques have used semantic similarity [1, 2], they have various weaknesses. For example, they do not explicitly and systematically consider usages of query terms, which would lower the quality of search result clustering.

In this paper, we propose and evaluate several methods to assign search results to a type of categories called *definition categories* (DCs). DCs are obtained based on both the possible meanings and the usages of the terms and/or phrases in a query using the techniques proposed in our previous paper [5].

H. Gao et al. (Eds.): WAIM 2012, LNCS 7418, pp. 185–197, 2012.

Unlike the work in [4] which considered only single-term queries, in this paper we consider both single-term and multi-term queries. Furthermore, the categories used in this paper are defined based on both the possible meanings and usages of the terms and/or phrases in the query using WordNet and Wikipedia [5]. The categories are ranked based on both the importance of the meanings/usages of each term/phrase in the query and the relationships between them. Moreover, we also introduce a new method which uses query rewriting technique to categorize SRRs.

This paper has the following contributions:

1. We introduce three new automatic real-time grouping methods. The first one is based on a query-rewriting technique (QRW). This method selects some query expansion terms (QET) and submits these QETs along with the original query to search engine(s) to retrieve related results for each category (DC). The second method (E3C) is extension to the CCC algorithm first introduced in [4]. E3C is designed to improve assigning SRRs that have low similarities with a DC to the right DC. The third method is a hybrid of the first two methods. It decides which of the first two methods to use in different situations in order to achieve better overall performance.
2. We perform extensive experiments to evaluate and compare the performance of our proposed algorithms. The experimental results indicate that some of the proposed algorithms have both good effectiveness and efficiency.

The rest of the paper is organized as follows. Section 2 reviews related work. Section 3 provides an overview of our approach. Section 4 presents the main steps of our approach. Section 5 reports experimental results. Section 6 concludes the paper.

2   Related Work

The general problem of document clustering and categorization has been studied extensively [8] and they will not be reviewed in this paper. Instead, we focus on related works that deal with the clustering and categorization of the search result records (SRRs) returned from search or metasearch engines.

Techniques for clustering web documents and SRRs have been reported in many papers and systems such as [2, 15, 17, 18]. There are also commercial search engines like yippy.com which clusters SRRS. However, these techniques perform clustering based on the syntactic similarity but not semantic similarity.

Some researchers used web directories like Yahoo directory or ODP to categorize/classify user queries. Mapping user queries to hierarchical sequences of topic categories was studied in [3]. The method in [9] maps user queries to categories using a user profile learned from the user's search history and a general profile derived from a concept hierarchy. The method in [16] classifies the search results into deep hierarchies using category candidates retrieved by query.

Another related area is *result diversification* (e.g., [13, 1]), which aims to select search results covering different meanings/usages and show them among the top-ranked results. These methods do not specifically cluster or categorize search results.

Query disambiguation [10] is relevant to identifying different meanings of a query for generating different categories. The issue of generating categories is not considered in this paper as our methods are based on already available categories.

Techniques for clustering and categorizing web documents using WordNet or other ontologies have also been extensively studied (e.g., [11, 14, 7, 15]) and some of them (e.g., [11, 14]) also tried to categorize SRRs based on the meanings of the query term.

Our approach differs from the above techniques significantly. First, our SRR grouping algorithm employs categorization, clustering and query rewriting techniques in a unique way. Second, our method also copes with SRRs that do not match any meaning/usage of the query term/phrase in WordNet or Wikipedia definitions. In other words, we utilize definitions provided by WordNet and Wikipedia but are not limited by them. Third, our approach also utilizes similarities that are computed using syntactical, semantic and common usage information.

3   System Overview

Our overall search result grouping system consists of the following main components:

1. Alternative Query Generation. For each user query Q, this step generates a set of *alternative queries* (AQs). All AQs contain the same set of query terms that appear in Q but contain different phrases. This step identifies different possible phrases comprised of the terms in Q as phrases are better at capturing the meanings of user queries. We will refer both query terms and phrases as **concepts**.

2. Definition Category Generation. A *definition category* (DC) is a combination of *meanings* or *usages* derived from the concepts of an AQ. This step generates all possible DCs for each AQ and it consists of three tasks: (a) *Usage generation*: Identify all possible meanings/usages for each concept in AQ using semantic dictionaries (WordNet and Wikipedia). (b) *Usage merging*: Merge similar meanings or usages for each concept into a single meaning/usage to reduce possible confusion to users. (c) *Definition category generation*: Generate DCs by combining one (possibly merged) meanings/usage from each concept in the AQ.

3. Definition Category Ranking. This step ranks the generated DCs. Each *DC* is generated from a specific alternative query AQ^*, from a specific meaning/usage of each concept in AQ^*, and from the combination of these specific meanings/usages. To rank the DCs, each DC is weighted in three aspects: (a) the importance of its AQ among all Aqs; (b) the importance of each DC among all DCs within each AQ; and (c) the document frequncy of each DC on the Web.

4. Submitting Query and Processing Results. For each user query, the top k ($k - 50$ in this paper) distinct results (duplicates are removed) are retrieved and are used as input to our SRR grouping algorithm (next component). Each result (SRR) usually consists of three different items: title, URL and snippet. Only the title and snippet of each SRR will be utilized to perform the grouping in our current approach. For each SRR, we first remove the stop words and stem each remaining word. Next, the SRR is converted as a vector of terms.

The above components were introduced in our previous works [4, 5] and will not be repeated in this paper. This paper focuses on the result grouping component as briefly reviewed below.

5. SRR Grouping Algorithms. We evaluate four major grouping algorithms in this paper (CCC, QRW, E3C and Hybrid). Our CCC and E3C algorithms consist of the following three steps to group SRRs: (i) Preliminary Categorization, (ii) Further Categorization, and (iii) Final Categorization. We explain these in more detail in section 4. The CCC algorithm here is similar to the CCC algorithm introduced in our previous work [4]. In [4], only WordNet was utilized to categorize SRRs and only single-term queries were considered. In this paper, we use both WordNet and Wikipedia to categorize SRRs and both single-term and multi-term queries are considered. The main difference between CCC and E3C is the similarity computation method used in the second step. The new method is not only based on the similarity between SRRs and DCs (categories) but also the similarities between un-categorized SRRs and already categorized SRRs. We will explain this in more details in Section 4.

Query ReWriting (QRW) uses the query expansion terms QETs generated from DCs along with the original query as new queries to retrieve SRRs related to each DC. We will discuss this in more details in 4.3. Hybrid solution uses one of the QRW or CCC [4] approaches based on the type of the DCs generated from the submitted query. Our experiments (section 5) show that this approach improves performance.

4   SRR Grouping Algorithms

We present four algorithms to group SRRs (CCC, E3C, QRW, Hybrid) in this section.

4.1   Algorithm CCC

Algorithm CCC in this paper consists of three major steps:

Step1: Categorize SRRs by assigning each SRR to the most similar DC if the similarity is greater than a threshold T_1. Temporary categories are obtained based on the current assignments and the remaining SRRs form another temporary category.

Step2: Further categorize the remaining SRRs by assigning each such SRR to the most similar temporary category using another threshold T_2.

Step3: Categorize/Cluster the set *RS* of remaining SRRs. Three alternative solutions were introduced for this step in [4]: (i) *Largest Frequency of Use* (*LF*): Assign RS to the cluster with the most common meaning; (ii) *Largest Category*: Assign RS to the cluster (from Step 2) with the largest size; (iii) *Clustering*: cluster the results in RS.

The CCC algorithm in this paper differs from that in [4] in the following aspects:

1. DCs used in this paper are based on both WordNet and Wikipedia definitions; however the one in [4] was just based on only WordNet definitions.
2. In this paper, we train thresholds in different steps of CCC to achieve the best performance. In [4] the thresholds were manually selected.

3. In this paper, the *LF* method in [4] is replaced by an HWC method, which assigns RS to the DC with the highest weight. *LF* is based on the largest *frequency of use* of the query term's synset in WordNet. HWC is based on the highest weighted (ranked) DC (definition category) obtained using both WordNet and Wikipedia (see [5] for details).

4.2   E3C (Extended CCC)

Algorithm **E3C** has the same three major steps as CCC except that the second step (i.e., Further Categorization) in E3C is implemented differently. In CCC, sometimes we couldn't assign a related SRR to the correct DC since we couldn't find high similarity between that SRR and the DC due to the lack of sufficient common words. To address this issue, we modify the second step in CCC. This is explained below.

Let's assume that there are similarity scores among documents and also between each document in a set of documents (SRRs) and a DC C_1. It is possible that an SRR R_1 has zero or very low similarity with C_1, but is still sufficiently similar to an SRR R_2 which is very similar to C_1 and has already been assigned to C_1. In this case we may assign R_1 to C_1 via R_2.

We categorize SRRs based on their $V(R, C_k)$ values (to be defined shortly) with DCs. Specifically, for each SRR R, find the DC C_k that has the highest $V(R, C_k)$ value to R among all DCs. If the value of $V(R, C_k)$ is very low, we postpone assigning R to a later step. This is to prevent assigning an SRR to a DC with a very low similarity. When more information becomes available later, we will try to assign this SRR again. If the $V(R, C_k)$ values between R and two DCs are very similar, it is easy to assign R to a wrong cluster. In this case, we also postpone assigning R to a later step.

We continue this method until one of the following conditions becomes true:

- All SRRs have been categorized.
- We cannot assign at least one new SRR to any DC.
- A pre-set number of iterations have been reached. In this case we don't continue assigning un-assigned SRRs to DCs and go to Step 3 of CCC.

We introduce two approaches to calculate $V(R, C_k)$ values. In the first approach, the value of $V(R, C_k)$ is calculated by

$$V(R, C_k) = Sim(R, C_k) + \sum_{i=1}^{n} Sim(R, d_i) P(d_i \mid C_k)$$

where $P(d_i \mid C_k) = \dfrac{Sim(d_i, C_k)}{\sum_{j=1}^{m} Sim(d_i, C_j)}$ and $1 \le k \le m$ and $Sim(R, C_k)$ is the similarity

between R and the kth DC, $Sim(R, d_i)$ is the similarity between R and d_i which is the *i*th already categorized SRR in DC C_k, m is the number of DCs built for a submitted query, $P(d_i \mid C_k)$ is the probability of d_i belonging to C_k (the kth DC), $Sim(d_i, C_k)$ is the similarity between d_i, which is the *i*th already categorized SRR in DC C_k, and C_k.

We consider these values in this method. First, the similarity between R and C_k. Second, the total similarities between R and the other SRRs in C_k ($1 \le k \le m$). This will be determined by calculating the similarity between R and each SRR d_i in C_k

multiplied by the probability of d_i belonging to C_k. By using this probability, we are more in favor of those assigned SRRs that have higher similarities to a DC over those with lower similarities to the same DC.

The reason behind this approach is, if there are similarities between an SRR R and SRRs already categorized in a DC C_k and the total value of these similarities and the similarity between R and C_k is above a threshold (e) and is the highest among all DCs, then there is a good chance that R belongs to C_k.

The second approach determines the value of $V(R, C_k)$ by using:

$$V(R, C_k) = Max\left(Max_{i=1}^n (Sim(R, d_i) * P(d_i \mid C_k)), \alpha * Sim(R, C_k)\right)$$

where $0 \leq \alpha \leq 1$. In this method we determine $V(R, C_k)$ by finding the maximum value among the similarity between R and SRR d_i in C_k multiplied by the probability of d_i belonging to C_k and the similarity between R and C_k. The logic behind this method is if an SRR R_1 has high similarity with an SRR R_2 which has already been categorized to a DC C_1, and also this R_2 has high similarity to DC C_1, then there is a good chance that this R_1 (uncategorized yet) belongs to C_1, compared to a situation that the same SRR R_1 has similarity to more SRRs (e.g., R_3, R_4) in another DC C_2, but the similarity between R_1 and R_3 (or R_4) multiplied by the probability of their (R_3 or R_4) belonging to C_2 is lower compared to the first case when R_1 is similar to R_2 and R_2 has been categorized into C_1.

4.3   Query ReWriting (QRW)

The last two algorithms send original queries and retrieve the first n SRRs (first 50 SRRs in this work). In those algorithms, usually high weighted DCs get relevant SRRs and DCs with lower weights don't receive any SRRs. In order to get relevant SRRs for low-weighted DCs, we need to retrieve many SRRs (in many cases thousands). To address this issue, we introduce the QRW method.

In this method, we first find a set of query expansion terms (QETs) for each DC (category), and send them along with the original query to a search engine to retrieve relevant SRRs for each DC. This will guarantee that there won't be any DCs without SRRs. Since our DCs are built based on Wikipedia and WordNet meanings/usages definitions, there is a very low chance to not retrieve any relevant SRRs for any DC. This method has the following steps: Send the original query to the DC generator, retrieve DCs, find query expansion terms (QET) for each DC, and finally send each DC's QETs along with the original query as a new query to a search engine to retrieve relevant SRRs for each DC. We now explain how to find the QETs for DCs (the first two steps have been discussed in Section 3 and our previous paper [5]). In order to find the best expansion terms for DCs we use two sources: Wikipedia and WordNet.

Due to space limitation, we cannot provide the details of our QET generation algorithm in this paper. The basic idea is sketched below. Given a DC as input, this method aims to obtain query expansion terms to represent the DC. QETs for a DC are a set of terms/phrases that summarizes the DC. We first use Wikipedia and WordNet to generate candidate QETs. Wikipedia provides some useful information for each concept (like meanings/usages definitions, categories, disambiguation page and etc.), which can be used to generate candidate QETs. WordNet provides information like synonyms, hypernyms, and etc. that can also be used as candidate QETs. These candidate QETs will be ranked for each DC based on their similarities and the top-ranked candidate QETs is then chosen as the final QETs for the DC.

4.4   Hybrid Method

We observed that algorithm E3C outperforms the QRW method for certain types of DCs while the opposite is true for other types of DC's (more details will be given shortly). We introduce a hybrid solution to take advantage of the strengths of both types of approaches to enhance the performance of search result grouping. The following are the main differences between E3C and QRW.

1. QRW focuses only on information retrieved from Wikipedia and WordNet, while E3C uses information retrieved from WordNet, Wikipedia and SRRs.
2. For DCs that do not contain a single phrase or for DCs with weak relationships between their concepts and meanings/usages, it is often difficult to determine a single meaning/usage which can express the intention of the DC. As a result, it is more difficult to find good QETs for these DCs. For this type of queries, E3C is more applicable than QRW.
3. Using Wikipedia and WordNet to categorize results places more emphasis on what meanings/usages are covered by a dictionary/encyclopedia for a concept. On the other hand, using SRRs to categorize the results emphasizes more on what contents are indexed by a search engine.
4. The importance of different meaning/usage recognized by a dictionary/ encyclopedia and SRRs can be different. A good balance between them can help the system categorize and rank SRRs with better accuracy and users' satisfaction. We try to achieve this goal by exploring different grouping algorithms.
5. Although Wikipedia is a dynamic source, SRRs are more dynamic and up-to-date. Just using Wikipedia to find possible meanings/usages may miss some usages (like persons' names). Using all sources (Wikipedia, WordNet and SRRs) makes our system capable of addressing this concern.
6. In CCC/E3C, depending on the number of SRRs retrieved, we may have some DCs with no SRRs assigned. In order to have at least some SRRs in each DC, we may need to retrieve thousands of SRRs. QRW does not suffer from this problem.

We differentiate four types of DCs. Each type of DC determines how well we can "guess" the real intention of a user by determining a meaning/usage for a user query. Based on our experiments, the QRW algorithm performs better for DCs for which we can find a meaning/usage, which can express the whole DC compared to other algorithms (e.g., when there is a definition page for the DC in Wikipedia which can be used as the DC's representative). On the other hand, E3C performs better for DCs for which we cannot find a single meaning/usage to represent those DCs (e.g., there is no definition page in Wikipedia which can be used as the DC's representative when sending the corresponding AQ to Wikipedia). Based on which algorithm (E3C or QRW) we use, different SRRs will be categorized and they will be categorized differently. For E3C, SRRs from the original query will be assigned to the similar

DCs, but for QRW, SRRs retrieved by queries formed by QETs along with the original query for each DC will be assigned to the corresponding DC.

In the hybrid algorithm, we use a method to select one of the algorithms (E3C or QRW) based on the type of DC generated from a query submitted by a user. We define four types of DCs. We also define different cases for each type of DC. These cases are used to determine the type of a DC by examining different definitions (if exist) in each DC to see if there is a single meaning/usage which can be used on behalf of other meanings/usages in a DC. Each query may consist of different concepts and each concept may have different meanings/usages. Therefore each DC may contain different meanings/usages. We try to examine each of these meanings/usages to see if there is relationship between them so we can consider all meanings/usages in a DC that are related to each other.

We classify DCs into the following four types:

Type 1: Single term Queries with single meaning.

> **Case 1:** A DC generated from an AQ (alternative query explained in section 3 step 1) which is a single term query with a single meaning/usage.

Type 2: DCs with no ambiguity. This type of DCs has only one meaning/usage (the DC contains only one meaning/usage) from both WordNet and Wikipedia after merging similar meanings/usages [5].
We recognize a DC as *Type 2* if one of the following cases is true:

> **Case 1:** A DC generated from an AQ which is a single term query with multiple meanings/usages from WordNet/Wikipedia. This means we can see multiple definition pages (there are multiple meanings/usages for this concept) when we submit this query to Wikipedia/WordNet. This type of AQs will generate multiple DCs, but each DC refers to one meaning/usage directly from WordNet/Wikipedia. All DCs generated from this AQ are *Type 2* in this case.
>
> **Case 2:** A DC generated from an AQ which is a valid phrase [5] with a single or multiple meanings/usages from Wikipedia/WordNet. Each DC refers to one meaning/usage directly from WordNet/Wikipedia. All DCs generated from this AQ are *Type 2* in this case.

Type 3: DCs with multiple definitions, but there is one definition which can be used to represent all other definitions in a DC. This type of DCs has multiple meanings/usages in WordNet or Wikipedia; however we can find one meaning /usage that can be used to represent all other meanings/usages in a DC. For example, if all meanings of a DC are synonyms, then one can be used to represent the others. As another example, if one meaning is a hypernym of other meanings, then this meaning can represent the others. We have identified 12 cases in which one meaning/usage can represent all other meanings/usages. But due to space limitation, they are not included here. These cases plus many examples for these cases can be found in a technical report [6].

Type 4: DCs with ambiguity: There is no single meaning/usage which can be used to represent all other meanings/usages in a DC. We recognize a DC as *Type 4* if it is not recognized one of the above three types.

The QRW method tends to generate longer queries for DCs in *Type 4* compared to DCs of *Types* 1, 2 and 3. The reason is that the system couldn't recognize one single meaning/usage for this type of DC. This type of DC has multiple meanings/usages. Each meaning/usage in a DC generates a set of expansion terms (explained in section 4.3). This usually will result in long and inaccurate QETs (query expansion terms). Furthermore, these QETs may retrieve irrelevant SRRs. Therefore, the Hybrid method uses the E3C algorithm for *Type 4* DCs and uses QRW algorithm for DCs of *Types 1, 2 and 3*.

5   Evaluation

5.1   Query Set and Performance Measures

Our query set contains 50 queries [6] with 25 from TREC (2003 and 2005) and 25 from AOL query logs. We sample 50 queries conditioned on: (a) the query set should have queries with different lengths, (b) the query set should have a mixture of queries with/without phrases, and (c) the query set should have queries with ambiguities. The reason for using queries from different sources is to have queries with different lengths and also have enough queries with ambiguities.

We evaluate four SRR grouping algorithms (CCC, E3C, QRW and Hybrid). For all algorithms, we use the *recall, precision* and *F1 measure* as the performance measures. For the SRR grouping algorithms, the recall and precision are defined in [4].

5.2   Performance of CCC and E3C in Different Steps

The CCC algorithm has three major steps to assign (categorize) SRRs to DCs. We evaluate the performance of each step. For Step 3, different methods are introduced to categorize unassigned SRRs. We also study the performance of each of these methods here. We evaluate the following four different methods for Step 3: Clustering, Largest Cluster (LC), Highest Weighted Cluster (HWC) and SIM (classification).

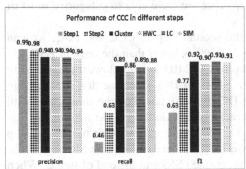
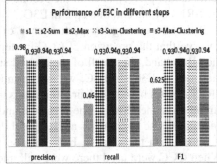

Fig. 1. Performance of CCC   **Fig. 2.** Performance of E3C

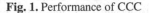

Fig. 1 shows the performance of CCC in different steps. We can observe that the precision of step 1 (p=0.99) is higher than the later steps. The reason is that we try to only assign those SRRs that have very high similarities to DCs. This will decrease the chance of assigning SRRs to wrong DCs. Among different methods for Step 3, clustering method performs the best considering both precision (0.94) and recall (0.89). This is mainly due to the fact that clustering method can group the SRRs beyond the discovered DCs while the other methods force the SRRs that do not match any DCs into incorrect categories. On the other hand, the recalls of earlier steps are very low compared to the later ones. The recalls for step 1, step 2 and step 3 are 0.46, 0.63 and > 0.85 (depending on what method was chosen for step 3). This is due to the fact that in the last step we try to categorize all SRRs and this will increase the recall value compared to earlier steps.

The E3C algorithm has three major steps to assign SRRs to the DCs built for a submitted query. We evaluate the performance of each step of the E3C algorithm here (Fig. 2). Step 1 in E3C is similar to that in CCC. In step 2 of E3C, we introduced two different approaches (Section 4.2). From Fig. 2, we can observe that the precision of step 1 (0.98) is higher than the later step. The reason is the same as that for algorithm CCC. In step 2, the second approach (i.e., when maximum value of $V(R, C_k)$ was chosen to assign an SRR to a DC) performs slightly better than the first approach (p=0.94 vs. p=0.93). We observe that for most cases both approaches categorize SRRs the same way. For cases that there were differences, the second approach tends to perform better. The reason is that if an SRR is very similar to an already assigned SRR to a DC, there is good chance that this SRR belongs to this DC, compared to a situation that the same SRR has moderate similarity with more SRRs in another DC, but the similarity with each of them is much lower compared to the one in another DC. On the other hand, the recall of earlier step (step 1) is very low compared to the later steps. The recalls for step 1, step 2-first approach (Sum) and step 2-second approach (Max) are 0.46, 0.93 and 0.94, respectively. This is due to the fact that in the second step we are less conservative and we give more chances to categorize SRRs to similar DCs since we have more information for each DC in this step (i.e., the SRRs assigned in Step 1 can be utilized). The performance in the third step stays almost the same as the performance in the second step since we were able to categorize most of the SRRs by the end of step 2 in E3C.

When comparing the performance of the second step in CCC and E3C, we can observe that the precision of CCC in step 2 (0.98) is better than the ones in E3C. On the other hand, the recall of the second step of CCC (0.63) is the lowest compared to the ones in E3C. In E3C, we assign more SRRs to DCs since we don't just assign SRRs to DCs based on their similarities to DCs but also based on the similarities with already assigned SRRs. The precision and recall of E3C in the second step for the first and the second approaches are 0.93 and 0.94. The precision of CCC in the second step is better than the ones in E3C because we only assign very similar SRRs to DCs. When CCC were used there were still uncategorized SRRs at the end of step 2. When E3C was used, we were able to categorize most of the SRRs at the end of step 2. In E3C there were few cases that we assigned SRRs to wrong DCs due to the fact that they either belong to DCs that are not generated by the DC generation algorithm in [5] or we assign them to the wrong DCs due to the lack of common words. Furthermore, in the second step of CCC we can have different thresholds. In the second step of CCC, we can be conservative and keep the value close to the value in the first step ($T_1 \geq T_2$) by decreasing the threshold value very little. In this case we only assign very similar SRRs to DCs. On the other hand, we can be more liberal by decreasing the threshold value more significantly from its original value in the first step in order to assign more SRRs to DCs. We observe that CCC performs better when more conservative approach is chosen. For example the precision, recall and F1 of CCC when clustering is chosen for more conservative approach are 0.94, 0.89 and 0.91, respectively, while for more liberal approach (the threshold value is much lower comparing to the conservative approach ($T_1 >> T_2 > 0$)), these values are 0.94, 0.79 and 0.86, respectively. We consider an approach to be more conservative when T_2 is closer to T_1, and more liberal when T_2 is closer to zero.

5.3   Overall Performance of All Algorithms

We introduced four algorithms to group SRRs in this paper (CCC, E3C, QRW and Hybrid). We study the performance of each of these algorithms here. Figure 3 shows the overall performance of different algorithms discussed in this paper.

To study the performance of CCC in this section, we consider the performance of the algorithm when clustering method is chosen in step 3, since it is the best among all the options. Similarly in E3C we consider the Max method in the second step since it performs better than the Sum method.

QRW performs better than CCC (F1 in QRW is 0.94 vs. F1 in CCC is 0.91). Hybrid method, which is combination of E3C and QRW algorithms, performs the best among different algorithms (precision is 0.96 and recall is 0.94). This is due to the fact that the system predicts what method performs better based on the submitted query. On the other hand, the Hybrid method needs more time to determine the types of the queries (its complexity is linear).

E3C performs better than CCC especially when we consider the recall (the recall for E3C is 0.94 vs. 0.89 for CCC, the precision for both algorithms are 0.94). This is more due to the fact that we were able to categorize more SRRs correctly for higher weighted categories. Based on our observation, E3C also is a faster algorithm (fewer iterations), although its complexity stays the same. Hybrid still performs slightly better than E3C (F1 is 0.95 for Hybrid vs. 0.94 for E3C).

The results show that all the four algorithms studied in this paper achieved significantly higher accuracy than a *Pure Clustering* method (this method just clusters all SRRs without considering DCs. The number of clusters is set to the number of DCs that are generated for other grouping algorithms. K-means is used in our implementation. After clustering, the clusters are compared to DCs and each cluster is assigned to the most similar DC.) The recall of Pure Clustering is 0.75 compared to the recalls of CCC, QRW, Hybrid and E3C which are 0.89, 0.93, 0.94 and 0.95, respectively. The precision for all algorithms are greater or equal to 0.94.

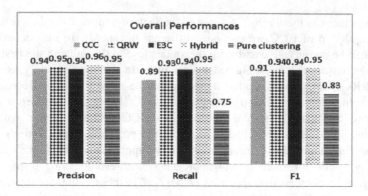

Fig. 3. Overall Performances

Each of the algorithms we have introduced in this paper has its own advantages and disadvantages. CCC, E3C and Pure Clustering are able to cover important web pages returned by a search engine since they all use the original query to retrieve results from a search engine. On the other hand, QRW may miss some important relevant results since we modified the original query and the modified query (QETs along with original query) is used to retrieve results for each specific DC from a search engine.

Furthermore, in CCC and E3C, we may not be able to find relevant SRRs for all DCs when only some top-ranked SRRs are retrieved per query (e.g., 50 SRRs). On average only 67% of the top 5 DCs can have assigned SRRs from the top 50 retrieved results in our dataset. This is due to the fact that those DCs were less popular and there were no relevant SRRs for those DCs among the top 50 results when the original queries were submitted. In QRW, we can find relevant SRRs for all DCs.

6   Conclusion

In this paper we investigated the problem of how to group the search result records (SRRs) from search engines. We previously have generated and ranked all possible categories of any user queries according to their match with the expected intention of the user. These categories can be used to categorize SRRs returned from search engines in response to user queries. By grouping the SRRs based on the different meanings/usages of the query term, it makes it easier for users to identify relevant results from the retrieved results. In this paper we focused on grouping SRRs and studied different grouping algorithms. Specifically, we proposed four algorithms that combine categorization, clustering and query expansion techniques. Our novel grouping algorithms use three different sources (Wikipedia, WordNet and SRRs) fully automatically to categorize the SRRs in a unique way. The categories built by our method are more meaningful and distinguishable than those by existing techniques since we build our categories based on different meanings/usages of queries' terms. We cover all possible meanings and usages of any terms and phrases by using WordNet and Wikipedia. We can also cluster uncategorized SRRs for those SRRs that do not belong to any categories built based on Wikipedia and WordNet. Our experimental results indicated that our SRR grouping algorithms are effective and highly accurate.

References

1. Agrawal, R., Gollapudi, S., Halverson, A., Ieong, S.: Diversifying search results. In: Proc. 2nd ACM Intl. Conf. on Web Search and Data Mining (2009)
2. Carpineto, C., Osinski, S., Romano, G., Weiss, D.: A survey of web clustering engines. ACM Computing Surveys 41(3), Article No. 17 (2009)
3. He, M., Cutler, M., Wu, K.: Categorizing Queries by Topic Directory. In: WAIM Conference, pp. 278–284 (2008)
4. Hemayati, R., Meng, W., Yu, C.: Semantic-Based Grouping of Search Engine Results Using WordNet. In: Dong, G., Lin, X., Wang, W., Yang, Y., Yu, J.X. (eds.) APWeb/WAIM 2007. LNCS, vol. 4505, pp. 678–686. Springer, Heidelberg (2007)
5. Hemayati, R.T., Meng, W., Yu, C.: Identifying and Ranking Possible Semantic and Common Usage Categories of Search Engine Queries. In: Chen, L., Triantafillou, P., Suel, T. (eds.) WISE 2010. LNCS, vol. 6488, pp. 254–261. Springer, Heidelberg (2010)
6. Hemayati, R., Meng, W., Yu, C.: Categorizing Search Results. Technical report (2012), `http://cs.binghamton.edu/~rtaghiz1/`
7. Hotho, A., Staab, S., Stumme, G.: WordNet Improves Text Document Clustering. In: ACM SIGIR Semantic Web Workshop (2003)
8. Jain, A.K., Murty, M.N.: Data Clustering: A Review. ACM Computing Surveys (1999)
9. Liu, F., Yu, C., Meng, W.: Personalize Web Search by Mapping User Queries to Categories. In: ACM CIKM Conference (2002)
10. Liu, S., Yu, C., Meng, W.: Word Sense Disambiguation in Queries. In: ACM CIKM Conference, pp. 525–532 (2005)
11. de Luca, E., Nürnberger, A., von-Guericke, O: Ontology-Based Semantic Online Classification of Documents: Supporting Users in Searching the Web. University of Magdeburg, Universitätsplatz 2, 39106 Magdeburg, Germany, AMR (2004)
12. Pitler, E., Church, K.: Using word-sense disambiguation methods to classify web queries by intent. In: Conference on Empirical Methods in NLP, vol. 3 (2009)
13. Santos, R.L.T., Macdonald, C., Ounis, I.: Intentaware search result diversification. In: SIGIR (2011)
14. de Simone, T., Kazakov, D.: Using WordNet Similarity and Antonymy Relations to Aid Document Retrieval. In: Recent Advances in Natural Language Processing, RANLP (2005)
15. Song, M.-H., Lim, S.Y., Kang, D.-J., Lee, S.-J.: Ontology-Based Automatic Classification of Web Documents. In: Huang, D.-S., Li, K., Irwin, G.W. (eds.) ICIC 2006, Part II. LNCS (LNAI), vol. 4114, pp. 690–700. Springer, Heidelberg (2006)
16. Xing, D., Xue, G., Yang, Q., Yu, Y.: Deep Classifier: Automatically Categorizing Search Results into Large-scale Hierarchies. In: Int'l. Conf. on Web Search & Data Mining (2008)
17. Zamir, O., Etzioni, O.: Grouper: A Dynamic Clustering Interface to Web Search Results. In: World Wide Web Conference (1999)
18. Zeng, H., He, Q., Chen, Z., Ma, W.: Learning To Cluster Web Search Results. In: ACM SIGIR (2004)

Optimal Sequenced Route Query Algorithm Using Visited POI Graph

Htoo Htoo[1], Yutaka Ohsawa[1], Noboru Sonehara[2], and Masao Sakauchi[2]

[1] Graduate School of Science and Engineering, Saitama University
[2] National Institute of Informatics

Abstract. Trip planning methods including the optimal sequenced route (OSR) query become a critical role to find the economical route for a trip in location based services and car navigation systems. OSR finds the shortest route, starting from an origin location and passing through a number of locations or points of interest (POIs), following the prespecified route sequence. This paper proposes a fast optimal sequenced route query algorithm from the current position to the destination by unidirectional and bidirectional searches adopting an A* algorithm. An OSR query on a road network tends to expand an extremely large number of nodes, which leads to an increase in processing time. To reduce the number of node expansions, we propose a visited POI graph (VPG) to register a single found path that connects neighboring POIs. By using a VPG, duplicated node expansions can be suppressed. We also perform experiments to show the effectiveness of our method compared with a conventional approach, in terms of the number of expanded nodes and processing time.

1 Introduction

The optimal sequenced route (OSR) query method has been proposed in recent years. It has been used for several trip planning applications, such as location-based services (LBS) and car navigation systems. The OSR finds the minimum-length route, starting from an origin location and passing through a number of locations or points of interest (POIs), following a prespecified route sequence.

Fig. 1 shows an example of an OSR query. You are currently at the "current position," and the final destination of your trip is home, which is labeled "destination." During the trip, you want to stop at a bank to withdraw money, and next at a Chinese restaurant to have dinner, and then at a movie theater, and finally return home. Although there may be many banks, restaurants, and movie theaters in the area, the OSR query chooses one from each category according to the specified sequence in order to minimize the total cost of the trip. The cost can be measured by several criteria, including distance, total trip time, safety, and the ease of the drive. However, in this paper we measure the cost according to the total trip distance.

The thick line in Fig. 1 shows the result of the OSR query, which provides the shortest distance route. In some applications, multiple result routes may be presented, so that the user can select the route he prefers. This case is called the "requested k shortest OSR (k-OSR)".

H. Gao et al. (Eds.): WAIM 2012, LNCS 7418, pp. 198–209, 2012.

Fig. 1. Optimal Sequenced Route

The OSR query was first proposed by Sharifzadeh et al.[10]. They proposed several algorithms to find the k-OSR in both vector (based on the Euclidean distance) and metric (based on the road-network distance) spaces. Among them, the progressive neighbor expansion (PNE) is the only algorithm that can be applied to road networks.

On a road network, the nearest neighbor (NN) object calculated by the Euclidean distance is not always the NN calculated on a road network [9]. The computation cost can drastically differ between these two distance measurements. The Euclidean distance between two points can be easily calculated; however, for the distance on a road network, we need to find the shortest path that connects two points. To find the shortest path, Dijkstra's algorithm [2] and the A* algorithm [5] are usually used. However, these algorithms consume a large amount of CPU power compared to the Euclidean distance-based search. In addition, a Euclidean distance-based algorithm can use the simple spatial index structures, such as, R-tree [4], to narrow the search space. In fact, the R-LORD algorithm proposed by Sharifzadeh et al. [10] employed R-tree for this purpose. Moreover, a spatial index based on the Euclidean distance is not effective for road network distance-based queries[9].

In this paper, we propose efficient algorithms for an OSR query on road networks. In an usual trip planning, a final destination is normally provided. For example, a home or an office can be a final destination of a trip. In this regard, in our trip planning method, the starting (usually the current position) and the destination positions of the trip are provided explicitly. When the destination is specified explicitly, we can adopt an efficient A* algorithm and the bidirectional search [6] for an OSR query.

The contributions of this paper are the following:

- To propose a visited POI graph (VPG), in order to reduce the number of node expansions by inhibiting duplicated node expansions.
- To present an efficient unidirectional search algorithm for a k-OSR query using the VPG. In addition, to present a bidirectional search algorithm to start the search from the current and destination points, in order to achieve a stable search time.
- To prove that the proposed method performs 100 times faster than the PNE algorithm [11] by conducting extensive experiments.

2   Related Work

The OSR query was first proposed by Sharifzadeh et al.[10]. They proposed several algorithms for an OSR query to operate on the Euclidean distance. Among them, the light optimal route discoverer (LORD) first finds a greedy route which is composed by the successive nearest neighbor search. The greedy route is found by performing a consecutive NN search from the starting point to the last visiting category. The search area is restricted by the length of this greedy route. Then, the LORD finds the optimal route in the reverse order (from the last category to the starting point), by narrowing the search area. The authors also proposed a more efficient algorithm called the R-LORD (R-tree-based LORD). However, these algorithms cannot be adapted directly to the road-network distance. Hence, for road network distance query, they proposed another algorithm named progressive neighbor exploration (PNE).

During almost the same time, Li et al. proposed the trip planning query (TPQ) [7].The TPQ is similar to an OSR query; however, the visiting order of the POI is not specified in the TPQ. Because of this free visiting order, the complexity of the TPQ is NP-hard, as in the traveling salesman problem. Therefore, Li et al. proposed several types of approximation algorithms. However, these algorithms cannot be directly applied to road networks, because of the heavy burden of the NN search. For TPQ on a road network, Li et al. proposed the minimum distance query (MDQ) algorithm. Basically, the MDQ expands nodes on the road network successively, finding the NN POI in the same way as by Dijkstra's algorithm. This causes duplicated node expansion, and the calculation time increases, especially when multiple trip-plan routes (k-TPQ) are requested.

Chen et al. [1] proposed another type of route query called the multi-rule partial sequenced route (MRPSR) query. This query generalizes both the OSR and the TPQ. For example, suppose we want to visit a bank, a restaurant, and a movie theater in that visiting order. A user may want to visit a bank before visiting both the restaurant and the movie theater because he needs to withdraw some money. However, the order of visiting to the movie theater and the restaurant can be exchanged. In this case, the visiting order is specified as a semi-ordered set, which can be represented as a directed graph. They called this graph an activity on vertex (AOV) network.

3   OSR Query Applying A* Algorithm

When we plan a trip, the starting position is obvious. In general, the current position acquired by GPS can be taken as the starting position, or the user specifies the starting position explicitly. In most situations, the final destination is also decided. Then, a trip planning query can be invoked with the starting (S) and the final destination (E) positions. In this case, we can adopt an A* algorithm for the efficient search of the TPQ. We can also use bidirectional search [8] for this purpose. Our algorithm proposed in this section uses both these methodologies, because they can reduce the calculation cost, which is mainly due to the considerable node expansion on the road network. We first describe an OSR query using the single-source A* algorithm (unidirectional search). We then develop it into a bidirectional search.

3.1   OSR Query by A* Algorithm

Let U_i be a category of the POI to be visited, and M be a sequence of U_i to specify the visiting order. That is, $M = \{U_1, U_2, \ldots, U_m\}$, and here m is the length of M ($m = |M|$). Our OSR query finds k optimal sequenced routes from the starting point S to the destination E, visiting each POI belonging to $U_i(1 \leq i \leq m)$ one after another, according to the given sequence M. The partial sequenced route (PSR) is the shortest route from the starting point S to one of the POIs in U_i, by passing through the POIs one after another choosing from U_j ($1 \leq j < i$) on the way according to the given sequence. SR is the total routes from S to E, visiting the POIs according to the given sequence M. To simplify the algorithm explanation, we assume that the POI is on a road-network node. However, this restriction can be easily resolved [9].

The A* algorithm has been applied to find the shortest route given S and E [5]. Here, we apply it to the k-OSR query. The A* algorithm evaluates the favorable node n to be expanded next by the cost $C = d(S, n) + h(n, E)$. Here, $d(x, y)$ is the distance from node x to y moving on the road network, and $h(y, z)$ is a heuristic function between y and z. Because we evaluate the cost of the OSR by the route length, we use the Euclidean distance between y and z as the value of $h(y, z)$.

Fig. 2 shows an example of a search using the A* algorithm. In this example, the search starts from S, and then finds P_1^1 belonging to U_1. From this POI, a new search targeted at the U_2's POI starts. In parallel, the search starts from S, finds P_2^1 belonging to U_1, and then another new search starts from the POI.

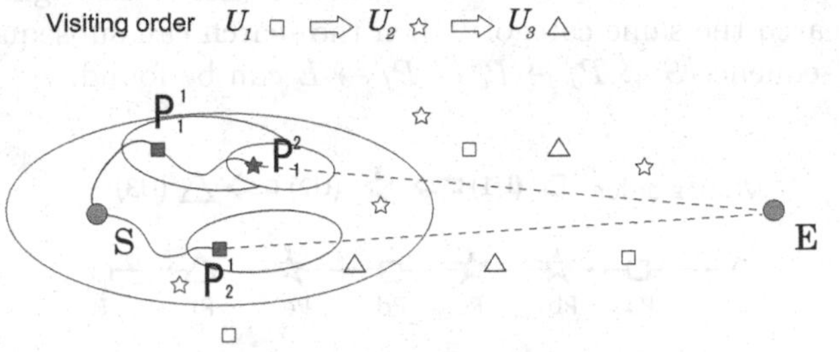

Fig. 2. Outline of POI search

As mentioned above, the A* algorithm decides the next-expanded node on the road network (n_a) by an extracted record from the priority queue (PQ), which gives the minimum cost $d(S, n_a) + h(n_a, E)$. For each road segment connected to n_a, the cost is calculated to compose a new record, as shown in Eq. (1), and it is then inserted back into the PQ:

$$C = d(S, n_a) + d(n_a, n_b) + h(n_b, E)$$

Here, n_b shows the opposite-side node to n_a of the road segment.

$$< C, U_i, L, n_b, n_a, P_{prev}, org > \tag{1}$$

In the record, C is the abovementioned cost, U_i is the next POI category to be visited, and L is the distance on the road network from S to n_b (i.e., $L = d(S, n_b)$). P_{prev} is the last-visited POI that belongs to U_i. n_a is necessary in the record to restore the PSR by backtracking from n_b to S. For the backtracking, Eq.(1) is recorded in a hash table indexed by n_b, after it is removed from the PQ. The term org is the origin of the PSR, i.e., S or E. This term is not always necessary for a unidirectional search where the origin of the search is predetermined; that is, a unidirectional search can start from either S or E. Both S and E can be the origins of a bidirectional search. Hence, both search origins are required for the bidirectional search.

Repeating the node extraction from the PQ and the node expansion, the search area is gradually enlarged. The search is terminated when the item n_b of the record extracted from the PQ is E. This terminating condition is the same for the typical A* algorithm.

Every time a POI P^i belonging to U_i is found, we start a new search targeting U_{i+1}, and simultaneously, we need to continue the search for another POI that belongs to the same U_i, ignoring P^i. Fig. 3 explains this necessity. This figure shows a situation where the POIs that belong to several categories of POI are arranged in a line. The search starting from S first finds P_a in U_1. Then, the next search targeting a POI in U_2 starts, and the search finds P_b. Next, a search targeting U_3 starts from P_b, and then finds P_f. Finally, the search reaches E and is then terminated. By this search, the sequence $S \to P_a \to P_b \to P_f \to E$ has been found. However, there are other OSRs that have the same length. For example, the other sequences $S \to P_a \to P_c \to P_f \to E$, and $S \to P_d \to P_e \to P_f \to E$ have the same length. If we invoke another search ignoring P_a and continue to search the same category, then the search can subsequently find S, and thus the sequence $S \to P_d \to P_e \to P_f \to E$ can be found.

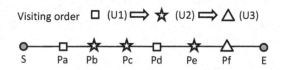

Fig. 3. OSR query setting border category

In a typical shortest-path search, Dijkstra's algorithm and the A* algorithm use a close set (CS) to avoid multiple node expansions. Once a node is expanded, it is registered to the CS, and the node will not be expanded again. On the other hand, an OSR query requires multiple CSes. Each CS records an expanded node from an individual source of searching (e.g., S or P_j^i). This characteristic of an OSR query causes multiple node expansions; that is, a node on a road network is expanded several times. For example, the search paths targeting U_1 started from

S to find the POIs P_1^1 and P_2^1, belonging to U_1, then new searches targeting U_2 start from both of them. These two searches are executed independently. Therefore, a node that has been expanded by another search can be expanded again, which causes a rapid increase in processing time. This also happens with the PNE when it adopts an incremental k-NN on a road network. We will deal with this problem using the bidirectional search in Section 4.

3.2 Bidirectional Search

The unidirectional search described above can be extended to a bidirectional search in a straightforward manner. A bidirectional search starts from S and E simultaneously under the control of one PQ. The record of Eq. (1) is put into the same PQ, which is independent of the origin of search. The search starting from S tries to find the POI according to the predetermined POI sequence M, and the search starting from E tries to find the POI in the reverse order of M; that is, $E \rightarrow U_m \rightarrow U_{m-1} \rightarrow \ldots \rightarrow U_1 \rightarrow S$.

The search is terminated when the search paths from both origins meet at a POI. Every time a search encounters the next POI to be visited, the arrival to the POI from another origin is checked. Suppose the POI is P^c belonging to U_C, and both PSRs from S to P^c and from E to P^c have been found; we can then obtain a complete SR by combining these two PSRs at P^c. When we need up to the k-th-shortest OSRs, we can obtain them by repeating node expansions until the k-th-shortest OSR is found. The abovementioned bidirectional search appears simple when only one shortest OSR is requested. However, when multiple k-OSRs are requested, there are some problems to be considered. In Section 4, we explain the problems and propose a solution.

When one of the categories in M has POIs with a very dense distribution, several independent searches will start from each POI belonging to the category. This causes an enormous number of node expansions, because the node expansions take place independent of each other such that the computation takes a long time. To avoid this effect, Fujii et al. [3] proposed a method to set a midway category (MC), named *bidirectional search with midway category* (BSWMC). The MC is selected from the POI category that has the highest density. When a search reaches a POI in the MC, no new search targeted at the next category starts. At this time, a PSR from S or E to a POI belonging to the MC is found. There is no additional search to find the next category starting from a POI in MC. Meanwhile, node expansions from another origin are advanced until one of them reaches the POI from the other side. At this time, a complete OSR is found.

The BSWMC method is suitable when we know the density of the POI in each category. In general, however, we cannot know the POI density. Even if we can conjecture the density, the POI is apt to be distributed with bias. Therefore, we need to improve the efficiency without setting the MC.

4 Suppressing Duplicated Node Expansion

Both the abovementioned approaches start a new node expansion from a POI belonging to U_i toward a POI belonging to U_{i+1}, every time a POI belonging to U_i is found. In Fig. 4, a search for a POI belonging to U_1 starts from S, and then the search finds P_1^1 and P_2^1 in that order. New searches for a POI belonging to U_2 start from P_1^1 and P_2^1. Then, the search that started from P_1^1 finds P_1^2 as the second-visited POI, at which a further new search starts for a POI belonging to U_3. Later, the search starting from P_2^1 reaches P_1^2, and then another new search for a POI belonging to U_3 starts. However, these two searches will consequently find the same path (PSR) from P_1^2 to E, as is shown in Property 1. Therefore, we need to suppress this redundant node expansion. A similar duplication also happened in the PNE when it adopted the incremental network expansion (INE).

Property 1. *k-PSRs starting from a POI belonging to U_i to E, obeying a pre-specified visiting sequence are determined uniquely.*

Proof. We deal with a time-invariant road network, and the constellation of the POIs is fixed. If k-PSR from a POI position belonging to U_i to E will not be changed at query time, then they are determined uniquely. Therefore, this supports the property.   □

Despite Property 1, a unidirectional search always starts a new search when a new POI is found. This causes duplicated node expansions, which find the same PSR. By reducing these duplicated node expansions, an efficient k-OSR query can be conducted.

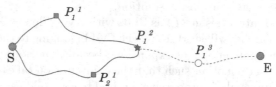

Fig. 4. Search path arrival to the same POI from multiple search paths

Property 2. *Consider searching an SR, the search started from S, finding POIs in order, then reaches a POI (P_j^i) belonging to U_i. Let the PSR from S to P_j^i be R, and the length be L_R. To advance this search, a new search targeted at the next category U_{i+1} starts from P_j^i. Consider another search path R' whose length is $L_{R'}$ has reached P_j^i. Then, we have the following relation between L_R and $L_{R'}$:*

$$L_R \leq L_{R'}$$

Proof. The priority queue (PQ) returns a node to be expanded according to the expected path length in ascending order. When R is returned from the PQ prior to R', the following relation stands:

$$L_R + h(P_j^i, E) \leq L_{R'} + h(P_j^i, E) \tag{2}$$

The heuristic distance value $h(P_j^i, E)$ is the same in both the left and right terms, so we obtain the following relation:

$$L_R \leq L_{R'} \tag{3}$$

Consequently, the equality in Eq.(3) is true when R and R' have the same length.   □

From these two properties, we can suppress duplicated node expansions. When P_1^3 is the nearest POI belonging to U_3 from P_1^2, it is determined uniquely independent when the paths reached P_1^2 from S. Then, when the search already started from P_1^2, which is the head of the PSR $S \to P_2^1 \to P_1^2$, to start a further node expansion from the same POI that is the head of another path is not useful.

In Fig. 4, we explained that a late-arrival PSR to P_1^2 (e.g., $S \to P_1^1 \to P_1^2$) cannot be the shortest path among SRs through P_1^2, because the first-arrival PSR to P_1^2 (e.g., $S \to P_2^1 \to P_1^2$) is shorter than the other late-arrival PSR. Then, when we need to find only one shortest OSR, we do not need to consider the late-arrival PSR at any POI. However, when we need to find k-OSR ($k > 1$), late arrival PSRs could be a part of the SRs. In this case, we need to consider the late-arrival PSR. Simultaneously, we need to consider suppressing the duplicated node expansions, which gives the same result.

To cope with this problem, we use the visited POI graph (VPG) as shown in Fig. 5. In the graph, nodes are visiting POIs (and two terminal points S and E) and edges are paths connecting neighboring POIs. An edge in VPG can be shared by plural OSR routes. For example, OSR routes R_1 and R_2 are sharing the link $P_2^3 P_3^3$ and $P_3^3 E$. Therefore, if we calculate for a link once, we can avoid the calculation for the same link again. This reduces the calculation cost of road network distance considerably, especially, when m is large.

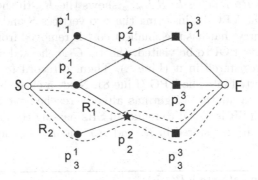

Fig. 5. Visiting POI graph

Hereafter, we refer to the unidirectional search with the VPG as the USVPG, and the bidirectional search with the VPG as the BSVPG. Algorithm 1 shows the BSVPG algorithm. The USVPG algorithm is almost similar to the BSVPG algorithm, although it is simpler. Line 1 initializes PQ with S and E. R is the result set returning k-OSR. The size of R is bounded by k, i.e., the number of requested SRs. Then $R.i$, the number of found SRs in R, does not exceed k.

Algorithm 1. Bi-directional Search with VPG

1: $PQ \leftarrow \{S, E\}$, $R \leftarrow \emptyset$, $R.L_{max} \leftarrow \infty$, $R.i \leftarrow 0$
2: Initialize VPG.
3: **loop**
4:     $e \leftarrow deleteMin(PQ)$
5:     **if** $e.C > R.L_{max}$ **then**
6:         **return** R
7:     **end if**
8:     **if** $e.nb$ is in the next visiting POI category **then**
9:         **if** $e.nb$ is an element of VPG **then**
10:             Add the link $(e.P_{prev} \rightarrow e.nb)$ to VPG.
11:             **if** PSRs from the opposite origin have reached $e.nb$. **then**
12:                 $R \leftarrow$ generateSR$(e.nb, VPG, R)$
13:             **end if**
14:         **else**
15:             Add the link $(e.P_{prev} \rightarrow e.nb)$ to VPG.
16:             **for all** road network node neighboring e **do**
17:                 compose Eq.(1) for U_{next}, enqueue it into PQ.
18:             **end for**
19:         **end if**
20:     **end if**
21:     **for all** road network node neighboring $e.nb$ **do**
22:         compose Eq.(1) for U_i, enqueue it into PQ.
23:     **end for**
24: **end loop**
25: **return** R

$R.L_{max}$ is the maximum SR length found so far; it has the value ∞ while $R.i$ is less than k. When $R.i$ reaches k, $R.L_{max}$ shows the $R.i$-th shortest SR length. Line 2 initializes the VPG by inserting the two vertices S and E.

In Line 4, the entry that has the smallest cost is removed from the PQ. When the entry is the next POI to be visited, the VPG is checked as to whether the POI has already registered in it (Line 8). Then, a record $(e.P_{prev} \rightarrow e.nb)$ is composed and inserted into the VPG (Line 8). Next, a check is performed as to whether a PSR from another origin has already reached the POI (Line 8). In this case, generated SR is called to make all SRs passing through the POI $e.nb$. The resulting SRs are registered in R, and $R.i$ and $R.L_{max}$ are altered by the results (Line 12).

Algorithm 2. Generatesr(e, VPG, R)

1: Trace VPG from e to the origin composing all PSR, then insert the result into $PSR1$.
2: Trace VPG from e to the reverse origin composing all PSR, then insert the result into $PSR2$.
3: Make all SR passing through e by making direct product between $PSR1$ and $PSR2$, then add them into R.
4: Renew $R.L_{max}$ and $R.i$ by the current condition.
5: **return** R

Lines 15 to 18 are always executed when $e.nb$ is the next visiting POI and the PSR from the opposite side has not reached $e.nb$. The record $(e.P_{prev} \to e.nb)$ is inserted into the VPG, and then the POI that belongs to the next visiting POI category is searched (Line 17).

The steps below Line 21 are always executed because of the reason described in Section 3.1. Then, when e is a POI, the search advances in two ways, one is to search the POI that belongs the next visiting POI category (Line 17), and the other is to search the POI that belongs to the same POI category (Line 22). When e is not a POI, usual node expansion is continued. This SR search is repeated until the number of found SRs exceeds the requested number k, and the cost of e becomes greater than $R.L_{max}$ (Line 23). After the latter condition has been satisfied, no shorter SR than k-th shortest SR in R will be created.

5   Experimental Results

To evaluate the performance of the proposed methods, we conducted extensive experiments. We used digital road-network data published by the Geospatial Information Authority of Japan (GSI), in the area of Saitama city, Japan. The road map consists of 25,586 road segments. We implemented the algorithms using the C# language. The computer used was a Core 2 Quad 2.40GHz CPU with 4GB RAM, and the operating system was Windows Vista.

We generated several sets of POIs in a pseudo-random sequence, with varying distribution density (p), which means the existence probability of a POI on a road segment (a road segment means a polyline between two intersections, or between an intersection and a dead end). For example, there are approximately 25 POIs in the subject area in the case of $p = 10^{-3}$, and approximately 256 POIs when $p = 10^{-2}$.

Fig. 6 shows the results between the unidirectional search (USVPG) and the bidirectional search (BSVPG). (a) compares their processing time among the unidirectional searches that start from S (USVPG-S), E (USVPG-E), and BSVPG. The density of the POI increases from the first POI (0.001) to the last POI (0.01). In these cases, the USVPG-S is the fastest, the USVPG-E is the slowest, and the BSVPG is moderate. When the density distributions of POIs are not equal, the unidirectional search starting from the less-dense side always outperforms the other.

Fig. 6(b) shows the result of another pattern. In this case, the middle category has the highest density, and the third category has the second-highest density. The BSVPG outperforms the others in this case. The experiments using the other combinations show the same tendency. Consequently, when we estimate the density of each visiting POI category, we can choose the fastest strategy by starting the search from the less-dense side. In general, however, the estimation is not easy. Even if we can know the statistical density of each POI category, the density may vary depending on its location.

Fig. 7(a) and (b) compare the performance between BSVPG and PNE according to the processing time. The experiments were conducted under three

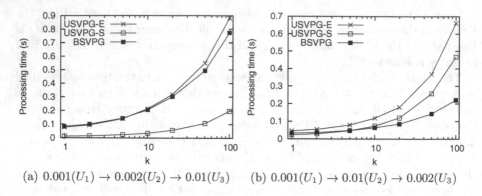

(a) $0.001(U_1) \rightarrow 0.002(U_2) \rightarrow 0.01(U_3)$ (b) $0.001(U_1) \rightarrow 0.01(U_2) \rightarrow 0.002(U_3)$

Fig. 6. Comparison between USVPG and BSVPG

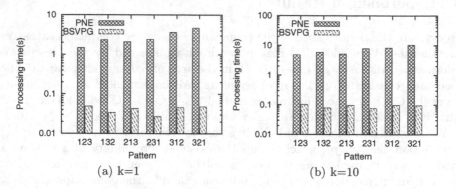

(a) k=1   (b) k=10

Fig. 7. Comparison between PNE and BSVPG

POI categories in which $p = 0.001$, $p = 0.002$, and $p = 0.01$, with shuffling of the order to be visited. In these figures, the pattern is shown by the three-digit number that corresponds to the three different POI densities: '1' corresponds to 0.001, '2' to 0.002, and '3' to 0.01. (a) shows $k = 1$ and (b) shows $k = 10$. As indicated by these results, the BSVPG always outperforms the PNE, and the processing time becomes stable with increasing k, independent of the POI density patterns.

6   Conclusion

In this paper, two fast algorithms called USVPG and BSVPG have been proposed to search the k-OSR for the road-network distance. Both algorithms are controlled by the A* algorithm. The BSVPG searches POIs from S and E simultaneously. This paper also proposes the VPG to reduce multiple node expansions, which is unavoidable in trip planning. This fact holds for all the existing trip planning methods that work on road-network distance measurements, for example, the OSR, TPQ, and MRPSR. Therefore, a strategy to reduce them can be the key to the fast trip planning. The VPG can be applied not only to the OSR, but also for several other types of trip planning that use road-network distance measurements.

Experimental results confirm that the presented algorithm can search the OSR approximately 100 times faster than the PNE. The USVPG and BSVPG largely contribute to the improvement in the VPG. The USVPG search, starting from the less-dense POI side, can find the OSR faster than the BSVPG, while the USVPG search starting from the other side will degrade. Therefore, in the case when the POI density cannot be known in advance, the BSVPG can be a good selection.

In this paper, we proposed an efficient method based on INE, however, incremental Euclidean restriction strategy [9] can also be applied for OSR search based on road network distance. On this strategy, two kinds of efficient algorithms are essential; one is an incremental OSR candidates generation method in Euclidean distance, and the other one is the efficient road network distance verification method for those candidates. Our future direction is to develop them.

Acknowledgments. This work was supported by Grants-in-Aid for Scientific Reaserch (KAKENHI) 24500107 and 23300337.

References

1. Chen, H., Ku, W.S., Sun, M.T., Zimmermann, R.: The multi-rule partial sequenced route query. In: ACM GIS 2008, pp. 65–74 (2008)
2. Dijkstra, E.W.: A note on two problems in connection with graphs. Numeriche Mathematik 1, 269–271 (1959)
3. Fujii, K., Htoo, H., Ohsawa, Y.: Fast optimal sequenced route query method by bi-directional search. Technical report, Technical Report of IEICE, ITS-2010-05 (2010) (in Japanese)
4. Guttman, A.: R-Trees: a dynamic index structure for spatial searching. In: Proc. ACM SIGMOD Conference on Management of Data, pp. 47–57 (1984)
5. Hart, P.E., Nilsson, N.J., Raphael, B.: A formal basis for the heuristic determination of minimum cost paths. IEEE Transactions of Systems Science and Cybernetics SSC-4(2), 100–107 (1968)
6. Ikeda, T., Hsu, M.-Y., Imai, H., Shimoura, S.N.H., Hashimoto, T., Tenmoku, K., Mitoh, K.: A fast algorithm for finding better routes by AI search techniques. In: 1994 Vehicle Navigation & Information System Conference, pp. 291–296 (1994)
7. Li, F., Cheng, D., Hadjieleftheriou, M., Kollios, G., Teng, S.-H.: On Trip Planning Queries in Spatial Databases. In: Medeiros, C.B., Egenhofer, M., Bertino, E. (eds.) SSTD 2005. LNCS, vol. 3633, pp. 273–290. Springer, Heidelberg (2005)
8. Ohsawa, Y., Fujino, K.: Simple trip planning algorithm on road networks without pre-computation. IEICE Transactions on Information and Systems J93-D(3), 203–210 (2010) (in Japanese)
9. Papadias, D., Zhang, J., Mamoulis, N., Tao, Y.: Query processing in spatial network databases. In: Proc. 29th VLDB, pp. 790–801 (2003)
10. Sharifzadeh, M., Kalahdouzan, M., Shahabi, C.: The optimal sequenced route query. Technical report, Computer Science Department, University of Southern California (2005)
11. Sharifzadeh, M., Kolahdouzan, M., Shahabi, C.: The optimal sequenced route query. The VLDB Journal 17, 765–787 (2008)

A Packaging Approach for Massive Amounts of Small Geospatial Files with HDFS

Jifeng Cui, Yong Zhang, Chao Li, and ChunXiao Xing

Department of Computer Science and Technology
Tsinghua University
Beijing 100084, China
cjfhly@163.com, {zhangyong05,xingcx}@tsinghua.edu.cn,
lichao00@tsinghua.org.cn

Abstract. The efficiency of dealing with massive small geospatial files deeply affects the performance of Web Geography Information System (WebGIS). The Hadoop Distributed File System (HDFS) is scalable to satisfy the requirement of massive data files storage, but not efficient in dealing with small files. In this paper, we proposed a method to pack a group of small files into one large logical file, and set up Hilbert spatial index inside the block with their spatial adjacency relation. The experimentation proved that this method reduces the size of block indices and increases the speed to search and retrieve the massive small spatial files.

Keywords: Hadoop Distributed File System(HDFS), massive small geospatial files, Hilbert spatial index, packaging approach.

1   Introduction

Modern science and technology provide many advantages for obtaining geospatial information, and the development of Internet of Things (IoT) requires massive information management. For example, satellites collect petabytes of geospatial data every day, while remote sensors and urban sensing activities are accumulating data at a comparable faster pace[1]. The massive geospatial files needs the support of corresponding storage technology and information retrieval technology to serve the WebGIS application. The massive network storage system has attracted more attention to solve this problem. Depending on cluster, grid and distributed file system, data storage and information retrieval services with a uniform interface[2]are provided to cooperate the different storage equipments and software together .

The Hadoop Distributed File System (HDFS) is a distributed file system designed to be deployed on low-cost hardware[3] with high fault-tolerance. HDFS is widely used in research for its open source and advanced architecture. It mainly settles the massive storage problem and data consistency. However, for massive small spatial files, HDFS does not perform well. In addition, lacking of consideration of the spatial relations among file objects, the efficiencies of storage utilization and data retrieval are low for spatial data, which cannot satisfy WebGIS.

H. Gao et al. (Eds.): WAIM 2012, LNCS 7418, pp. 210–215, 2012.

In this paper, according to the spatial adjacency between geospatial files, we proposed a method to pack a group of small files into one large logical file, and set up block inner index. The experimentation proved that this method reduces the size of block indices and increases the speed to search and retrieve the massive small spatial files.

2   Related Work

In distributed file systems, metadata is used to describe the storage file's content and location, which plays an important role for data locating and filename searching. Currently, distributed file systems place the metadata on the metadata server. The typical structure of metadata servers is master/slave, and map-reduce is used as its computing model to locate the file[4].

There are two types of file systems used in data intensive applications[7], general parallel file system and distributed file system. A general parallel file system is designed for High Performance Computing applications which run on large clusters with the needs of high scalability and concurrent storage I/Os. Examples of general parallel file systems include Sun's LustreFS[11] and the open source Parallel Virtual file system (PVFS) [10].

Distributed file system is widely used in Internet services. Google File System (GFS) [12], Hadoop Distributed File System(HDFS) [8] and Amazon Simple Storage Service (3S) [16]are typical paragons which support cloud computing environment. Unfortunately, these three file systems can not satisfy users' requirements on processing massive numbers of small files. Whereas, in the cloud environment, lots of enterprises would like to publish their files, most of which are consisted of small files. Dissatisfactory performance of handling small files becomes a bottleneck of GFS, HDFS and 3S, in progress towards cloud applications.

Researches on small file storage based on HDFS can be classified into two categories: general solutions and special solutions to meet the demand of specific applications. The former includes Hadoop Archives (HAR), SequenceFile and MapFile[8]. A HAR is a file archiving facility that packs files into HDFS blocks, which contains metadata. A SequenceFile provides a persistent data structure for binary key-value pairs. It uses file names as the key and file contents as the value to support compressing and decompressing at record level or block level. A MapFile is a type of sorted SequenceFiles with an index to permit lookups by key. It consists of an index file and a data file. The data file stores key-value pairs as records, which are sorted in key order. As for special solutions, recent papers[13][14][15] are proposed some attracting solutions we present below.

3   Solution for Data Disposal

We should first access the metadata server to locate the files before accessing files. Since the massive number of small files makes the metadata table very large, the metadata server's performance may be the system's bottleneck. Therefore, we take two steps to relieve the bottleneck. This strategy manages the metadata based on the analysis of the geospatial information storage and applications, and develops an engine on HDFS for accessing geospatial files. The engine can help us manage the massive amount of small files effectively.

We present the metadata catalog strategy and design index structure.Firstly, we pack a group of small files, which locates in the same area, into one large file, and order the small files by the geography map splitting criteria. The order of each file denotes the file's location in this area.   For example, we split the region 1 into 12*12 tiles, and give a number with row and column for every tile. Therefore, the order of every file in the group has a unique code. We order them by the Hilbert coding rule. With this method every small file has an order in the group. We store their sizes and offsets in block into the index file, which is stored in the data node.   The indices in the blocks and the storage for these groups are shown in Fig 1.

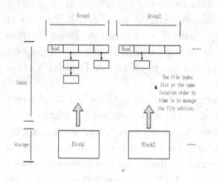
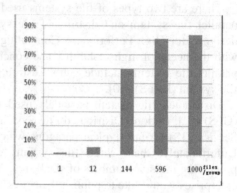

Fig. 1. Organization of i ndices for small files

Fig. 2. The space efficiency

After the packing step, we set up the order of the group with hashing method, so that the file index will have two levels. The first level is the group index, and the second is the files' order in the group. We only store the group id and block-id in the group index, in order to guarantee that the small files' index file can be found when needed.

In the experiment, we use the hashing algorithm to get the index file, and use the HDFS to store the packed files and index files.

4   The Experimentation Results and Analysis

We set up a HDFS system and conduct experimentations. There are three DataNodes and one NameNode in our HDFS: A IBM Server (2 Intel CPU 3. 20GHz, 2 GB memory and 1 TB disk), acts as NameNode;   DELL Servers (2 Intel CPU 2. 00GHz, 1 GB memory and 500GB disk),   act as DataNodes. The hadoop file system's version is 0.20.2,there are 178560 files all together with the total size is 47. 5GB involved in the experiments.

In our HDFS, every file is stored in one or more blocks. If we did not pack the small files into one logical file, the space of storage would be wasted. In addition, a proper block size should be set to minimize the number of blocks so as to improve the efficiency of metadata searching. We calculate the storage efficiency for different number of files group. The files' size is 1024K for every group, and the block size is 64M. The ratio of space utilizing is shown in the average value as fig 2.

As what we can assume from the graph, if the size of group does not reach the size of block, the ratio is increasing with the number of files. When the size of group exceeds it, the ratio of space utilization is only affected by the times of the group size dividing the block size.

We put the files into a list of directories. Each directory is a group of files, which are packed into one block. One group has 12 files or the times of 12 files because of the quad-splitting of spatial index. The time of writing files in HDFS is shown below. Y-axis is the time cost in minutes; x-axis is the number of files. The time here also includes the packing time.

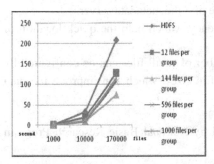

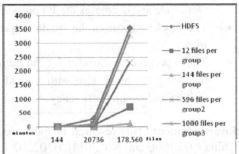

Fig. 3. Time of writing small files   Fig. 4.   Time of reading small files

As shown in Fig 3, the small files packing method is effective for time saving, especially for the groups with large numbers of small files in one block. We test the time of creating index for every group of files. Result show that it only took about 443 seconds to pack files and create the file index for 17856 files. Obviously, the time is comparatively short for such number of files comparing with the writing time.

We read files one by one and get the file content from the file list in memory. In experiments, we read files' group first, get the file's offset from the index files, and then obtain the file's content from the group. With the different number of files in each group, we use the same strategy as writing. The time of reading the files is shown below. Y-axis is time cost in seconds; X-axis is the number of files.

As shown in the figure, the time of reading files from HDFS in groups is shorter than that not in groups. Yet it is not obvious when the number of files exceeds 1000. In addition, the group number has little influence for reading.

Packing the files with spatial adjacency into one logical file can improve the small files accessing efficiency, but the number of small files for one logical file is limited,

When the size of logical file is larger than a block size, the writing speed of small files descends, and the reading speed for each small file slows down quickly. When the size of logical file is three times of block size, the efficiency is almost the same with the strategy without group. Therefore, we should select appropriate number of small files and the size of logical file relative to block size. A certain space for some small files' different backup is required to be reserved as well, which will improve data accessing.

5   Conclusions

In this paper, we introduce the metadata group strategy with HDFS, which uses a two-steps index for geospatial small files metadata management. The experiment shows that this strategy is effective for geospatial small files' management in the following aspects:

- The two level index model, which distributes the index of NameNode to DataNode and decreases the IO pressure of the metadata server.
- The block index uses the spatial adjacency, which help the quick location of files
- The appropriate parameter for the number of small files can be configured by balancing the block size with logical file size, which can improve the data writing and reading efficiency.

Acknowledgment. This research is sponsored by National Basic Research Program of China (973 Program) No. 2011CB302302

References

1. Yanga, C., Goodchildb, M., Huanga, Q., Nebertc, D., Raskind, R., Xue, Y., Bambacusf, M., Faye, D.: Spatial cloud computing: how can the geospatial sciences use and helpshape cloud computing? International Journal of Digital Earth 4(4), 305–329 (2011)
2. Siddhisena, B., Warusawithana, L., Mendis, M.: Next generation multi-tenant virtualization cloud computing platform. In: Advanced Communication Technology (ICACT), pp. 405–410 (2011)
3. Armbrust, M., Fox, A., et al.: Above the Clouds: A Berkeley View of Cloud Computing, Technical ReportNo. UCB/EECS-2009-28, University of California at Berkley (2009)
4. Dean, J., Ghemawat, S.: MapReduce: Simpli_ed Data Processing on Large Clusters. In: OSDI (2004)
5. Dick, M.E.: Leveraging P2P overlays for Largescale and Highly Robust Content Distribution and Search. In: VLDB 2009, p. 1059 (2009)
6. Yang, C.P., Raskin, R., Goodchild, M.F., Gahegan, M.: Geospatial Cyberinfrastructure: Past, present and future. Computers, Environment and Urban Systems 34(4), 264–277 (2010)

7. Amirian, P., Alesheikh, A., Bassiri, A.: Interoperable Exchange and Share of Urban Services Data through Geospatial Services and XML Database, Complex. In: 2010 International Conference on Complex, Intelligent and Software Intensive Systems, pp. 62–68 (2010)
8. Zhang, J., You, S., Gruenwald, L.: Indexing large-scale raster geospatial data using massively parallel GPGPU computing. In: Proceedings of the 18th SIGSPATIAL International Conference on Advances in Geographic Information Systems (GIS 2010), pp. 450–453. ACM, New York (2010)
9. Hadoop archives,
   `http://hadoopapache.org/common/docs/current/`
   `hadoop_archives.html`
10. Lopes, P.A., Medeiros, P.D.: pCFS vs. PVFS: Comparing a Highly-Available Symmetrical Parallel Cluster File System with an Asymmetrical Parallel File System. In: D'Ambra, P., Guarracino, M., Talia, D. (eds.) Euro-Par 2010, Part I. LNCS, vol. 6271, pp. 131–142. Springer, Heidelberg (2010)
11. Von Laszewski, G.: Concurrency and Computation: Practice and Experience. Special Issue: Grid Computing. High Performance and Distributed Application 22(11), 1433–1449 (2010)
12. Ghemawat, S., Gobioff, H., Leung, S.-T.: The Google File System. In: SOSP 2003, Bolton Landing, NewYork, USA, pp. 29–43 (October 2003)
13. Liu, X., Han, J., Zhong, Y., Han, C., He, X.: Implementing WebGIS on Hadoop: A Case Study of Improving Small File I/O Performance on HDFS. In: IEEE International Conference on Cluster Computing and Workshops, CLUSTER 2009, pp. 1–8 (2009)
14. Dong, B., Qiu, J., Zheng, Q., Zhong, X., Li, J., Li, Y.: A Novel Approach to Improving the Efficiency of Storing and Accessing Small Files on Hadoop: a Case Study by PowerPoint Files. In: 2010 IEEE International Conference on Services Computing, pp. 65–72 (2010)
15. Jiang, L., Li, B., Song, M.: The Optimization of HDFS Based on Small Files. In: Proceedings of IC-BNMT 2010, The 3rd IEEE International Conference on Broadband Network& Multimedia Technology, pp. 912–915 (2010)
16. DeCandia, G., Hastorun, D., Jampani, M., Kakulapati, G., Lakshma, A., Pilchin, A., Sivasubramanian, S., Vosshall, P., Vogels, W.: Dynamo: amazon's highly available key-value store. In: Proceeding SOSP 2007 Proceedings of Twenty-First ACM SIGOPS Symposium on Operating Systems Principles, vol. 41(6), ACM, New York (2007)

Adaptive Update Workload Reduction for Moving Objects in Road Networks*

Miao Li, Yu Gu, Jia Xu, and Ge Yu

Northeastern University, ShenYang, LiaoNing 110819, China
limiao@research.neu.edu.cn, {guyu,xujia,yuge}@ise.neu.edu.cn

Abstract. Location-Based Service (*LBS*) in road networks has many dazzling applications. Under the context of road networks, while many approaches have been proposed to speed up the location-based queries, to the best of our knowledge, none of them pays attention to the updating protocol, which is a vital aspect directly impacting the system performance. In this paper, we focus on designing adaptive updating protocol to reduce the updating workload between moving objects and the database server. We build an effective motion model, called Road-Network Safe Range (*RNSR*) for each object. The *RNSR* enables large space tolerance for the moving objects. Extensive experiments using real-world dataset justify that our proposals apparently cut down the system updating workload while still guarantee a certain query accuracy.

1 Introduction

Nowadays, more and more location-based services (*LBSs*) which rely on the locations of the moving objects have facilitated our daily life. Compared with the *LBS* problem in an ideal Euclidean space, *LBS* under the context of road network is more practical with many dazzling applications, such as the traffic management which is illustrated in the following example.

Example. By knowing the location and velocity of each vehicle, we can predict their potential locations after a period of time. Based on those predicted locations, drivers on the road can be guided to avoid the traffic jam.

In this paper, we focus on the *Predicted Range Queries*. While a variety of works have been proposed to speed up location-based queries [1] [2], the updating protocol has not been sufficiently studied. In [3] and [4], Chen et al. proposed adaptive updating protocol to reduce the updating workload on an ideal Euclidean space. However, under the context of road network, the moving trajectory of each object is constrained by the road shape, which can help us better design the updating protocol. We summarize our contributions as follows:

First,we design a *Location-based Updating Protocol* for the road networks. The protocol utilizes the trajectory similarities between neighboring objects to

---

* The National Natural Science Foundation of China (61003058); the Fundamental Research Funds for the Central Universities(N100704001,N110404006).

H. Gao et al. (Eds.): WAIM 2012, LNCS 7418, pp. 216–221, 2012.

estimate the updating frequency of the system. Second, we propose a *Velocity-based Updating Protocol* which employs the velocity information of each object to estimate the system updating frequency. We perform extensive experiments using real-world datasets to verify our two proposals.

2   Related Work

There are many papers which have given solutions for the safe region selection and optimization [5] [6] [7]. However, all these methods are designed and implemented for the ideal Euclidean space.

In literature [3], Chen et al. proposed an updating protocol which is constructed on the basis of the safe region (called $STSR$). The $STSR$ is built for every moving object to approximate its location. Under the setting of $STSRs$, a larger $STSR$ introduces less active updates but more passive updates, and a smaller $STSR$ causes less passive updates but more active updates. Based on such intuition, a cost model is built to estimate both of the system total updating cost and the composition of active updates and passive updates. In [8], a updating protocol was designed for solving the similarity pair queries for the application of discovering alliances for online game players. Its updating protocol is also constructed on the basis of safe regions. Although this approach is adaptive under different parameter settings, it did not consider the velocity information of each object which may help to improve the cost model for better safe region adjustment.

Amongst all existing works, the work [3] by Chen et al is most relevant to ours, since we both consider supporting the predicted range queries and minimizing the updating workload.

3   Preliminaries

We assume there are n moving objects $O\{o_1, o_2, ..., o_n\}$,being monitored in the system. The time axis is sliced into snapshots at discrete times $T\{0, 1, ..., t, ...\}$. Based on the motion properties of the moving objects, we introduce the Road-Network Safe Range $(RNSR)$ to describe the states of the moving objects.

Each moving object o_i has a $RNSR$, which is denoted by $R(o_i) = \{k(path_j), LS, VLS, t_r, t_e\}$. Here, $k(path_j)$ is the slope of the $path_j$ with o_i on; LS is a line segment on the road network; VLS indicates the both endpoints' velocity of the LS, t_r is the reference time; and t_e is the expiration time $(t_e > t_r)$.

On the basis of [3], we extend the basic update protocol. Our updating protocol also considers the following types of update, namely the active update and the passive update.

Active Update. If the object o_i moves out of its $R(o_i)$ (in the same path), it will incur the active update.

Passive Update. A passive update is issued when the database conducts a query. At the query time t, when the predicted $R(o_i)$ partially intersects with the query line segment, it cannot determine whether the moving object o_i is in the query line segment. Thus the server will send a probing message to the object o_i. As long as the server receives the message, and then o_i gives its exact location to the server and updates the predicted line segment.

4   Optimization Techniques

We consider that the moving objects on the same path have similar speed. Each object should review whether its current location is still in its $RNSR$ in each time period. If the object o_i does not stay in the $R(o_i)$, o_i will send an update message to the server in order to inform its location and reassign a new $RNSR$ to o_i.

Based on the above analysis, we can infer two different probabilities for the active update.

Location-Based Active Update Probability Calculation. The probability of inconsistency before the expiration time t_e for an object o_i is denoted as $P_a(R(o_i))$. We define the moving objects intersecting with $R(o_i)$ as similar records $R_{NN}(o_i)$ to o_i. However, the similar records $R_{NN}(o_i)$ can not get the active update probability of the object o_i accurately. Therefore, we give the covered records $R_k(o_i)$ from $R_{NN}(o_i)$ to solve this problem as follows.

Definition 1: If the $R(o_i)$ completely covers its similar records $R_{NN}(o_i)$ and its current location, these records are called covered records R_k. The value of R_k is denoted by $| \{R_k(o_i) \in R_{NN}(o_i) \mid R(o_i) \supseteq R_k(o_i)\} |$.

We get the active update probability from the historical perspective as $P_a(R(o_i))$
$= 1 - \frac{|\{R_k(o_i) \in R_{NN}(o_i) | R(o_i) \supseteq R_k(o_i)\}|}{|R_{NN}(o_i)|}$.

Velocity-Based Active Update Probability Calculation. We introduce two parameters λ and ϵ. An initial $R(o_i)$ is initialized to a region centralized at the object o_i with a length of 2λ and ϵ denotes the speed increment. The two endpoints' velocities of a line segment is $< v - \epsilon, v + \epsilon >$. Given the current velocity of the object o_i, we can calculate the active update probability following formula, $P_a(R(o_i)) = \prod_k p_k \cdot (1 - p_s)$. Here, p_s is the probability that o_i and similar records have the same velocity range; p_k is the probability that o_i and similar records have the different velocity ranges.

Passive Update Probability. According to the type of the active update, the passive update probability is also divided into the following two cases: probability based on location and probability based on velocity.

(1)Probability Based on Location

When the query line segment is sent to the database on the road network, the passive update is issued. That is, the predicted line segment partially overlaps the query line segment. We will estimate the number of the objects per meter. If the $RNSR$ completely covers the query line segment or does not cover the query line segment, we do not need to update the $RNSR$.

Assume the update time of the $R(o_i)$ is t_u, and the length of the object o_i on the road network is Len with $Len(o_i) = 2\lambda \cdot (t_u - t_r)$. During this period, the probability of a passive update is: $P_p(R(o_i)) = \min\{\frac{Len(o_i) \cdot K}{(t - t_r) \cdot L}, 1\} \times (1 - P_a(R(o_i)))$

Here L is the whole length of the path which o_i belongs to; K is the expected number of queries happening at each timestamp.

(2) Probability Based on Velocity

Probability based on velocity is similar to the probability based on location. The length of the object o_i at the update time t_u on the path is $Len(o_i) = \sum_{t=t_u}^{t_e} [2\epsilon \cdot (t - t_r) + 2\lambda]$. The probability of a passive update under this case is the same to that of the probability based on location.

Cost Model. According to our analysis, the active update cost of the objects is $Cost_{P_a} = \sum_{i=1}^{N} P_a(R(o_i))$, where N is the active update number of the objects. And the passive update cost is $Cost_{P_p} = \sum_{i=1}^{N} P_p(R(o_i)) \times 2$, where 2 indicates: the probing message is sent to the object and an update message is sent back from the object. So the total cost is $Cost_{total} = Cost_{P_a} + Cost_{P_p}$.

5   Optimization of $RSNR$

In this section, we present two methods to assign a suitable $RNSR$ for each object and to calculate the updating probability and cost.

5.1   Optimization Based on Greedy Algorithm

As the solving process of Greedy algorithm is simple and fast, we uses this method combined with a United Space for optimization to find the suitable $RNSR$ for each moving object. Now, we give the formal a definition of the United Space.

Definition 2: United Space (US) is a line segment, which is composed by the intersection of similar records on a path and this intersection must cover the location of the object o_i.

Define the minimum value of $u_i.LS$ as $u_{min}.LS$ using Greedy algorithm. If the record u_i has been chosen, we no longer consider it. Thus, we select the minimum $u_i.LS$ as $u_{min}.LS$ every time and add it to $RNSR.LS$ until there is no element in the US or the workload can not be accepted by the system. Finally, we can obtain the initial $RNSR$.

5.2   Optimization Based on TS

On the basis of Greedy algorithm, better $RNSR$ can be found for each object by $TABU$ Search (TS) algorithm, which absorbs inferior solution to jump out of local optimal state.

First, we use the Greedy Algorithm to determine the initial $RNSR$ for each object. Second, we discuss the algorithm based on TS. The aim of this part is to extend the $RNSR.LS$ by selecting records similar to object o_i (details are illustrated in Algorithm1). To judge whether the similar records have intersection with the United Space elements, we should calculate the workload of new $Cost_{total} R(o_i)$ until its workload can not be accepted by the system (i.e. the termination condition.).

Algorithm 1. Active Update based on TS

1. Search the United Space
2. Search the initial $RNSR.LS$
3. D_{old}=the initial $RNSR.LS$
4. **for** the current $P_{total}(D_{old}) \leq \beta^*$ **do**
5.   Include $min\{D_{old} + u_i.LS\}$ into $RNSR.LS$
6. **if** $RNSR.LS$ is the initial $RNSR.LS$ **then**
7.   **for** the current $P_{total}(D_{old}) \leq \beta^*$ **do**
8.   Include $min\{D_{old} + u_i.LS\}$ into $RNSR1.LS$
9.   **if** $RNSR1.LS$ is the initial $RNSR1.LS$ **then**
10. $D = the initial RNSR.LS$
11.   **else**
12. $D = RNSR1.LS$
13. **else**
14. $D = RNSR.LS$
15. **for** the current $P_{total}(D) \leq \beta^*$ and U_{AL} is not NULL **do**
16.   Include $min\{D_{old} + u_i.LS\}$ in D
17. **if** find the new D then **then**
18.   return the optimal $RNSR.LS = D$
19. **else**
20.   **for** the current $P_{total}(D) \leq \beta^*$ and U_{AL} is not NULL **do**
21.   Include $min\{D_{old} + u_i.LS\}$ in $D1$
22.   **if** find the new $D1$ **then**
23.   return the optimal $RNSR.LS = D1$
24.   **else**
25.   return the optimal $RNSR.LS = D$

---

6   Experimental Study

In this section, we evaluate the updating cost of our proposed algorithms. We use the real dataset is provided by $R - tree$ Portal[1].

In this paper, our algorithms are implemented in C++ and run on a PC with 2.4GHz Inter(R) Core(TM)2 Duo CPU, 2G RAM and 160G disk.

Figure 1 illustrates that different settings of ϵ in $Velocity - U$ lead to different numbers of active update. As can be seen from the figure, the greater ϵ is, the fewer the number of active update is. This is because as the ϵ becomes larger, $R(o_i).LS$ is also increased, and thus the object o_i does not need to update for some time. And the number of active update has stabilized over time. In Figure 2, we can see that different λ can also result in different total costs.

Figure 3 indicates the system cost when the objects move in different intervals of speed. We can see that all these three methods obey the variation tendency that the faster speed can lead to a higher total cost. And, as the speed value increases, the total cost of update will increase linearly. Also, it can be inferred that the total cost of $Velocity - U$ is smaller than the other two methods easily. Figure 4 makes a comparison of $Location-UG$, $Location-UT$ and $Velocity-U$

---

[1] http://www.rtreeportal.org/index.php

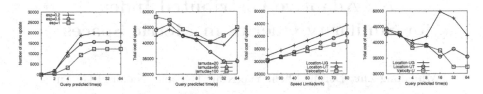

Fig. 1. Different ϵ **Fig. 2.** Different λ **Fig. 3.** Result   **Fig. 4.** Result

by using different query predicted time. We can observe that the total update cost will be generally decreased with the longer query predicted time.

7   Conclusion

In this paper, we present an extended updating protocol for the moving object databases under the context of the road networks. We use $RNSR$ (Road-Network Safe Range) for each object to quantify its movement characteristics. Based on the updating protocol, two methods are designed for reducing the system workload. Extensive experiments are conducted to verify that both of our proposals are more effective and feasible than the basic updating protocol.

References

1. Zhang, Z., Yang, Y., Tung, A.K.H., Papadias, D.: Continuous k-means monitoring over moving objects. IEEE Trans. Knowl. Data Eng. 20(9), 1205–1216 (2008)
2. Hu, H., Xu, J., Lee, D.L.: A generic framework for monitoring continuous spatial queries over moving objects. In: SIGMOD Conference, pp. 479–490 (2005)
3. Chen, S., Ooi, B.C., Zhang, Z.: An adaptive updating protocol for reducing moving object databases workload. PVLDB 3(1), 735–746 (2010)
4. Zhang, Z., Cheng, R., Papadias, D., Tung, A.K.H.: Minimizing the communication cost for continuous skyline maintenance. In: SIGMOD Conference, pp. 495–508 (2009)
5. Amir, A., Efrat, A., Myllymaki, J., Palaniappan, L., Wampler, K.: Buddy tracking - efficient proximity detection among mobile friends. Pervasive and Mobile Computing 3(5), 489–511 (2007)
6. Küpper, A., Treu, G.: Efficient proximity and separation detection among mobile targets for supporting location-based community services. Mobile Computing and Communications Review 10(3), 1–12 (2006)
7. Xu, Z., Jacobsen, H.-A.: Adaptive location constraint processing. In: SIGMOD Conference, pp. 581–592 (2007)
8. Yiu, M.L., Leong Hou, U., Saltenis, S., Tzoumas, K.: Efficient proximity detection among mobile users via self-tuning policies. PVLDB 3(1), 985–996 (2010)

An Adaptive Distributed Index
for Similarity Queries in Metric Spaces

Mingdong Zhu, Derong Shen, Yue Kou, Tiezheng Nie, and Ge Yu

College of Information Science & Engineering, Northeastern University, China
dr.zhumd@gmail.com, {shenderong,kouyue,nietiezheng,yuge}@ise.neu.edu.cn

Abstract. As the amount of data is growing rapidly, efficient and scalable index structures for managing large-scale data are attracting more and more attention. To efficiently query and manage the data in metric spaces, an adaptive distributed index, MT-Chord, is proposed. MT-Chord integrates Chord based routing protocol and M-tree based index structure to support efficient similarity query processing in metric spaces. Each index node has multiple replicas for load-balance and a cost model is presented to dynamically tune the number of replicas based on the query and update pattern at the granularity of each index node. MT-Chord is a truly scalable, efficient and adaptive distributed index structure for query processing in metric spaces, which is verified by our extensive experimental studies on three real-life datasets extracted from different data sources.

Keywords: distributed index, metric space, similarity query.

1 Introduction

Driven by growing demand for efficiently querying and managing large-scale data, the distributed storage systems have received considerable attention. And a lot of systems are proposed, such as Google's BigTable, Yahoo's Pnuts and Amazon's Dynamo, they store data with key-value model and have scalability and fault-tolerance. But they only support simple key-value queries instead of similarity queries such as range queries and KNN queries. For example, it often happens that a traveler wants to find the restaurants from which the distance is less than 1 km or the two of the nearest restaurants. That's to say, the user needs to issue a range query with $range = 1$ km or a KNN query with $k = 2$. To support these queries in the large-scale dataset with high-dimensional or unstructured data, an efficient distributed index structure is pressing needed.

As for querying high-dimensional or unstructured data, such as locations, images, micro-blogs, protein sequences, and so on, there are basically two kinds of index structures: vector based index structures(spacial access methods) [1, 2] and distance based index structures [3,4]. To obtain scalability these index structures are extended by integrating with p2p overlay, and many distributed index structures are proposed. For example, [5] integrates R-tree and CAN overlay, [6]

H. Gao et al. (Eds.): WAIM 2012, LNCS 7418, pp. 222–227, 2012.

integrates quad-tree and Chord overlay, and [7] combines GHT and Chord overlay. Unfortunately, existing vector based distributed index structures cannot efficiently query high-dimensional data [5]. What's more, some complex unstructured data cannot be precisely represented by the vector model. Existing distance based distributed index structures [7–10] have to select pivots in advance, and if the distribution of data changes after updating, the index structure needs to be totally rebuilt or the performance cannot be guaranteed. That's to say, they are static rather than dynamic.

Hence, a novel distributed index, MT-Chord, is proposed to support range queries and KNN queries for large-scale complex data. MT-Chord integrates Chord [11] based routing protocol and M-tree [3] based index structure by disseminating the M-tree index nodes to Chord overly. The contributions of this paper are summarized as follows. (1) A novel distributed index structure, MT-Chord, is presented. (2) A cost model and an adaptive tuning algorithm are put forward which can dynamically definite the optimal number of replicas for each index node and tune it accordingly. (3) Results of extensive experimental studies on real-life datasets demonstrate the efficiency and scalability of the proposed index structure and algorithms.

The rest of this paper is organized as follows: Section 2 presents related works and Section 3 shows the structure of our index structure. Section 4 proposes the cost mode. Section 5 presents the experimental studies and we conclude in Sect. 6.

2 Related Work

To satisfy with similarity queries for large-scale data, various index structures are proposed. For example, [6] integrates quadtree index with Chord overlay to enable more powerful accesses to data in p2p networks. [5] integrates R-tree and CAN overlay to process multi-dimensional data in a cloud system. [12] combines B-tree and BATON overlay to provide a distributed index which has high scalability but incurs low maintenance. They both choose a part of local index nodes to build global index node by computing the cost model. In metric spaces, [13] partitions the data space into clusters and selects a reference point for each cluster, and every data object is assigned a one-dimensional key according to the distance to its clusters reference object. [9] integrates the idea of [13] and Chord overlay to distribute the storage space and parallelize the execution of similarity search. [8] proposes a mapping mechanism that enables to actually store the data in well-established structures such as the B-tree. [7] proposes a distributed index, GHT* index, which can exploit parallelism in a dynamic network of computers by putting a part of the index structure in every network node.

3 MT-Chord Index Structure

MT-Chord index is a type of distributed M-tree index built on a shared-nothing cluster which is organized by the Chord overlay. The key problem of MT-Chord index lies in query processing and determining the optimal number of replicas of each index node, which are detailed in Sect. 4. In this section the structure of MT-Chord is presented.

Each index node has multiple replicas, and MT-Chord is built by mapping each replica to a Chord node using consistent hashing. Figure 1 depicts some mappings between index nodes and Chord nodes. As shown in Fig. 1, the data are logically organized by M-tree index, while they are physically stored in the Chord overlay. Every index node has at least three replicas. When a new index node is created during object insertion, it is assigned a node ID (nid) and initially it has three replicas whose replica ID (rid) are 0, 1 and 2 respectively. Each replica is mapped to the corresponding Chord node by a hashing function taking nid and rid as parameters. For example, in Fig. 1, the index node 4 has 4 replicas, and these replicas are mapped to Chord node b, d, g and h through a hashing function H("(nid, rid)"). As for the hashing function, SHA-1 can be adopted, because it has the capability to map two index nodes with similar ID numbers to totally different Chord nodes [6].

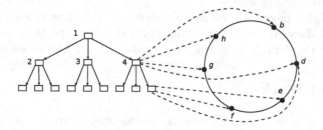

Fig. 1. Some mappings between logical M-tree and physical Chord overlay

Considering query processing, let rn_i denote the number of replicas of the index node i, where i is the node ID. These replicas are not only used to guarantee the robustness of MT-Chord in case of node failure, but also used to balance the query load. For example, if a client issues a query to a index node i, firstly the client randomly chooses a value $v \in [0, rn_i - 1]$, and then sends the query to the Chord node corresponding to H("(i, v)"), say, n. If n's location is in the client's cache, the query will be directly sent, otherwise the location should be acquired by using Chord routing protocol in advance. In the similar way, the query should be sent from n to one or more child nodes chosen by computing the distance between the query and the child nodes, recursively, until leaf nodes are reached and the query is answered.

Intuitively, more replicas can increase query throughput by balancing query load, however, at the same time, they increase the cost of update because more data need to be updated. Hence, a balance should be found. Generally for a relatively stable tree index, the index nodes close to root node are more likely to be queried than be updated while for the index nodes close to the leaves it's just the opposite, which make the intent of adaptive replication feasible.

To determine the optimal number of replicas for each index node, periodically an algorithm of cost estimation is called as a basis for increasing or decreasing replicas, which is detailed in Sect. 4.

4   Adaptive Index

Now we consider the cost of replication in the distributed environment. We do this by estimating the number of messages which are relevant to replicas of index nodes in the network. For our index structure, the number of messages is linearly proportional to the number of distance computation, hence it can reflects both network and computation cost.

In query processing, if the number of messages received by a Chord node exceed its processing capacity, the excess messages will be discarded and be reissued. So we can estimate the query cost as: $cost_q = (\frac{qn_i}{rn_i} - cap) \cdot H_i$, where qn_i is the number of queries in a time unit through index node i, cap is the number of messages which the Chord node can process before a timeout happens and H_i is the hight of the index node i. When index nodes split or merge and index structure updates, synchronization is needed to keep replicas consistent, and A slightly mordified Paxos protocol is proposed to keep replicas consistent during update. One synchronization needs $csyn \triangleq 4 \cdot (4-1) + rn_i - 3 = rn_i + 9$. messages. We can divide the cost of update into two parts: the cost of splitting and the cost of merging. When an index node needs split, it needs two synchronizations of replicas and one notification to its parent. Similarly, when an index node needs merge, it needs two synchronizations of replicas and two notifications to its parent. So the cost of update is: $cost_u = [ps(2 \cdot csyn + n) + pm(2 \cdot csyn + 2 \cdot n)] \cdot un$, where un is the number of updates in a time unit, ps and pm is the probability of splitting and merging respectively, which can be estimated by using the random walk theory [12]. Finally, the total cost of index node i can be summarized as: $cost_i = (\frac{qn_i}{rn_i} - cap) \cdot H_i + [ps \cdot (2 \cdot csyn + n) + pm \cdot (2 \cdot csyn + 2 \cdot n)] \cdot un$. And when $rn_i = opt_i = \sqrt{\frac{qn_i}{2 \cdot un \cdot (ps+pm)}}$, $cost_i$ takes the minimum value.

After the optimal number of replicas is figured out, the current number of replicas will be set to the value.

5   Experiment

Three real-life datasets are used in our experiences. YouKu dataset: 500,000 images are obtained from YouKu videos which involve different categories such as film, music, education, and so on. MicroBlog dataset: 1,000,000 micro blogs are downloaded from Sina and Tencent which are short messages with at most 140 characters. DNA dataset: 8,130,000 protein sequences of length 64 are extracted from the largest, the smallest, and the median chromosome of human MT-Chord is compared with RT-CAN [5]. PeerSim is used to simulate a Chord overlay. All experiments are conducted with JAVA 1.6.0_25 and Intel(R) Core(TM) i7 Quad CPU 870 @2.93GHz and 8G RAM.

Performance of Queries. In this experiment, we evaluate the performance of range queries. Figure 2 shows the throughput with the different computing nodes for the three datasets. With increasing of the computing nodes, the throughput grows linearly, which confirms the scalability of MT-Chord. And the performance of MT-Chord is better than that of RT-CAN. In Fig. 3, different radii are adopted, and as expected, with increasing of the query radius the performance degrades.

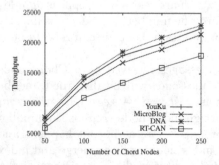

Fig. 2. Performance of query

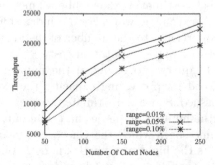

Fig. 3. Effect of range

Performance of Updates. In our algorithm, the process of update is similar to the range query with radius set to 0 and the involved computation cost is considerably smaller, so generally its performance is better than that of the range query. As shown in Fig. 4, the performance of updates in MT-Chord keeps good because of the adaptive tuning algorithm and is better than that in RT-CAN.

Effect of Dimensionality. The dimensionality of YouKu dataset varies from 20 to 100 to evaluate the performance of our index. As shown in Fig. 5 the performance of our index is better than that of RT-CAN and is not as sensitive to dimensionality as RT-CAN, although the performance is slightly degrade with increasing the dimensionality.

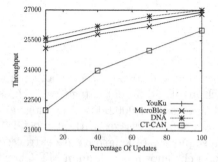

Fig. 4. Performance of updates

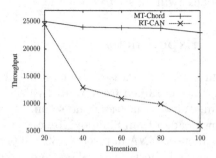

Fig. 5. Effect of dimensionality

6   Conclusion

MT-Chord is a distributed M-tree built on top of the Chord p2p overlay. As far as we know, MT-Chord is the first distributed and adaptive index in metric spaces which doesn't have to choose pivots before constructing the index. MT-Chord index can dynamically tune the number of replicas at the granularity of the index node. A cost model are proposed to estimate the cost. Extensive experiments are conducted, which verify the efficiency and scalability of MT-Chord.

Acknowledgements. This research is supported by the State Key Program of National Natural Science of China (61033007), the National Natural Science Foundation of China (60973021), and the Fundamental Research Funds for the Central Universities(N100704001).

References

1. Chen, G., Vo, H.T., Wu, S., Ooi, B.C., Özsu, M.T.: A Framework for Supporting DBMS-like Indexes in the Cloud. In: Proc. of VLDB, pp. 702–713 (2011)
2. Cao, Y., Chen, C., Guo, F., Jiang, D., Lin, Y.: A Cloud Data Storage System for Supporting Both OLTP and OLAP. In: Proc. of ICDE, pp. 291–302 (2011)
3. Ciaccia, P., Patella, M., Zezula, P.: M-Tree: An Efficient Access Method for Similarity Search in Metric Spaces. In: Proc. of VLDB, pp. 426–435 (1997)
4. Chiueh, T.C.: Content-based Image Indexing. In: Proc. of VLDB, pp. 582–593 (1994)
5. Wang, J., Wu, S., Gao, H., Li, J., Ooi, B.C.: Indexing Multi-dimensional Data in a Cloud System. In: Proc. of SIGMOD, pp. 591–602 (2010)
6. Tanin, E., Harwood, A., Samet, H.: Using a Distributed Quadtree Index in Peer-to-Peer Networks. VLDB J. (VLDB) 16(2), 165–178 (2007)
7. Batko, M., Gennaro, C., Savino, P., Zezula, P.: Scalable Similarity Search in Metric Spaces. In: Proc. of DELOS, pp. 213–224 (2004)
8. Novak, D., Batko, M.: Metric Index: An Efficient and Scalable Solution for Similarity Search. In: Proc. of SISAP, pp. 65–73 (2009)
9. Novak, D., Zezula, P.: M-Chord: A Scalable Distributed Similarity Search Structure. In: Proc. of INFOSCALE, pp. 181–190 (2006)
10. Falchi, F., Gennaro, C., Zezula, P.: Nearest Neighbor Search in Metric Spaces Through Content-Addressable Networks. Inf. Process. Manage. 43(3), 665–683 (2007)
11. Stoica, I., Morris, R., Karger, D.R., Kaashoek, M.F., Balakrishnan, H.: Chord: A Scalable Peer-to-Peer Lookup Service for Internet Applications. In: Proc. of SIGCOMM, pp. 149–160 (2001)
12. Wu, S., Jiang, D., Ooi, B.C., Wu, K.: Efficient B-tree Based Indexing for Cloud Data Processing. In: Proc. of VLDB, pp. 1207–1218 (2010)
13. Jagadish, H.V., Ooi, B.C., Tan, K., Yu, C., Zhang, R.: iDistance: An Adaptive B+-Tree Based Indexing Method for Nearest Neighbor Search. ACM Trans. Database Syst. 30(2), 364–397 (2005).

Finding Relevant Tweets

Deepak P.[1] and Sutanu Chakraborti[2]

[1] IBM Research - India, Bangalore
[2] Indian Institute of Technology, Madras
deepak.s.p@in.ibm.com, sutanuc@iitm.ac.in

Abstract. When a user of a microblogging site authors a microblog post or browses through a microblog post, it provides cues as to what topic she is interested in at that point in time. Example-based search that retrieves similar tweets given one exemplary tweet, such as the one just authored, can help provide the user with relevant content. We investigate various components of microblog posts, such as the associated timestamp, author's social network, and the content of the post, and develop approaches that harness such factors in finding relevant tweets given a query tweet. An empirical analysis of such techniques on real world twitter-data is then presented to quantify the utility of the various factors in assessing tweet relevance. We observe that content-wise similar tweets that also contain extra information not already present in the query, are perceived as useful. We then develop a composite technique that combines the various approaches by scoring tweets using a dynamic query-specific linear combination of separate techniques. An empirical evaluation establishes the effectiveness of the composite technique, and that it outperforms each of its constituents.

1   Introduction

Twitter[1] has now become a popular way of sharing breaking news, personal updates and spontaneous ideas and has been observed to function as a social sensor [1]. The *home timeline* of a twitter user is a stream of updates from users that she has chosen to follow. A list [2] denotes a group of users, and the stream associated with each list comprises tweets from users in the group. Home timelines and streams could contain tweets of users outside their purview whose tweets are *re-tweeted* (i.e., forwarded) by users that are being monitored by the stream. Despite numerous third-party applications, accessing content outside those generated or re-tweeted by users among lists or followees[2] is limited to keyword or hashtag[3] based search apart from other obvious mechanisms like visiting twitter user profiles to see tweets they have authored. All of these require some effort on the part of the user.

When a user authors a tweet, it provides valuable information about the topic that the user is currently interested in. We argue that tweets from the public

---

[1] https://twitter.com/
[2] If A follows B, we say that B is a followee of A.
[3] http://en.wikipedia.org/wiki/Hashtag

H. Gao et al. (Eds.): WAIM 2012, LNCS 7418, pp. 228–240, 2012.

timeline, i.e., even those outside the home timeline, that pertain to the topic of the authored tweet could provide information that pertain to the topic of interest (as embodied by the exemplary tweet just authored) of the user. In this paper, we address this problem of *finding tweets relevant/related to a particular tweet*. Retrieving similar/related entities to a *query* entity is a classical retrieval problem that has been widely addressed in several other domains such as general relational data [3] and program code [4] among others.

To the best of our knowledge, finding *related or relevant* tweets has not yet been looked into, in past research. Tweets are short messages limited to *140* characters with metadata such as timestamp, author handle and original author in case of re-tweets. The *140* character restriction poses a major challenge to traditional text search engines that are not robust to noise common in tweet data such as deliberate omissions of relevant words and mispellings. Our specific contributions are as follows:

- We propose various techniques to assess the relevance of a tweet wrt a query tweet that harness aspects such as time of the query tweet, the social network of the query tweet's author, and content-based similarity.
- An extensive empirical study of such techniques on real-world twitter data that reveals that content based techniques are most effective and that extra information not already present in the query tweet is generally welcome to the user.
- We then propose a composite approach for tweet relevance assessment that leverages the goodness of the various proposed techniques and illustrate empirically that the composite technique fares best.

2   Problem Definition

Given a tweet Q, a set of tweets T and k, we would like to identify an ordered set of k tweets from T, $T_Q = [t_1, t_2, \ldots, t_k]$, such that the tweets in T_Q are similar/relevant to Q. T is a set of tweets from the public timeline and could contain tweets outside the home timeline of the author of Q. In particular, we would like to develop a scoring function, $S(Q, t)$ that would be used to score every tweet $t \in T$ wrt Q, the score being directly related to estimated relevance of t to Q. Informally, T_Q contains the top-k tweets according to the $S(.,.)$ function ordered in a non-increasing order of their scores. In the rest of the paper, we will use D to denote a large corpus of tweets (that are not in T) that we will use to compute corpus-based stats to denote generic twitter behavior such as word idfs to be used in some of the techniques that we describe. We will use traditional IR quality metrics in our evaluation and will elaborate on them in a later section.

3   Scoring Tweets

We will now outline various intuitive scoring functions for tweets; the scoring functions that we develop exploit one of (1) time, (2) author and social network or (3) content of the tweets in question.

Table 1. Scoring Functions Overview

Type	Technique	Scoring Function
Time	TS	$\mathcal{S}_{TS}(\mathcal{Q}, t_i) = -1 \times |timestamp(t_i) - timestamp(\mathcal{Q})|$
Social Network	SC	$\mathcal{S}_{SC}(\mathcal{Q}, t_i) = \frac{|\mathcal{N}(Author(\mathcal{Q})) \cap \mathcal{N}(Author(t_i))|}{|\mathcal{N}(Author(\mathcal{Q})) \cup \mathcal{N}(Author(t_i))|}$
	GD	$\mathcal{S}_{GD}(\mathcal{Q}, t_i) = -1 \times GraphDist(Author(\mathcal{Q}), Author(t_i))$
Content	TC	$\mathcal{S}_{TC}(\mathcal{Q}, t_i) = \frac{\Sigma_w (tf.idf(\mathcal{Q})[w] \times tf.idf(t_i)[w])}{\sqrt{\Sigma_w (tf.idf(\mathcal{Q})[w])^2} \times \sqrt{\Sigma_w (tf.idf(t_i)[w])^2}}$
	QS	$\mathcal{S}_{QS}(\mathcal{Q}, t_i) = \Sigma_{w \in \mathcal{Q}}(f(t_i, w) \times idf(w))$
	ED	$\mathcal{S}_{ED}(\mathcal{Q}, t_i) = \prod_{w \in \mathcal{Q}}(1.0 + \sum_{w' \in t_i} sim(w, w') \times idf(w'))$
	WC	$\mathcal{S}_{WC}(\mathcal{Q}, t_i) = \prod_{w \in \mathcal{Q}}(1.0 + \sum_{w' \in t_i} p(w'|w) \times idf(w'))$
	RC	$\mathcal{S}_{RC}(\mathcal{Q}, t_i) = \prod_{w \in \mathcal{Q}}(1.0 + \sum_{w' \in t_i} p'(w'|w) \times idf(w'))$
	WS	$\mathcal{S}_{WS}(\mathcal{Q}, t_i) = \prod_{w \in \mathcal{Q}}(1.0 + \sum_{w' \in t_i} lesk(w', w) \times idf(w'))$

3.1 Time-Based Scoring (TS)

Taking cue from *reverse chronoligical ordering*, the standard presentation mode in most feedreaders and *twitter* website, time-based scoring uses temporal proximity as a proxy of relevance. This leads to an intuitive scoring function, $\mathcal{S}_{TS}$ in Table 1; since the absolute time-difference is inversely related to relevance, the -1 ensures that the scoring is directly related to the relevance.

3.2 Author Social Network Based Scoring

In most datasets that are dumps of just tweets, social network connections (e.g., *follows, likes* type relationships) are unknown. Thus, we outline a graph structure between twitter users based on the tweets using the following notion of edge:

$$A \leftrightarrow B \Rightarrow sentTweetTo(A, B) \vee sentTweetTo(B, A) \qquad (1)$$

$sentTweetTo(.,.)$ denotes that the first user addressed (i.e., sent to or mentioned) the second in a tweet. This, unlike the *follows* relationship, is undirected; we stick with this formulation for tweet relevance assessment since existence of either way of communication is intuitively indicative of being interested in similar topics.

Shared Connections (SC). The immediate neighborhood of a user in a social network largely defines her interests, since those are whom she has directly interacted with (according to our way of inducing links). The Jaccard similarity[4] between the set of immediate neighbors between the author of the Q and the candidate tweet t_i can then be used as the scoring function S_{SC}; $\mathcal{N}(Author(t))$ denotes the set of immediate neighbors of the author of the tweet t according to Eq. 1. Among tweets that have no shared connections with the query, recency based ordering is employed.

Graph Distance (GD). Jaccard similarity does not differentiate between tweets whose authors do not have any shared connections with the author of Q. However, tweets from users who are just a few hops away may be more relevant than those who are further away; a scoring function that assesses relevance of tweets as inversely related to the distance between authors would incorporate such a notion. Such an intuitive scoring function S_{GD} uses $GraphDist(.,.)$, the minimum number of hops between authors, in its formulation.

3.3   Content-Based Scoring

The *140* character restriction in Twitter leads to deliberate infusion of various kinds of noise such as missing vowels, unnatural abbreviations and omissions of less informative words [5]. Omissions of words as well as usage of different word variants both aggravate the sparsity problem [6], and hence we explore various kinds of noise-robust processing and similarity measures (e.g., ontology-based techniques etc.) that can uncover latent similarities in estimating relevance of tweets. The dataset that we work with had URLs in only as little as 0.7% of the tweets, and thus, we omit considering URLs and content of their web pages.

tf.idf Cosine (TC). *tf.idf* is a simple and common scoring function that is used for text processing [7]. S_{TC} denotes scoring according to the cosine similarity of *tf.idf* vectors between tweets; the IDF being computed over $\mathcal{D}$ (Ref. Section 2).

Query Centric Similarity (QS). Consider the query tweet $Q =$ *"blasts in mumbai !!! :O"* and two candidate tweets $t_1 =$ *"mumbai blasts again, omg"* and $t_2 =$ *"mumbai blasts kill at least 10 people"*. S_{TC} would score t_1 higher due to it having a larger *fraction* of query words. However, besides being relevant to Q, t_2 is seen to provide additional information, whereas t_1 is mostly redundant wrt the query tweet. Since the extra information contained in t_2 may be actually useful, we would like to score t_2 at least as much as t_1[5]. A query centric similarity measure that does not discount the score for additional information in a

---

[4] http://en.wikipedia.org/wiki/Jaccard_index

[5] Such a consideration is not very relevant for scenarios for which *tf.idf* cosine similarity is traditionally used, which is that of comparing reasonably long documents such as newswire articles. This is because, very similar/duplicate/identical content is statistically unlikely due to the length of the documents. However, since tweets are short text snippets containing a few words, identical content is highly likely and hence needs special consideration.

candidate tweet would remedy this problem and score both t_1 and t_2 identically. Such a scoring function is given in $\mathcal{S}_{QS}$, where $f(t_i, w)$ denotes the frequency of w in the tweet t_i.

Edit Distance Based Similarity (ED). Twitter's character restriction leads to authors frequently resorting to vowel dropping and usage of uncanny abbreviations; *parliament* is often abbreviated to *parlmnt* or *prlmnt* whereas *atlntc* is used to refer to *atlantic*. Such shortening of words is detrimental to similarity measures that rely on occurences of the same word in two tweets to quantify relevance. Levenstein distance [8] quantifies the distance between two tokens as the number of character edits required to transform one string to the other [9]. Such similarities could potentially be robust to various kinds of noise; e.g., (*atlntc* and *atlantic* have a low Levenstein distance of 2). We use $sim(w_1, w_2)$ to denote $1 - edfrac(w_1, w_2)$ where $edfrac(w_1, w_2)$ denotes the edit distance between w_1 and w_2 measured as a fraction of the shorter word. We design a scoring function $\mathcal{S}_{ED}$ that aggregates the similarity between each word in the query with its edit distance based similarity with each word in the candidate. The addition of 1.0 to the inner sum is a standard practice to ensure that one of the inner sums evaluating to zero does not lead to an overall zero score.

Word Co-occurences (WC). Edit distance, while being robust to mispellings, is unaware of semantic relatedness between words. For example, *Christmas* and *Yule*[6] would have a low score despite the semantic relatedness being evident due to co-occurence in tweet data (an example tweet reads *'Yule would be the perfect day to take a day off and do my Christmas baking'*). In exploiting such word co-occurences, we use conditional probabilities first outlined in [10]. The conditional probability of occurence of a word w' given a word w is:

$$p(w'|w) = \frac{|\{d|d \in \mathcal{D} \wedge w \in d \wedge w' \in d\}|}{|\{d|d \in \mathcal{D} \wedge w \in d\}|}$$

where $\mathcal{D}$ denotes the large corpus of tweets. Such similarities are then aggregated in usual style to yield the scoring function $\mathcal{S}_{ED}$.

Reply Correlations (RC). Replies to a user's tweets may intuitively be treated as relevant to a tweet. Since reply tweets are not tagged with the tweet being replied to, we heuristically estimate replies to the author within two hours of authoring a tweet as being replies to the tweet. An example tweet, reply pair thus extracted reads as [*I am actually really excited for Friday what gunna happen to the government, I just hope there is no election*]. Given a set of such $[t, r]$ pairs, we outline a similar conditional probability formualtion that estimates the probability of a word w' occuring in a reply to a tweet containing w:

$$p'(w'|w) = \frac{|\{[t, r]|[t, r] \in \mathcal{TR} \vee w \in t \vee w' \in r\}|}{|\{[t, r]|[t, r] \in \mathcal{TR} \vee w \in t\}|}$$

---

[6] http://en.wikipedia.org/wiki/Yule

Such word-correlations across question-answer pairs were used in [11] in building translation models for question answer forums. $p'(w'|w)$ values are then aggregated to form the $\mathcal{S}_{RC}$ scoring function.

WordNet Similarity (WS). Semantic relatedness between words may be assessed using ontologies such as Wordnet[7]. Among the various similarity measures proposed for quantifying pair-wise similarity between words in WordNet [12], we found the *Lesk* measure [13] to be most effective in estimating tweet relevance; $lesk(.,.)$ estimates the similarity of a pair of words as being proportional to the extent of overlap of their dictionary definitions. $\mathcal{S}_{WS}$ outlines a scoring function that uses the $lesk(.,.)$ similarity measure between words.

4   Empirical Evaluation

Dataset and Experimental Setup: In the absence of twitter datasets with relevance judgements (such as LETOR[8] document datasets), we create our own dataset from a large twitter corpus of 977252 tweets obtained from Fundacion Barcelona Media[9]. Upon choosing 50 random tweets to represent queries, for each query tweet, we collect 200 tweets from the immediate past, but restrict to those that have a common non-stopword with the chosen query tweet. The common non-stopword restriction is applied since choosing tweets using only the recency parameter would invariably lead to mostly irrelevant tweets, leaving us with very few non-zero relevance judgements for the same labeling effort. We sought the help of human annotators to judge the *binary relevance* (either as relevant or non-relevant) of each of the $10k$ (i.e., 200×50) tweets to their respective query tweet. This effort was sizeable and was spread over 2-3 humans, and altogether took roughly 25 hours; this leads to an average of 9 seconds for labeling each tweet. For each query $\mathcal{Q}$, the $\mathcal{T}$ set is formed by the 200 selected tweets[10]. The corpus, after removal of the 10000 potential candidates, and 50 queries, forms $\mathcal{D}$, a dataset of 967202 tweets. The inability to undertake large-scale labeling due to the human effort involved constrains us to work with the set fo 50 queries.

Evaluation Measures: We use popular Information Retrieval evaluation measures [14] such as Mean Reciprocal Rank (MRR), Mean Average Precision (MAP), Normalized Discounted Cumulative Gain (NDCG) and Precision (P) to evaluate the effectiveness of the techniques. The NDCG, MAP and P measures are computed on the top-k results; we choose $k=10$ consistently, unless otherwise mentioned. We refer the interested reader to [14] for details of these measures, owing to lack of space. Each of these measures are plotted with their 95% confidence intervals in the charts we present. Since we work with a small set of queries, we present analysis of statistical significance when appropriate; our evaluation of statistical significance uses randomization tests [15] with p-value at < 0.05.

---

[7] http://wordnet.princeton.edu/
[8] http://research.microsoft.com/en-us/um/beijing/projects/letor//
[9] http://caw2.barcelonamedia.org/node/7
[10] Please contact the first author for a copy of the dataset.

Table 2. Time Difference and Precision

Time Difference	Precision in %
< 10 secs	30.00%
< 20 secs	31.11%
< 40 secs	26.97%
< 60 secs	23.45%
< 80 secs	13.74%
All	12.39%

Table 3. Social n/w Distance and Precision

Distance b/w Authors	Precision in %
1	66.67%
2	25.13%
3	14.27%
4	13.88%
5	13.74%
All	12.39%

Table 4. Top-10 Pairs According to WC

(w, w')	$p(w'|w)$
(spears,britney)	0.0542
(degrees,number_token)	0.0493
(peanut,butter)	0.0481
(going,to)	0.0472
(minutes,number_token)	0.0435
(searching,for)	0.0431
(happy,!)	0.0429
(check,out)	0.0423
(anyone,?)	0.0414
(suggestions,?)	0.0389

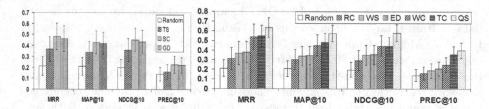

Fig. 1. Time and Author based Techniques

Fig. 2. Content based Techniques

4.1 Time and Author Based Techniques

We plot the various metrics for Time and Author based techniques along with an approach that randomly retrieves tweets from $\mathcal{T}$. Since $\mathcal{T}$ comprises of tweets that have at least one common word, *Random* measures how well such a lexical constraint works in assessing relevance. TS is seen to be 75% better than *Random* whereas SC and GD provide up to 35% gains over TS. The performance of TS is indicative of the temporal coherence of topics in tweets, twitter being considered as a mode of real-time communication. Table 2 analyzes the correlation between temporal closeness and relevance of tweets where high accuracies of up to 30% are observed when only tweets within 20 seconds are considered. For the social network based techniques, we found it interesting that twitter users who have as many as 2-3 hops in between still exhibit some similarity in the topics of discussion. An analysis of social network distance and precision in Table 3 suggests that very high accuracies of up to 67% may be achieved when only immediate neighbor's tweets are considered.

4.2 Content Based Techniques

An evaluation of the various content-based techniques appear in Figure 2. **RC** is seen to fare the worst at a mere 50% better than *Random*. This was found to be due to the high conditional probabilities assigned to exclamation words and wishes (e.g., *lol, omg, congrats* etc), replies to tweets being short and often comprising such words. We found that 30% replies had a question mark. In analyzing **WS** that fared marginally better than RC, we examined the top word pairs according to *lesk* metric, and found [*weekend, week*], [*north, south*] etc to be among them. Tweets that talk about *north* are unlikely to talk about *south*, despite these being semantically similar words; thus, semantic word similarity measures are found to have some apparent drawbacks in assessing tweet relevance. In a similar analysis of **ED** that was found to be competetive with WS, we find that there were as many as 9 spurious matches (e.g., [*protest, promise*], [*breaking, being*], [*against, again*] and [*chocolate, coconut*]) among the top-20 pairs assessed as maximally similar wrt edit distance based similarity. Word similarity assessments for **WC** were found to be much more accurate, leading to an improved overall performance. The top-10 pairs according to the $p(.|.)$ measure (used in WC) as listed in Table 4 are seen to model word similarity fairly accurately, thus explaining the better performance of WC. **TC**, the traditional document similarity metric, is seen to surprisingly perform slightly better than WC; this indicates that the semantic similarities estimates of WC, ED etc are offset by spurious matches produced by them. **QS** is seen to be the best performing technique, indiciating that additional information embodied in words that do not occur in the query are indeed perceived as relevant by the user. This validates our hypothesis that twitter users are likely to be interested in new information.

Statistical Significance (p < 0.05): TC's precision was found to be statistically significant over many techniques it outperforms, whereas QS provides statistical significant results on *all measures* over all techniques except for WC and TC. *QS is found to be statistically significant over TS on MAP and NDCG*, while being significant over WC on all measures except MRR (Ref. Table 8).

5   A Composite Technique

Though content based techniques are seen to be most effective in assessing tweet relevance, that the different techniques are designed differently could mean that the techniques may have some orthogonality among them. TS is likely to perform well for extremely time-sensitive topics like an ongoing sporting contest, whereas social network based scoring may perform better during times when the common interest in a social network peaks (e.g., when the common interest is related to politics, activity would peak when there is an election). Thus, a composite technique that is able to identify scenarios where specific techniques (e.g., content-based, time-based etc.) are likely to be more effective and rely on them highly for such cases, is likely to perform better than the separate methods.

Correlation Analysis: We now analyze whether there is indeed some orthogonality among the techniques using a simple correlation analysis. For each technique, we create a vector whose i^{th} value denotes the precision (@10) obtained for the ith query. The Pearson co-efficient[11] among such precision vectors for the top-5 techniques are tabulated in Table 5 with high values in boldface. The correlation co-efficient ranges between -1.0 (inverse correlation, i.e., high precision indexes of one technique corresponding to low precision in the other) and 1.0 (denoting direct correlation). However, since correlation co-efficients can only uncover *linear* relations, inferences based on them need to be taken with a pinch of salt. The techniques may be grouped into two groups of correlated techniques, one with TS and SC, and the other comprising the content based techniques. Any pair from across these groups have a low correlation of 0.5-0.6; though this is higher than 0 indicating independence, that they are far enough from 1.0 provides some hope of orthogonality between them that may be exploited.

Table 5. Correlation Analysis

Table 6. Distributions with Similar Entropy

r	TS	SC	WC	TC	QS
TS	1.0	**0.85**	0.64	0.59	0.52
SC	**0.85**	1.0	0.67	0.68	0.61
WC	0.64	0.67	1.0	**0.80**	0.68
TC	0.59	0.68	**0.80**	1.0	**0.88**
QS	0.52	0.61	0.68	**0.88**	1.0

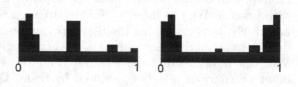

A Weighting Score: For a scoring function, query combination $[\mathcal{S}(.), \mathcal{Q}]$, we first linearly normalize the score of $\mathcal{S}(\mathcal{Q}, t_i)$ to between 0 and 1 by scaling[12]. We use $\mathcal{S}_{\mathcal{N}}(.)$ to denote this normalized version. If this were to represent the true relevance distribution, it would have a low entropy with all the relevant tweets scoring at 1.0 and others scoring at 0.0. In the absence of relevance judgements, we could just prefer low-entropy distributions of $\mathcal{S}_{\mathcal{N}}(.)$ since entropy is inversely related to the uniformity (randomness) of the scoring. Entropy, however, is unable to differentiate between distributions in Table 6. Among those, we would intuitively prefer the right distribution since that has more objects scored close to 1.0. We use these intuitions to define a weighting for a $[\mathcal{S}(.), \mathcal{Q}]$:

$$w(\mathcal{S}, \mathcal{Q}) = average(top\text{-}k(\{\mathcal{S}_{\mathcal{N}}(\mathcal{Q}, t) | t \in \mathcal{T}\})) - entropy(\{\mathcal{S}_{\mathcal{N}}(\mathcal{Q}, t) | t \in \mathcal{T}\})$$

where the *top-k* denotes the *top-k* values in the input set, and the *average*(.) and *entropy*(.) denote the average and entropy of the distributions respectively. Both these terms are in the range $(0, 1)$ and the average of the *top-k* terms (being the average of the highest k terms in a normalized distribution) is likely to be larger than the entropy. This motivates the subtraction-based construction.

---

[11] http://en.wikipedia.org/wiki/Pearson_product-moment_correlation_coefficient

[12] For the content-based techniques that use the product formulation, we score tweets using the log of the respective formula, prior to normalization.

An Integrated Approach: The weighting score can be used to linearly combine multiple scoring functions, $\{S_1, \ldots, S_p\}$, as follows:

$$S_{\{S_1,\ldots,S_p\}}(Q,t) = \sum_{i=1}^{p} \begin{cases} 0.0, & if \ w(S_i, \mathcal{T}) \leq 0.0 \\ w(S_i, \mathcal{T}) \times S_{\mathcal{N}i}(Q,t), otherwise \end{cases}$$

we use a cut-off of 0.0 thereby not allowing those (Q, S_i) combinations that have a weighting score evaluating to negative to influence the scoring.

Table 7. Composite Technique Evaluation

Evaluation Measure	Best Among Components	S_c	% Impr. Recorded
MRR	0.633 (QS)	0.663	4.7%
MAP@10	0.575 (QS)	0.603	4.9%
NDCG@10	0.578 (QS)	0.606	4.8%
PREC@10	0.398 (QS)	0.406	2.0%

Table 8. Statistical Significance

Technique Pair	Statistically (p < 0.05) Significant Metrics
QS over WC	map, ndcg, prec
QS over TC	map, ndcg
S_C over WC	map, ndcg, prec
S_C over TC	map, ndcg
S_C over QS	map, ndcg

Evaluation: We combine the top-5 content based techniques WS, ED, WC, TC and QS along with TS and SC in the above construction to form a combined technique S_c. Table 7 illustrates that S_c beats the the best performing component technique on each of the measures, albeit by small margins. This establishes the utility of the weighting score formulation in leveraging the strengths of the various techniques in building a technique that outperforms the components, and establishes the combination as the preferred technique for retrieving similar/relevant tweets. Table 8 shows that S_C is statistically significant over TC and QS on MAP and NDCG; it was found to be statistically significant on all metrics against techniques not listed in the table.

Computational Cost: Each of our separate techniques are in $\mathcal{O}(|\mathcal{T}|l^2p)$ where l denotes the number of tokens in a tweet, and p denotes the number of characters in a token (for edit distance calculations that are linear); this is so since token-pair similarities (e.g., $lesk(.,.)$ and $p(.|.)$) may easily be pre-computed (so are IDF values). Normalization (for $S_{\mathcal{N}}$) and final scoring each take $\mathcal{O}(|\mathcal{T}|)$ in series, leading to an overall complexity of $\mathcal{O}(|\mathcal{T}|l^2p)$. It may be noted that l is often 15-20 at max since tweets are limited to 140 characters.

6 Related Work

The problem of finding similar microblog posts (e.g., tweets) is, to the best of our knowledge, has not received much attention yet. We provide a brief overview of literature under two separate heads.

Microblog Processing: Retrieval related tasks that have been attempted on microblog data include identifying re-tweetable tweets [16], non-query specific ranking of tweets [17] to replace the reverse chronological ordering, diversity-conscious retrieval of tweets related to a topic (e.g., *oil spill*) [18] and mechanisms for gathering user input on tweet ranking [19]. Retrieving extrinsic content such as news articles [20] and RSS feed entries [21] in correlation with tweet information has also been of interest.

Non-content based approaches in Retrieving Entities: Information retrieval and top-k processing [3] have become pervasive. Approaches that focus on user profiles or temporal information have been of interest in social media processing [22][23]. While immediate followees have been found to be useful to recommend to a user to follow [24], [25] assesses familiarity based evidence to be more useful than network proximity. Another aspect of similarity is that pertaining to locations [26]; geo-tagging, however, is not yet very popular on twitter. Incorporating temporal information in collaborative filtering has been of utility in retrieving timestamped entities [27].

7   Conclusions and Future Work

We analyzed the problem of finding similar/relevant microblog posts, to a query post. Similarity search in microblog posts, to the best of our knowledge, has not received enough attention. Microblog posts in the popular microblogging service, Twitter, are short text snippets with associated metadata such as the timestamp and author handle. Towards retrieving relevant tweets, various intuitive techniques that separately exploit content and metadata such as timestamp and author social network were developed. An evaluation on real-world data confirms that content-based techniques are most effective and that tweets containing new information while being lexically similar with the query are perceived to be very useful. Based on a correlation analysis of the techniques, we then formulated a weighting score that heuristically estimates the effectiveness of specific techniques for given queries and used that to build a composite scoring function that assesses relevance using a query-specific linear combination of the different schemes. An empirical evaluation illustrates the statistically significant superiority of the composite technique over the component techniques, and establishes it as the technique of preference for assessing tweet relevance.

Many recommendations are accompanies by an interpretable reason e.g., gmail[13] explains why it marked email threads as interesting by including an explanation indicating the reason. We would like to develop techniques to derive interpretable explanations to accompany suggestions of *relevant tweets*. Further, as and when geo-encoded tweet data are available, we intend to explore the utility of geo-information in assessing tweet relevance.

---

[13] http://www.gmail.com

References

1. Sakaki, T., Okazaki, M., Matsuo, Y.: Earthquake shakes twitter users: real-time event detection by social sensors. In: WWW, pp. 851–860 (2010)
2. Wu, S., Hofman, J.M., Mason, W.A., Watts, D.J.: Who says what to whom on twitter. In: WWW, pp. 705–714. ACM, New York (2011)
3. Deshpande, P.M., Deepak, P., Kummamuru, K.: Efficient online top-k retrieval with arbitrary similarity measures. In: EDBT, pp. 356–367 (2008)
4. Krinke, J.: Identifying similar code with program dependence graphs. In: WCRE, p. 301. IEEE Computer Society, Washington, DC (2001)
5. Subramaniam, L.V., Roy, S., Faruquie, T.A., Negi, S.: A survey of types of text noise and techniques to handle noisy text. In: AND, pp. 115–122 (2009)
6. Allison, B., Guthrie, D., Guthrie, L.: Another Look at the Data Sparsity Problem. In: Sojka, P., Kopeček, I., Pala, K. (eds.) TSD 2006. LNCS (LNAI), vol. 4188, pp. 327–334. Springer, Heidelberg (2006)
7. Steinbach, M., Karypis, G., Kumar, V.: A comparison of document clustering techniques. In: KDD Workshop on Text Mining (2000)
8. Levenshtein, V.I.: Binary codes capable of correcting deletions, insertions, and reversals. Technical Report 8 (1966)
9. Wang, W., Xiao, C., Lin, X., Zhang, C.: Efficient approximate entity extraction with edit distance constraints. In: SIGMOD, pp. 759–770 (2009)
10. Sanderson, M., Croft, W.B.: Deriving concept hierarchies from text. In: SIGIR, pp. 206–213 (1999)
11. Xue, X., Jeon, J., Croft, W.B.: Retrieval models for question and answer archives. In: SIGIR, pp. 475–482 (2008)
12. Pedersen, T., Patwardhan, S., Michelizzi, J.: Wordnet: Similarity - measuring the relatedness of concepts. In: AAAI, pp. 1024–1025 (2004)
13. Banerjee, S., Pedersen, T.: An Adapted Lesk Algorithm for Word Sense Disambiguation Using WordNet. In: Gelbukh, A. (ed.) CICLing 2002. LNCS, vol. 2276, pp. 136–145. Springer, Heidelberg (2002)
14. Robertson, S., Zaragoza, H.: On rank-based effectiveness measures and optimization. Inf. Retr. 10, 321–339 (2007)
15. Smucker, M.D., Allan, J., Carterette, B.: A comparison of statistical significance tests for information retrieval evaluation. In: Proceedings of the Sixteenth ACM Conference on Conference on Information and Knowledge Management, CIKM, pp. 623–632 (2007)
16. Uysal, I., Croft, W.B.: User oriented tweet ranking: a filtering approach to microblogs. In: CIKM, pp. 2261–2264 (2011)
17. Duan, Y., Jiang, L., Qin, T., Zhou, M., Shum, H.Y.: An empirical study on learning to rank of tweets. In: COLING, pp. 295–303 (2010)
18. De Choudhury, M., Counts, S., Czerwinski, M.: Identifying relevant social media content: leveraging information diversity and user cognition. In: HT (2011)
19. Sarma, A.D., Sarma, A.D., Gollapudi, S., Panigrahy, R.: Ranking mechanisms in twitter-like forums. In: WSDM, pp. 21–30 (2010)
20. Chen, J., Nairn, R., Nelson, L., Bernstein, M.S., Chi, E.H.: Short and tweet: experiments on recommending content from information streams. In: CHI (2010)
21. Phelan, O., McCarthy, K., Smyth, B.: Using twitter to recommend real-time topical news. In: RecSys, pp. 385–388. ACM, New York (2009)
22. Pennacchiotti, M., Gurumurthy, S.: Investigating topic models for social media user recommendation. In: WWW (Companion Volume), pp. 101–102 (2011)

23. Diaz, F., Metzler, D., Amer-Yahia, S.: Relevance and ranking in online dating systems. In: SIGIR, pp. 66–73. ACM, New York (2010)
24. Hannon, J., Bennett, M., Smyth, B.: Recommending twitter users to follow using content and collaborative filtering approaches. In: RecSys, pp. 199–206 (2010)
25. Guy, I., Jacovi, M., Perer, A., Ronen, I., Uziel, E.: Same places, same things, same people?: mining user similarity on social media. In: CSCW, pp. 41–50 (2010)
26. Lee, M.-J., Chung, C.-W.: A User Similarity Calculation Based on the Location for Social Network Services. In: Yu, J.X., Kim, M.H., Unland, R. (eds.) DASFAA 2011, Part I. LNCS, vol. 6587, pp. 38–52. Springer, Heidelberg (2011)
27. Ding, Y., Li, X., Orlowska, M.E.: Recency-based collaborative filtering. In: Proceedings of the 17th Australasian Database Conference, ADC, vol. 49, pp. 99–107 (2006)

Fgram-Tree: An Index Structure Based on Feature Grams for String Approximate Search[*]

Xing Tong and Hongzhi Wang

Department of Computer Science and Technology
Harbin Institute of Technology
{hitxingt,whongzhi}@gmail.com

Abstract. String approximate search is widely used in many areas. Indexing is no doubt a feasible way for efficient approximate string searching. However, the existing index structures have a common weakness that they do not obey the nature of the index which is a function by mapping different data to different index items, similar data to similar index items, in order to query easily. In this paper, we propose a new type of string indexing structure called Fgram-Tree, which is based on feature grams to build itself and filter strings. It obeys the two maps by placing similar strings into the same node, different strings into different nodes that could greatly improve the efficiency of index. Our index is able to support for different types of search. Compared to other index, it provides high scalability and fast response time.

Keywords: Fgram-Tree, index structure, string approximate search.

1   Introduction

Approximate string handling in the application of computer systems plays an important role and is widely used in the database, information retrieval and other areas. However, approximate string matching also brings technical challenges. Firstly, it is not trivial to measure the difference between two strings. Secondly, even though many string measured functions have been proposed, it is costly to calculate these measured functions.

Based on the above discussion, for efficient string approximate search, a natural way is to build the appropriate indexing structure for the strings. Currently, the most extensive index structure for approximate string matching is inverted table structure that will split the string into grams, and measure the string by edit distance metric [3-10]. Even though gram-based method can process approximate string matching efficiently in many cases, it has many weaknesses. First, it cannot effectively deal with the data update. Second, it has to introduce many collection operations when we use the inverted table to do the query, which increases the complexity of the query.

There are still a lot of non-inverted list indexing structures supporting approximate string search. For example, [2] proposed edit distance tree structure that could hash

---

[*] This paper was partially supported by NGFR 973 grant 2012CB316200 and NSFC grant 61003046, 6111113089. Doctoral Fund of Ministry of Education of China (No. 20102302120054).

H. Gao et al. (Eds.): WAIM 2012, LNCS 7418, pp. 241–253, 2012.

each string into a number, and insert it into a B+ tree structure, which can support the data update well, but this structure emphasis on the ordering of the string too much and cannot pay full attention to the similarity of strings, therefore a large number of similar strings cannot be in the same leaf node, and the result is that the filter effect is not obvious, getting many alternative leaf nodes.

In summary, existing methods have drawbacks. We attempt to address these problems and design an index structure for efficient approximate string matching. Inspired by hashing index and B+ tree, if similar strings are mapped into the same entry in the index, during the search, they can be accessed by once probing. Thus, the design of our index is to cluster the string set based on the similarity and build index with each cluster as an item. Such that similar strings can be accessed in batch in the index.

We choose a tree structure as the skeleton of our index, since such structure supports the data updating. Each node in the index corresponds to a set of similar strings. To represent the nodes in the index, for each node, we extract some grams from the strings in the corresponding string set, which are called feature grams. With feature grams, when some new strings are added to the set, they are added to the nodes with similar strings and distinguished from strings in other nodes. With this consideration, it is crucial to select the feature grams. To choose the feature grams, we use a cluster method to cluster the strings together and extract the feature grams from the center of each cluster. Our index is able to support threshold-based search, top-k search.

The contributions of this paper include:

1 We present a new type of string indexing structure that supports a variety of types of similarity string search efficiently.
2 An effectiveness and efficient index building algorithm with complexity is proposed.
3 Extensive experimental results on real datasets show that our indexing scheme achieves comparable performance against other solution on search operations.

The rest of the paper is organized as follows. Section 2 discusses the necessary background and gives some formal problem definitions. Section 3 presents the basic principles of the Fgram-Tree. Section 4 presents details of the index building process and algorithm complexity. Section 5 presents a comprehensive evaluation of the proposed techniques and Section 6 concludes the paper.

2   Preliminaries

In this section we present some preliminary knowledge regarding string processing as well as the basic problem definition.

In the literature of approximate string matching, edit distance is commonly used to measure the similarity of two strings. We use edit distance to measure string similarity in our query process operations.

Definition 1. (Edit Distance)
The edit distance between two strings s_i and s_j is the minimum number of single character edit operations (insertion, deletion, and substitution) that are needed to transform s_i to s_j. We denote the edit distance between s_i and s_j as $ed(s_i, s_j)$.

Figure 1 shows a simple database table including 5 distinct strings and its general index structure. $ed(s_0, s_1)$ is 1, since s_0 is transformed to s_1 with a substitute operation from single character 'y' to 'l'.

Next, we give the formal definitions of string approximate search with respect to edit distance.

Definition 2. (Threshold-based Search)
Given a query string q, a string set D and threshold θ, find all strings in D with edit distance no larger than θ.

In Figure 1, a threshold-based search query q ="Joe" with θ= 1 will return strings s_0, s_1 and s_4, whose edit distances to q are no larger than 1.

Definition 3. (Top-k Search)
Given a query string q and a string set D, find k strings in D with edit distance no larger than any other strings in D.

In Figure 1, a top-k search query q ="Janet" with $k = 2$ will return strings s_2 and s_3, which are more similar to q than any other strings in D.

When we do the approximate search, the algorithm of counting edit distance between two strings runs in $O(|s|^2)$ time for strings of length $|s|$, based on a standard dynamic programming method. As a result, when we deal with large amount of strings, we use various filtering techniques to prune the number of strings before counting edit distance. We split a string into grams.

Definition 4. (Ngram Split)
Ngram split of a string is a set, that is composed by all the substrings with length N.

For example, the 2-gram split for "Joey" is {Jo, oe, ey}.

Some filters use the fact that if $ed(s_1, s_2) \leq \theta$, then the lengths $|s_1|$ and $|s_2|$ should differ by at most θ, and the number of common n-grams of them should be at least Num_c:

$$Num_c = Max\{|s_1|, |s_2|\} + 1 - (\theta + 1) * n \qquad (1)$$

Formula (1) is easy to understand. One single character edit operation could involve n grams at most, so θ operations involve $\theta * n$ grams at most and a string contains $|s| - n + 1$ grams. As a result, the subtraction between two numbers is the least common n-grams.

3   Fgram-Tree

This section introduces the index structure and discusses query process methods including two kinds of approximate search queries.

3.1   Index Structure

We propose Fgram-Tree, a novel type of string indexing structure in a tree structure with each node representing a set of similar strings. Fgram-Tree pays more attention to the treatment of similar strings and supports threshold-based search, top-k search. We propose the formal definition of Fgram-Tree.

Definition 5. (Fgram-Tree)

A Fgram-Tree is a tree structure where each leaf node is represented as a triple (bs, cbs, ids) and each intermediate node is a tuple (bs, cbs), where bs is a set of ngram splits of strings contained by all lower nodes and occurrences of each gram, cbs is the center of bs to represent bs (More details about cbs will be described in Section 4), and Ids is the set of the strings attached to the node.

We use an example to illustrate our index. General index structure is shown in Figure 1. Similar strings s_0, s_1, s_4 are stored in $LNode_1$, and strings s_2, s_3 are stored in $LNode_2$. To reduce the storage overhead and the computing complexity, we store bs and cbs as bitmaps. Corresponding to each gram there are one bit in bs and a frequency of that gram. To facilitate the presentation, we do not use the bitmap form but the collection way in Figure 1, and we will discuss the problem by the collection way in the rest of the paper.

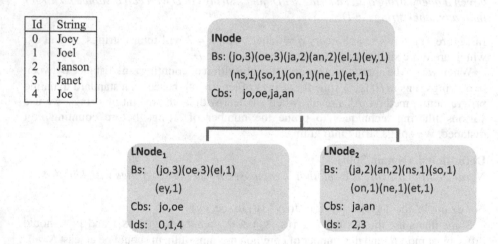

Id	String
0	Joey
1	Joel
2	Janson
3	Janet
4	Joe

INode
Bs: (jo,3)(oe,3)(ja,2)(an,2)(el,1)(ey,1)
    (ns,1)(so,1)(on,1)(ne,1)(et,1)
Cbs: jo,oe,ja,an

$LNode_1$
Bs: (jo,3)(oe,3)(el,1)
     (ey,1)
Cbs: jo,oe
Ids: 0,1,4

$LNode_2$
Bs: (ja,2)(an,2)(ns,1)(so,1)
     (on,1)(ne,1)(et,1)
Cbs: ja,an
Ids: 2,3

Fig. 1. An example of string dataset and its general index structure

3.2   Query Process

In this section, we will discuss the algorithms for threshold-based search and top-k search based on our index.

Before the introduction of query process supported by Fgram-Tree, we give two filter conditions which are used in the query process and related to the similarity of strings. The basic idea of query process is that they are used to prune excess nodes impossible to contain similar strings with the query.

Condition 1: According to Formula (1), if the number of elements in intersection between the query string q and bs of node c is smaller than Num_c, node c must not contain a string similar with the query. Obviously, if the number of common n-grams bs sharing with q is smaller than Num_c, any string belonging to node c must not share such many grams with q, either.

Condition 2: If there is no intersection between a query q and cbs of node c, c must not contain a string meeting the query. We use center-based clustering method to construct our index and cbs is just the center for each cluster, thus it must intersect with q otherwise q does not belong to that cluster. We will discuss this condition specifically in Section 4.3.

Threshold-Based Search
Firstly, we discuss threshold-based search. We use these two conditions to prune nodes to obtain the leaf nodes as few as possible. And then every string s in the obtained node set is verified to check whether s satisfies the constraint in the query. We give pseudo code of threshold-based search in Algorithm 1.

Algorithm 1. Threshold-basedSearch (string q, tree node N, threshold θ)

1: **if** N is the leaf node **then**
2:   **for** each $s \in N$ **do**
3:     **if** Verify(q, s, θ) **then**
4:       Add s in search result
5: **else**
6:   **for** each child $c \in N$ **do**
7:     **if** CommonGramBs(q, c, θ) && CommonGramCbs(q,c) **then**
8:       Threshold-basedSearch(q,c,θ)

Algorithm 1 traverses the nodes meeting Condition 1 and Condition 2 in the index recursively (Line 6-8) and verifies the strings in the visited the leaves (Line 1-4) by the verification algorithm only in $O(\theta|s|)$ mentioned in [1]. The strings satisfying the constraint are added to the final results. The function CommonGramBs() and CommonGramCbs() correspond to Condition 1 and Condition 2 respectively.

Top-K Search
Next, we discuss top-k search method. To take advantage of the feature of our index that similar strings are located in the same node in our index, we could firstly locate the leaf node where the most similar string with query q is by setting threshold 0 and choose k most similar strings in that node as the initialization result. Then we visit other nodes in the same way and update the result if a string with smaller threshold than the largest one in the result is found. Thus, we only need to access the index once. The top-k search algorithm is as follows:

Algorithm 2. Top-kSearch (string q, tree node N, threshold θ, result heap H)

1: **if** N is the leaf node **then**
2:   **for** each $s \in N$ **do**
3:     Geteditdistance(q, s)
4:     Insert s into H
5:     **if** |H|>k **then** pop top entry
6: **else**
7:   **for** each child $c \in N$ **do**
8:     **if** CommonGramBs(q, c, θ) && CommonGramCbs(q, c) **then**
9:       Top-kSearch(q,c,θ,H)

In Algorithm 2, we locate the nodes which contain the most similar string with q with the same pruning conditions as Algorithm 1(Line 7-9). We use a max-heap to keep the current top-k similar strings and directly calculate the edit distance in Geteditdistance()(Line 1-5). When a leaf node meets the conditions, for the strings in that leaf, we calculate their edit distances with q and insert them into the heap. If the number of elements in max-heap is more than k, we pop the strings with the largest distance.

4   Construction of Index Structure

According to our discussion in Section 3, in order to filter more nodes in the same level, strings in the same node should be similar while those in different nodes should be not similar. Therefore, we design a center-based clustering method with each tree node equivalent to a cluster. *Cbs* in the index structure is just the center of our cluster. We design a center-based algorithm for the construction of such index in this section. At first, we introduce the framework of our clustering method in Section 4.1. Since center initialization, node selection and center update are basic operations for the clustering, we discuss them in Section 4.2, Section 4.3 and Section 4.4, respectively. At the end of this section, we analyze the time complexity of the index construction.

4.1   Overall Method

We show of overall framework of our method in this section by using a recursive approach to make a hierarchical clustering. In each level, firstly we initialize a suitable center *cbs* for each cluster. Secondly, we iteratively choose a center for every string and update the centers until the centers do not change. We show the overall index construction in Algorithm 3.

Algorithm 3. RecurMakeNode (tree node N, int r,,int k)

1:   **if**(the size of strings in $N>r$) **then**
2:       InitNode(N)
3:       **while**(true)
4:           **for** each string s of N **do**
5:               child node c=NodeChoice(s,N)
6:           **for** each child node c of N **do**
7:               SetCenter()
8:           **if** each child $c'cbs$ in N does not change **then**
9:               **break**
10: **else**   **return**
11: **for** each child node c of N **do**
12:       **RecurMakeNode** (c,r,k)

The input of Algorithm 3 is a tree node N which contains all the strings waiting to be clustered,   the number k of clusters and the size r of strings in a child node. In order to take full advantage of the memory page size, k and r are by memory limit. The output is the index based on tree with root N. And if the size of strings in a child node c is greater than r, the algorithm splits c (Line 2-9). Otherwise, it visits other nodes recursively. In Line 2, the center is initialized with InitNode(). Line 3-9 is the iterative process. The condition of ending iteration is that both the grams and their frequencies of all children nodes' cbs do not change (Line8-9). Line 4-5 chooses the suitable child node for s using NodeChoice(). Line 6-7 updates cbs of every child node using the method SetCenter().

Next, we will discuss the details about InitNode(), NodeChoice() and SetCenter() in Section 4.2, Section 4.3 and Section 4.4, respectively.

4.2   Center Initialization

In this section, we propose the method to initialize the center to accelerate the iteration rather than randomly generate centers.

Center-based clustering method is to select a center on behalf of the characteristics of each cluster. Clearly, two strings are similar if they have many common grams. Then we can extract some grams from every strings in the cluster to form a gram collection as the center, which share grams with each string. Moreover, we should extract the grams that similar strings share, because these grams could make similar strings located in the same cluster.

Based on above discussion, we pick some shared grams from bs as our center. And according to the idea of vote, the frequency of these shared grams should be higher than that of other grams. Thus, we choose some grams of high frequency to initialize our cluster center cbs. We use the trie structure[5] to initialize the center by picking some high frequencies grams from the trie in InitNode().

We develop an efficient three-step algorithm in Algorithm 4 to achieve the goal. Assume there are k centers and initialize for cbs_i ($i=1,2...k$).

In the first step, we construct a trie and select k grams in the leaf node belonging to different strings with the highest frequency and add them into $initgram_i$ ($i=1,2...k$) respectively (line 6-7).

Algorithm 4. InitNode (tree node N)

```
1:   center cbs_i of N's every child (i=1,2...k)
2:   initgram_i ← φ(i=1,2...k)
3:   Heap H ← φ
4:   TrieNode root ← MakeTrie()
5:   traverse the trie, H ← k grams whose frequency is highest
6:   for each initgram_i do
7:       initgram_i ← H[i]
8:       for each substring s of H[i] with length n do
9:           if(a gram begins with s in trie) then
10:              initgram_i ← gram
11:      for each gram of initgram_i do
12:          cbs_i ← the standard grams of gram
```

In the second step, to each *initgram*, we find all of grams in the leaf node starting with n common characters with the gram in *initgram*, espacially that n is the length of grams in *cbs* and add these grams to *initgram*. (line 8-10)

In the last step, to each *initgram*, every gram in *initgram* is split into standard grams. And we add them to the node center *cbs*. (line 11-12)

The main computation cost of Algorithm 4 is the construction of the trie with $O(N)$, since when the trie is constructed the cost of collecting high frequency grams(Line 7-12) is mainly trie traversal with time complexity $O(N)$.

4.3   Node Choice

In this section, we discuss the determination method of the clusters the strings belonging to in the iteration.

The goal is that strings in the same node should be similar but not similar to strings in other nodes. It means that strings in the same node should share more grams while less grams with those in different nodes. Thus, the smaller the intersection size of any centers is, the less common grams are in different nodes. Specifically, if we insert a string s into one node, the intersection size may become large. As a result, when we make node choice for a string, we should minimize the incremental intersection size after it is inserted.

To illustrate this problem clearly, we discuss the change of the intersection between two centers. Two $cbs_{1,2}$ divide the gram set into four subsets r_1, r_2, r_3, r_4. $cbs_1 = r_1 \cup r_3$,$cbs_2 = r_2 \cup r_3$. r_3 is the set of shared grams between cbs_1 and cbs_2, while r_1 and r_2 present their unique grams collections. r_4 is the set of all other grams not belonging to $cbs_{1,2}$. Obviously, $|r_3|$ is the intersection size and our goal is minimizing it after choosing a node for ngram split gs of a string. We give the choosing method as follows.

If $gs \cap r_1 = gs_1, gs \cap r_2 = gs_2$, we choose cbs_i where $|gs_i| = Max(|gs_1|, |gs_2|)$.

For the classification, we have two possible ways. The first is that gs is assigned to cbs_1. Such that $|r_3| = |r_3| + |gs_2|$. The second is that gs is assigned to cbs_2. Such that $|r_3| = |r_3| + |gs_1|$. Thus if $|gs_1| > |gs_2|$, we choose cbs_1 and otherwise we choose cbs_2. Therefore, we choose the center which has the most common grams with gs during the filtering of common grams with other centers. For the convenience of discussion, we define $|gs_i|$ as single covered degree *sc-degree*.

Based on the above discussion, we compute *sc-degree* for every center and select the center whose *sc-degree* is the largest. In the first step, we maintain the intersection between any two *cbs* represented by bitmap in a two-dimensional array to support the operation filtering of common grams with other centers, therefore it runs in $O(k)$ time where k is the number of clusters. In the second step, we select the center with the largest *sc-degree*. Obviously, the complexity of NodeChoice() is $O(k^2)$.

While querying a string q, CommonGramCbs() in Algorithm 1,2 is used to judge whether there is a intersection between the ngram split gs of q and *cbs*. Because of our node choice method, gs must have a intersection with *cbs* which greatly enhanced filter function of the index.

4.4   Center Update

In this section, we introduce the update method of center cbs during the index construction.

In each iteration if a gram is chosen as the center, it must be contained in the ngram splits of two strings. Therefore, we should choose the grams from those with frequencies no less than two. If we apply such update method, the number of grams in the center will be very large which may cause a new problem that many grams would lead to excessive iterations.

Grams in each center must be constituted by each ngram split of the string waiting for clustering. According to the node choice method mentioned in Section 4.3, to a string s, due to the center containing the overlapping part with ngram split of s, s will be chosen by this center. Then we can say that this overlap controls string s. Assuming that the number of elements in each overlapping set has taken to the minimum, then it is obvious that the center size will be minimal.

Next, we formally explain this process by drafting some definitions and get the conclusion in Theorem 1 that a large number of grams would not increase the iteration times.

Definition 6. (Control Effect)
To a ngram split gs of s, if common(cbs_i, cover)$\geq$common(cbs_j, cover)(j=1,2...k, j$\neq$i) where Common(A,B) is sc-degree between set A and B and cover is the subset of gs, then cover determines which center s belongs to and every gram in cover has control effect to gs.

Definition 6 illustrates a control effect existing in a string s that means we may only compare a part of ngram split of s with the center to select the appropriate cluster rather than to compare the whole ngram split collection. Every center should be composed by covers and when each |*cover*| takes the min value, the center size will be minimal. Then, we define min cover set as well as the min center.

Definition 7. (Min Cover Set)
Mcs is a min cover set , if its arbitrary subsets are not cover sets.

Definition 8. (Min Center)
Min center Mcenter={$mcs_1, mcs_2 ... mcs_{|S|}$}, mcs_i is the min cover set of s_i, $s_i \in S$, where S presents all of strings in one node.

Definition 8 shows the composition of the minimal center. Then we use above definitions to prove a theorem that implies larger center size would not increase the iteration times.

Theorem 1. *When Mcenter converges, any center Lcenter with larger size satisfying Lcenter $\supseteq$ Mcenter must also converge.*

Proof. Suppose that all the strings in one node is S, to each ngram split of string s in S, we divide it into two sets min cover set (*mcs*) and remaining set (*rs*). According to Definition 6, *mcs* could control any subset *sub* of *rs*, as a result, when *Mcenter* converges, every mcs_i (i=1,2...|S|) converges, then sub_i must converge. Obviously,

$Mcenter = \{mcs_1, mcs_2 \ldots mcs_{|S|}\}$ and the center $Lcenter = \{mcs_1 \cup sub_1, mcs_2 \cup sub_2 \ldots mcs_{|S|} \cup sub_{|S|}\}$, therefore $Lcenter$ converges.   □

Theorem 1 shows that due to the control effect, when $Mcenter$ converges, $Lcenter$ must also converge. In another word, their iteration times are the same. And because the size of $Mcenter$ is smaller than $Lcenter$, the iteration times would not change with the increasing of center size.

We use SetCenter() to update the center in Algorithm 3. And because our method is based on operations of bitmap, the complexity of SetCenter() is $O(1)$ as well as judging convergence.

4.5   Complexity Analysis

In this section, we analyze the time complexity of index construction.

Algorithm 3 visits each node recursively. In every recursion, it contains two phases, initialization and iteration. The complexity of the initialization phase is the establishment of the trie with $O(N)$, where N is the number of strings. In iteration phase, because updating center and judging convergence run in constant time, the main consuming is choosing node with cost $O(k^2)$, where k is the number of clusters. Iteration phase has 2 levels of iterations with m and N times respectively, so the cost of the second phase is $O(mk^2N)$ where m is the iteration times when the center converges. As a result, in every recursion Algorithm 3 runs in $O(mk^2N)$ and in the whole recursive process Algorithm 3 runs in $O(mk^2NlogN)$. Because m would not increase with the bit size already mentioned in Section 4.4 and m is far smaller than N in reality as well as k is a constant by memory limit, then the time complexity of Algorithm 3 is $O(NlogN)$.

5   Experimental Evaluation

In this section we evaluate the performance of the index construction and the query respect to threshold-based and top-k search on edit distance with a real dataset.

Our data are from DBLP(http://dblp.uni-trier.de/xml/). We extracted the author element from the article list as the data set. The maximum length of strings is 56 and average length is 22.

We chose Bed-Tree[2] for comparison. [2] proposed an edit distance tree structure that hash each string into a number, and insert it into a B+ tree structure. As a result, it is able to work in both of memory and external memory. And in our experiment, Bed-Tree takes 2-gram, 4 string bucket, 200 strings in one node, and gram counting order. We will measure the performance of queries for Bed-Tree and Fgram-Tree both in memory and external memory.

We classified all the characters into two categories: 26 English letters and the other non-English letters as one character, then a string could be split into 2-grams up to 729, therefore we chose 729 bits for each bs and cbs. The page size is set to be 4kB. In order to take full advantage of the page size, each leaf node stored up to 200 strings and each internal node stored 20 bs as well as their cbs, i.e, $r =200$ and $k=20$ in Algorithm 3.

We compile all the programs in Windows7 using jdk 6.0 with java. The experiments are run on a Intel(R) Core(TM)2 Duo CPU E7500 2.93GHz with 2 GB main memory and 7200 RPM disk drive.

5.1   Experimental Results of Index Construction

In this section, we presents the variation of iteration times with the size of center firstly, and then shows the time of building index.

In our algorithms, we selected grams whose frequency is no less than 2 as the center of each node. Figure 2(a) shows the variation of iteration times with the size of center. In this experiment, we use a data set with 10000 randomly selected strings. To each center size, we repeated 10 times to cluster these strings, and took the average value of 10 times as the iteration times. As can be seen from Figure 2(a), when the size of center is greater than 15, the iteration times is essentially the same, which validates Theorem 1. The center is slower to converge when the median is less than 10 because few bit size cannot guarantee the control effect.

Figure 2(b) shows the variation of construction time with string size. We select the number of strings ranging from 0.01 million to 2 million. According to our analysis, the complexity of Algorithm 3 is $O(NlogN)$ that is consistent with the growth trend of the curve in Figure 2(b).

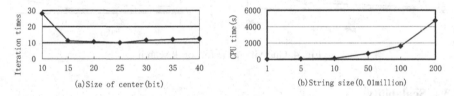

(a) Size of center (bit)   (b) String size (0.01million)

Fig. 2. The time of building index

5.2   Experimental Results for Query Processing

During the inquiry experiment, we used one million strings from author property of DBLP and randomly selected 100 strings as query objects, then got the average querying time.

We had test queries in two settings, one is the index is located in disk while the other is the index is in main memory. In the first type of experiments, we measured the performance of queries when all data required for answering a query needed to be retrieved from disk. In the second setting, the experiment represents the other extreme in which all data required to answer a query is already in memory.

Figure 3 and Figure 4 present the experimental results of threshold-based search and top-k search respectively. An observation from the figure is that the efficiency increases with the value of the abscissa. When the threshold or k is relative small, the efficiency of both methods is similar. But when they get large, the efficiency of Bed-Tree gets slower than our method. It is because the pruning ability of Bed-Tree with high level is weak, it will generate many strings from different leaves for edit distance computation.

From the experimental results, Fgram-Tree consistently outperforms Bed-Tree both for raw disk performance and for a fully memory. The most important reason is that Bed-Tree does not locate similar strings in the same node. As a comparison, many similar strings lie in different nodes in Bed-Tree and thus it has to generate more nodes for edit distance computation than Fgram-Tree. Therefore, the Bed-Tree is costly.

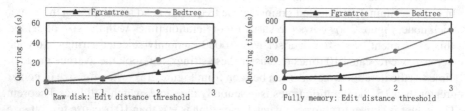

Fig. 3. Querying time for threshold-based search

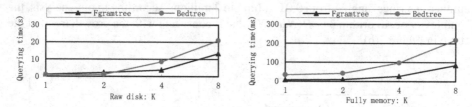

Fig. 4. Querying time for top-k search

5.3   Scalability

We varied the number of indexed strings on author property of DBLP to evaluate the scalability of our techniques. The experimental results are shown in Figure 5 and Figure 6. The results show that Fgram-Tree offers better scalability than Bed-Tree both in threshold-based search and top-k search for raw disk or fully memory. Because we used filter conditions CommonGramBs() and CommonGramCbs() simultaneously in our query process algorithms, we could filter more nodes unsatisfied with the query even though large amount of data. As a result, we got a better result.

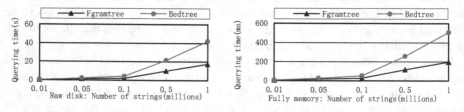

Fig. 5. Threshold-based search scalability with edit distance threshold 3

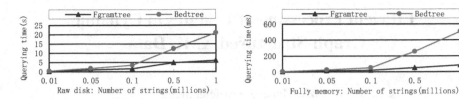

Fig. 6. Top-k search scalability with top8

6   Conclusion

In this paper we propose a general tree-based index structure to support a broad class of string approximate queries with respect to edit distance. We map similar strings in the same index node and different strings in different nodes to accelerate the query processing. To make the index support the query effectively, we design a center-based clustering approach, which can locate similar strings into the same node with the time complexity $O(NlogN)$. The experimental results show our indexing scheme achieves comparable performance against other solutions on threshold-based and top-k queries.

References

1. Sakoe, H., Chiba, S.: Dynamic programming algorithm optimization for spoken word recognition, pp. 159–165 (1990)
2. Zhang, Z., Hadjieleftheriou, M., Ooi, B.C., Srivastava, D.: Bed-tree:an all-purpose index structure for string similarity search based on edit distance. In: SIGMOD (2010)
3. Hadjieleftheriou, M., Koudas, N., Srivastava, D.: Incremental maintenance of length normalized indexes for approximate string matching. In: SIGMOD, pp. 429–440 (2009)
4. Yang, X., Wang, B., Li, C.: Cost-Based Variable-Length-Gram Selection for String Collections to Support Approximate Queries Efficiently. In: SIGMOD (2008)
5. Li, C., Wang, B., Yang, X.: Improving performance of approximate queries on string collections using variable-length grams. In: VLDB (2007)
6. Xiao, C., Wang, W., Lin, X.: (Ed-Join)–an efficient algorithm for similarity joins with edit distance constraints. In: VLDB (2008)
7. Xiao, C., Wang, W., Lin, X.: (PPjoin)Efficient similarity joins for near duplicate detection. In: Proceedings of the International World Wide Web Conference Committee (2008)
8. Hadjieleftheriou, M., Chandel, A., Koudas, N., Srivastava, D.: Fast indexes and algorithms for set similarity selection queries. In: ICDE (2008)
9. Behm, A., Ji, S., Li, C., Lu, J.: Space-constrained gram-based indexing for efficient approximate string search. In: ICDE (2009)
10. Arasu, A., Ganti, V., Kaushik, R.: Efficient exact set-similarity joins. In: VLDB, pp. 918–929 (2006)

Efficient Processing of Updates in Dynamic Graph-Structured XML Data[*]

Lizhen Fu and Xiaofeng Meng

Information School, Renmin University of China, 100872 Beijing, China
`fulizhen303@163.com, xfmeng@ruc.edu.cn`

Abstract. When the ID/IDREF relationship is considered, an XML document needs to be modeled as an ordered graph more naturally than an ordered tree. Then it becomes more difficult to process the updates of XML document. This paper studies the incremental maintenance of the document order and reachability relationship in Graph-structured XML. We propose an extended interval labeling scheme to label the document order and reachability relationship in XML. We identify the main reason for the inefficiency of updates of the labels. To accelerate the processing of updates, we design a novel index, called XUI. Based on the index, we propose an efficient update method, called UOGX. Our experimental evaluation illustrates the space efficiency and update performance of the proposed labeling.

Keywords: Graph-Structured XML, Interval labeling, Incremental • Maintenance Algorithm.

1   Introduction

As XML is gaining unqualified success in being adopted as a universal data exchange format, particularly in the World Wide Web, the problem of managing and querying XML documents poses interesting challenges to database researchers. Previous works always model the XML as a directed tree. However, in many applications, an XML document needs to be modeled as a directed graph more naturally than a tree. For example, the XML document of the relationship of chapters and authors adapts to graph structure since one author may write more than one chapter and one chapter may have more than one author. A fragment of an XML document about books is shown in Fig.1. Obviously, the Graph-structured XML document can be represented in tree structure by duplicating the element with more than one incoming paths. But it will result in redundancy. If the information in Fig.1 is represented with a tree-structured XML document, the element "author" will be duplicated. XML standard

---

[*] This research was partially supported by The National Science and Technology Major Project of Key Electronic Devices, High-end General-purpose Chips and Fundamental Software Products Foundation of China under Grant No. 2010ZX01042-002-003; the Natural Science foundation of China under Grant No 61070055, 91024032, 91124001, 60833005; the Fundamental Research Funds for the Central Universities, and the Research Funds of Renmin University under Grant No. 11XNL010, 10XNI018.

H. Gao et al. (Eds.): WAIM 2012, LNCS 7418, pp. 254–265, 2012.

uses ID and IDREF types to avoid redundancy. ID represents a unique ID name for the attribute that identifies the element within the context of the document. The IDREF type allows the value of one attribute to be an element elsewhere in the document provided that the value of the IDREF is the ID value of the referenced element. Taking ID/IDREF into account, XML data should be modeled as a graph. Unlike other graph data (e.g. graph data in biological networks, social networks, and so on), there is an ordering, document order, defined on all the nodes in XML data. So XML data should be modeled as an ordered directed graph in this paper.

To process the queries (e.g. XQuery[1]) in XML database efficiently, a common method is that assign each node a label in advance and process queries using these labels. For Graph-structured XML, when the XML is static, it is easy to extend some existing labeling schemes to label them, like the approach in [2]. However, when the XML becomes dynamical, in other words, XML data changes over time, ID/IDREF relationships make the maintenance of labeling schemes of XML data more complex. The difficulty of this problem is that we not only need maintain the document order but also need maintain the reachability relationship. Although there have existed some methods to process updates of labeling for tree-structured XML, they can't be used to process updates of labeling of Graph-structured XML. We will give the detailed reasons in section 3.

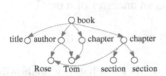
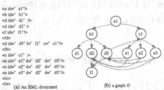

Fig. 1. An order directed graph   **Fig.2.** An Example of Graph-structured XML

In response to these, this paper studies the incremental maintenance of labeling of XML modeled as an ordered directed graph. We identify the main reason for the inefficiency of updates of labeling. Then, we propose an efficient and general update approach for dynamic XML. The main contributions of this paper include:

— We model the graph-structured XML data as ordered directed graph and propose an updatable extended interval labeling for XML data.
— Illustrate inefficient cases in updates of the extended labeling and propose a simple approach to deal with them.
— To get more efficient, we design an effective indexing, called XUI. Based on this indexing, we propose two novel algorithms for deletion and insertion of edges.
— We conduct comprehensive experiments to demonstrate the benefits and performances of our labeling scheme and updating algorithms.

The remainder of the paper is organized as follows: Section 2 presents summarizing the preliminaries. In section 3, we analyze inefficient cases in updates and present two naïve methods. Section4 introduces the novel indexing XUI. The efficient updating algorithms are presented in Section 5. In Section 6, we present an experimental evaluation of the proposed labeling and our update algorithms. Related work is discussed in Section 7 and Section 8 concludes this paper.

2   Preliminaries

2.1   Graph-Structured XML Data

Before giving the definition of Graph-structured XML Data, we first introduce the document order of XML. The document order is an order defined on all the element nodes in the document corresponding to the order in which the first character of the XML representation of each element occurs in the XML representation of the document after expansion of general entities.

Definition 1(Graph-Structured XML Data): With IDREF/ID in an XML document representing reference relationship, an XML document can be considered as an ordered directed graph, denoted as $G=(V, E, r)$. Each element node is mapped to a node in V. Nesting relationships and reference relationships of elements are mapped to two kinds of edges in E, nesting edge and reference edge. $r \prec$ is a partial order on V. For two nodes u and v in V, $u \prec v$ *iff* the order of v is more than u's in the document order.

From definition 1, we can see the reference edges can't affect the document order of a node in G. In this paper, we use $T(G)$ to denote a spanning tree of G, composed of all nodes in V and all nesting edges. A reachability relationship $v \to u$ represents that there is a path in G from v to u, in other words, v is an ancestor of u on G.

2.2   Extended Interval Labeling Scheme

In 1989, [2] proposed an interval-based labeling to label reachability relationships on directed acyclic graph. However, it didn't concern the document order between nodes. Since 2000, several interval-based labeling schemes have been proposed for tree-structured XML which use an interval to represent the order of nodes, like work [11] In addition, there have been a lot of approaches to maintain this kind of labeling, such as CDQS [8], Float-point [9], and so on.

Table 1. Labeling of G

Node	a1	b1	c1	d1	d2	d3	e1	e2	e3	f1	C
Id	[1,20]	[2,11]	[12,19]	[3,4]	[5,8]	[9,10]	[13,14]	[15,16]	[17,18]	[6,7]	[9,10][12,19]
ISet	[1,20]	[2,11]	[5,8] [9,10] [12,19]	[3,4]	[5,8]	[5,8] [12,19] [9,10]	[5,8] [9,10] [12,19]	[5,8] [9,10] [12,19]	[5,8] [9,10] [12,19]	[6,7]	[5,8] [9,10] [12,19]

Based on previous works, we design an extended interval labeling scheme which is updatable for XML. The basic idea of our labeling is to assign an special interval *Id* [*start, end*] for each node *n* in *G* to identify the node and represent the order of nodes and assign an interval set *ISet* to record nodes which can be reached by *n* in *G*. There are two steps to construct it. First, assign each node in *T(G)* an interval as *Id*, according to approach in [11]. Second, each strongly connected component (SCC)in *G* is contracted to one node to convert *G* to a DAG *D*; in a reverse topology order, compute an interval set like [2] for each node in *D*; all of *ISet* of nodes in the same SCC are set as the interval set of SCC. Table 1 is the labels of nodes on G in Fig.2.

According to above construction steps, it is easy to understand the two following Lemmas. Lemma2 is proved according to lemma1.Due to limitations on space, we don't give proofs.

Lemma1: Given a node n in G, any two intervals in $n.ISet$ can't intersect. Similarly, given C a strongly connected component, any two intervals in $n_C.ISet$ or $n_C.ID$ can't intersect.

Lemma2: Given two nodes u and v in G, u can reach v iff $v.ID$ is contained by some interval in $u.ISet$; $u \prec v$ iff $v.ID.start < u.ID.start$.

3   Analyses of Updates of Extended Interval Labeling

In XML, updates (updates of a node and an ID/IDREF attribution) can be transformed into a sequence of insertions and deletions of edges. So insertions and deletions of edges are the core operations of updates. In our paper, we only discuss insertions and deletions of edges. We focus on how to maintain the reachability relationship and the document order. Note that insertion and deletion of an element will cause insertion and deletion of its subelements. So, the update of nesting edges can't change reachability relationships of left nodes. For simplify, we only consider the insertions and deletions of edges causing by inserting or deleting a leaf node in XML. In fact, it is easy to extend our method to process the case causing by inserting or deleting a internal node. Firstly, consider the insertion of an edge (u, v). There are two cases: 1) Insert a reference edge; 2) Insert a nesting edge. In the remainder of the paper, $A(u) = \{a|a \rightarrow u\}$, $D(v) = \{d|v \rightarrow d\}$.

Case1. See Fig. 3(a). Consider the insertion of a reference edge (u, v), and suppose that a can't reach d before inserting. According to lemma2, a's label must be updated after inserting. Because the inserted edge is a reference which can't affect the order of nodes, $a.ID$ needn't be changed. Case1 is the bottleneck of insertions.

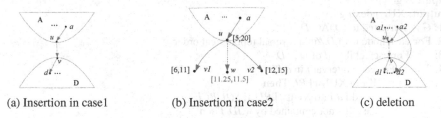

(a) Insertion in case1   (b) Insertion in case2   (c) deletion

Fig. 3. Illustration of insertion of an edge

Case2. Inserting a leaf element in XML will cause an insertion of a nesting edge in G. See Fig. 3(b). Insert a nesting edge (u, w), where $v1$ and $v2$ are children of u, w is inserted between $v1$ and $v2$. In this case, nodes in $A(u)$ and nodes (like $v2$) after w are affected. Luckily, Float-point [9] can avoid relabeling problems in tree-structured XML when a new node is inserted. See Fig.3.(b).

According to above analysis, we propose a naïve approach to process the insert of an edge (u, v) as follow:

1. Insert a nesting edge: label (v)// label v as described in case2;
2. Insert a reference edge:

For each node $w \in A$

    If w cannot reach v: $w.ISet = \cup \ v.ISet$, merge intersected intervals in $w.ISet$

Next, discuss a deletion of an edge (u, v) in G. If the deleted edge is a reference edge, it can't change the structure of $T\ (G)$. So such deletion will not affect ID labels of nodes. After deleting a nesting edge (in other words, deleting a leaf node), the order and reachability of left nodes can't be changed. According to lemma 2, their labels needn't be changed too. So, we focus on the deletion of a reference edge. Shown as Fig 3(c), after deleting the edge (u,v), $a1$ in $A(u)$ can't reach $d1$ in $D(v)$, but $a2$ in $A(u)$ can still reach $d2$ in $D(v)$. So, if a can still reach d, it is wrong to remove $v.ISet$ from $a.ISet(a \in A(u))$ directly. $a.ISet$ must be reconstructed. A naïve approach to process the deletion of an edge (u, v) is as follow:

1.   remove $v.ISet$ from $u.ISet$
2.   **For** each node $w \in A(u)$// in a reversal topological order, including u
3.   **For** each child d of w in G
4. $w.ISet = \cup \ d.ISet$, merge the intersected intervals in $w.ISet$
5.   **IF** $w.Iset$ does not change **Then** Return

4   XUI Index

Above naïve approaches need lots of times of merging two interval sets and once of computing A which needs to visit the labels of all nodes once. A deletion results in calculating the reversal topological of A (the time Complexity is $O(|V|)$). These largely increase the time of processing the updates. To improve the update efficiency, we specially design a new indexing, called XUI. In this section, we will introduce it in detail.

Algorithm 1. Construction of XUI indexing

Input: A graph G
Output: XUI indexing
01: G is converted to a DAG D
02: **For** each node w in D //in a reversal topological order
03:    **For** each child d of w in D
04:        **For** each interval i in $d.ISet$
05:            **If** i in XUI[w].RL **Then**
06:                add d into XUI[w].RL[i].$SOURCE$
07:            **Else If** i not contained by $w.ID$ **Then**
08:                add (i, d) to XUI[w].RL

In designing the index, we consider both update efficiency and lookup efficiency of the index. Hash Index is a good choice. So we design XUI based on Hash Index. The structure of XUI is shown in Fig.4. In hash table, the number of a node n in G is set as a key and each node points to an interval list RL. In RL, each record consists of three parts: $START, END$, and $SOURCE$, where $START$ and END represent an interval, and $SOURCE$ is a nodes set composed by children of n. Let LS the interval set of RL. For a node u in G, $LS_u = \{i | i \in children(u).ISet \ and \ i \nsubseteq u.ID \}$. For each record, $SOURCE$ stores u's children whose label contains the interval related with the $SOURCE$. Algorithm 3 illustrates the construction procedure of XUI.

5   Incremental Maintenance Algorithms for Labeling

Based on the analysis in Section 3, we propose a novel messaging-based method, called UOGX, which makes full use of the indexing XUI to speed up the processing of updates. The main idea of our method drives from a discovery: 1) after deleting an edge $(v1, v2)$, if some node on the paths from a to $v1$ can still reach $d \in D(v2)$, then a must be able to reach d; 2) after inserting $(u1, u2)$, If the reachability relationship between some node on paths from $a' \in A(u1)$ to $u1$ and $d' \in D(u2)$ is not changed, then the reachability relationship between a' and d' mustn't be changed. According to this discovery, we propose a bottom-up method to re-label nodes. We first define the structure of message queue, followed by the introduction of algorithms for insertion and deletion.

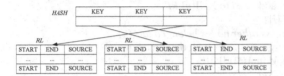

Fig. 4. The structure of XUI   **Fig. 5.** Message Queue

Data structure (*Message Queue*): Message Queue consists of a sequence of message subset (*MS*) from the children of corresponding node $n \in D$. A piece of message includes three parts: interval (*start* and *end*), type of operation (type, insertion or deletion) and the source of message (*from*). See Fig.5. To facilitating message processing, we merge messages in *MS* according to their intervals and type in algorithms.

5.1   Insertions for Extended Interval Labeling

In this section, we will describe how to use message queue to process the insertion of an edge in detail.

In algorithm 2, lines 01-02 deal with Case2 in section 3. Other lines deal with Case1. If u and v are in the same SCC *C1*, then the label and index of *C1* can't be affected. So nothing needs to do (Line 03). Otherwise, the labels of some nodes in $A(u)$ may be changed. Lines 04-25 process this case. We first process the node u in lines 06-13. As we know, in *ISet*, each interval represents all nodes on a sub-tree of $T(G)$. For case1 (line07), if an interval i is not be contained by some interval of $u.ISet$, then the corresponding nodes of i can't be reached by u before inserting. So $u.ISet$ must be updated. In line 08, update u's label and index item. In lines 09-11, send a piece of message to u's presents. If i is contained by $u.ISet$, then the corresponding nodes of i can be reached by u before inserting. So we needn't change u's label and only need to update the corresponding index item. According above discovery (2), the label of any node in $A(u)$ needn't be changed. So, we needn't send messages to their parents (lines12-13). If $i \subseteq u.ID$, we needn't change u' index item, as shown setion4. Then, we process u's ancestor nodes in D (lines 15-23) until the message queue is empty. For each ancestor node, the process is the same as u's. At last, we should remark that if u or v is in some SCC *C'* and *C'* can reach v before deleting d, anything won't be changed. Otherwise, replace u or v with C' and deal with it by algorithm 5.

Algorithm 2. Insert-edge

Input: a DAG D, an edge (u, v) to be inserted, a label set S ,XUI indexing
Output: a new label set S'
01: **If** the edge is a nesting edge **Then**
02:          Label(v) and Return
03: **If** u and v are in the same *SCC* **Then   Return**
04: **Else** $Q = \emptyset$; //Q is a message queue //*if u is in a SCC C*
 05:       **For** each interval i in $v.ISet$ // *process the node u*
06:          **Case**1: $i \not\subseteq j \in u.ISet$ // *all interval can't intersect (Lemma1)*
07:             add i to $u.ISet$ and add(i, v) to XUI[u].RL
08:             **For** each parent w of u // *send message to u's parent nodes*
09:                **If** $MS[w]$ not in Q **Then** Q.Enqueue $(MS[w])$// $MS[w]$ *is message subset of w*
10:                   add(i, *insertion,u*) to $MS[w]$
11:          **Case**2: $i \subseteq j, j \in u.ISet$ and $i \not\subseteq u.ID$
12:             **If** $i \notin XUI[u].RL$ **Then** add (i, v) to XUI[u].RL.
13:             **Else** add v to XUI[u].$RL[i]$
14:       **While** Q not empty **Do** // *process other nodes belonged to A(u)*
15:       Q.Dequeue $(MS[v'])$ and merge messages with the same interval
16:       **For** each   piece of message M in $MS[v']$
17:          **Case**1:$M.interval \not\subseteq j \in v'.ISet$
18:             add $M.interval$ to $v'.ISet$ and add($M.interval$, $M.from$ to XUI [v'].RL
19:             **For** each parent w of v'
20:                **If** $MS[w]$ not in Q **Then** Q.enqueue $(MS[w])$
21:                   add($M.interva$, *insertion,,* v') to $MS[w]$
22:          **Case**2: $M.interval \subseteq j, j \in v'.ISet$ and $M.interval \not\subseteq v'.ID$
23:             **If** $M.interval \notin k, k \in XUI[v'].RL$ **Then**
24:             add $(M.interval, M.from)$ to XUI[v'].RL
25:             **Else**   add $M.from$ to XUI [v'].$RL[M.interval]$

Complexity: In algorithm2, the dominating steps are lines 15-26. For each node v' in MQ, we performs XUI updates once and at most *max* times (the maximum |$v.ISet$|) reachability tests in lines 18-25. We need less than $Max*|E|$ times index updates and reachability tests, where $|E|$ is the total number of edges in D. Hence, in the worse case, the overall time complexity is O ($|E|$).The main space cost is that of message queue. Suppose m the maximum space cost of a piece of message in MS. The cost of space is less than $m*Max* |V|$, where $|V|$ is the total number of nodes in D. Hence, in the worse case, the overall space complexity is O ($|V|$).

Example 1: D (Fig.6 (a)) is the corresponding directed acyclic graph of G (Fig.2(b)). Its XUI indexing and labeling are shown in Fig.6 (b) and Table1. Consider an insertion of a reference edge $(c1, d1)$ in G. First, replace $c1$ with C. In processing C, because [3,4] in $d2.ISet$ is not contained by $C.ISet$, add [3,4] to $C.ISet$, update XUI[C] and send messages to $b1$and $a1$. The results are shown in fig.6(c).Then, $MS(b1)$ is out of MQ. Because [3, 4] is contained by $b1.ID$, we needn't update $b1$'s label and index, and send message to its parents (see Fig.6 (d)). Finally, MS ($b1$) is out of MQ. Similarly, we needn't do anything to deal with $a1$.

5.2   Deletions for Extended Interval Labeling

Using XUI indexing, it is easy to check the *reachability relationship* between two nodes after deleting. In this section, we introduce how to use message queue and XUI indexing to process the deletion of an edge from the bottom up.

Algorithm 3. Delete-edge

Input: a DAG D, an edge (u, v) to be deleted, a label set S ,XUI indexing
Output: a new label set S'
01: **If** (u, v) is a nesting edge **Then** // *if u and v are in same SCC, then we need do nothing*
02:        delete v's label and delete v' item from XUI
03: **Else**   //*(u, v) is a reference edge;*
04:     **For** each interval i in $v.ISet$ //$Q \leftarrow \emptyset$ Q is a message queue
05:        **If** $i \not\subset u.ID$ **Then**
06:           delete v from XUI[u].RL[i] and delete j from XUI[u]. RL //XUI[u]. $RL[j]=\emptyset$
07:           **If** $i \in u.ISet$ and $i \notin$ XUI[u]. RL **Then**
08:              delete i from $u.ISet$
09:              **For** each parent w of u
10:                 **If** $MS[w]$ not in Q **Then** $Q.Enqueue$ ($MS[w]$)
11:                    add(i, $delete,u$) to $MS[w]$// $MS[w]$ *is message subset of w*
12:              **If** $i \supset k \in$ XUI[u]. RL **Then**
13:                 add $Maximum(k)$ to $u.ISet$
14:                 add($Maximum(k)$, $insert,u$) to $MS[w]$// *for each parent w of u*
15:     **While** Q not empty **do**
16:        $M[v'] \leftarrow Q.Dequeue()$
17:        For each message M in $MS[v']$ //*firstly, process insertion messages*
18:           **Case**1: $M.type$ is insertion
19:              add ($M.interval$, $M.from$) to XUI[v']. RL
20:           **Case**2: $M.type$ is deletion
21:              **If** $M.interval \not\subset v'.ID$ **Then**
22:                 delete $M.from$ from XUI[v'].RL[$M.interval$]
23:                 delete i from XUI[v']. RL //XUI[v']. $RL[i]=\emptyset$
24:                 **If** $M.interval \in u.ISet$ and $M.interval \notin$ XUI[u]. RL **Then**
25:                    delete $M.interval$ from $v'.ISet$
26:                    **For** each parent w of v'
27:                       **If** $MS[w]$ not in Q **Then** $Q.Enqueue$ ($MS[w]$)
28:                          add($M.interval$, $delete,v'$) to $MS[w]$
29:                    **If** $M.interval \supset k \in$ XUI[u]. RL **Then**
30:                       add $Maximal(M)$ to $v'.ISet$
31:                       add($Maximal(M)$, $insert,u$) to $MS[w]$ //*for each parent w of v'*

In algorithm 3, Lines 01-02 deal with the deletion of a nesting edge. Others process reference edges. Lines 04-14 process u. In line 5, if i is contained by $u.ID$, then nodes represented by i still can be reached by u. So u's label needn't be changed. According to the property of XUI indexing, we needn't change u's index too. From discovery (1), we know all u's ancestors needn't relabel too. So we needn't send message to u's ancestors. Otherwise, update u's index in line 06. In line 07, $i \in u.ISet$ and $i \notin$ XUI[u]. RL means that all u's children can't reach node n ($n.ID=i$), after deleting. So u must be relabeled (line 8) and send messages to its parents (lines 9-11). $i \supset k \in$ XUI[u]. RL means that u can also reach n' descendant. Hence, we relabel u and send messages to u' parents in lines13-14. $i \notin$ XUI[u]. RL means that some child of u is still able to reach n, so we need do nothing, according to discovery (1). Then process u's ancestors until the message queue is empty (lines 15-31). To relabel node v' correctly, we first process insertion messages in $MS[v']$. If the type of a piece of message is insertion, then we only need update v'index. In a MS, if there is a piece of insertion message, then there must be a piece of deletion message with the same interval. Therefore, if it needs to relabel v', we will do it in case2 (lines 20-31). Lines 20-31 are similar with lines05-14. In Lines30-31, the function *Maximal* (*M*) returns all intervals of XUI[*u*]. *RL*.which are contained by *M. interval* and not contained by other interval in XUI[*u*]. *RL*. Finally, we should remark that if *u* or *v* is in some *SCC* *C'* and *C'* can still reach *v* after deleting *d*, anything won't be changed. Otherwise, replace u or v with *C'* and deal with it by algorithm 3.

Complexity: In algorithm3, the dominating steps are lines 15-31. We perform lines16-31 at most |*E*| times. In lines16-31, the dominating operations are once XUI lookup and at most *max* times reachability tests, where the maximum of |v.*ISet*| is *max*. Hence, we at most need *max**|*E*| times reachability tests and |*E*| times XUI lookups. The overall time complexity is O (|*E*|), in the worst case. The main space cost is that of message queue. Suppose *m* the maximum space cost of a piece of message in *MS*. Hence, the cost of space is at most $m* d*max*$ |*V*|, where *d* is the maximum fan-out. The overall space complexity is O (|*V*|) in the worse case.

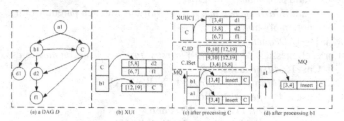

Fig. 6. An illusion of insertion

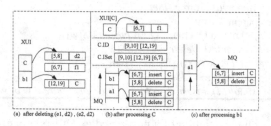

Fig.7. An illusion of deletion

Table 2. Data set

	Density(E/V)	Size	DAG size
G1	1.0	3000	3000
G2	1.5	3000	2456
G3	2.0	3000	1353
G4	2.5	3000	734
G5	3.0	3000	421
XMark	1.2	8518	7801

Example 2: Consider deletion sequence (*e1, d2*), (*e2, d2*), (*e3, d2*) in the graph *G* presented in Fig.2 (b). We use the labels depicted in table1 and the index shown in Fig.6 (b). After deleting (*e1, d2*) and (*e2, d2*), all indexes aren't changed (the result is shown in Fig.7 (a)) and all labels aren't changed, which concise with the fact that the reachability relationships do not change after deletions. In processing (*e3, d2*), we first replace *e3* with *C*. After deleting (*e3, d2*), in XUI[*C*].*RL* [5, 8], SOURCE related with [5,8] becomes empty, so the interval [5, 8] must be delete from *C. ISet* and XUI[*C*].*RL*. Because there is an interval [6, 7] in XUI[*C*], insert [6, 7] to *C. ISet* and send two pieces of messages to *b1* and *a1* (see Fig.7 (b)). Then *MS* [*b1*] is out of queue and process *b1*. Because [6, 7] and [5, 8] are contained by *b1.ID*, nothing needs to be done (see Fig.7(c)). Similarly, *a1* is dealt with as same as *b1*. These meet the fact that *C* can't reach *d2* and can also reach *f1*, and *a1* and *b1* can still reach *d2* and *f1*, after deleting the three edges.

6   Experimental Result

In this section, we present the results and analysis of part of our extensive experiments on the algorithms in this paper. All experiments were carried out using a Intel^R Core^{TM} Duo CPU type E8300 with two cores at 2.83GHz. We implemented all our algorithms with C++. The dataset we tested were the XMark benchmark [10] and generating random graphs. XMark can be modeled as a graph with complicated schema and circles. To check the effect of density, we generated 5 random graphs with different density. Their parameters are shown in Table2.

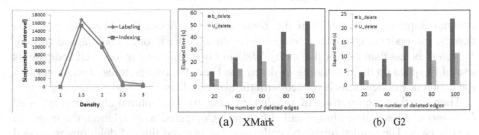

<table>
<tr><td>(a) XMark</td><td>(b) G2</td></tr>
</table>

Fig. 8. Size of labeling and indexing   **Fig.9.** Deletion performance

We are particularly interested in the following issues: the size of labeling and indexing and updates performance. In following, we will discuss in detail.

6.1   Size of Labeling and Indexing

We performed an experiment with five different files in table2 and compared the increase of the size of labeling and indexing with the increase of their density. The results are depicted in Fig 8. From the results, we can see, firstly, the size of labeling and indexing increases sharply with the increase of density, and in 1.5, the maximum is reached. Then the size becomes to decrease with the increase of density. Because the DAG size will decrease when the density becomes larger, lot of nodes can share the same labels. In addition, we can get that the size of indexing is related closely with the size of labeling. When the density is very small (for example, the structure of graph is close to a tree structure), our labeling scheme is very good.

6.2   Performance of Updates

To check the performances of insertion and deletion, we randomly chose 100 edges in graph. First, we deleted these edges from the graph, and then inserted them into graph.

In the experiment, we run our result in XMark data set and G2 because they have more labels. The size of labeling of XMark is 54697 and the size of its indexing is 46878.The results are depicted in Fig.9 and Fig.10. As we known, insertions and deletions of nesting edges are very simple, so we only present the results of reference edges.

First, consider the deletion. In Fig.9, the basic deletion algorithm is called *b_delete* and our optimal algorithm is called *U_delete*. From these results, we can know our

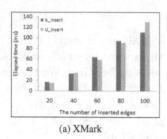 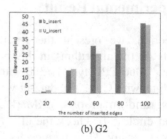

(a) XMark   (b) G2

Fig. 10. Insertion performance

optimal approach is better than the basic approach. Specially, when the graph is denser, the optimal approach is much faster than the basic one. The elapsed time of *U_delete* is less than half of that of *b_delete* in Fig.9 (b). We can see XUI indexing is useful to speed up the processing of deletions. From the two pictures, we can get that the efficiency of deletions is related with the size of labels.

Then consider the insertion of an edge. See Fig.10. The optimal algorithm is called *U_insert* and basic algorithm is called *b_insert*. Compared with deletion, the insertion is very easy. The results verify the collusion. The elapsed time of deletions is several hundred times of that of insertions with same edges. In some cases, the elapsed time of *U_insert* is larger than the *b_insert*, because *U_insert* needs to update the index. However, using XUI indexing, *U_delete* can improve the performance of deletions and processing of deletion is the bottleneck of updates. So, it is worth to do that.

7   Related Work

We classify the related works into two categories, labeling schemes for graph data and incremental maintenances of interval labeling scheme.

As we known, testing reachability relationships is a fundamental operator in queries on directed graphs. So far, a lot of labeling schemes have been proposed for this problem. 2-hop [5], optimal tree [2], optimal chain [4] are the well-known labeling schemes. Recently, Jin et al. proposed a labeling schemes 3-hop [6] for sparse graphs and dense graphs, and van Schaik et al. proposed a memory efficient labeling scheme PWAH [7] for very large graphs. However, these approaches can't be used to label Graph-structured XML directly. There are two reasons: 1) Most of them are not suitable to be extended for labeling the document order. 2) Most of them are difficult to be maintained. In this paper, we choose a based-interval labeling to extend, because interval labeling has made a great success in labeling ordered Tree–Structured XML.

On the other hand, there have been a lot of works to maintain interval labeling scheme for Tree-structured XML, such as CDQS [8], Float-point [9], and so on. All of them can be used to our approach to process the updates of nesting edges. To simplify, we choose Float-point in this paper. Float-point uses real values for the" start" and "end" of intervals, so we can insert many real values between any two different real values.

To the best of our knowledge, there haven't been studies on processing of update of Graph-structured XML. In this paper, we focus on how to maintain the reachability relationship and the document order in Graph –structured XML.

8   Conclusion and Future Work

In this paper, we model XML with ID/IDF as an ordered directed graph, and propose an updatable extended interval labeling. After analyzing the inefficiency of processing updates of this labeling, we give a basic solution. To improve the Performance of updates, we design a novel indexing XUI. Finally, we present an efficient approach to deal with the updates of extended interval labeling, based on the XUI indexing and message-passing queue. The main idea of this method comes from our discovery as described as section5.2. Through experiment and analysis on XMARK data sets and five generated random graph. By the experiment results, we confirmed the effectiveness and efficienty of our method.

In this paper, we deal with the problems on small data set, when the data becomes larger, how to update the labeling is a new challenge. In addition, due to limitations on space, we also give the algorithm for the case of the generation or destroy of a strongly connected component. Hence, in the future, we will consider these problems. and section5.2 without consideration of the generation or destroy of a strongly connected component.

References

1. XQuery, http://www.w3.org/TR/xquery/
2. Agrawal, R., Borgida, A., Jagadish, H.V.: Efficient management of transitive relationships in large data and knowledge bases. In: SIGMOD Conference 1989, pp. 253–262. ACM Press, New York (1989)
3. Wang, H., Li, J., Luo, J., Gao, H.: Hash-basesubgraph query processing method for graph-structured XML documents. PVLDB 1(1), 478–489 (2008)
4. Jagadish, H.V.: A compression technique to materialize transitive closure. ACM Trans. Database Syst. 15(4), 558–598 (1990)
5. Cohen, E., Halperin, E., Kaplan, H., Zwick, U.: Reachability and distance queries via 2-hop labels. In: Proceedings of the 13th Annual ACMSIAM Symposium on Discrete algorithms, pp. 937–946. ACM Press, New York (2002)
6. Jin, R., Xiang, Y., Ruan, N., Fuhry, D.: 3-HOP: a high-compression indexing scheme for reachability query. In: SIGMOD Conference 2009, pp. 813–826. ACM Press, New York (2009)
7. van Schaik, S.J., de Moor, O.: A memory efficient reachability data structure through bit vector compression. In: SIGMOD Conference 2011, pp. 913–924. ACM Press, New York (2011)
8. Li, C., Ling, T.W., Hu, M.: Efficient updates in dynamic XML data: from binary string to quaternary string. VLDB J 17(3), 573–601 (2008)
9. Amagasa, T., Yoshikawa, M., Uemura, S.: QRS: A Robust Numbering Scheme for XML Documents. In: The 19th Int. Conf. on Data Engineering (ICDE 2003), pp. 705–707. IEEE Press, New York (2003)
10. Schmidt, A., Waas, F., Kersten, M.L., Carey, M.J., Manolescu, I., Busse, R.: XMark: A benchmark forXML data management. In: VLDB (2002)
11. Zhang, C., et al.: On Supporting Containment Queries in Relational Database Management Systems. In: Proc. of ACM SIGMOD, pp. 425–436 (2001)

Extracting Focused Time for Web Pages

Sheng Lin[1], Peiquan Jin[1], Xujian Zhao[1], Jie Zhao[2], and Lihua Yue[1]

[1] School of Computer Science and Technology,
University of Science and Technology of China, 230027, Hefei, China
[2] School of Business, Anhui University, 230029, Hefei, China
`linsh@mail.ustc.edu.cn`

Abstract. Time plays important roles in Web search, because most Web pages contain temporal information and a lot of Web queries are time-related. In this paper, we concentrate on the extraction of the focused time for Web pages, which refers to the most appropriate time associated with Web pages. In particular, two critical issues are deeply studied. The first issue to extract implicit temporal expressions from Web pages, and the second is to determine the focused time among those extracted temporal information. For the first issue, we propose a new dynamic approach to resolve the implicit temporal expressions in Web pages. For the second issue, we present a score model to determine the focused time for Web pages. We conduct experiments on real data sets to measure the performance of our algorithms. The results show that our approach outperforms the competitor algorithms.

Keywords: Temporal Expressions Extraction, Web search, Focused Time.

1 Introduction

Temporal information plays an important role in many research areas such as information extraction, topic detection, question answering, query log analysis, and Web search. Temporal information usually appears in Web pages as temporal expressions, which are typically divided into two types, namely explicit expressions, e.g., March 7, 2012, and implicit expressions, e.g., Today. The various forms of temporal expressions impose some challenging issues to temporal information extraction within the scope of Web search:

(a) How to determine the right temporal information for implicit expressions contained in Web pages? Differing from the explicit expressions, which can be directly found in a calendar, the implicit expressions need a transformation process and usually a referential time is required.

(b) How to determine the focused time for a Web page? A Web page may contain a lot of temporal information, but which ones are the most appropriate times associated with the Web page? This is very important to temporal-textual Web search engines which support both terms-based and time-based queries.

For the first issue, the difficult part is to select the referential time which is used to resolve implicit expressions. For example, to determine the exact time of the implicit

H. Gao et al. (Eds.): WAIM 2012, LNCS 7418, pp. 266–271, 2012.

expression "Yesterday" in a Web page, we must know the date of NOW under the context. For the second issue, namely focused time determination, the difficult part is to develop an effective scoring technique to measure the importance and relevance of the extracted temporal information. For instance, suppose "April, 2011" and "17 April, 2011" are two extracted time words, "17 April, 2011" is contained in "April, 2011". Therefore, even "April, 2011" rarely appears in the Web pages, it will still be the focused time for the page in case that there are a great number of extracted time words contained by "April, 2011".

In this paper, the main contributions of the paper can be summarized as follows:

(a) We propose a new reference time dynamic-choosing approach to extract implicit temporal expressions in Web pages (see Section 3).

(b) We present a score model to determine the focused time for Web pages. Our score model takes into account both the frequency of temporal information in Web pages and the containment relationship among temporal information (see Section 4).

2   Related Work

GUTime is part of the TARSQI (Temporal Awareness and Reasoning Systems for Question Interpretation) toolkit (TTK) [1], which is the state-of-the-art tool for this natural-language processing task, it has a good performance in the extraction of explicit temporal expressions, but it does not perform very well in dealing with the implicit temporal expressions, especially in the case of lack of the document publication time. To improve the GUTime performance, we need to improve the reference choosing mechanism of GUTime.

Most of the works on temporal expression normalization do not give an effective reference time choosing method for implicit times in real texts. More specifically, the pioneer work by Lascarides [2] investigated various contextual effects on different temporal-reference relations. Then Hitzeman et al. [3] discussed the reference-choosing taking into account the effects of tense, aspect, temporal adverbials and rhetorical relations. Dorr and Gaasterland [4] presented the enhanced one in addition considering the connecting words. But they are theoretical in nature and heavily dependent on languages. Currently, the static choosing mechanisms [5, 6] for reference time are applied into some systems widely. Nevertheless, they are not adaptable to universal implicit times. Zhao [7] proposed a novel reference time dynamic-choosing mechanism which considers the global reference time and local reference time respectively.

In general, we can use the frequency of the temporal expressions to determine which the most relevant time is, but it does not take the relation among temporal expressions into consideration. There have been some studies on the extraction of the focused locations, such as the algorithm proposed by Zhang [8] and Web-a-where [9]. Inspired by them, we proposed a score model which takes the relation between temporal expressions into account to determine the focused time for a web page.

3   Extracting Temporal Expressions

The explicit temporal expressions can be recognized by many time annotation tools, such as GUTime, and they get high accuracy ratio in extraction. In this paper, we employ the GUTime tool to extract explicit temporal expressions. The biggest difference of recognition between the explicit and implicit temporal expressions is that the implicit temporal expressions need to determine a reference time, so choosing the right reference is the key to the identification of the implicit temporal expression.

In this paper, we classify temporal expressions two classes. One is called Global Time (GT) whose temporal semantics is independent of the local context, and takes the report time or publication time as the referent. Another one, Local Time (LT), makes reference to the narrative time in text above on account of depending on the current context. Table 1 gives some examples of GT and LT in real texts.

Table 1. Common Global Temporal Expressions and Local Temporal Expressions

Class	Sub-class	Examples	Class	Sub-class	Examples
	year	last year		year	that year
GT	month	next month	LT	month	October
	day	this Friday		day	the second day

In our approach, there is a reference time table which is used to hold full reference time for the whole text, and we need to update and maintain it dynamically after each normalizing process. The time table consists of two parts: Global Reference Time (GRT) and Local Reference Time (LRT).

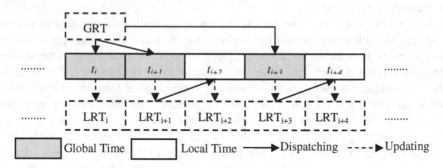

Fig. 1. Interaction between reference times and target times

In Fig. 1, we notice that different classes of time dynamically and automatically choose references based on their respective classes rather than doing it using the fixed value or the inconsiderate rule under the static mechanism.

4   Determining the Focused Time

In this section, we propose a novel and efficient algorithm to determine the most relevant time of an article. We consider two aspects when calculating the score of a temporal expression, namely the term frequency of the temporal expression and the relevance between temporal expressions. Here, we define the score of a temporal expression as a sum of an explicit score and an implicit score. The explicit score is related to the term frequency of a temporal expression, and accordingly the implicit score is related to the contribution made by all its children expressions. Table 2 shows the parent-child relationships among all the six time granularities we consider.

Table 2. The parent-child temporal relationships

time granularity	parent time granularity
DAY	MONTH
MONTH	QUARTER
QUARTER	HALF
HALF	YEAR
YEAR	DECADE

The explicit score $ES(T_i)$ is defined as the term frequency of T_i in the article. As compared to implicit temporal expressions, the explicit temporal expressions are more accurate in the extraction. So we add a weighting factor d to the implicit temporal expressions. The explicit score of T_i is defined as formula (1).

$$ES(T_i) = TF_{ETE}(T_i) + d * TF_{ITE}(T_i) . \tag{1}$$

Here, $TF_{ETE}(T_i)$ refers to the term frequency of the explicit temporal expressions which are recognized as T_i. $TF_{ITE}(T_i)$ refers to the term frequency of the implicit temporal expressions which are calculated as T_i. d is the weighting factor.

The implicit score $IS(T_i)$ is related to all the scores of its children, we denoted as $C_1, C_2, \ldots , C_n$ respectively, and we use the letter N to represent how many children unit T_i contains. For example, if the granularity of T_i is MONTH, then the value of N is 30 because a month contains about 30 days. Here, we use the factor α to represent how much contribution the children of T_i make. So the implicit score $IS(T_i)$ can be defined as formula (2).

$$IS(T_i) = \frac{1}{\alpha \times N} \sum_{i=1}^{n} S(C_i) . \tag{2}$$

5 Experiment

5.1 Dataset

Two real data sets are chosen in the experiments. The first small data set consists of 3,148 People's Daily news articles published in January, 1998. This data set is used to measure the performance of temporal expression extraction. The data collection contains 21,176 manually annotated Basic Temporal Expressions.

The second large data set consists of 1,812,933 English news articles crawled from the New York Times website. In this paper, we made an exhaustive statistics on the temporal information of the New York Times articles and use mean reciprocal ranking (MRR) to measure the effectiveness of the focused time extraction algorithm.

5.2   Evaluation on Temporal Expressions Extraction

For evaluating our algorithm objectively, we compare the experiment data with other two methods on the same testing corpus. The first compared method applies the publication time or report time (PT/RT) as the unique referent for normalization. The nearest narrative time (NNT) by Lin [6] is taken as the second compared method. Table 3 presents the results. It shows that our method exceeds the compared ones evidently.

Table 3. Results of temporal expressions extraction

Method	Average referent updating/article	Accuracy
PT/RT	0	68.42%
NNT	7.8	76.19%
DRT	4.2	83.55%

5.3   Evaluation on Focused Time Determination

We extract all the news articles in our corpus. It contains a total of 6,455,985 temporal expressions, and it includes 3,763,923 (58%) explicit temporal expressions and 2,692,062 (42%) implicit temporal expressions. More specifically, it mainly contains 3,157,156 DAY expressions (49%), 2,317,796 YEAR expressions (36%) and 959,747 MONTH expressions (15%).

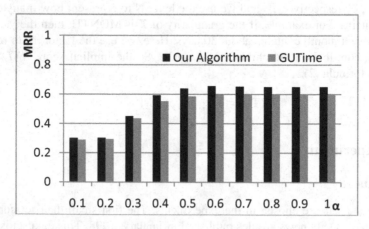

Fig. 2. MRR values of our score algorithm

To evaluate the effectiveness of the focused time extraction algorithm, we randomly select 50 news articles, and each of them contains more than 10 temporal expressions. The parameter d in our algorithm is set to 0.5, the parameter α ranges from 0.1

to 1.0, and the step is 0.1. For each value of α, using mean reciprocal ranking (MRR) evaluation standard, we estimate the ranking result of temporal expressions in each news article. Fig. 2 shows that the reference time dynamic-choosing mechanism works well in GUTime, and it also tells us that our algorithm gets the best performance when α is set to 0.6.

6   Conclusions

In this paper, we present an approach to determine the focused time for web pages. In particular, we apply the reference time dynamic-choosing mechanism to the temporal expressions extraction tool GUTime, which makes it more effective in recognizing. In addition, we make an exhaustive statistics on the temporal information of the New York Times articles. The reference time dynamic-choosing mechanism cannot be integrated into GUTime duo to the restriction of application architecture of GUTime. So rewriting the GUTime and making full use of the temporal information, particularly the focused time, in web pages are our next research focuses in the future.

Acknowledgement. This work is supported by the National Science Foundation of Anhui Province (NO. 1208085MG117), and the USTC Youth Innovation Foundation.

References

1. Verhagen, M., Pustejovsky, J.: Temporal Processing with the TARSQI Toolkit. In: Proceedings of the 22nd International Conference on Computational Linguistics (Coling 2008), pp. 189–192 (2008)
2. Lascarides, A., Asher, N., Oberlander, J.: Inferring Discourse Relations in Context. In: Proceedings of the 30th Meeting of the Association for Computational Linguistics, pp. 1–8 (1992)
3. Hitzeman, J., Moens, M., Grover, C.: Algorithms for Analyzing the Temporal Structure of Discourse. In: Proceedings of the 7th European Meeting of the Association for Computational Linguistics, pp. 253–260 (1995)
4. Dorr, B., Gaasterland, T.: Constraints on the Generation of Tense, Aspect, and Connecting Words from Temporal Expressions. Technical Report CS-TR-4391, UMIACS-TR-2002-71, LAMPTR-091, University of Maryland, College Park, MD (2002)
5. Jang, S.B., Baldwin, J., Mani, I.: Automatic TIMEX2 Tagging of Korean News. ACM Transactions on Asian Language Information Processing 3(1), 51–65 (2004)
6. Lin, J., Cao, D.F., Yuan, C.F.: Automatic TIMEX2 tagging of Chinese temporal information. Journal of Tsinghua University 48(1), 117–120 (2008)
7. Zhao, X.J., Jin, P.Q., Yue, L.H.: Automatic temporal expression normalization with reference time dynamic-choosing. In: Coling 2010, pp. 1498–1506 (2010)
8. Zhang, Q., Jin, P., Lin, S., Yue, L.: Extracting Focused Locations for Web Pages. In: Wang, L., Jiang, J., Lu, J., Hong, L., Liu, B. (eds.) WAIM 2011 Workshops. LNCS, vol. 7142, pp. 76–89. Springer, Heidelberg (2012)
9. Amitay, E., Har'El, N., Sivan, R., Soffer, A.: Web-a-where: geotagging Web content. In: Proc. of SIGIR, Sheffield, United Kingdom, pp. 273–280 (2004)

Top-Down SLCA Computation Based on Hash Search

Junfeng Zhou[1], Guoxiang Lan[1], Ziyang Chen[1], Xian Tang[2], and Jingfeng Guo[1]

[1] School of Information Science and Engineering,
[2] School of Economics and Management, Yanshan University, Qinhuangdao, China
{zhoujf,guoxianglan,zychen,txianz,jfguo}@ysu.edu.cn

Abstract. In this paper, we focus on efficient processing of a given XML keyword query based on SLCA semantics. We assign to each node an ID that equals to its visiting order when traversing the XML document in deep-first order, based on which we construct two kinds of indexes. The first index is an inverted list L of IDDewey labels for each keyword k, where each IDDewey label $l \in L$ represents a node v that directly contains k, l consists of node IDs corresponding to all nodes on the path from the document root to v. The second index is a hash table, which records, for each pair of node v and keyword k, the number of *occurrence* of k in the subtree rooted at v. Based on the two indexes, we propose an algorithm, namely TDHS, that takes the shortest inverted IDDewey label list as the working list and computes all SLCA results in a *top-down* manner based on hash search. Compared with existing methods, our method achieves the worst case time complexity of $O(m \cdot |L_1^{ID}|)$ for a given keyword query Q, where $|L_1^{ID}|$ is the number of distinct node IDs in the shortest inverted IDDewey label list of Q. Our experimental results verify the performance advantages of our method according to various evaluation metrics.

1 Introduction

Keyword search over XML data has attracted a lot of research efforts [1–3, 5–13] in the last decade, where a core problem is how to efficiently answer a given keyword query. Typically, an XML document can be modeled as a node-labeled tree T, and for a given keyword query Q, lowest common ancestor (LCA) is the basis of existing XML keyword search semantics [5, 6, 10, 13], of which the most widely followed variant is smallest LCA (SLCA) [8, 10]. Each SLCA node v of Q on T satisfies that v is an LCA node of Q on T, and no other LCA node of Q can be v's descendant node. The meaning of SLCA semantics is straightforward, i.e., smaller trees contain more meaningfully related nodes.

To facilitate SLCA computation on XML data, existing methods [2, 8–12] usually assign to each node v a unique ID that is either node ID (underlined number in Fig. 1) that is compatible with the document order [12], or Dewey label [8, 10, 11], or its variant, such as JDewey [2] and IDDewey [9] (Dewey labels consisting of node ID). For simplicity, we do not differentiate a node, its ID and the corresponding IDDewey unless there is ambiguity. For example, when we say node 3, it denotes node x_2 in Fig. 1 with ID 3 and IDDewey label 1.2.3. Based on the adopted labeling scheme, inverted lists are built for all keywords for fast SLCA computation.

H. Gao et al. (Eds.): WAIM 2012, LNCS 7418, pp. 272–283, 2012.

Table 1 shows the comparison of these algorithms, from which we have the following observations: (1) HS [9], LPSLCA [11] and FwdSLCA/BwdSLCA [12] are better than Stack [10], IL [10], IMS [8] and JDewey [2] according to their time complexity; (2) Among HS [9], LPSLCA [11] and FwdSLCA/BwdSLCA [12], LPSLCA (Fwd-SLCA/BwdSLCA) needs to afford $\log|L_m|(\log|L_m^{ID}|)$ cost to check whether a given node directly or indirectly contains a keyword by probing other inverted lists. On the contrary, the HS algorithm takes the shortest list L_1 as the working list and sequentially processes all IDDewey labels of L_1. In each iteration, it picks from L_1 an IDDewey label l and checks whether nodes represented by IDs of l contain all keywords of the given query in their subtree. By maintaining a hash mapping between each pair of node and keyword, the checking of whether a node contains a certain keyword in its subtree becomes a hash search, instead of probing other inverted list. Therefore, HS removes the $\log|L_m|(\log|L_m^{ID}|)$ factor from its time complexity. However, HS still suffers from much redundant computations when the number of results is much less than the length of the shortest inverted list, no matter the shortest inverted list is short or long, as shown by the following example.

Table 1. Worst case time complexities of different algorithms on SLCA computation, where $|L_1|(|L_m|)$ is the length of the shortest (longest) inverted list consisting of Dewey/JDewey/IDDewey labels, $|L_1^{ID}|(|L_m^{ID}|)$ is the number of distinct node IDs in the shortest (longest) inverted IDDewey label list.

Algorithm	Time Complexity	Labeling Scheme
Stack [10]	$O(d \cdot m \cdot (\sum_1^m |L_i|))$	
IL [10]	$O(d \cdot m \cdot |L_1| \cdot \log|L_m|)$	Dewey
IMS [8]	$O(d \cdot m \cdot |L_1| \cdot \log|L_m|)$	
LPSLCA [11]	$O(m \cdot |L_1| \cdot \log|L_m| + d \cdot m \cdot |L_1| \cdot \log\frac{|L_m|}{|L_1|})$	
JDewey [2]	$O(d \cdot m \cdot |L_1| \cdot \log|L_m|)$	JDewey
FwdSLCA [12]	$O(m \cdot |L_1^{ID}| \cdot \log|L_m^{ID}|)$	Node ID
HS [9]	$O(m \cdot \log d \cdot |L_1|)$	IDDewey
TDHS	$O(m \cdot |L_1^{ID}|)$	

Example 1. Consider processing $Q = \{a, b\}$ on D in Fig. 1. As $|L_a| = 200 < |L_b| = 1000$, the HS algorithm takes L_a as the working list. In each iteration, it sequentially picks an IDDewey label l from L_a, and processes all node IDs of l to compute a candidate SLCA node by probing the hash table. For this query, HS needs to check, for each label of a_1 to a_{100}, whether node 2 and node 3 contain keyword b in their subtrees; similarly, HS needs to check, for each label of a_{101} to a_{200}, whether node 105 and node 106 contain keyword b in their subtrees. Even though there are only two qualified SLCA nodes, i.e., x_1 (node 2) and x_3 (node 105), after processing this query, HS probes the hash table 400 times, which is unnecessary and time-consuming in practice.

The main reason for the redundancy problem of HS lies in that it processes each ID-Dewey label individually, without noticing that some IDs are repeatedly appearing in many different IDDewey labels in the same inverted IDDewey label list. For example, consider query $Q = \{a, b\}$ again. In L_a of Fig. 1, node a_1 to a_{100} share three IDs, i.e., 1,2 and 3, and node a_{101} to a_{200} share three IDs, i.e., 1, 105 and 106. Even though it does not need to check whether node 1 contains b by using binary search on each

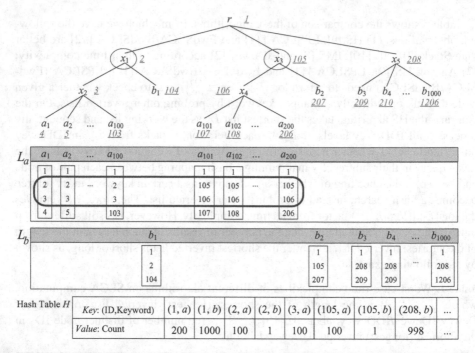

Fig. 1. A sample XML document D, where only nodes that directly or inderectly contain a and b are kept for explanation. L_a and L_b are the two inverted lists of IDDewey labels for $Q = \{a, b\}$, where all IDDewey labels in the two lists are shown in document order.

IDDewey label, it needs to repeatedly check the satisfiability of node 2, 3, 105 and 106 for 100 times, respectively.

To realize *fine-grained optimization*, we propose an efficient algorithm, namely TDHS, that computes all candidate SLCA nodes in a *top-down manner* to accelerate the SLCA computation. Intuitively, our method takes all nodes in the set of inverted IDDewey label lists as leaf nodes of an XML tree T, and checks whether it contains all keywords of the given query. The "top-down" processing strategy means that if T contains all keywords, T must contain at least one SLCA node, we then remove the root node of T and get a forest $\mathcal{F}_T = \{T_1, T_2, ..., T_n\}$ of subtrees corresponding to the set of child nodes of T's root node. Based on $\mathcal{F}_T$, we check whether each subtree contains all keywords. If no subtree in $\mathcal{F}_T$ contains all keywords, it means that T is a smallest tree that contains all keywords, then we directly output T's root node as an SLCA node; otherwise, for each subtree in $\mathcal{F}_T$ that contains all keywords, we just need to *recursively* compute its subtree set until no subtree in a subtree set contains all keywords.

To check whether there exists some subtrees of T that contain all keywords, our method records in a hash table H, for each pair of node v and keyword k, the number of occurrence of k in the subtree rooted at v, as shown in Fig. 1. During processing, our method takes the shortest inverted IDDewey label list L_1 (corresponding to k_1) as the working list, and checks whether a node represented by each *distinct* node ID is a qualified SLCA node, rather than repeatedly processing an ID as HS does when it is contained by many IDDewey labels. Specifically, Our method computes all *common ancestor* (*CA*) nodes in a top-down way. After processing a node v, it firstly gets the number of occurrence of k_1 in subtree T_v by one probe operation on the hash table, then skips all IDDewey labels that contain v's ID to get its next sibling node. Therefore, our method avoids the redundant probe operations on the hash table, and achieves the worst-case time complexity of $O(m \cdot |L_1^{ID}|)$, where $|L_1^{ID}|$ is the number of distinct node IDs in the shortest inverted IDDewey label list for a given keyword query.

Example 2. Continue Example 1. To process query $Q = \{a, b\}$ on the XML document in Fig. 1, our method takes L_a as the working list and computes CA nodes in a top-down way. It firstly checks whether node x_1 is a CA node by probing the hash table one time, then checks whether x_1's child node, i.e., x_2, is a CA node by another probe operation on H. Since x_2 is not a CA node, we directly skip all IDDewey labels containing 3 (x_2's ID) by the third probe operation on H. As x_2 does not have a sibling node, the processing of x_1 stops and we output x_1 as an SLCA node. Similarly, we just need to afford three probe operations on H to process x_3 and x_4, and output x_3 as an SLCA node. As a comparison, to process Q, our method just needs to probe H 6 times, while HS needs 400 probe operations on H.

The rest of the paper is organized as follows. In Section 2, we introduce background knowledge. We introduce our TDHS algorithm in Section 3. In Section 4, we present the experimental results, and conclude our paper in Section 5.

2   Background Knowledge

2.1   Data Model

We model an XML document as a node-labeled tree, where nodes represent elements or attributes, while edges represent direct nesting relationship between nodes in the tree. If a keyword k appears in the node name or attribute name, or k appears in the text value of v, we say v directly contains k. Fig. 1 is a sample XML document.

The positional relationships between two nodes include Document Order ($\prec_d$), Equivalence (=), AD (ancestor-descendant, $\prec_a$), PC (parent-child, $\prec_p$), Ancestor-or-self ($\preceq_a$) and Sibling relationship. $u \prec_d v$ means that u is located before v in document order, $u \prec_a v$ means that u is an ancestor node of v, $u \prec_p v$ denotes that u is the parent node of v. If u and v represent the same node, we have $u = v$, and both $u \preceq_d v$ and $u \preceq_a v$ hold.

2.2   Query Semantics

For a given query $Q = \{k_1, k_2, ..., k_m\}$ and an XML document D, inverted lists are often built to record which nodes directly contain which keywords. We use L_i to denote the inverted list of k_i, of which all nodes are sorted in document order. Let $S = \{v_1, v_2, ..., v_n\}$ be a set of nodes, $lca(S) = lca(v_1, v_2, ..., v_n)$ denotes the lowest common ancestor (LCA) of all nodes in S.

The LCAs of Q on D are defined as $LCA(Q) = LCA(L_1, L_2, ..., L_m) = \{v | v = lca(v_1, v_2, ..., v_m), v_i \in L_i (1 \le i \le m)\}$. E.g., the LCA nodes for $Q = \{a, b\}$ on D in Fig. 1 include r, x_1 and x_3.

Compared with LCA, SLCA [8, 10] defines a subset of $LCA(Q)$, of which no LCA in the subset is an ancestor of any other LCA, which can be formally defined as $SLCA(Q) = \{v | v \in LCA(Q) \text{ and } \nexists v' \in LCA(Q), \text{ such that } v \prec_a v'\}$. In Fig. 1, although r is an LCA node, r is an ancestor of x_1 and x_3, thus the set of SLCAs for $Q = \{a, b\}$ on D in Fig. 1 are x_1 and x_3.

For SLCA computation, researchers have proposed many algorithms [2,8–12], which have been discussed in Section 1.

3   The Algorithm for SLCA Computation

3.1   Data Organization

We assign to each node an ID that equals to its visiting order when traversing the XML document in deep-first order, as shown by the italics numbers in Fig. 1, based on which we construct two kinds of indexes.

The first index is an inverted list L of IDDewey labels for each keyword k, where each IDDewey label $l \in L$ represents a node v that directly contains k, l consists of node IDs corresponding to all nodes on the path from the document root to v. E.g., L_a and L_b in Fig. 1 are the two inverted IDDewey label lists of keyword a and b, respectively.

The second index is a hash table, which records, for each pair of node v and keyword k, the number of occurrence of k in the subtree rooted at v, which is shown by the "Count" value. H in Fig. 1 shows partial content of the hash table, from which we know that there are 100 occurrence of keyword a in the subtree rooted at node 2 by using "$(2, a)$" as a key to probe H, which can be denoted as $100 = H[(2, a)]$; on the contrary, using "$(3, b)$" as a key to probe H, we know that the subtree rooted at node 3 does not contain keyword b, which can be denoted as $(3, b) \notin H$.

3.2   The TDHS Algorithm

For a given query $Q = \{k_1, k_2, ..., k_m\}$, we always assume that $0 < |L_1| \leq |L_2| \leq ... \leq |L_m|$, the case where at least one IDDewey label list is empty can be easily processed before line 1. We omit it for simplicity. As shown by Algorithm 1, our method takes the shortest inverted IDDewey label list L_1 as the working list, which is associated with a "$cursor$" pointing to some IDDewey label of L_1. Let l be the ith IDDewey label of L_1, id_v the jth ID of l denoting node v, we have $l = L_1[i]$, $id_v = l[j]$ and $id_v = L_1[i][j]$. In the following discussion, we use $|l|$ to denote the number of IDs of l, and $|L_1|$ the number of IDDewey labels of L_1.

We use T to denote a subtree rooted at a certain node. $T.root$ represents the IDDewey label of the root node of T, $T.start(T.end)$ denotes, in L_1, the position of an IDDewey label, which corresponds to the first (last) node in T that directly contains k_1. As shown by Algorithm 1, initially, we set node 1 as the root node of T (line 1), and all nodes of L_1 as leaf nodes of T (line 2 and 3). In line 4, we recursively process T by calling procedure processSubTree(T).

In line 5 to 17, we recursively process subtree T and output $T.root$ as an SLCA node if no child node of $T.root$ is a CA node. In line 5, we set $flag$ with default value

"TRUE" to denote that $T.root$ is an SLCA node. In line 6 to 7, we make "$cursor$" point to the first IDDewey label of L_1, such that the length of $L_1[cursor]$ is greater than that of $T.root$. In line 8 to 16, we repeatedly check whether each child node of $T.root$ is a CA node. In each iteration, we get the ID id_v corresponding to a child node v of $T.root$ in line 9, then we check whether node v is a CA node in line 10. If function isCA(id_v) returns TRUE, it means that v is a CA node, which also means that $T.root$ is not an SLCA, thus we set $flag$ with the "FALSE" value (line 11). In line 12, we get the subtree T' rooted at v, and recursively process T' in line 13. If function isCA(id_v) returns FALSE in line 10, we firstly get the number of occurrence of k_1 in the subtree rooted at v in line 15, then skip all IDDewey labels containing id_v in line 16. After processing all child nodes of $T.root$, in line 17, if $flag$ = TRUE, it means that no one of $T.root$'s child nodes is a CA node, then we directly output $T.root$ as a qualified SLCA node.

In line 18 to 22, to get the subtree T' rooted at v, we firstly get the number of occurrence of k_1 in T' by probing the hash table H using (id_v, k_1) (line 18), then set the IDDewey of v as $T.root.id_v$ in line 19, where $T.root$ is the IDDewey of v's parent node. In line 20, we set the value of $T'.start(T'.end)$, which corresponds to the first (last) node in T' that directly contains k_1.

In line 23 to 25, we check whether a given node v is a CA node, which needs to repeatedly check, in line 23 to 24, whether $k_i(i > 1)$ appears in the subtree rooted at v by using (id_v, k_i) as the key to probe the hash table H at most $m - 1$ times.

Example 3. Consider processing query $Q = \{a, b\}$ on the XML document in Fig. 1, Algorithm 1 takes L_a as the working list. As node r is the document root node, our method recursively finds child CA nodes of node r. The first CA node found by our method is x_1 by using $(2, b)$ as the key to probe the hash table. After that, our method checks whether x_2 is a $\overline{\text{CA}}$ node, which can be done by another probe operation on the hash table using $(3, b)$ to find whether there are occurrence of b in the subtree rooted at x_2. Since $(3, b)$ does not appear in the hash table in Fig. 1, x_2 is not a CA node, our method skips all IDDewey labels containing 3 to find x_2's next sibling node. Note that the number of skipped nodes equals to the number of occurrence of keyword a in the subtree rooted at x_2, which is 100 and can be found by another probe operation on the hash table using $(3, a)$. As x_2 does not have other sibling nodes, the skipping operation moves the "$cursor$" to the IDDewey of a_{101}, and the processing of x_1 is stopped, then x_1 is outputted as an SLCA node. The following processing is similar, our method needs to check the satisfiability of x_3 and x_4. For each one of them, we need to use one probe operation to check whether it is a CA node, and another probe operation for x_4 to skip IDDewey labels containing 106. In summary, to process $Q = \{a, b\}$, our method needs to check the satisfiability of x_1, x_2, x_3 and x_4 by probing the hash table four times, and skipping useless IDDewey labels by probing the hash table two times. The total number of probe operations invoked by our method is 6.

Now we analyze the complexity of TDHS. Assume that for a given query $Q = \{k_1, k_2, ..., k_m\}$, the set of IDDewey label lists satisfy $0 < |L_1| \leq |L_2| \leq ... \leq |L_m|$. During processing, our method takes the shortest inverted IDDewey label list L_1 (corresponding to k_1) as the working list, and checks whether a node represented by each *distinct* node ID is a CA node, rather than repeatedly processing an ID as HS does when it is contained by many IDDewey labels. Therefore, the total number of processed node IDs equals to the number of distinct IDs in all IDDewey labels of L_1, which is denoted by L_1^{ID}. For each ID id_v corresponding to node v, we need at most $m-1$ probe operations to check whether v is a CA node, and at most one probe operation to skip IDDewey labels containing id_v. That is, for each id_v, our method needs at most m probe operations on the hash table H. Therefore, the worst-case time complexity of our method is $O(m \cdot |L_1^{ID}|)$.

Algorithm 1. TDHS(Q) /*$Q = \{k_1, k_2, ..., k_m\}, 0 < |L_1| \le |L_2| \le ... \le |L_m|$*/

```
 1   T.root ← 1
 2   T.start ← 1
 3   T.end ← |L_1|
 4   processSubTree(T)
```

Procedure processSubTree(T)

```
 5   flag ← TRUE
 6   if (|L_1[T.start]| > |T.root|) then cursor = T.start
 7   else cursor = T.start + 1
 8   while (cursor ≤ T.end) do
 9       id_v ← L_1[cursor][|T.root| + 1]
10       if (isCA(id_v)) then
11           flag ← FALSE
12           T' ← getSubTree(T, id_v)
13           processSubTree(T')
14       else
15           count ← H[(id_v, k_1)]
16           cursor ← cursor + count
17   if (flag = TRUE) then output T.root as a qualified SLCA result
```

Function getSubTree(T, id_v)

```
18   count ← H[(id_v, k_1)]
19   T'.root ← T.root.id_v
20   T'.start ← cursor
21   T'.end ← T'.start + count − 1
22   return T'
```

Function isCA(id_v)

```
23   foreach (k_i ∈ Q, i > 1)do
24       if ((id_v, k_i) ∉ H) then return FALSE
25   return TRUE
```

---

4   Experimental Evaluation

4.1   Experimental Setup

Our experiments were implemented on a PC with Intel(R) Core(TM) i5 M460 2.53GHz CPU, 2 GB memory, and Windows 7 as the operating system. The algorithms used for comparison include the IL [10], IMS [8], JDewey [2], LPSLCA [11], FwdSLCA [12] and HS [9] algorithms. All algorithms were implemented using Microsoft VC++, all results are the average time by executing each algorithm 1000 times on hot cache. We did not make comparison with Stack [10] because Stack has been verified not as efficient as other existing methods [8–12].

We used XMark (582MB) dataset for our experiment. We have selected 30 keywords classified into three categories according to their occurrence frequencies (i.e. $|L_{IDDewey}|$ line in Table 2): (1) low frequency (100-1000), (2) median frequency

Table 2. Statistics of keywords used in our experiment

Keyword	tissue	baboon	necklace	arizona	cabbage	hooks	shocks	patients	cognition	villages
$L_{IDDewey}$	384	725	200	451	366	461	596	382	495	829
Keyword	male	takano	order	school	check	education	female	province	privacy	gender
$L_{IDDewey}$	18441	17129	16797	23561	36304	35257	19902	33520	31232	34065
Keyword	bidder	listitem	keyword	bold	text	time	date	emph	incategory	increase
$L_{IDDewey}$	299018	304969	352121	368544	535268	313398	457232	350560	411575	304752

(10000-40000), (3) high frequency (300000-600000). Based on these keywords, we generated 18 queries as shown in Table 3. Index sizes are listed in Table 4, where Dewey/JDewey/IDDewey is for IL, IMS, JDewey and LPSLCA; Node ID is for Fwd-SLCA, while IDDewey+Hash Table is for HS and TDHS.

Table 3. Queries on 582MB XMark dataset, $|L_{min}|$ denotes the length of the shortest IDDewey label list for a query, N_S is the number of qualified SLCA results, $R_S = N_S/|L_{min}|$ denotes the result selectivity.

ID	Keywords	$|L_{min}|$	N_S	$R_S(\%)$	Freq.
QX1	villages,hooks	461	9	1.95	Low
QX2	baboon,patients,arizona	382	1	0.26	
QX3	cabbage,tissue,shocks,baboon	366	9	2.46	
QX4	shocks,necklace,cognition,cabbage,tissue	200	9	4.5	
QX5	female,order	16700	570	3.41	Med
QX6	privacy,check,male	18428	29	0.16	
QX7	takano,province,school,gender	17129	107	0.62	
QX8	school,gender,education,takano,province	17129	107	0.62	
QX9	bold,increase	304706	34136	11.2	High
QX10	date,listitem,emph	304969	43777	14.35	
QX11	incategory,text,bidder,date	299018	1	0.0003	
QX12	bidder,date,keyword,incategory,text	299018	1	0.0003	
QX13	incategory,cabbage	366	224	61.2	Random
QX14	province,bold,increase	33520	427	1.27	
QX15	listitem,emph,arizona	451	1	0.22	
QX16	bold,increase,hooks,takano	461	6	1.3	
QX17	emph,arizona,villages,education	451	1	0.22	
QX18	check,bidder,date,baboon	742	1	0.13	

Table 4. Comparison of index sizes

Dataset	Dewey/JDewey/IDDewey	Node ID	IDDewey+Hash Table
XMark(582MB)	2.3GB	817MB	3.0GB

4.2   Performance Comparison and Analysis

For a given query, we define the *result selectivity* as the size of the results over the size of the shortest inverted list. The metrics for evaluating these algorithms include: (1) running time, and (2) number of probe operations on the hash table, which is only used for the HS and TDHS algorithms. The reason we use the second metric is that for all compared algorithms, only HS and TDHS are based hash search, which do not need the comparison operation between IDDewey labels. For other algorithms, we only compare their running time.

Fig. 2 shows the running time of different algorithms for query QX1 to QX18. Table 5 shows the number of probe operations of HS and our TDHS algorithms on the

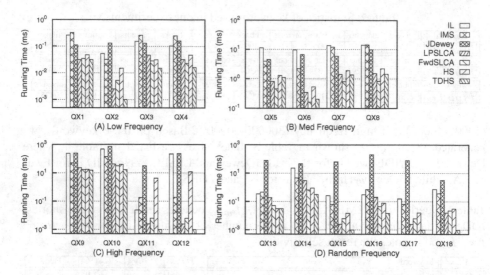

Fig. 2. Comparison of running time for SLCA computation on XMark dataset (log-scaled)

hash table for these queries, which can be used to facilitate the understanding of the performance difference between HS and TDHS. From Fig. 2 and Table 5, we have the following observations.

(1) our method outperforms the HS algorithm for most of these queries. The reason lies in that for each distinct ID in IDDewey labels of the shortest inverted IDDewey label list, our method processes it without redundant probe operations on the hash table, which is further verified by Table 5. E.g., for query QX2, QX4, QX6, QX11, QX12, QX15 QX16, QX17 and QX18, the number of probe operations invoked by our method is much less than that of the HS algorithm. For query QX11 and QX12, our method outperforms HS by up to two orders of magnitude. The lower the selectivity, the more benefits can be brought by our method.

(2) compared with LPSLCA and FwdSLCA, our method can work better than them for many queries, such as QX2 to QX4, QX10 to QX12, and QX14 to QX18, this is because when checking wether a node is a CA node, our method does not need to probe other inverted lists based on binary search, instead, we simplify this operation with a simple hash search, which can be done more efficiently. Even though, our method suffers from inflexibility when compared with these two algorithms, such as QX5. The reason lies in that our method takes the shortest inverted IDDewey list as the working list, and it needs to verify the satisfiability of CA node for IDs of IDDewey labels in this list. As a comparison, LPSLCA and FwdSLCA may skip more useless IDs by using IDs of other lists to probe the shortest list.

(3) our method is much more efficient than IL, IMS and JDewey, because our method only processes IDs of the shortest inverted list, while IL, IMS and JDewey need to make repeatedly comparison between Dewey labels to check their positional relationships and compute their LCA node, which is costly in practice.

Besides the above observations, for existing methods, we have the follow results: (1) IL could perform better than IMS in some cases, such as QX1, QX3, QX4, QX11, QX13

Table 5. Comparison of the number of probe operations on hash table, where $N^{ID}_{L_{min}}$ is the number of IDs in all IDDewey label of the shortest inverted list, while $N^{DistID}_{L_{min}}$ is the number of *distinct* IDs in all IDDewey labels of the shortest list.

Query	QX1	QX2	QX3	QX4	QX5	QX6	QX7	QX8	QX9
$N^{ID}_{L_{min}}$	3869	3306	3016	1704	116429	113081	100338	100338	1542891
HS	1899	764	3361	2383	61044	31937	153856	198913	1057672
$N^{DistID}_{L_{min}}$	2674	2309	2071	1198	68049	62162	53324	53324	683370
TDHS	942	12	772	450	35506	5005	34608	34726	482506
Query	QX10	QX11	QX12	QX13	QX14	QX15	QX16	QX17	QX18
$N^{ID}_{L_{min}}$	2353169	1196072	1196072	3016	173900	2255	3869	2255	6013
HS	2216869	299018	897054	1060	46858	451	3601	451	2659
$N^{DistID}_{L_{min}}$	574800	353220	353220	2071	105795	1355	2674	1355	4014
TDHS	310463	2	4	916	7500	2	862	2	350

and QX16, but the performance gain is usually much less than that of IMS on IL, such as QX5, QX6, QX14 and especially QX12, because for these queries, the set of Dewey label lists for each one has different keyword distributions, and IL is not as flexible as IMS on utilizing various keyword distributions to accelerate the computation. (2) JDewey is usually beaten by IL and IMS, especially for queries with Dewey label lists having huge difference on their lengths, such as QX11, and QX13 to QX18, this is because JDewey needs to process all lists of each level from the leaf to the root; and for all lists of each level, after finding the set of common nodes, it needs to recursively delete all ancestor nodes in all lists of higher levels, which is very expensive in practice. (3) LPSLCA and FwdSLCA outperform HS for many queries, because HS suffers form much redundant probe operations on hash table. Besides, all LPSLCA, FwdSLCA and HS are much better than IL and IMS, the reason lies in that LPSLCA, FwdSLCA and HS do not need to repeatedly check the positional relationships and compute LCA node for two Dewey labels.

Besides the 18 queries in Table 3, we randomly generated 174406 queries with 2, 3, 4 and 5 keywords based on the 30 keywords of Table 2, which contains all possible combinations of these keywords, that is, $174406 = C^2_{30} + C^3_{30} + C^4_{30} + C^5_{30}$. Based on these random queries, we record their average running time based on different result selectivities, which provides us a way to better understand different algorithms.

Fig. 3 shows the impacts of result selectivity on performance of these algorithms. Again, from the four figures in Fig. 3 we know that the average performance of our TDHS algorithm is better than other existing methods, the more keywords involved in a given query, the more benefits can be got by our method. Compared with the HS algorithm, our method performs much better when the result selectivity is low. Compared with IL, IMS, LPSLCA and FwdSLCA, the performance gain increases when the number of keywords involved in a query increases.

Note that in Fig. 3 (A) and (B), with the increase of the result selectivity, the average time used by all methods for result selectivity in [40,100] is less than that in [30,40), which can be explained by Fig. 4, where for queries with 2 and 3 keywords, the number of results decreases with the change of result selectivity from [30,40) to [40,100], which means that the performance of all algorithms is also affected by the number of results.

We shown the scalability from two aspects based on Fig. 3: (1) fixing the number of keywords and varying the result selectivity, which is just explained in the previous

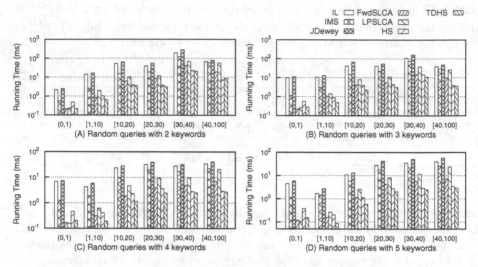

Fig. 3. Comparison of running time with different result selectivity

paragraphs, (2) fixing the result selectivity and varying the number of keywords, which can be got from the four sub-figures of Fig. 3 and is omitted by limited space. The general trend can be stated as: the performance of IL, IMS, JDewey, LPSLCA and FwdSLCA will be better with the increase of the number of keywords, this is because the performance of IL, IMS, JDewey, LPSLCA and FwdSLCA can utilize the positional relationships between keyword nodes to skip useless nodes, the more keywords involved, the more possibility for these algorithms to make optimization. Compared with IL, IMS, JDewey, LPSLCA and FwdSLCA, HS can work better with the increase of result selectivity and the number of processed keywords in a given query. Compared with HS, our TDHS algorithm can work better when the result selectivity is low.

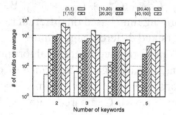

Fig. 4. Average number of results for each selectivity

Fig. 5. Running time of Q10 on different XML documents

Further, we shown in Fig. 5 the scalability when executing Q10 on XMark dataset from 116MB to 1745MB (15x). The query time of these algorithms grows sublinearly with the increase of the data size. Also, the query time for TDHS is consistently about 30 times faster than IL, 9 times faster than IMS, 40 times faster than JDweey, 2.5 times faster than LPSLCA, 2 times faster than FwdSLCA and 2.3 times faster than HS. For other queries, we have similar results, which are omitted due to space limit.

5   Conclusions

In this paper, we proposed an efficient algorithm, namely TDHS, that computes all qualified SLCA nodes in a *top-down* way based on *hash search*. Our method records in a hash table H, for each pair of node v and keyword k, the number of occurrence of k in the subtree rooted at v. During processing, our method takes the shortest inverted IDDewey label list L_1 as the working list, and checks whether a node represented by each *distinct* node ID id_v is a qualified SLCA node, rather than repeatedly processing id_v as HS does when it is contained by many IDDewey labels. As a result, our method avoids the redundant probe operations on the hash table, and achieves the worst-case time complexity of $O(m \cdot |L_1^{ID}|)$, where $|L_1^{ID}|$ is the number of distinct node IDs in L_1. Experimental results verify the performance advantages of our method according to various evaluation metrics.

Acknowledgment. This research was partially supported by grants from the Natural Science Foundation of China (No. 61073060, 61040023, 61103139), the Fundamental Research Funds of Hebei Province (No. 10963527D), and the Hebei Science and Technology research and development program (No. 11213578).

References

1. Bao, Z., Ling, T.W., Chen, B., Lu, J.: Effective xml keyword search with relevance oriented ranking. In: ICDE (2009)
2. Chen, L.J., Papakonstantinou, Y.: Supporting top-k keyword search in xml databases. In: ICDE (2010)
3. Bao, Z., Ling, T.W., Chen, B., Lu, J.: Effective xml keyword search with relevance oriented ranking. In: ICDE (2009)
4. Guo, L., Shao, F., Botev, C., Shanmugasundaram, J.: Xrank: Ranked keyword search over xml documents. In: SIGMOD Conference (2003)
5. Li, G., Feng, J., Wang, J., Zhou, L.: Effective keyword search for valuable lcas over xml documents. In: CIKM (2007)
6. Li, Y., Yu, C., Jagadish, H.V.: Schema-free xquery. In: VLDB (2004)
7. Liu, Z., Chen, Y.: Identifying meaningful return information for xml keyword search. In: SIGMOD Conference (2007)
8. Sun, C., Chan, C.Y., Goenka, A.K.: Multiway slca-based keyword search in xml data. In: WWW (2007)
9. Wang, W., Wang, X., Zhou, A.: Hash-Search: An Efficient SLCA-Based Keyword Search Algorithm on XML Documents. In: Zhou, X., Yokota, H., Deng, K., Liu, Q. (eds.) DASFAA 2009. LNCS, vol. 5463, pp. 496–510. Springer, Heidelberg (2009)
10. Xu, Y., Papakonstantinou, Y.: Efficient keyword search for smallest lcas in xml databases. In: SIGMOD Conference (2005)
11. Zhou, J., Bao, Z., Chen, Z., Lan, G., Lin, X., Ling, T.W.: Top-Down SLCA Computation Based on List Partition. In: Lee, S.G., Peng, Z., Zhou, X., Moon, Y.-S., Unland, R., Yoo, J. (eds.) DASFAA 2012, Part I. LNCS, vol. 7238, pp. 172–184. Springer, Heidelberg (2012)
12. Zhou, J., Bao, Z., Wang, W., Ling, T.W., Chen, Z., Lin, X., Guo, J.: Fast slca and elca computation for xmlkeyword queries based on set intersection. In: ICDE (2012)
13. Zhou, R., Liu, C., Li, J.: Fast elca computation for keyword queries on xml data. In: EDBT (2010)

Top-K Graph Pattern Matching:
A Twig Query Approach '

Xianggang Zeng[1], Jiefeng Cheng[1], Jeffrey Xu Yu[2], and Shengzhong Feng[1]

[1] Shenzhen Institutes of Advanced Technology, Chinese Academy of Sciences, China
{xg.zeng,jf.cheng,sz.feng}@siat.ac.cn
[2] The Chinese University of Hong Kong, Hong Kong, China
yu@se.cuhk.edu.hk

Abstract. There exist many graph-based applications including bioinformatics, social science, link analysis, citation analysis, and collaborative work. All need to deal with a large data graph. Given a large data graph, in this paper, we study finding top-k answers for a graph query, and in particular, we focus on top-k cyclic graph queries where a graph query is cyclic and can be complex. The capability of supporting top-k cyclic graph queries over a data graph provides much more flexibility for a user to search graphs. And the problem itself is challenging. After investigating a direct yet infeasible solution, we propose a new twig query approach. In our approach, we first identify a spanning tree of the cyclic graph query, which is used to generate a list of ranked twig answers on-demand. Then we identify the top-k answers for the graph query based on the twig answer list. In order to find the best twig query in solving a given cyclic graph query, cost-based optimization for twig query selection is studied. We conducted extensive performance studies using a real dataset, and we report our findings in this paper.

1 Introduction

With the rapid growth of World-Wide-Web and new data archiving/analyzing techniques, there exists a huge volume of data available in public, which is graph structured in nature including bioinformatics, social science, link analysis, citation analysis, and collaborative network. *Graph pattern matching* is long investigated in database study. It traditionally stands for subgraph isomorphism problem [25,23], which determines whether a small graph pattern is exactly contained in another graph, or graphs in a large graph collection (as the data). Its main application is the so called frequent subgraph mining, which has been extensively studied for the last decade [29,28,22,30].

In recent year, there are an increasing number of applications which need to deal with large standalone graphs, such as link analysis, social networks and bioinformatics. it is important to know patterns existed in a single large graph [31]. Graph pattern matching also is no longer limited to subgraph isomorphism. Usually, the conditions on the matched instances in the data graph generalize to the label requirements and the structural requirements, which are succinctly represented by a query graph [3,5,24,26,32]. The graph pattern matching of this type has many applications: It is used in the join processing for managing large *XML* documents [26]. In life sciences, graph pattern matching can be used for protein interaction networks comparison and protein structure

H. Gao et al. (Eds.): WAIM 2012, LNCS 7418, pp. 284–295, 2012.

matching [24]. In software engineering, graph pattern matching is used for dependance-related code search over the system dependence graph of the program [27]. Graph pattern matching is also central to quering massive RDF repository using SPARQL [33]. Other applications include interactive graph visulization [19], computing service discovery [7] and 3D object matching [8].

In this paper, we study top-k graph pattern matching problem which is to find the top-k answers for a graph pattern query over a large data graph. Particularly, to be distinguished from all existing work, our focus is on finding top-k answers of a cyclic graph pattern query. A naive solution is to find all possible answers from the underneath graph using an existing approach [5,26,32], and rank all the answers by the answer weight in order, and report the first k answers with the smallest total weight. However, there can be an enormous number of answers in the underneath graph. The cost of exhaustively enumeration for all of them can be prohibitive. Another solution is to use top-k join [16] to progressively compute the top-k answers rather than the complete set of answers. However, the top-k join solution requires a large amount of memory or have to use costly nested-loop join processing [1]. Other standard top-k processing such as TA [9] can not be directly applied because it needs sorted object lists, where the object to be searched are respectively sorted by each attribute in different sorted object lists. Such conditions can be not easily met in our problem setting.

Finding top-k matches for a cyclic graph query is challenging. However, linear cost algorithms in terms of time and space exist for finding top-k answers for twig queries, as introduced in [12]. It is possible operate on a ranked list of twig answers, in order to find top-k answers of a graph queries. Therefore, the overall processing can be efficient and scalable because those ranked lists of twig answers can be obtained very efficiently. Even if a large number of twig answers on the ranked list have to be enumerated before finding the requested top-k answers of the graph query, the cost of such processing increases marginally. Based on this motivation, in this paper, we propose a efficient solution for top-k graph pattern matching.

Our Contributions: Our contributions are as follows: (1) We propose a new top-k graph pattern matching problem and investigate a baseline solution based on the existing top-k processing technique (Section 2). (2) We propose a new twig query approach which is efficient and scalable in terms of different value scales of k (Section 3); (3) Since not each twig answer in a list corresponds to an answer of the graph query, different twig lists can result different cost in solving a given cyclic graph query. We propose cost-based optimization to select the best twig query ((Section 3.1); (4) We conducted extensive performance studies using a real dataset, and we confirm the efficiency of our proposed approach (Section 4).

We discuss related work in Section 5 and Section 6 concludes this paper.

2   Problem Statement

We discuss top-k graph matching for a given graph query over a large data graph. The data graph is defined as a weighted node-labeled graph $G_D = (V, E, \Sigma, \text{label}, W_e)$. Here, V is a set of nodes. E is a set of edges that can be directed or undirected. Σ is a set of node labels, which are usually far less than all nodes in G_D. label is a mapping

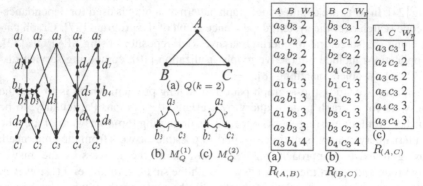

Fig. 1. Graph G_D Fig. 2. Query/Answers   Fig. 3. The Sample Storage

function which assigns each node, $v_i \in V$, a label $X \in \Sigma$, and label(v_i) is hence the label of node v_i. Given a label $X \in \Sigma$, the extent of X, denoted as $\text{ext}(X)$, is the set of all nodes in G_D that are X-labeled. The weight function $W_e(u,v)$ assigns a weight to every edge $(u,v) \in E$. The shortest distance from a node u to a node v, denoted $W_p(u,v)$, is the minimum total weight along a path from u and v, in G_D. And a shortest path from u to v is a path from u to v with the minimum total weight $W_p(u,v)$. In the following, we use $V(G)$ and $E(G)$ to denote the set of nodes and the set of edges in a graph G.

Fig. 1 shows a simple data graph, G_D, in which all edges are weighted 1. There are 4 labels, $\Sigma = \{A, B, C, D\}$. In Fig. 1, we use the small letter x with a unique number i to signify an X-labeled node $x_i \in \text{ext}(X)$.

A graph query $Q = (V(Q), E(Q))$ is an undirected and connected graph, where $V(Q)$ consists of labels in Σ, and $E(Q)$ is a set of edges between two nodes in $V(Q)$. To simplify exposition, we assume unique labels in $V(Q)$ in the following discussion, and it is straightforward to extend our approach to the case of repeated labels. A graph query Q is a *tree query* (or *twig query*) if it is cycle-free. Otherwise, it is a *cyclic query* (or *graph query*). In the following of this paper, tree query and twig query (cyclic query and graph query) are used interchangeably. Fig. 2(a) shows a graph query with 3 labels. For this query, Fig. 2(b) and Fig. 2(c) illustrate two answers, $\langle a_2, b_2, c_2 \rangle$ and $\langle a_3, b_3, c_3 \rangle$, that can be found in the data graph G_D (Fig. 1).

To answer a graph query Q is called the **graph pattern matching problem**, or **GPM** for short. An answer of Q, denoted M_Q, over G_D, is an n-ary node-tuples, $\langle v_1, v_2, \cdots, v_n \rangle$, where $v_i \in V(G_D)$ and $n = |V(Q)|$. There exists a one-to-one mapping, $\lambda: Q \to M_Q$ that satisfies two kinds of conditions specified by Q, namely, label condition and structural condition. (1) The label conditions specified by Q are satisfied, i.e., for every $X \in V(Q)$, there is a node $x \in M_Q$ labeled by X in G_D; and (2) the structural conditions indicated by Q are satisfied, i.e., for every edge $(X,Y) \in E(Q)$, there is a connected path in G_D between the two corresponding nodes $x, y \in M_Q$.[1]

GPM asks for all answers for a given query. However, the total number of answers in a large graph can be enormous. To return a large number of answers to a user can be

---

[1] In a directed graph G_D, the directed path either from x to y or from y to x can satisfy an undirected edge $(X,Y) \in E(Q)$.

overwhelming for the user to digest, and the computation overhead is also prohibitive. Therefore, in this paper, we focus on efficiently finding top-k answers of a graph query.

Top-k Answers: The top-k answers for Q are determined by a score function. Numerous score functions are discussed in the literature [21,14,2,15,20,12], which are usually based on *node scores* and *edge scores*. The node score is used to reflect the node importance of $v \in M_Q$, while the edge score is used to reflect the connection strengths of (u, v), where $u, v \in M_Q$ and $(\lambda^{-1}(u), \lambda^{-1}(v)) \in E(Q)$.

For simplicity, we consider the edge score only with the following equation. However, our approach is extendable to include the node score.

$$score(M_Q) = \Sigma_{(A,D) \in E(Q)} W_p(u, v) \tag{1}$$

where $(A, D) = (\lambda^{-1}(u), \lambda^{-1}(v))$ is a query edge in Q and $u, v \in M_Q$, Therefore, for each query edge of Q, there is a corresponding score component of M_Q. Eq. 1 is the sum of $|E(Q)|$ several edge score components. Intuitively, the smaller distance between two nodes in G_D indicates a closer relationship between them. Therefore, an answer with a smaller score of Eq. 1 is regarded to be better. In other words, an answer M_Q tends to be ranked higher if $score(M_Q)$ is smaller.

k-GPM Problem: Consider a graph query Q against a large graph G_D. A k-GPM problem finds the top-k answers of Q. Therefore, for k-GPM query Q, its answer is a list: $(M_Q^{(1)}, M_Q^{(2)}, \cdots, M_Q^{(k)})$, such that any $score(M_Q^{(i)})$, $1 \le i \le k$, is no greater than that of any other answers of Q. In this paper, we study the problem of k-GPM for a cyclic query Q.

Based on those path lengths in Fig. 2(b) and Fig. 2(c), the top-2 matches are $M_Q^{(1)} = \langle a_3, b_3, c_3 \rangle$ and $M_Q^{(2)} = \langle a_2, b_2, c_2 \rangle$, where $score(M_Q^{(1)}) = 3$ and $score(M_Q^{(2)}) = 6$.

2.1 A Direct Solution Based on Top-k Join

A storage Scheme. Like the existing work [12,13], we materialize the edge transitive closure of a data graph G_D, and store it in tables. The reasons why we proceeds our discussion with such a storage scheme bare (a) it speeds the queries by exempting the burden to search for the large number of required shortest paths at query time; (b) although the transitive closure is very large in size, a compressing scheme such as 2-hop covers [6,4] easily works on it for better space consumption; (c) it supports efficient search of trees [12,13] or even graphs (as will discussed below) for large graphs.

In detail, a table $R_{(A,D)}$ stores information for all shortest paths from A-labeled nodes to D-labeled nodes, which can be implemented as a database relations containing three columns essentially: A, D and distance. Here, columns A and D are for A-labeled nodes and D-labeled nodes, respectively. The column distance is for the corresponding distance $W_p(a, d)$ where a and d are A-labeled and D-labeled nodes in the same tuple. Below, we use $R_{(A,D)}$ to refer to this table. There can be $|\Sigma|^2$ tables, each corresponding to a different pair of labels in Σ. Later, we use t to signify a tuple in $R_{(A,D)}$, while t.distance denotes $W_p(a, d)$ for $a, b \in t$. A table supports two ways to access it: the sequential access, which scans the table sequentially, and the random access, which retrieve a tuple using given a and b. Fig. 3 shows $R_{(A,B)}$, $R_{(B,C)}$ and $R_{(A,C)}$ for the data graph G_D in Fig. 1.

A top-k Join Solution. The top-k join algorithms [16] can progressively compute top-k joins of several tables without computing all joins of those tables. Particularly, we briefly describe the adaption of a representative [16], called the *hash ripple join*, or simply HRJN.

Suppose there are $|E(Q)| = l$ edges in Q and each edge is identified by a number. Let the i-th edge in $E(Q)$ be (X, Y), and we use R_i to denote $R_{(X,Y)}$. A multi-way join on R_1, R_2, $\cdots$, and R_l can be used to compute answers of Q. Here, R_i and R_j are joined together, if a common query node X appears as an end node in both the i-th and the j-th edges of $|E(Q)|$. And the join is based on the equality condition on the corresponding common X columns in table R_i and table R_j. The top-k join algorithm [16] requires that R_1, R_2, $\cdots$, R_l are sorted in the ascending order of all $W_p(a, d)$ in their `distance` columns. HRJN sequentially scans those tables on disk. The tuples already scanned into memory are referred as *seen tuples*, while those not scanned yet are *unseen tuples*. For each table R_i, a hash index is built for those seen tuples of R_i. In detail, during the sequential scan, when an unseen tuple t from R_i is accessed, HRJN probes all hash indexes to compute all valid join combinations of t between all seen tuples of R_j, $i \neq j$. In this way, HRJN progressively joins R_1, R_2, $\cdots$, and R_l and a buffer is used to maintain temporary top-k anwsers that have been found. The HRJN can stop early when the the upper bound of those top-k answers in the buffer is even smaller than the lower bound of all *unseen* answers.

3   The New Twig Query Approach

Finding top-k matches for a cyclic graph query is challenging. However, linear cost algorithms in terms of time and space exist for finding top-k answers for twig queries, as introduced in [12]. It is possible operate on a ranked list of twig answers, in order to find top-k answers of a graph queries. Therefore, the overall processing can be efficient and scalable because those ranked lists of twig answers can be obtained very efficiently. Even if a large number of twig answers on the ranked list have to be enumerated before finding the requested top-k answers of the graph query, the cost of such processing increases marginally. We first briefly review the top-k twig query processing, then we discuss our new twig query approach for k-GPM.

Gou et al. shows in [12] that a linear cost algorithm exists in terms of time and space in order to efficiently find the top-k answers of a twig query T over G_D, where G_D is stored using the aforementioned storage scheme with edge transitive closure. To process a given twig query T, the bottom-up strategy starts with the smallest *subtrees* of T and then considers larger subtrees till T is fully considered. The time and space requirement is linear to all data inputs [12], namely $O(\sum_{(X,Y)\in E(T)} |R_{(X,Y)}|)$. Specifically, the time cost is $O(\sum_{(X,Y)\in E(T)} |R_{(X,Y)}|)$ for the first (top-1) answer. After the first answer, only a fixed amount of time Δ, which is independent of the data size, is required to output the second (top-2) answer, the third (top-3) answer and so on. Interesting readers can refer to [12] for more knowledge. It is important to note that this top-k twig pattern matching does not directly applied for a cyclic query Q, because we cannot decompose the cyclic graph pattern in the same manner as that of the bottom-up strategy.

Our Approach. Given a k-GPM query Q, the processing consists of two closely related tasks:

Task-1: select a *best* twig query based on Q to construct the ranked list of these twig query answers progressively;

Task-2: process the ranked list to find the top k answers for the k-GPM as soon as possible.

The two tasks are executed simultaneously: In *Task-1*, the answers of the twig queries are generated on demand, as long as they are requested by *Task-2*. While *Task-2* computes the answers of Q based on those twig query answers. When all answers of the k-GPM are sucessfully found, it stops all processing.

In *Task-1*, those twig queries are obtained by considering the spanning trees of the graph structure of Q. We call such twig queries the *t-queries* of Q. Thus, any one spanning tree of Q can be a t-query of Q. A t-query returns the answers in the non-descending order of their weights. Therefore, for each t-query, its ordered answers form a ranked list. For easy representation, this ranked list of a t-query is called a *t-list*. Similar to a sequential scan over a ranked list, a t-list is constructed and processed progressively: In a t-list, all twig answers seen that far are examined; the last answer is the latest answer returned by the t-query, which is consumed by *task-2* immediately when it is generated. New answers of the t-queries are generated on demand and appended to the t-list, as long as they are requested by *Task-2*.

Task-2 tries to extend each twig answer to an answer of Q. Consider a twig answer. If it can be sucessfully extended to an answer of Q, the set of nodes in the two answers must be identical. Thus, the twig answer can be a partial answer of Q, where we only need to find the additional edges among those nodes for Q. Those edges represents connections among those nodes in the data graph, which are required to satisfy Q. To this end, we consider all *missing query edges*, which are those edges appeared in Q, but do not exist in the t-query. For each missing query edge, say (A, D), note that we already have two corresponding nodes a and d in the twig answer. So we only need to look up in $R_{(A,D)}$ to see if there is a record for the required shortest path between a and d. If the record can be successfully found, the required shortest path between a and d exists; so we add an edge between a and d to this partial answer, otherwise, this twig answer cannot be extended to an answer of Q and is discarded immediately. Finally, if the records for all missing query edges are found, we successfully obtain an answer of Q with all required shortest paths.

Algorithm. Algorithm 1 begins with the computation for an optimal t-query T (Line 3). Section 3.1 will address the details of find the best t-query with the minimum estimated cost. At Line 4, the corresponding t-list of T is initialized. Therefore, a call of the function $S.next()$ will return the lastest available answer in S; this call also tells S to generate the next answer for T (Line 6). All answers in the t-list are discarded immediately when it is examined as follows: we try to extend the latest answer of T (returned by $S_T.next()$), denoted $M_T^{\perp}$, to an answer of Q. If $M_T^{\perp}$ can be successfully extended to

---

Algorithm 1. KGPM_Twig

Input: A cyclic graph query Q.
Output: The answers for the k-GPM query of Q.

```
1  begin
2  |   let B be a buffer for the temporary top-k answers of Q;
3  |   select the optimal t-query T;
4  |   initialize the t-list S for T;
5  |   repeat
6  |   |   M_T^⊥ ← S.next();
7  |   |   if extendable(M_T^⊥, M_Q) then
8  |   |   |   update the top-k buffer B with M_Q;
9  |   |   |   ϖ ← the largest score of all answers in B ;
10 |   |   ω ← lbound(M_T^⊥);
11 |   until (ϖ ≤ ω and |B| ≥ k) ;
12 |   return B;
13 end
```

---

a match of Q, a function $extendable(M_T^\perp, M_Q)$ returns true, where the result is passed to M_Q by reference. The top-k buffer $\mathcal{B}$ is updated with M_Q (Line 8). $\overline{\omega}$ is the largest cost for a match in $\mathcal{B}$ and hence is also updated at Line 9. Line 10 derives a lower bound $\underline{\omega}$ that far for all unseen answers of Q with a bounding function $lbound()$ based on all latest answers of the t-queries. At last, we stop the whole processing once the stop condition of Line 12 is satisfied, which suggests we all already have that answer for the k-GPM in the buffer.

Stop Condition. Note that T returns answers ranked by $\Sigma_{(A,D) \in E(T)} W_p(u,v)$. Particularly, $lbound(M_T^\perp)$ directly returns the score of $M_T^\perp$, namely

$$lbound(M_T^\perp) = score(M_T^\perp) \tag{2}$$

The above equation is based on the fact that the score of any unseen answer of Q should be at least greater than the score of the last answer of T. To understand it, notice that for an answer, M_Q, where $score(M_Q) \leq score(M_T^\perp)$, it contains an corresponding answer of T, which must be ranked before $M_T^\perp$, hence is identified before $M_T^\perp$. Therefore, all such M_Q should be examined already. To see how the stop condition works, note that $lbound(M_T^\perp)$ (or $\underline{\omega}$) is growing larger and larger as more and more twig answers are identified; in the meanwhile, $\overline{\omega}$ will not grow as the processing proceeds; it finally equals to the score of the k-th answer of Q. Therefore, the stop condition $\overline{\omega} \leq \underline{\omega}$ can be satisfied. Note that $\underline{\omega}$ can be futhered tightened by considering the smallest score components corresponding to those missing query edges, which can be easily collected over the base tables offline.

Example 1. For the example cyclic query in Fig. 2(a), we use a t-query tree T_1 (Fig. 4(a)) of Q to demonstrate the twig query approach. We construct S_{T_1} shown in the table on the left of Fig. 4(b). When the first twig answer, $\langle a_3, b_3, c_3 \rangle$, comes, whose score is 2, we look up the weight of (a_3, c_3) in $R_{(a,c)}$ and it is 1. Thus the first answer of Q is obtained and its overall score is 3. It is buffered and $\overline{\omega}$ is set to be 3. And we can obtain the

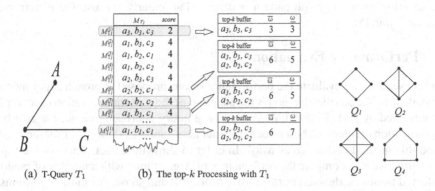

(a) T-Query T_1 (b) The top-k Processing with T_1

Fig. 4. An Example

Fig. 5. Tested Graph Queries

current $\underline{\omega}$ also as 3, which is the sum of the twig answer score and the smallest possible score for the query edge (A, C). Since the size of the buffer is below the required number, the stop condition is not satisfied. Then, we move to the second twig answer and it turns out not extendable. Thus, we repeat the precessing and find another Q answer upon the 7-th twig answer. The stop condition is satisfied upon the 11-th twig answer.

3.1 Cost-Based Optimization for T-Query Selection

Given a graph query, there can be many different t-queries for it. Each t-query can give a different cost in solving the k-GPM. Therefore, it is important to select the optimal t-query given a graph query Q. We discuss a cost-based selection to find the t-query with the smallest cost will be used. We mainly consider the time as the cost to be optimized. Note that [12] already shows the time requirement of a twig query is $O(\sum_{(X,Y)\in E(T)} |R_{(X,Y)}|)$ for the first (top-1) matches, and a fixed amount of time for the second (top-2) match, the third (top-3) match and so on. Therefore, the cost to solve a k-GPM with the t-query T can be estimated as below,

$$cost(T) = c \cdot \sum_{(X,Y)\in E(T)} |R_{(X,Y)}| + (1 - c) \cdot N \qquad (3)$$

where c is a constant coefficient to tune the weight of two kinds of cost. Now the central issue for optimizing k-GPM is how to estimate the number of twig answers which are consumed before the stop condition is satisfied, namely, N in the above equation.

In order to estimate the value of N, we assume that all answers of a graph query Q are evenly distributed in S_T. Moreover, we assume the first k answers of Q obtained by processing S_T are the k answers for the k-GPM. In this way, N can be simplified as below:

$$N = k \cdot sel_T$$

where sel_T is the average number of twig answers of T that are needed to obtain each answer of Q. Let N_Q and N_T denote the total number of answers of Q and T respectively. With obove assumption, there is $sel_T = \frac{N_T}{N_Q}$. N_Q and N_T can be estimated based on existing work on graph pattern matching. Particularly, we use the pattern match estimation in [5].

4   Performance Evaluation

In this section, we evaluate the performance of our proposed approach experimentally. Specifically, the baseline method is the adaption of the top-k join solution (Section 2.1), represented by join, for k-GPM. For our twig query approach, we use the cost-based optimization to select the best t-query in order to instantiate Algorithm 1 for the k-GPM. And this method is denoted as twig*. In order to show the effectiveness of the t-query selection, we also compare the performance of Algorithm 1 with a number of randomly selected t-queries, denoted as random1, random2 and so on. All those algorithms are implemented using C++. The value of c in Eq. 3 is set as 0.2. We show the elapsed time and required memory for the four cyclic queries Q_1, Q_2, Q_3, and Q_4, according to these k values: 10, 50, 100, 150 and 200. The structures of these queries are shown in Fig. 5.

We experimented on the real dataset, DBLP[2]. We construct a "co-authorship graph" based on the data. This graph contains 840,688 nodes (authors) and 3,078,263 edges. An edge between two nodes indicates that the two corresponding authors have co-authored one or more papers. As discussed in [18], co-authorship graphs capture many key features of social networks. We treat this graph as a social network, and use the method described in [18] to compute its edge weight. We assign node labels based on text clustering algorithms. In detail, we use the paper titles as the text feature for all authors. One author's text feature can come from multiple titles from that author. We group all authors into 100 clusters. For each author, we assign the cluster ID as its node label.

We conducted all experiments on a PC with a 3.4GHz processor, 180G hard disk and 2GB main memory running Windows XP.

Compare to the Baseline Method. This test is to compare the performance of our twig query approach (twig*) with the baseline method, namely the top-k join solution for top-k graph matching (join). Fig. 6 shows this test. In general, twig* outperforms join noticeabley. For all queries and most k values, join needs much more time and memory, than twig*. Moreover, the time and memory for join increase significantly as k increases. But both the time and memory of twig* increase quite slower. For example, in Fig. 6 (c) (Q_3), when k increases from 10 to 200, the elapsed time of join increases from $1,005$ milliseconds to $180,142$ milliseconds. In contrast, the time of twig* is from $4,013$ to $4,648$ milliseconds. join even cannot finish Q_4 in 2 hours when k equals 100 or a larger value (Fig. 6 (d)). Only when k is as small as 5 and 10, there are some cases that join can outperform twig*. However, join become quite slow when k takes a relatively larger value. It is because join has to perform a huge number of joins with those seen tuples in memory when k is 10, 50 or above. Moreover, it has to keep many seen tuples in memory so as k increase, the memory consumption of join also becomes larger and larger as k increases. For example, in Fig. 6 (h), join needs 34 and 56 megabytes of memory for Q_4 when $k = 10$ and $k = 50$. twig* only needs 16 megabytes of memory for all k values.

---

[2] http://dblp.uni-trier.de/xml/

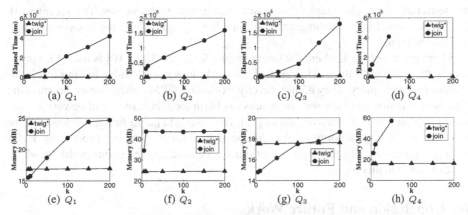

Fig. 6. Compare with Top-k Join Solution

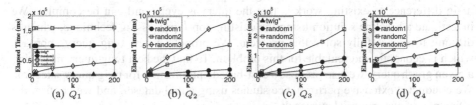

Fig. 7. Cost-based Optimization for T-query Selection

Performance of our Cost-based Optimization. Fig. 7 shows the elapsed time of twig* as compared to a number of randomly selected t-queries. We can see that twig* is the fastest t-query in solving k-GPM of all graph queries. Take Q_4 as an example, twig* spends 5,920 milliseconds to 6,369 milliseconds, while random1 needs 6,112 milliseconds to 6,589 milliseconds and random2 needs 9,481 milliseconds and 35,277 milliseconds. The effectiveness of our cost-based optimization can be successfully verified.

5 Related Work

Graph pattern matching is a long investigated topic for database applications [25,23]. Two lines of work can be identified in terms of the underlying graphs to be searched. One is a large collection of small graphs of hundreds of nodes, where the frequent subgraph mining is studied extensively in recent years, such as [29,28,22,30] to name a few.

Our work belongs the other line, which deals graph searching over a large standalone data graph [2,14,3,5,26,12,32,31]. We can further categorize the work in this line work according to whether there are user-given query graph. [2,14,31] do not have a user-given query graph. [2,14] belong to the so-called keyword search over graphs, which can find the top-k connected trees in the graph such that all user-given keywords are included in the trees. We omit the large number of other studies in this direction for we are searching for cyclic graphs that match user-given graph queries. The recent work [31] even studied frequent subgraphs within a large graph, where the answer structure are also not fixed.

There are many work where the query graph [3,5,24,26,32,11,10] is used. Our problem is the same with them. However, these work does not consider finding top-k answers of a graph query. They do not directly work on k-GPM either, because computing all answers at first and then sort the results to obtain the k best ones can be very expensive. Our solution is closely related to [12]. However, [12] is to find top-k answers for twig queries. A large amount of work has been done for top-k query processing. Refer to [17] for a survey. The top-k join algorithm [17] can be applied to our problem, which we examined in this paper.

6   Conclusion and Future Work

In this paper, we propose a new top-k graph pattern matching problem, in which the main difference to existing work is that the query is cyclic and can be complex. We investigate a baseline solution using the existing top-k processing technique, which is difficult to successfully solve this problem. We propose a new twig query approach which is economic and scalable for this problem. To find the best twig query in solving a given graph query, we also propose a cost-based optimization for twig query selection. We conducted extensive performance studies using a real dataset, and we confirm the efficiency of our proposed approach.

Acknowledgments.   This work is supported by NSFC (Grant No. 61103049) and Shenzhen Research Fund (Grant No. JC201005270342A). Xianggang Zeng is supported by Shenzhen New Industry Development Fund (Grant No. CXB201005250021A). Shengzhong Feng is supported by Special Funds of The Chinese Academy of Sciences (Grant No. XDA06010500).

References

1. Agrawal, P., Widom, J.: Confidence-aware join algorithms. In: ICDE (2009)
2. Bhalotia, G., Hulgeri, A., Nakhe, C., Chakrabarti, S., Sudarshan, S.: Keyword searching and browsing in databases using BANKS. In: ICDE (2002)
3. Chen, L., Gupta, A., Kurul, M.E.: Stack-based algorithms for pattern matching on DAGs. In: VLDB (2005)
4. Cheng, J., Yu, J.X.: On-line exact shortest distance query processing. In: EDBT (2009)
5. Cheng, J., Yu, J.X., Yu, P.S., Wang, H.: Fast graph pattern matching. In: ICDE (2008)
6. Cohen, E., Halperin, E., Kaplan, H., Zwick, U.: Reachability and distance queries via 2-hop labels. In: Proc. of SODA 2002 (2002)
7. Corrales, J.C., Grigori, D., Bouzeghoub, M.: BPEL Processes Matchmaking for Service Discovery. In: Meersman, R., Tari, Z. (eds.) OTM 2006. LNCS, vol. 4275, pp. 237–254. Springer, Heidelberg (2006)
8. Demirci, M.F.: Graph-based shape indexing. In: Machine Vision and Applications (2010)
9. Fagin, R., Lotem, A., Naor, M.: Optimal aggregation algorithms for middleware. In: PODS (2001)

10. Fan, W., Li, J., Luo, J., Tan, Z., Wang, X., Wu, Y.: Incremental graph pattern matching. In: SIGMOD (2011)
11. Fan, W., Li, J., Ma, S., Tang, N., Wu, Y., Wu, Y.: Graph pattern matching: From intractable to polynomial time. In: VLDB (2010)
12. Gou, G., Chirkova, R.: Efficient algorithms for exact ranked twig-pattern matching over graphs. In: SIGMOD (2008)
13. He, H., Wang, H., Yang, J., Yu, P.S.: BLINKS: ranked keyword searches on graphs. In: SIGMOD (2007)
14. Hristidis, V., Papakonstantinou, Y.: Discover: keyword search in relational databases. In: VLDB (2002)
15. Hwang, H., Hristidis, V., Papakonstantinou, Y.: ObjectRank: a system for authority-based search on databases. In: SIGMOD (2006)
16. Ilyas, F., Aref, G., Elmagarmid, K.: Supporting top-k join queries in relational databases. The VLDB Journal 13(3), 207–221 (2004)
17. Ilyas, I.F., Beskales, G., Soliman, M.A.: A survey of top-k query processing techniques in relational database systems. ACM Comput. Surv. 40(4), 1–58 (2008)
18. Kempe, D., Kleinberg, J., Tardos, E.: Maximizing the spread of influence through a social network. In: KDD (2003)
19. Koenig, P.-Y., Zaidi, F., Archambault, D.: Interactive searching and visualization of patterns in attributed graphs. In: Graphics Interface Conference (2010)
20. Liu, F., Yu, C., Meng, W., Chowdhury, A.: Effective keyword search in relational databases. In: SIGMOD (2006)
21. Page, L., Brin, S., Motwani, R., Winograd, T.: The PageRank citation ranking: Bringing order to the web (1998) (submitted for publication)
22. Haichuan, S., Ying, Z., Xuemin, L., Xu, Y.J.: Taming verification hardness: an efficient algorithm for testing subgraph isomorphism. In: VLDB (2008)
23. Shasha, D., Wang, J.T.L., Giugno, R.: Algorithmics and applications of tree and graph searching. In: PODS (2002)
24. Tian, Y., Patel, J.: TALE: A tool for approximate large graph matching. In: ICDE (2008)
25. Ullmann, J.R.: An algorithm for subgraph isomorphism. J. ACM 23(1) (1976)
26. Wang, H., Li, J., Luo, J., Gao, H.: Hash-base subgraph query processing method for graph-structured XML documents. In: VLDB (2008)
27. Wang, X., Lo, D., Cheng, J., Zhang, L., Mei, H., Yu, J.X.: Matching dependence-related queries in the system dependence graph. In: ASE (2010)
28. Williams, D., Huan, J., Wang, W.: Graph database indexing using structured graph decomposition. In: ICDE (2007)
29. Yan, X., Yu, P.S., Han, J.: Graph indexing: a frequent structure-based approach. In: SIGMOD (2004)
30. Yuan, Y., Wang, G., Wang, H., Chen, L.: Efficient subgraph search over large uncertain graphs. In: VLDB (2011)
31. Zhu, F., Qu, Q., Lo, D., Yan, X., Han, J., Yu, P.S.: Mining top-k large structural patterns in a massive network. In: VLDB (2011)
32. Zou, L., Chen, L., Özsu, M.T.: Distance-join: Pattern match query in a large graph database. In: VLDD (2009)
33. Zou, L., Mo, J., Chen, L., Özsu, M.T., Zhao, D.: gstore: Answering sparql queries via subgraph matching. In: VLDB (2011)

Dynamic Graph Shortest Path Algorithm*

Xueli Liu and Hongzhi Wang

Harbin Institute of Technology
{xueli.hit,whongzhi}@gmail.com

Abstract. Shortest paths computation in graph is one of the most fundamental operation in many applications such as social network and sensor network. When a large graph is updated with small changes, it is really expensive to recompute the new shortest path via the traditional static algorithms. To address this problem, dynamic algorithm that computes the shortest-path in response to updates is in demand. In this paper, we focus on dynamic algorithms for shortest point-to-point paths computation in directed graphs with positive edge weights. We develop novel algorithms to handle the single-edge updating and the batch edge updating. We prove that our algorithms can compute the shortest paths for updated graph in time polynomial to the size of updated part of the graph. We experimentally verify that these dynamic algorithms significantly outperform their batch counterparts in response to small changes, using real-life data and synthetic data.

1 Introduction

The shortest-path problem is a routine graph problem in a variety of real-word applications, e.g,routing in a road network, routing/data harvesting in sensor networks[1]. It is often defined in terms of all-pairs shortest path, single-source shortest-path, and point-to-point shortest path. In practise,there is an industrial demand for computing point-to-point shortest path(P2P) on dynamic large-scale network such as road network whose edges are dynamic changed with the traffic condition, sensor network whose sensor may not work effectively. Given a graph $G = (V, E)$, a source node $s \in V$, a terminal node $t \in V$, and a list of changes ΔG, e.g, edge deletions, edge insertions and edge weight updating, the dynamic shortest path algorithm is to compute the update shortest path from s to t in $G \oplus \delta G$. We call an algorithm which handle only the edge insertion as incremental algorithm. The decremental algorithm solves the edge deletion.

As discussed in [2], the traditional computation complexity analysis for the static algorithms is no longer suitable for the dynamic algorithms. It is clearly that the cost of a dynamic algorithm only depends on the update size in the computation. Using worst-case analysis analysis and a function with the size of problem input to express it is not informative. Instead, one can define an

* This paper was partially supported by NGFR 973 grant 2012CB316200 and NSFC grant 61003046, 6111113089. Doctoral Fund of Ministry of Education of China (No.20102302120054).

H. Gao et al. (Eds.): WAIM 2012, LNCS 7418, pp. 296–307, 2012.

adaptive parameter $|changed|$, which captures the size of the changes in the input and output, to analysis the dynamic algorithms. This parameter indicates that the updating cost is inherent to the dynamic algorithm itself. An dynamic algorithm is said to be bound if its computation cost depends only on the size of $|changed|$, and not on the size of the entire input. i.e, it can be expressed by a function of $|changed|$. This paper, we define the set of nodes whose input value and output value change as $|changed|$. We use the parameter $|AFF|$ to denote the number of nodes in $|changed|$ and $||AFF||$ to denote the number of edges incident on some nodes in $|changed|$.

Contributions. In this paper we focus on dynamic point-to-point shortest-path problem(P2P) for directed graphs with positive edge weights. We provide effective dynamic algorithms for unit update, i.e, a single-edge deletion or insertion, and bath updates, i.e, a list of edge deletions and insertions mixed together. We show that all the dynamic algorithms are bound. Using both real-life data and synthetic data, we experimentally evaluate the efficiency and scalability of our dynamic algorithms. We find that our dynamic algorithms perform significantly better than their static counterparts, even in worst case. When data graphs are changed up to 40%, our algorithms consistently outperform the static algorithm.

Organization. Section 2 states the basic definition and data structure. The dynamic shortest-path problems for single-edge update and bath update are studied in section 3,4, respectively. Section 5 represents experimental results. This paper ends with conclusion in Section 6.

Related Work. There are some work concerning the dynamic shortest path problem. IEven [3] and Rohnert [4] presented algorithms for maintaining shortest paths on directed graphs with arbitrary real weights. Their algorithms required $O(n^2)$ per edge insertion; Ramalingam and Reps [5], [2], Frigioni et al. [6], [7] introduced a batch dynamic shortest path algorithms with arbitrary real weights, they use the model that the running time of their algorithm is analyzed in terms of the output change rather than the input size, but they only handle the single edge update. As far as we know, there is no algorithm which the worst case is always better than recomputing the new solution from scratch.

There are some works on special case to compute the shortest path problem. [8] depicted an incremental shortest path algorithm for directed graphs with positive integer weights less than C: the amortized running time of their algorithm is $O(Cnlogn)$ per edge insertion. [9] designed a fully dynamic algorithm with mortized time $O(n^{9/7}log(nC))$ for all pair shortest path problem on planar graphs with integer weights. A fully dynamic algorithm for single-source shortest paths in planar directed graphs in $O(n^{4/5}log^{13/5}n)$ mortized time per operation was presented in [10].

The existing method which related to ours is proposed in [5]. This paper proposed a fully-dynamic algorithm called SWSF-FP which handled the batch update at the same time. Our partial work batch update is closely to the SWSF-FP, but we maintain it on a special shortest path tree to reduce the unnecessary computation.

2   Problem Statement and Data Structure

2.1   Problem Statement

Let $G = (V, E)$ be a directed graph with a non-negative length function len: $V \times V \to R^+ \bigcup \{\infty\}$, where V is the vertex set, $E \subseteq V \times V$ is the edge set. Let $s(t) \in V$ be an arbitrary but fixed source(destination), $(v_1, v_2, ...v_s)$ be a shortest path from v_1 to v_s . With $dist(v)$ we denote the length of a shortest $s - v - path$ in G for any $v \in V$. i.e, $dist[v]$ is the shortest distance from s to t.

Given a shortest-path P from s to v, a batch update $\triangle G$ to G, the dynamic shortest-path problem is to find the new shortest-path P' from s to t without recomputing from the original graph.

2.2   Data Structure

To avoid of recomputing the P' from scratch, we introduce some auxiliary data structures.

The basic idea of finding new shortest path in dynamic graph is to store the single source shortest path graph(SP), then update the SP to obtain the result. There is lots of unnecessary computation in the update process. To decrease the redundant cost, we propose a new structure which is called adaptive shortest-path subgraph. Before we define the adaptive shortest-path subgraph, we introduce the concept of SSP edge: An edge in the graph is said to be an SSP edge iff it occurs on some shortest path from the source vertex s to the other vertex $v \in G$. Thus, an edge (u, v) is an SSP edge iff $dist[u] = len(u, v) + dist[v]$. Adaptive shortest-path subgraph are defined as followed:

Definition 1. *A subgraph T is said to be an adaptive shortest-paths subgraph for a given graph G with source node s if*
   (1)for each $v \in V(T)$, there is a path from s to v.
   (2)every edge in T is a SSP edge.

We maintain a distance vector D containing $d(v)$ for each node v in $ASP(G)$. After each update, D has to be updated accordingly. For every each vertex $v \in V(G)$, we use $out[x]$ to denote the outgoing vertices of x, $in[x]$ to denote the ingoing vertices of x.

For each node v in $ASP(G)$, we maintain a list $pre[v]$ to denote the previous node of v in the shortest-path from s to v. (Considering the shortest-path between two nodes is not unique, so $pre[v]$ is a list.) $Succ[v]$ represent the list of successor node of v .

As pointed out in section 1, instead of analyzing the cost of the dynamic algorithms in terms of the size of the entire input, we analyze them by the size of $|changed|$. The notion of affected areas is introduced to characterize $changed$.

Affected Areas. Let G be the graph before the edge update, G' be the graph after the edge update, $ASP(G)$ be the adaptive shortest-path subgraph of G, $ASP(G')$ be the new adaptive shortest-path subgraph of G'. The $Affecte areas$ of G by the edge update is the difference between $ASP(G)$ and $ASP(G')$. We use $|AFF|$ to denote the number of vertices in Affected areas, $\|AFF\|$ to denote the number of edge incident to the vertices in Affected Area.

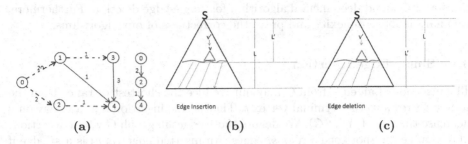

Fig. 1. ASP(G) and affected area

Example 1. Consider the weighted directed graph G shown in the left of Fig1.(a). Without loss of generality, we set 0 as the source node and 4 the destination node. The shortest-path from 0 to 4 is the indirect root $0 \rightarrow 2 \rightarrow 4$. The right graph of Figure 1 depicts the adaptive shortest-path graph $ASP(G)$ of G and the left figure1 with dotted line part is the shortest-graph $SP(G)$ of G. Check the ASP(G) against $SP(G)$, we obtain that the size of $SP(G)$ is smaller than the size of $SP(G)$.

Proposition 1. *The size of $ASP(G)$ is always less than or equal to the size of $SP(G)$.*

All the auxiliary structure are summarized in table 1.

Table 1. Auxiliary Data Structure

$d[v]$	The shortest distance from source s to v;
D	Distance vector containing d[.] of all nodes in graph G;
$SP(G)$	The shortest path graph for graph G;
$ASP(G)$	The adaptive shortest path graph for graph G;
$pre[v]$	The previous node list of v in $ASP(G)$.
$Succ[v]$	The successor node list of v in $ASP(G)$.
$in[v]$	The previous node list of v in graph G.
$out[v]$	The successor node list of v in graph G.
$|AFF|$	The number of affected nodes when update $D[]$.
$\|AFF\|$	The number of affected edge incident to the affected nodes when update $D[]$.

The preconditions of the algorithm is in following:

(1) ASP(G), which store all the shortest-path form s to vertex v which $d[v] \leq d[t]$.

(2) A priority queue Q, a vertex $v \in Q$ if $v \in V(ASP(G))$ and $v[Succ] = \emptyset$.

3 Unit Update

In this section, we provide a bound incremental algorithm for single-edge insertions and a bound decremental algorithm for single-edge deletion. Furthermore, we analysis the complexity and prove the correctness of our algorithms.

3.1 Single-Edge Insertion

The insertion of an edge from G may only reduce the shortest distance from the source vertex s to the terminal vertex t. That is, the insertion operate may only decrease the size of $ASP(G)$. We denote those edges in graph G whose insertions may change the shortest-path as si edges. An inserted edge (u, v) is a si edge if $u \in V(ASP(G))$. one can verify the following:

Proposition 2. *Given a graph G and a shortest-path from the source node s to the terminal node t, only the insertions of si edge in G may reduce the shortest distance from s to t.*

Example 2. Recall the *Example* 1, if we insert an edge $(2, 1)$ with weight 3, which is not a si, the shortest path is not changed and the adaptive shortest path graph of the new graph $G' ASP(G')$ holds the line.

Based on Proposition 1, we provide algorithm *InsertEdge* shown to compute the shortest-path from s to t when the edge $e = (v', v)$ is inserted into G.

Algorithm 1. Procedure InsertEdge

Input : Graph G, shortest-path $P(s, t)$, edge $e = (v', v)$ with weight w to be
added to G.
Output: The updated shortest-path $P'(s, t)$.
Insert e into G;
Heap$Q := \emptyset$;
If e is not an si edge, return; mindist $\leftarrow d[v'] + w$;
If $d[v] >$ mindist$d[v]$ InsertHeap$(v, d[v])$ and d[v]$\leftarrow$ mindist;
while $Q \neq \emptyset$ **do**
> $u =:$DeleteMinHeap();
> cut the nodes whose dist are larger than dist[v] and return if u==t;
> **if** $d[u] < d[t]$ **then**
> > **for** *every vertex* $x \in out(u)$ **do**
> > > **if** $w(u, x) + d[u] < d[x]$ **then**
> > > > $d[x] = w(u, x) + d[u]$; AdjustHeap(x,d[x]);

According to the *Prev*[t], return P';

In the first step, we insert the edge into the original graph G. Then we judge wether the edge e is a si edge. If not, return P and the $ASP(G)$ is not changed. There are two situations when $e \in si$: (1) the insertion of e indeed reduces the shortest distance from s to t. (2)the shortest path is remain P. In the first case, when update the $ASP(G)$, we need to cut the node and edge whose dist is larger than $dist[v]$. In the other case, we need update the affected node the $ASP(G)$. A node v is an affected node when the update $d[v]$ is less than original $dist[v]$, these nodes may only appear in the subtree of v'.

Correctness and Complexity. When the inserted edge(v', v) affect the $ASP(G)$, two kinds of node may be affected. One is the subtree nodes of v, the distance of this kind of nodes decrease, the other kinds is the nodes in subtree nodes of v in $G \backslash ASP(G)$, this kind of nodes may be added into the $ASP(G)$. Our algorithm traverse the outgoing edge of v, then relax these edges and update corresponding *prev* and *Succ*. This step we update the new distance of nodes in $ASP(G)$ until its distance is larger than new $d[t]$, also add the nodes who is not in the $ASP(G)$ but update distance lower than $d[t]$. It ensures the correctness of updating $ASP(G)$.

Let L be the distance of P, L' be the distance of P'. The affected area of InsertEdge is shown in Fig1.(b): let δ be the nodes in in smaller triangle, $|\delta|$ be the number of δ and $||\delta||$ be the number of edges related to nodes in δ. This triangle includes the subtree nodes of v. The update of this area node takes $O(\delta \log \delta)$ time to update. Denote the number of diagonal marked area as m, we need delete the node in this area from ASP(G). It takes $O(m)$ time. The entire complexity of this algorithm is $O(m + \delta \log \delta)$. If the insertion edge did not change the shortest path, $m = 0$. Let $|AFF| = m + \delta$, we can verify that this algorithm is bound by the affected nodes.

3.2   Single-Edge Deletion

In contrast to edge insertions, the deletion of an edge may only increase the shortest distance from s to t. i.e, the old shortest-path from s to t may be not connected. We identify those edges in G whose deletions affect the shortest-path, referred to as *sd* edges, as follows. An edge (u, v) in G is a *sd* edge for a shortest-path P if $u \in V(P), v \in V(P)$. It suffices to consider *sd* edge for edge deletions:

Proposition 3. *Given a graph G and a shortest-path from the source node s to the destination node t, only the deletions of sd edge in G may increase the shortest distance from s to t.*

After delete the edge in the shortest-path P, we first check the *prev* list from terminal vertex t. Going through the *prev* list, if the vertex s is reached, the algorithm return Path P, the shortest-path is not change. We only need to update the subtree rooted at v in $ASP(G)$. If not, we have to find another shortest-path from G. According to the Proposition 2, we only need to extend the $ASP(G)$ until we meet the vertex t. In the worst case, the path from s to t is not connected after the deletion operation. In this case, the $ASP(G)$ is equal to $SP(G)$, that is to say, we have to compute all the shortest-path from single source s to the other vertex. In the process of updating $ASP(G)$, firstly, we should identify the affected area. The affected area consists of vertices whose $d[]$

Algorithm 2. Procedure DeleteEdge

Input : Graph G,the shortest-path $P(s,t)$, and an edge $e = (v^{'},v)$ in G is to
be deleted .
Output: The updated shortest-path $P'(s,t)$
delete edge e from G;
Heap Q:= $\emptyset$;
if $e \in sd$ then
    Identify the affected area;
    // update the $ASP(G)$
    for *every vertex* $u \in affctedarea$ do
        $d[u]=$min$d[x] + len(x,u)|x$is not in AffectedArea and $x \in in(u)$;
        InsertHeap(u,d[u])If $d[u] \neq \infty$;

    while $Q \neq \emptyset$ do
        $u =:$DeleteMinHeap();
        if $d[u] < d[t]$ then
            for *every vertex* $x \in Succ[u]$ do
                if $d[x] < d[u] + len(x,u)$ then
                    $d[x] = d[u] + len(x,u)$;Update ASP(G);
                    DecreaseHeap(x,d[x]);

    if $d[t] = \infty$ then
        Heap Q=all the leave nodes of the $ASP(G)$;
        extend the $ASP(G)$ until reach t;
    search pre[t] to find $P^{'}$;

---

changes. We use a stack to find them. Suppose v is an affected node, we check
the nodes in $Succ[v]$ and put them into the stack. Then we handle them one by
one. If the *prev* of node x in $Succ[v]$ is empty after deleting v from $prev[x]$, we
add x into the affected area.

Correctness and Complexity. According to the updated ASP(G), we can
obtain the update shortest-path $P^{'}$. The process of the DeleteEdge algorithm
ensures every path in the ASP(G) is a shortest-path, so, $P^{'}$ is the shortest-path
from s to t. The size of the affected vertices is the number of vertices in the
smaller triangle and diagonal marked area in Fig1.(c)(edge deletion). We denote
the number as $|AFF|$. The number of associate edge as $||AFF||$. We adopt
Fibonacci heap to update the ASP(G), so the complexity of the algorithm is
$O(|AFF| + ||AFF||log||AFF||)$.

4   Batch Update

Given a series of edge operations $+e_1, -e_2, -e_3, +e_4, ..,$ $'+'$ $('-')$ represent the
insertion(deletion) operation. It is cost when we update the shortest-path one by
one because of the turbulence of the ASP(G).To avoid redundant computation
and improve the efficiency of the update, We introduce a batch update algorithm
for point-to-point shortest path problem.

The batch update algorithm is based on two principle. (1)remove redundant edge operation as much as possible. (2)handle multiple edge operations simultaneously rather than one by one.

Algorithm 3. Algorithm BatchUpdate

Input : Graph G,the shortest-path $P(s,t)$, and the batch update $\triangle G$.

Output: The updated shortest-path $P'(s,t)$

update G; MinDelta(ASP(G),$\triangle G$);

for *every edge* $e = (v',v)$ *in* $\triangle G$ **do**

    $new[v] = min\{d[p] + len(p,v)|p \in in(v)\}$;

    if $d[v]! = new[v]$ **then**

        InsertHeap($v, min\{new[v], dist[v]\}$);

if *the* $\triangle G$ *affect* P **then**

    $d[t]=\infty$;

while $Q!=\emptyset$ **do**

    $u =$:FindAndDeleteMin(Q);

    if $d[u] < new[u]$ **then**

        $d[u]=\infty$;

        for *every vertex* $x \in Succ(u)$ **do**

            $new[x] = min\{d[p] + len(p,x)|p \in in(x)\}$;

            key=min{d[x],new[x]};

            if $d[x]! = new[x] and key < d[t]$ **then**

                InsertHeap(x,key);

    if $d[u] > new[u]$ **then**

        $d[u] = new[u]$;

        if $u==t$ **then**

            return;

        for *every vertex* $x \in out(u)$ **do**

            $new[x] = min\{new[v], new[u] + len(u,x)\}$;

            $key = min\{d[x], new[x]\}$;

            if $d[x]! = new[x] and key < d[t]$ **then**

                InsertHeap(x,key);

return P' ;

In algorithm BatchUpdate, the procedure MinDelta minimize the size of $\triangle G$ by reducing the nodes who don't affect the $ASP(G)$. i,e, the inserted edges which are not si edges and the deleted edges which are not sa edges. When updating $ASP(G)$, we use $new[v]$ to store the tempt distance of shortest path from s to v. Let key=min{d[v],new[v]}, the key value sort the sequence of nodes to be process. This order can set all the affected nodes on correct values.

Theorem 1. *In the Batch update algorithm, every affected vertices at most processed two times.*

Proof. When all the $\triangle G$ are edge deletions, no vertices can have shorter distances. According to the algorithm, $d[v]$ will be first assigned ∞ and then back to its correct value. Therefore, the affected vertices are insert into heap and extracted from heap twice. Similarly, when all the input updates are edge insertion, no vertices can have longer distances. $d[v]$ is assigned his $new[v]$ value and will not be processed again.Consequently, the affected vertices are processed only once.

Correctness and Complexity. When the iteration procedure abort, all the ASP(G) are update correctly. Let AFF be the affected area nodes, $|AFF|$ be the number of AFF and $||AFF||$ be the affected edge associated vertices in AFF. The complexity of our algorithm is $O(|AFF| + ||AFF||log||AFF||)$.

5   Experiments

In this section, we present an experimental study using both real-life and synthetic data, we conduct three sets of experiments to evaluate (1) the effectiveness of our algorithms compared to Dijikstra and DynamicDijikstr, (2)the efficiency and scalability of batch updates algorithms, (3) the performance and advantage of our algorithm compared to the scratch method and MFP [11] method.

Experimental Setting. We used two real-life datasets to evaluate the effectiveness of our algorithms, and synthetic data to change graph characteristics, for an in-depth analysis.

(1)Real-life data. Both of the real-life datasets come from websites. (a)The first is taken from website [1], it describes the North American Road Network with 174k nodes and 179k edges and San Joaquin County road network with 18K nodes and 23k edges. Each road was represented by the nodeID and his location information, each edge was associated a float number denoting the distance of two nodes. We extract the North American Road network data snapshots based on the road location,each has 18k nodes and 20k edges. (b)the second datasets came from a crawled LiveJournal social network graph. With node denotes person and edge denotes friendship relation. The size of this datasets is 5M with 75knodes and 508k edges. We extract snapshots based on the age of people, each consisting of 20k nodes and 80k edges.

(2)synthetic data..We use a generator to produce directed weight graph with two parameters, the number of nodes and the average degree. The parameter density denotes the density of edge in the graph.

Implementation. We implemented the following in C++: (1)InsetEdge and DeleteEdge (2)BatchUpdate (3)the Dijkstra algorithm.(4)DynamicDijkstra [11]. (5)MFP.

All experiments were executed on a machine with an Intel Pentium@2.93GHz CPU and 2GB of memory running Windows XP.

---

[1] http://www.cs.fsu.edu/~lifeifei/SpatialDataset.htm

Experimental Result

1.The Efficiency and Scalability of Single-Edge Update Algorithms

We use real-life and synthetic data to evaluate the efficiency of our algorithms. If the source and terminal nodes were fixed, we select arbitrary an edge for 10 times to compute the average running time and the affected area.

Next we present our findings. Fig2.(a) shows the efficiency of our algorithms

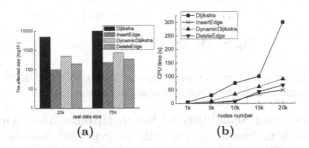

(a)   (b)

Fig. 2. The efficiency and scalability of single-edge update

using real-life data. The data size is 20k and 175K. Both of these two cases, our algorithms perform better. This is because our unit algorithm save a lot of time recomputing the shortest path graph. The affected size represents the same character.

Figure2.(b) shows the scalability of our algorithms. We generate graph with fixed average degree 500. The size of graph vary 1K nodes to 20k nodes, each increases by 5k nodes. In contrast to the DynamicDijikstra algorithm, the Insert-Edge algorithms always perform good. Its average of the running time changes in a small range. In practise, when we insert an edge into a big graph, the probability of the edge changing exiting shortest path is very small. Even in the shortest path changed case, the InsertEdge only need to compute a tiny subtree which was introduced in Fig1.(b), while the DynamicDijikstra have to update all the subtree in the entire shortest path graph. However, the procedure DeleteEdge sometimes defeated by DynamicDijikstra algorithm. The reason for that is the worst case occurs or the updated d[t] far outweight the old d[t]. In the worst case, the DynamicDijikstra algorithms just update the subtree of node of the tailing deleted edge. But the DeleteEdge must extend the adaptive shortest path graph. Let L be the old shortest distance from source vertex to terminal vertex, L' be the updated value. Consider the following situations: (1)L=L', that is, the distance of shortest path do not change. (2)L'=∞. i,e. there is no path from the source vertex to the end vertex. We compare these two cases in Fig3.(a). $|L - L'| = 1$ represents L'=∞ . In best case(L=L'), InsertEdge algorithm has decisive advantage. While in the worst case, the DynamicDijikstra performs better.

We also study the effects of the graph density. We fix the graph size as 2k with average degree 100 and vary the average degree from 100 to 1000, separated by 200. It can be found in Fig3.(b) that dealing with dense graph need more time

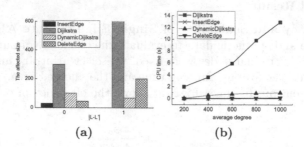

Fig. 3. The affection of parameters

for Dijikstra algorithms, the dynamic algorithms is affected little. In fact,the dense graph reduce the probability of worst case occurring.

2.The Efficiency and Scalability of Batch Update Algorithms

We compare batch update algorithm with MFP in both real-life dataset and synthetic dataset. The change size set as 10% of the edge number. Every setting is implemented 10 times to obtain the average value. In most cases, the performance of BatchUpdate is superior to the MFP. The cause is the affected nodes that BatchUpdate update less than MFP in general case. However, in worst case, BatchUpdate needs some extra cost to extend the adaptive shortest graph to a single source shortest graph. The result shows the average time. We can see that algorithm BatchUpdate gains advantage over MFP and Dijikstra algorithms. The scalability of BatchUpdate is depicted in Fig4.(b),we set the size of graph as above described the scalability of unit update.The figure shows that our algorithms scale well.

Finally, we change the size of ΔG to present the adaptability of our algorithms. We increase the updating edge number by 10% percent of all edges number, from 10% to 50%.And the graph size is 2K nodes with average degree 100. The result is shown in Fig2.(c), the BatchUpdate algorithm perform well until 40% edges changed.

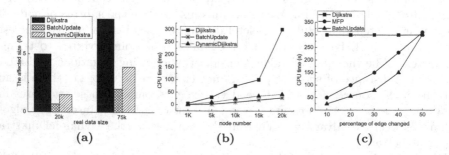

Fig. 4. The experimental study of batch update

6 Conclusion

This paper study the dynamic algorithms of Point-to-Point shortest path problem. We propose a new method using Adaptive shortest path graph to reduce the size of update cost comparing to the single source shortest path problem.Using this method, we present the unit update and batch update algorithms. We state that all the algorithms are bound by the change of input and output size. Finally, we experimentally demonstrate our algorithms perform better than recompute the shortest path from static algorithm and the dynamic algorithm maintaining the shortest path graph. Unfortunately, There is no dynamic algorithm always performing better than recomputing from scratch. It is an open issue to deserve study.

References

1. Bauer, R., Wagner, D.: Batch dynamic single-source shortest-path algorithms: An experimental study. Experimental Algorithms 2, 1–20 (2009)
2. Ramalingam, G., Reps, T.: On the computational complexity of dynamic graph problems. Theoretical Computer Science 158(1-2), 233–277 (1996)
3. Even, S., Gazit, H.: Updating distances in dynamic graphs. J. Algorithms (1985)
4. Rohnert, H.: A dynamization of the all-pairs least cost problem. In: Mehlhorn, K. (ed.) STACS 1985. LNCS, vol. 182, pp. 279–286. Springer, Heidelberg (1984)
5. Ramalingam, G.: An incremental algorithm for a generalization of the shortest-path problem. J. Algorithms (1991)
6. Franciosa, P.G., Frigioni, D., Giaccio, R., Sapienza, L.: Semi-Dynamic Shortest Paths and Breadth-First Search in Digraphs. Search 2(20244) (1997)
7. Frigioni, D., Marchetti-spaccamela, A., Nanni, U.: Fully Dynamic Algorithms for Maintaining Shortest Paths Trees 1. Journal of Algorithms (201), 251–281 (2000)
8. Ausiello, G., Italiano, G.F.: Incremental algorithms for minimal length paths. J. Algorithms (1991)
9. Henzinger, M.R.: Faster Shortest-Path Algorithms for Planar Graphs 23, 3–23 (1997)
10. Fakcharoenphol, J., Rao, S.: Planar graphs, negative weight edges, shortest paths, and near linear time. Foundations (2002)
11. Chan, E.P.F., Yang, Y.: Shortest Path Trees Computation in Dynamic Graphs. Most, 1–45

A Framework for High-Quality Clustering Uncertain Data Stream over Sliding Windows[*]

Keyan Cao[1], Guoren Wang[1,2], Donghong Han[1], Yue Ma[1], and Xianzhe Ma[1]

[1] College of Information Science & Engineering, Northeastern University, China
[2] Key Laboratory of Medical Image Computing (NEU), Ministry of Education
caokeyan@gmail.com

Abstract. In recent years, data mining over uncertain data stream has attracted a lot of attentions along with the imprecise data widely generated. In many cases, the estimated error of the data stream is available. The estimated error is very useful for the clustering process, since it can be used to improve the quality of the cluster results. In this paper, we try to resolve the problem of clustering uncertain data stream over sliding windows. The tuple expected value and uncertainty are considered meanwhile in the clustering process. We therefore propose the algorithm based on Voronoi diagram to reduce the number of expected distance calculation over sliding windows. Finally, our performance study with both real and synthetic data sets demonstrates the efficiency and effectiveness of our proposed method.

Keywords: Uncertain data stream, clustering, data mining.

1 Introduction

In recent years, a large amount of uncertain data stream, such as sensors data stream is generated. Analyzing and mining such kinds of data stream have becoming a hot topic. Mining uncertain data stream is much more difficult than mining tasks over deterministic data stream.

Clustering is one of the most important tasks in data mining field. Most of traditional clustering methods in an uncertain data set treat the distances between tuples as the unique factor to cluster. However, when the data is uncertain with error, not only consider the expected value, but also consider the uncertainty of the data. The presence of uncertainty can significantly affect the results of clustering. In the clustering process, the data with higher uncertainty may be treated differently from those which have lower uncertainty [1] [2].

In this paper, we adopt the sliding window model and Voronoi diagram to cluster uncertain data stream. Our main contributions are as follows:

(1) We provide a method to quantify tuple uncertainty in order to distinguish the uncertainty of different estimated error.

[*] This research was supported by the National Natural Science Foundation of China (Grant No. 61073063, 61173029, 60803026 and 61173030).

H. Gao et al. (Eds.): WAIM 2012, LNCS 7418, pp. 308–313, 2012.

(2) The cluster algorithm is proposed with Voronoi diagram over sliding window.

(3) We also implement a series of experiment evaluation, showing that our method is effective and efficient.

This paper is organized as follows: In the next section, we propose a metric to quantify uncertainty of tuple in Section 2. In section 3, we propose an algorithm for clustering uncertain data stream with Voronoi diagram over sliding windows, and elaborate the implementation details respectively. Section 4 we will discuss the experimental results. Finally, we give a brief conclusion and point out the future work in Section 5.

2   Uncertain Definition

In this section, the method which is introduced to quality of uncertainty is used for distinguishing the quality of uncertain data. We assume it is available that the estimated error of the data stream. Each point is composed of tuple value and estimated error of each dimension. In fact, it's an uncertainty deduction procedure. We can see that the smaller the range of tuple, the less uncertainty of it. If we can obtain certain data, its uncertainty will reduce to zero, then it transforms to a deterministic situation. Above analysis indicates that estimated error of a tuple is highly correlated with its uncertainty. We further define the tuple uncertainty as the function.

Definition 1. *(tuple uncertainty) tuple x_i is two-dimensional data, the tuple uncertainty of $x_i((x_i^1, \psi(x_i^1)), (x_i^2, \psi(x_i^2)))$, $\psi(x_i^1) > 0$, $\psi(x_i^2) > 0$, is defined as*

$$u(x_i) = \frac{\psi(x_i^1) \cdot \psi(x_i^2)}{\psi(x_i^1) \cdot \psi(x_i^2) + 1} \tag{1}$$

$\psi(x_i^d)$ $(1 \leq d \leq 2)$ *is the error value of the d-dimensional of x_i.*

Property 1. (low bound) $u(x_i) > 0$, uncertainty of tuple x_i is greater than zero.

Proof. because of $\psi(x_i^1) \cdot \psi(x_i^2) > 0$, and $\psi(x_i^1) \cdot \psi(x_i^2) + 1 > 1$, so

$$u(x_i) = \frac{\psi(x_i^1) \cdot \psi(x_i^2)}{\psi(x_i^1) \cdot \psi(x_i^2) + 1} > 0$$

Property 2. (upper bound) $u(x_i) < 1$, uncertainty of tuples x_i is always less than 1.

Proof. suppose

$$u(x_i) = \frac{\psi(x_i^1) \cdot \psi(x_i^2)}{\psi(x_i^1) \cdot \psi(x_i^2) + 1} \geq 1$$

the inequality further implies that

$$\frac{1}{\psi(x_i^1) \cdot \psi(x_i^2) + 1} \leq 0$$

obviously, this inequality does not hold, the assumption is not true. So the uncertainty of tuple $u(x_i)$ is always less than 1.

Property 3. Tuple uncertainty is only a function of estimated error. It does not depend on the actual values of tuple, but only on its estimated error for each dimension.

By Definition 1, the uncertainty of tuple is available. Through uncertainty of tuple, uncertainty of cluster can be obtained.

Definition 2. *(cluster uncertainty) the cluster is formed of n tuples $x_1, x_2, \cdots, x_n$, u_{x_i} represent the uncertainty of the x_i, the uncertainty of cluster is define as*

$$U = \frac{\sum_{i=1}^{n} u_{x_i}}{n}$$

3   Clustering Uncertain Data Stream

Based on the quantification of uncertainty, we propose the algorithm to cluster uncertain data stream with Voronoi diagram over sliding windows. Although in [1], Aggarwal and Yu proposed Error based Cluster Feature (ECF) to handle uncertain data stream. However, the influence of uncertainty has not been considered in the procedure of uncertain stream clustering, which affects the quality of cluster results. In this paper, we emphasize the influence that uncertainty to cluster results.

3.1   Problem Definition

We assume that uncertain data stream consists of a set of two-dimensional uncertain tuples $x_1, x_2 \cdots, x_i, \cdots, x_n$, arriving at time stamp $T_1, T_2, \cdots, T_i, \cdots, T_n$, for any $i < n$, $T_i < T_n$. The tuple x_i is expressed as $((x_i^1, \psi(x_i^1)), (x_i^2, \psi(x_i^2)))$. Estimated error of tuple for each dimension is $\psi(x_i^d)$ $(1 \le d \le 2)$ respectively.

3.2   Framework

The micro-clustering model was first proposed in [3] for large data sets, in order to cope with the uncertain stream model, a variety of micro-clusters are defined, named Micro-Cluster Feature (MCF)(Definition 3).

Definition 3. *(MCF): A Micro-Cluster feature(MCF) for a set of d-dimensional $(1 \le d \le 2))$ points with time stamps $T_1, T_2, \cdots, T_i, \cdots, T_n$, for any $i < n$, $T_i < T_n$, and error $\psi(x_i^d)$ is defined as the (2d+3) tuple $(\overline{CF2^x}(C), \overline{CF1^x}(C), n(C), U(C), t(C))$, $\overline{CF2^x}(C)$ and $\overline{CF1^x}(C)$ each correspond to a vector of d entries. The definition of each of these entries is as follows:*

• *For each dimension, the sum of squares of the data values is maintained in $\overline{CF2^x}(C)$. The value of j-th entry is $\sum_{i=1}^{n} (x_i)^2$.*

- *For each dimension, the sum of the data is maintained in $\overline{CF1^x}(C)$. The j-th entry of $\overline{CF1^x}(C)$ is equal to $\sum_{i=1}^{n} x_i$.*
- *The number of points in the cluster is maintained in $n(C)$.*
- *The uncertainty of cluster is maintained in $U(C)$.*
- *The time point of the most recent point is maintained in $t(C)$.*

We note that error based micro-clusters maintain the important additive property which is critical to its use in the clustering process.

Property 4. (MCF additive): Let C_1, C_2 denote two sets of points. $MCF(C_1 \cup C_2)$ is available, according to $MCF(C_1)$ and $MCF(C_2)$.

The values of entries $\overline{CF2^x}(C)$, $\overline{CF1^x}(C)$, and $n(C)$ are the sum of the corresponding entries in $MCF(C_1)$ and $MCF(C_2)$. The single temporal component $t(C_1 U C_2)$ is given by max $t(C_1), t(C_2)$. The additive property is an important one, when new data point arrived, MCF of the new cluster can be calculated easily and rapidly.

In the practical application, people often more care about the most recent distribution of the data stream, sliding window model can be used to better present the characteristics of current data stream. The algorithm we proposed is based on sliding window, can phase out outdated data in a timely manner, present the characteristics of current data stream really. In this section, we adopt the model to update uncertain data stream over sliding window was proposed in [4].

3.3   Clustering Algorithm

For the m points arrived early, we establish micro-cluster for each point, and build voronoi diagram. In Figure 1, when the next uncertain data x_i arrives, calculate the uncertainty of tuples, by Definition 1. At the same time, we determine whether the range of data point is within a cluster in the voronoi diagram. If it is true, calculate the distance d_{is} from expected value of x_i to centroid of the cluster, when $d_{is} < \delta$ (δ is the threshold) add uncertain data x_i to the cluster, update the MCF. Otherwise, we merge the two clusters, create a new cluster for the current data. If the range of data x_i is on the junction of n clusters, we calculate the distance from expected value of data x_i to centroid of the n clusters respectively, and select some of them ($d_{is} < \delta$), assuming the data are added to each cluster, calculate $\Delta U(C)$ of each cluster, pick up the cluster with maximum value of to absorb the data point x_i.

3.4   Update Voronoi Diagram

While a new data point is added to a cluster, the centroid of cluster may be changed, in Figure 2, point c_1 is centroid of the cluster, however, c_1' is the new centroid of the cluster after it was merged a new point, we update the Voronoi diagram, only need to recalculate perpendicular bisector from c_1' to centroid of adjacent cluster, and no need to change other parts of Voronoi diagram. This reduces the number of unnecessary calculations, to improve updated speed.

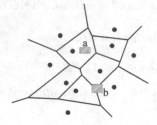

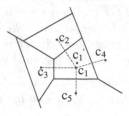

Fig. 1. Uncertain data point in voronoi diagram

Fig. 2. Update cluster

4 Experiment

In this section, we employed UMicro [5] as the competitive method. To evaluate the clustering quality, efficiency of the H-QCUDS and UMicro algorithm, both real and synthetic data sets are used in our experiments. Specifically, we show the algorithm that we proposed can get high quality clustering result. The method can achieve superior performances and it is scalable to fit for the stream requirement.

4.1 Clustering Quality

We use uncertainty mean (UM) to measure the level of uncertainty in clustering result. As [2], the UM is defined as the average uncertainty of overall online micro clusters. In general, the smaller UM value is, the greater quality of the clustering result.

Figures 3 shows the clustering quality comparison results. It is clear that, in each case, the H-QCUDS provided superior quality to UMicro under UM criterion. The high clustering quality of H-QCUDS benefit from that our algorithm pay more attention to the uncertainty of the data tuples.

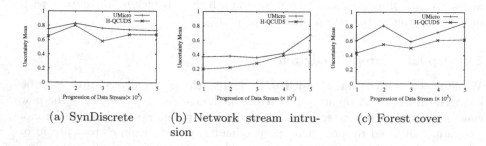

(a) SynDiscrete

(b) Network stream intrusion

(c) Forest cover

Fig. 3. Uncertainty of stream clustering

4.2 Efficiency Test

In Figures 4, we illustrate the efficiency of the clustering method on the different data sets. We can see that both the execution time of H-QCUDS and UMicro grow linearly as the stream proceeds, and H-QCUDS is more efficient than UMicro. It is clear that the H-QCUDS method is able to process thousands of points per second in each case. From this perspective, our H-QCUDS method is not only effective, but also a very efficient clustering method for uncertain data stream.

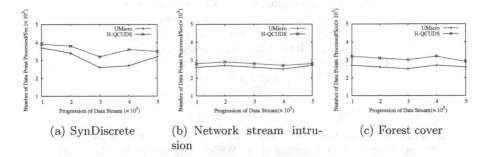

(a) SynDiscrete   (b) Network stream intrusion   (c) Forest cover

Fig. 4. Efficiency of stream clustering

5   Conclusions

In this paper, we proposed H-QCUDS, an algorithm used to cluster uncertain data stream with Voronoi diagram over the sliding window. A Micro-Clustering Feature (MCF) is presented to tracking the uncertain data streams. In order to improve clustering quality, H-QCUDS algorithm adopts a two phase stream clustering selection process. The experimental results show that the clustering quality of the H-QCUDS algorithm is higher than that of UMicro algorithm. Next steps of our work are taking more complex uncertain data stream into account, such as high-dimensional.

References

1. Aggarwal, C.C., Yu, P.S.: A survey of uncertain data algorithms and applications. Knowledge and Data Engineering 21(5), 609–623 (2009)
2. Zhang, C., Gao, M., Zhou, A.: Tracking high quality clusters over uncertain data streams. In: ICDE 2009, pp. 1641–1648 (2009)
3. Zhang, T., Ramakrishnan, R., Livny, M.: BIRCH: An efficient data clustering method for very large databases. In: SIGMOD 1996, vol. 25(2), pp. 103–114 (1996)
4. Chang, J.L., Cao, F., Zhou, A.Y.: Clustering evolving data streams over sliding windows. Journal of Software 18(4), 905–918 (2007)
5. Aggarwal, C.C., Yu, P.S.: A framework for clustering uncertain data streams. In: ICDE, pp. 150–159 (2008)

Bayesian Network Structure Learning from Attribute Uncertain Data

Wenting Song[1,2], Jeffrey Xu Yu[3], Hong Cheng[3], Hongyan Liu[4],
Jun He[1,2,*], and Xiaoyong Du[1,2]

[1] Key Labs of Data Engineering and Knowledge Engineering, Ministry of Education, China
[2] School of Information, Renmin University of China
`{songwt,hejun,duyong}@ruc.edu.cn`
[3] The Chinese University of Hong Kong
`{yu,hcheng}@se.cuhk.edu.hk`
[4] Department of Manage. Sci. & Eng., Tsinghua University
`hyliu@tsinghua.edu.cn`

Abstract. In recent years there has been a growing interest in Bayesian Network learning from uncertain data. While many researchers focus on Bayesian Network learning from data with tuple uncertainty, Bayesian Network structure learning from data with attribute uncertainty gets little attention. In this paper we make a clear definition of attribute uncertain data and Bayesian Network Learning problem from such data. We propose a structure learning method named *DTAU* based on information theory. The algorithm assumes that the structure of a Bayesian network is a tree. It avoids enumerating all possible worlds. The dependency tree is computed with polynomial time complexity. We conduct experiments to demonstrate the effectiveness and efficiency of our method. The experiments show the clustering results on uncertain dataset by our dependency tree are acceptable.

Keywords: uncertainty, Bayesian Network structure, dependency.

1 Introduction

In the past, researchers in data mining and machine learning area usually assume data is certain or precise, however that is not always the case. Data uncertainty arises in many applications such as sensor network and user privacy protection. Uncertain data can be divided into two categories by uncertainty source: artificial uncertain data and inherent uncertain data. People sometimes add noise to data for some purpose such as user privacy protection. As a result, the data become artificially uncertain. There are also inherently uncertain data. For example the scientific measurement techniques and tools are inherently imprecise and they are responsible for the generation of inherent uncertain data. Many researchers focus on uncertain data management and mining in database area and data mining area in recent years. In database area, there are three

---

* Corresponding author.

H. Gao et al. (Eds.): WAIM 2012, LNCS 7418, pp. 314–321, 2012.

models of uncertain data. The first one is tuple uncertainty model [1] [2]. Each tuple in a probabilistic database is associated with a probability which represents the likelihood the tuple exists in the relation. The second one is attribute uncertainty model [3]. In attribute uncertainty model, each attribute in a tuple is subject to an independent probability distribution. Correlated uncertainty model [5] is the third one. Attributes are described by a joint probability distribution.

Many data mining algorithms have been proposed to analyze uncertain data, for example, mining frequent patterns from uncertain transaction database [2], naïve Bayesian classifiers for correlated uncertain data [7] [8], and clustering uncertain objects [6]. However there are few works on data mining from attribute uncertain data, and to the best of our knowledge no work has focused on how to learn Bayesian Network (*BN*) structure from such data. Attribute independency is a common assumption in database and data mining area, but it is not always reasonable, because there are dependency relationships among attributes. Attribute uncertainty due to measurement error or inherent uncertainty shouldn't be the reason for independence assumption. The structure learning from attribute uncertain data can reveal the essential relationship between attributes.

In this paper, we propose the problem of *BN* structure learning from attribute uncertain data and an algorithm named *DTAU* to solve the problem. Experiments demonstrate the effectiveness of our proposed algorithm.

The rest of the paper is organized as follows. Section 2 discusses related work on Bayesian Network structure learning for uncertain data. Section 3 gives relevant definitions. Section 4 introduces our structure learning algorithm *DTAU*. Section 5 is experimental study and Section 6 concludes the paper.

2   Related Work

Bayesian Network (*BN*) is a powerful tool to represent joint probability distribution over a set of variables or attributes. A *BN* is made up of two components: a directed acyclic graph (*DAG*), whose nodes represent variables and a set of conditional probability tables (CPTs) which specifies the conditional distribution of each variable given its parent in the *DAG*. Given a *BN* structure (*DAG*), there have been many algorithms for parameter learning. However, sometimes the *BN* structures are unknown for lack of domain knowledge. Thus the *BN* learning problem is of great importance. It has been proved that *BN* structure learning problem is *NP* completed [9].

Many researchers have proposed approximation algorithms to solve the structure learning from certain data problem. These methods are divided into two categories. The methods in the first category are based on information theory which is used to measure dependency relationships between nodes [10]. The methods [11] [12] in the second category aim to maximize score function of the possible structure considering that each node has no more than *K* parents. As this problem is *NP*-hard when *K* is bigger than 1, heuristic rules based methods are usually used. For the attribute uncertain data, we make use of the information theory to solve the problem and assume the structure is a tree.

3   Problem Definition

In this section we describe some concepts about the problem of learning *BN* structure from attribute uncertain data. The term *observation* is a concept in *BN* learning from certain data problem.

Definition 1 (*Attribute*). An *attribute* X_i is a component or aspect of an *object O*. X_i can take any value in $D(X_i)$ which is the possible value domain of X_i. D_i represents the size of $D(X_i)$. The *attribute* X_i is represented by a node (a random variable) in the *BN* structure.

Definition 2 (*uncertain example*). An vector $ue = \{P_1(X_1), P_2(X_2), ..., P_m(X_m)\}$ is an *uncertain example* if each $P(X_i)$ *is* an probability distribution or probability density function over $D(X_i)$.

Definition 3 (*Uncertain observation*). Given an *attribute* X_i, an *observation* $P_i(X_i)$ of X_i ($1 \leq i \leq m$) in an uncertain example *ue* is an uncertain observation.

Definition 4 (*Attribute Uncertain training dataset*). An uncertain training dataset D is composed of uncertain examples, $D = \{ue_1, ue_2, ..., ue_n\}$.

In this paper we focus on the problem of learning a *BN* structure from attribute uncertain training dataset. We assume the structure of Bayesian Network is a tree.

In the following parts of the paper we study the problem of learning structure from discrete attribute uncertain data. If they is continuous, we can discrete them.

4   Bayesian Network Structure Learning Algorithm *DTAU*

In this section, we start with a brief introduction of a naïve *BN* structure learning algorithm based on the exponential possible worlds. Then we will explain why the naïve method is unacceptable. At last we will show our approximation method which takes polynomial time.

Definition 5 (*Possible world*). Given an uncertain dataset about m attributes, it generates possible worlds, where each world is a certain dataset about the m attributes which has the same size of examples with the uncertain one. Each *possible world* W_i is associated with a probability $Pr(W_i)$ that the world exists.

The naïve method is based on possible world over all attributes (*PWAA*). The idea is converting the attribute uncertain dataset to some certain datasets. Given an uncertain discrete dataset AUD with N examples and m attributes, first we compute every possible world W_i of AUD. Each observations of example e_{ij} in possible world W_i is a possible value in the domain of the corresponding attribute. Those possible worlds form a set W and its size is $\Pi^m_{i=1}(D_i)^N$. Second, we treat each possible world W_i as a certain training dataset, and then we learn a dependency tree under the corresponding training data set. The tree with the highest score can be recognized as the right one. The total number of trees (obviously containing the duplicates) is the number of the possible worlds. The score of a tree T_i is $\Sigma_{Wj.tree=Ti} Pr(W_j)$.

The number of possible worlds is exponential, so the solution presented above costs exponential time complexity to construct the dependency tree. We propose an algorithm named *DTAU* (Dependency Tree learning from Attribute Uncertain data) to construct a dependency tree without enumeration of possible worlds and reduce the enormous computation. Our idea is to make use of the attribute uncertain dataset directly.

For a traditional certain training dataset, a popular way to construct the dependency tree with the closest probability approximation is called Chow-Liu tree [10]. The kernel idea in [10] is how to compute the dependency between each two attributes. The dependency between nodes X_i and X_j is measured by mutual entropy $I(X_i, X_j)$. The computation is defined by Equation 1. The value of mutual entropy $I(X_i, X_j)$ is always positive or zero. The value is more close to zero, then the dependency between the two attribute is weaker. The zero value means they are independent. We propose the *DTAU* algorithm to learn a dependency tree from attribute uncertain data. The *DTAU* algorithm is consistent with Chow-Liu tree under certain training data. The key point in the *DTAU* algorithm is how to compute the dependency between each two attributes under attribute uncertain training data. Equation 2 shows an initial approximation of $I(X_i, X_j)$. The two equations are from [10].

$$I(X_i, X_j) = \sum_{x_i, x_j} P(X_i = x_i, X_j = x_j) \log \frac{P(X_i = x_i, X_j = x_j)}{P(X_i = x_i)P(X_j = x_j)} \tag{1}$$

$$I(X_i, X_j) \approx \sum_{s \in D(X_i), t \in D(X_j)} N(X_i = s, X_j = t) \log \frac{\frac{N(X_i = s, X_j = t)}{n}}{\frac{N(X_i = s)}{n} \frac{N(X_j = t)}{n}} \tag{2}$$

Equation 3, 4 and 5 shows how to approximate the frequency in equation 2 and we get the final approximation of $I(X_i, X_j)$ by equation 3, 4 and 5.

$$N(X_i = s, X_j = t) = \sum_{k=1}^{n} P_{ki}(X_i = s) P_{kj}(X_j = t) \tag{3}$$

$$N(X_i = s) = \sum_{k=1}^{n} P_{ki}(X_i = s) \tag{4}$$

$$N(X_j = t) = \sum_{k=1}^{n} P_{kj}(X_j = t) \tag{5}$$

From the equations above we can learn that if the probability $P_{ki}(X_i=s)$ is the highest for attribute X_i and the probability $P_{kj}(X_j = t)$ is the highest for attribute X_j, then the occurrence probability for pair (s, t) may be the highest. The idea behind the equation is that the independence assumption doesn't have effect on the overall dependency computation. In other words, if the two attributes are independent, the computation result is zero. If they are not, the computation result is positive. The equation shows the consistence with certain data. We prove the result and we don't describe the details of the proof for the limitation of space in this paper.

The *DTAU* algorithm is divided into three steps. The first step is to compute the dependency between each two uncertain attributes and construct a weighted undirected graph. Then we follow the tree construction method in the Chow-Liu tree algorithm. The second step is to get a maximum spanning tree by a greedy algorithm which is 2-ratio approximation of the optimal tree. The last step is to add the direction by width first traverse. The pseudo-code is given below:

Algorithm *DTAU*

Input: an attribute uncertain dataset AUD with n examples and m variables
Output: a dependency tree T

Major steps:

Construct a weighted undirected completed graph $G = (V, E), V = \{X1, X2, \ldots, Xm\}$;

for i =1 to m
   for j =1 to i
Compute dependency $I(X_i, X_j)$ by equation 1,2,3,4 and 5;
$e(X_i, X_j) = I(X_i, X_j)$
end for
end for
T = greedy_max_spanning_tree(G);
Select a random node X_k in T as beginning node;
Add direction for each edge by breadth-first traverse beginning with X_j;

The time complexity of *DTAU* algorithm is $O(nm^2)$, which is smaller than the naïve solution. We can confirm that if the difference between $I(X_i, X_j)$ *and* $I(X_i, X_k)$ satisfies a *t*-condition that the absolute value of the difference between $I(X_i, X_k)$ and $I(X_i, X_k)$ is bigger than t, the dependency tree created by the *DTAU* method is the same with the one in the possible world with the highest probability. Because if the *t*-condition is satisfied, the partial orders for all mutual entropy are the same. The partial orders can determine the structure. We prove this result and we don't describe the details of the proof or the computation of t for the limitation of space. Experiments show that our method performs well even when the *t*-condition can't be satisfied.

5   Experiments

As there hasn't been any public attribute uncertain dataset, the attribute uncertain datasets we use are generated from certain datasets artificially. We generate the attribute uncertain datasets by adding noise to the UCI machine learning datasets which are standard for traditional *BN* learning problems. We convert the Letter recognition and Balance datasets to attribute uncertain datasets. The noise addition strategy is described as follows. For each training example in the original certain dataset, we assign a probability p which is not smaller than a bound α to the corresponding attribute's observation in the original certain training example, and assign a low probability to other possible values for this attribute. Table 1 shows an example of the original certain dataset, where $D(Attribute\ 1) = \{a, b, c\}$ and $D(Attribute\ 2) = \{d, e, f\}$. Table 2 shows an uncertain dataset obtained after noise addition with α being 0.5. By this way we get attribute uncertain training datasets denoted by AU-Letter-α, and AU-Balance-α respectively.

Table 1. An example of certain data

Attribute 1	Attribute 2
a	d
b	e

Table 2. An example of noise addition

Attribute 1			Attribute 2		
a: 0.8	b: 0.1	c: 0.1	d: 0.5	e: 0.25	f: 0.25
a: 0.2	b: 0.6	c: 0.2	d: 0.15	e: 0.7	f: 0.15

For each uncertain dataset we generate, the uncertain observations in an attribute uncertain dataset are closer to the ones of the original certain dataset when α is closer to 1. The experiment on AU-Letter-0.5 shows that the partial orders of the dependency measure (information entropy) in attribute uncertain data is almost the same with the one in the certain dataset. Figure 1 shows the tree we learn from AU-Letter-0.5 and the Chow-Liu tree learned from the certain dataset.

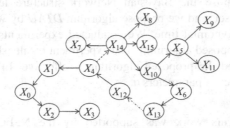

Fig. 1. Dependency tree from the dataset Letter and the correspondingly certain dataset

The two trees share the black solid edges. Edge<X_{13}, X_{12}> belongs to the tree from uncertain dataset and edge <X_4, X_{14}> belongs to the tree from certain dataset. We find that only edge <X_4, X_{14}> and edge <X_{13}, X_{12}> are different. The difference between $I(X_4, X_{14})$ and $I(X_{12}, X_{13})$ accounts for the bigger one of the two information entropy less than 1.3%.

We design experiments on AU-Balance to demonstrate the effectiveness of the dependency tree by clustering results. We use *DTAU* algorithm to learn a dependency tree from AU-balance data. Then we generate a certain sample dataset *i* for the uncertain training example ue_i in the dataset. Then we treat the sample dataset *i* as the training dataset and the dependency tree as *BN* structure to learn the joint probability distribution on all attributes in the uncertain dataset and then the uncertain example ue_i turns to be an uncertain object. We cluster those uncertain objects by algorithm *UK*-means [4] and the original certain objects by *K*-means and compare the two clustering results. We do experiments on the AU-Balance dataset with different values of α.

Table 3. Clustering precision under different parameters

Dataset	Cluster 0	Cluster 1	Cluster 2	precision
AU-Balance-0.6	6	161	103	43.2%
AU-Balance-0.8	21	115	98	37.44%
AU-Balance-0.9	5	119	156	44.8%
Balance certain	16	175	145	53.74%

We test three different values of parameter α, 0.6, 0.8 and 0.9. For each dataset, the cluster result is compared with the true class labels. Table 3 shows the results under different uncertain dataset. The numbers in column 2, column 3 and column 4 represent the size of correct examples in the corresponding cluster. The measure precision shows the percentage of correct clustered examples. From this table we can see that the precision for each of the three uncertain dataset is quite close to the certain one, for the certain one is always the possible world with the highest probability. The experiments show that the dependency tree generated by our method is acceptable and α is an important factor to the cluster results.

6   Conclusion

In this paper we propose the Bayesian Network structure learning problem on attribute uncertain dataset, and we propose algorithm *DTAU* by which we can learn a dependency tree in polynomial time. We conducted experiments to demonstrate the effectiveness of our proposed algorithm. The experiment results show the dependency trees are acceptable and the proposed algorithm is effective. In the future, we will further analyze the effect of parameters α.

Acknowledgement. This work was supported by the NSFC under Grant No. 70871068 and 71110107027 and the Major National Sci. and Tech. Project of China under Grant No. 2010ZX01042-002-002-03.

References

1. Dalvi, N., Suciu, D.: Effcient query evaluation on probabilistic databases. The VLDB Journal 16(4), 523–544 (2007)
2. Bernecker, T., Kriegel, H.P., Renz, M., Verhein, F., Zuefle, A.: Probabilistic frequent item set mining in uncertain databases. In: 15th ACM SIGKDD International Conference on Knowledge Discovery and Data Mining, pp. 119–128. ACM Press, Paris (2009)
3. Singh, S., Mayfield, C., Shah, R., Prabhakar, S., Hambrusch, S., Neville, J., Cheng, R.: Database support for probabilistic attributes and tuples. In: 24th IEEE ICDE International Conference on Data Engineering, pp. 1053–1061. IEEE Press, Cancún (2008)
4. Lee, S.D., Kao, B., Cheng, R.: Reducing UK-means to K-means. In: 7th IEEE ICDM Workshops International Conference on Data Mining Workshops, pp. 483–488. IEEE Press, Omaha (2008)
5. Gullo, F., Ponti, G., Tagarelli, A., Greco, S.: A Hierarchical Algorithm for Clustering Uncertain Data via an Information-Theoretic Approach. In: 8th IEEE ICDM International Conference on Data Mining, pp. 1053–1061. IEEE Press, Pisa (2008)
6. Günnemann, S., Kremer, H., Seidl, T.: Subspace Clustering for Uncertain Data. In: SIAM SDM SIAM Conference on Data Mining, pp. 385–396. SIAM Press, Ohio (2010)
7. He, J., Zhang, Y., Li, X., Wang, Y.: Naive Bayes classifier for positive unlabeled learning with uncertainty. In: SIAM SDM SIAM Conference on Data Mining, pp. 361–372. SIAM Press, Ohio (2010)

8. Ren, J., Lee, S.D., Chen, X., Kao, B., Cheng, R., Cheung, D.: Naive bayes classification of uncertain data. In: 9th IEEE ICDM International Conference on Data Mining, pp. 944–949. IEEE Press, Miami (2009)
9. Dalvi, N., Suciu, D.: Learning Bayesian networks is *NP*-complete. Lecture Notes In Statistics, pp. 121–130. Springer, New York (1996)
10. Chow, C., Liu, C.: Approximating discrete probability distributions with dependence trees. IEEE Transactions on Information Theory Journal 14(3), 462–467 (1968)
11. Heckerman, D., et al.: A tutorial on learning with Bayesian networks. In: Learning in Graphical Models, Michael I. Jordan, Massachusetts (1999)
12. Friedman, N., Nachman, I., Peér, D.: Learning Bayesian Network Structure from Massive Datasets: The "Sparse Candidate" Algorithm. In: UAI 14th Conference on Uncertainty in Artificial Intelligence, Madison, pp. 206–215 (1999)

Bandwidth-Aware Medical Image Retrieval in Mobile Cloud Computing Network[*]

Yi Zhuang[1], Nan Jiang[2], Zhiang Wu[3], Dickson Chiu[4], Guochang Jiang[5], and Hua Hu[6]

[1] College of Computer & Information Engineering, Zhejiang Gongshang University, P.R. China
[2] Hangzhou First People's Hospital, Hangzhou, P.R. China
[3] Jiangsu Provincial Key Laboratory of E-Business,
Nanjing University of Finance and Economics, P.R. China
[4] Dickson Computer Systems, HKSAR, P.R. China
[5] The Second Institute of Oceanography, SOA, Hangzhou, P.R. China
[6] School of Computer, Hangzhou Dianzi University, P.R. China
zhuang@zjgsu.edu.cn

Abstract. This paper proposes a bandwidth-aware content-based Medical Image retrieval method in Mobile Cloud computing environment, called the *MiMiC*. The whole query process of the *MiMiC* is composed of three steps. First when a doctor submits a query image I_q, a parallel image set reduction process is first conducted at a master node level. Then the candidate images are transferred to the slave nodes for a refinement process to obtain the answer set. Finally, the answer set is transferred to the query node. The proposed method including the adaptive load balancing scheme is specifically designed for solving the heterogeneity of the mobile cloud and an index-support image set reduction algorithm for reducing the data transfer cost in cloud environment. Additionally, we propose a bandwidth-conscious multi-resolution-based data transfer technique to further improve the query performance. The experimental results show that the performance of the algorithm is efficient and effective in minimizing the response time by decreasing the network transfer cost while increasing the parallelism of I/O and CPU.

Keywords: Medical image, multi-resolution, mobile cloud.

1   Introduction

Nowadays, with an explosive increase of medical multimedia data in hospital information management systems(HIMS), many applications require an efficient access

---

[*] This paper is partially supported by the Program of National Natural Science Foundation of China under Grant No. 61003047, No.71072172, No.61103229, No.60903053; the Program of Natural Science Foundation of Zhejiang Province under Grant No. Z1100822, No. Y1110644, Y1110969, No.Y1090165; the Science & Technology Planning Project of Wenzhou under Grant No. G20100202.

H. Gao et al. (Eds.): WAIM 2012, LNCS 7418, pp. 322–333, 2012.
© Springer-Verlag Berlin Heidelberg 2012

method to support content-based multimedia retrieval at a large scale. As one of the most important media types, management, query, and analysis of medical images plays a critical role in modern HIMSs. Although a considerable amount of related work has been carried out on medical image indexing and similarity query in high-dimensional spaces [1][2], most of them focus on a centralized way (i.e., single- PC-based) which cannot scale up well to large data volume. The query efficiencies of these centralized methods are unsatisfactory because the response time is linearly increasing with the size of the searched file. Therefore the design of high performance medical image query methods becomes a critically important research topic.

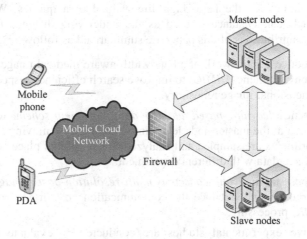

Fig. 1. The architecture of a mobile cloud environment

A Cloud, especially mobile Cloud (*Mc*) can be seen as a type of flexible computing infrastructure consisting of many computing nodes, which can provide resizable computing capacities to different users anywhere anytime. To fully harness the power of the *Mc*, efficient data management is needed to handle huge volumes of medical image data and support a large number of concurrent end users (e.g., doctors). Additionally, the *Mc* environment provides us with a location-based query that enables doctors to retrieve patient records and images conveniently. To achieve this, scalable, high-throughput, location-based querying and indexing schemes are generally required. However, as shown in Fig. 1, for *Mc*-based medical image query, exploring parallelism in the *Mc* to speed up the queries is a new research topic, which has received little attention so far. The challenges include three main aspects:

1) *High computation cost in medical image retrieval*: for a medical image, three characteristics are represented: *high pixel resolution, high-dimensional*, and *large-scale*. So the query cost of such medical images is very high.

2) *The mobility of users in the Mc*: most of the users (i.e., doctors) in the *Mc* are moving. That means the spatial position of each user varies with the variance of time. So how to perform an optimal data placement is also a challenging issue.

3) *The instability and heterogeneity of the MC*: the nodes in the *MC* are instable, that means, some nodes may be down or connected intermittently to the network. The bandwidth of any two nodes in the *MC* may be different according to the variance of time. There is no guarantee that the total response time of each query can be equal.

To address the above challenges, we propose an efficient distributed similarity (*MiMiC*) query processing technique in the mobile cloud environment. The *MiMiC* includes two enabling techniques, namely, *learning-based optimal data placement scheme* and *bandwidth-conscious multi-resolution-based image data transfer*. We have implemented the *MiMiC* method and extensive experiments indicate that our method is specifically suitable for the large high-dimensional data queries. Without loss of generality, Euclidean Distance is used as the underlying distance function in our research. The contributions of this paper are summarized as follows:

- We introduce a framework of a bandwidth-aware medical image retrieval in the mobile cloud environment (*MiMiC*) to improve search efficiency, especially for large-scale high-dimensional image repository.

- We present a *learning-based adaptive data placement scheme* to maximize the query parallelism at the master node level, in which doctors' moving trajectories with different departments are sampled and analyzed to get optimal placement positions of the medical image data with different department.

- We propose a *bandwidth-conscious multi-resolution-based image data transfer algorithm* to progressively reduce the communication cost in *MC* environment and speed up *MiMiC* processing.

- Extensive experimental studies are conducted to evaluate the efficiency, scalability, and robustness of our proposed algorithm.

The rest of paper is organized as follows. Related works are reviewed in Section 2. Preliminary work is given in Section 3. In Section 4, two enabling techniques, viz., the *adaptive learning-based data placement scheme* and the *bandwidth-conscious multi-resolution-based data transfer scheme* are introduced to facilitate a fast similar search over mobile cloud. In Section 5, we propose a *MiMiC* query processing algorithm. In Section 6, we perform comprehensive experiments to evaluate the efficiency of our proposed method. We conclude the paper in Section 7.

2   Related Work

Medical images have often been used for retrieval systems and the medical domain is often cited as one of the principal application domains for content-based access technologies in terms of potential impact. The famous two retrieval systems are the ASSERT system [1] on the classification of high resolution CTs of the lung and the IRMA system [2] for the classification of images into anatomical areas, modalities and view points. As both of them are based on a single PC environment, their processing scalabilities are limited, especially for a large volume of the medical images.

Much effort has been invested in designing distributed storage systems to manage large amounts of data, such as Google File System [3] (GFS), which serves Google's applications with large data volume. BigTable [3] is a distributed storage system for managing structured data of very large scales. Yahoo proposed PNUTS [5], a hosted, centrally controlled parallel and distributed database system for Yahoo's applications. These systems organize data into chunks, and then randomly disseminate chunks into clusters to improve data access parallelism. Some central servers working as routers are responsible for guiding queries to nodes that hold query results. Amazon's Dynamo [4] is a readily available key-value store based on geographical replication, and it can provide eventual consistency. MapReduce [6] was proposed to process large datasets disseminated among clusters.

Berchtold *et al.* [9] proposed a fast parallel similarity search in multimedia databases by providing a near-optimal distribution of data items among the disks. Furthermore, Sahin et al. [10] presented a similarity search using disk array. Recently, Peer-to-Peer(P2P)-based similarity search has been received attention increasingly. CAN [11] is the first system supporting multi-dimensional data. These works behave poorly when the data distribution is skewed. pSearch [13], a P2P system based on CAN, is proposed for document retrieval in P2P networks by rotating the dimensions in indexing. Another system also based on CAN proposed by Schmidt & Parashar [12] is highly inefficient for exact search. Panos *et al.* [14] proposed the multi-dimensional indexing schemes in the P2P network environment. Few work, however, have touched on medical image query in *Mc* environment due to the different query mechanism.

3   Preliminaries

3.1   Problem Formulation

The list of symbols used throughout the rest of this paper is first summarized in Table 1.

Table 1. Meaning of Symbols Used

Symbols	Meaning
Ω	a set of medical images
$\Omega(t)$	medical images from the t-th department and $t \in [1, \alpha]$
I_i	the i-th medical image and $I_i \in \Omega$
D	the number of dimensions
n	the number of medical images in Ω
I_q	a query image user submits
$\Theta(I_q, r)$	the query sphere with centre I_q and radius r
$d(I_i, I_j)$	the distance between two medical images
$\Omega'(t)$	the candidate image set from the t-th department and $t \in [1, \alpha]$
$\Omega''(t)$	the answer image set from the t-th department and $t \in [1, \alpha]$
α	the number of the departments(*or* master nodes *or* slave nodes)

DEFINITION 1. *A mobile cloud(MC) is a graph, which is composed of Node and Edge, formally denoted as MC=(N,E,T), where N refers to the set of nodes, E refers to a set of edges representing the network bandwidths for data transfer at time T, and T means the time.*

In the above definition, due to the instability and heterogeneity of the *Mc* environment, the bandwidth of any two nodes in *Mc* may be different and variant with the change of the time.

DEFINITION 2. *The nodes in MC, formally denoted as $N=N_q+N_m+N_s$, can be logically divided into three categories: the query node(N_q), master nodes(N_m), and slave nodes(N_S), where N_m is composed of α master nodes(N_m^i) and N_s is composed of α slave nodes(N_S^j), where N_m^i is the i-th master node, N_S^j denotes the j-th slave node, for $i=j\in[1,\alpha]$.*

As defined in Definition 2, in a MC, a doctor submits a query from the query node; the master nodes are responsible for storing the medical images(Ω) of different departments and their corresponding indexes; the slave nodes can receive the candidate images (Ω') obtained by image set filtering in the master nodes. Then the refinement processes (*distance computation*) of the candidate images in every slave node are conducted. Finally the answer set (Ω'') is sent back to the query node.

In addition, for easy illustration, suppose that one department corresponds to a master node and a slave node. So the number of master nodes(α) *or* the number of the slave nodes(α) equals to that of departments(α), respectively.

3.2   iDistance

iDistance [8] is a distance-based high-dimensional index method for similarity search. In this paper, we adopt it as an index to efficiently support medical image dataset reduction.

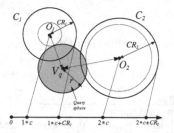

Fig. 2. Distance mapping to one-dimensional value

First, data space is partitioned into T clusters and a reference image O_j for each cluster C_j is selected. Each image is assigned a one-dimensional value according to the distance to its cluster's reference image. Having a large constant c to separate individual clusters, the index key value for a image $I_i\in C_j$ is $key(I_i)=j*c+d(I_i,O_j)$. Then, all images in cluster C_j are mapped to the interval $[j*c, (j+1)*c]$, as shown in Fig. 2. In this way, the problem of similarity search is transformed to an interval search problem. For a range query $\Theta(I_q,r)$, for each cluster C_j that satisfies the inequality $d(O_j,I_q)-r\leq R_j$, the images that are assigned to the cluster C_j and their key values belonging to the interval $[j*c+d(O_j,I_q)-r, j*c+d(O_j,I_q)+r]$ are retrieved. For these images I_i the actual distance to the query image is evaluated and thereafter, if the inequality $d(I_i,I_q)\leq r$ holds, I_i is added to the result set.

4 Enabling Techniques

To facilitate efficient similar query processing in mobile cloud environment, in this section, we introduce two enabling techniques: a *dynamic learning-based data placement scheme* and a *bandwidth-conscious multi-resolution-based transfer mechanism*. Their purposes are to minimize the transfer cost and maximize the query parallelism, respectively.

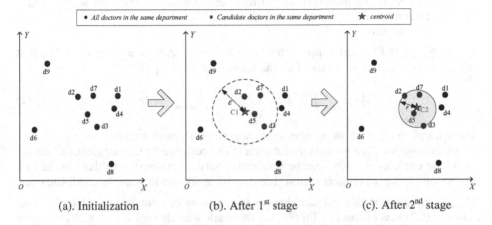

| • All doctors in the same department | • Candidate doctors in the same department | ★ centroid |

(a). Initialization   (b). After 1st stage   (c). After 2nd stage

Fig. 3. Learning-based trajectory mining

4.1 Learning-Based Data Placement Scheme in Master Nodes

For many existing distributed database systems, the study of data placement is critically important to the efficiency of distributed query processing. Different from traditional distributed systems, the *MC* is a mobile, wireless, and heterogeneous environment in which users do not stay at a fixed place. This may raise several challenges to data placement in such a dynamic environment as explained below.

4.1.1 Motivations

The motivations of the learning-based data placement scheme are based on the following key observations:

1) *Fast data access is very important for the distributed query in the MC environment. So the position of the data server that the doctors can access is critically important. That is, to minimize the total communication cost, the total distance between doctors and the server should be minimized.*

2) *For doctors in the same department, not all of them are always in a region (e.g., their department) all the time. That means several doctors may appear in other regions of the hospital due to some other duties. The position axis of the centroid, however, will not be dominated by a few doctors out of their department, but the most. It motivates us to devise a two-stage approach to obtain an optimal centroid position as illustrated in Fig. 3.*

4.1.2 Mining Doctors' Moving Trajectories

As mentioned above, the efficiency of fast data access is very important to the location-based query in the *MC*. So the basic idea of this approach is to find an optimal data placement in the *MC* by first mining doctors' moving trajectories. First, a doctor object can be modeled by a four-tuple:

$$d_i ::= \; < i, \; Dep, \; Loc, \; Tim > \tag{1}$$

where
- *i* means the doctor's ID.
- *Dep* refers to the department that the doctor d_i affiliated with.
- *Loc* is the axis location of d_i, formally denoted as: $Loc::=<x, y>$, where *x* and *y* refer to the axis values respectively.
- *T* is the time.

The 1ˢᵗ Stage. In Fig. 3(a), suppose that there are nine doctors in a department. We first calculate the centroid (C1) axis of all the nine doctors by Eq. (2).

$$C1.x = \frac{1}{|Dep|} \sum_{i=1}^{|Dep|} d_i.Loc.x \; , \quad C1.y = \frac{1}{|Dep|} \sum_{i=1}^{|Dep|} d_i.Loc.y \tag{2}$$

where $|Dep|$ means the total number of doctors in the department *Dep*.

Given centroid C1, a virtual circle region (*VCR*) centered by C1 and radius ε (see the dash blue circle in Fig. 3(b)) can be obtained, where ε is a threshold value. So the *VCR* can be seen as the department region. The doctors inside the *VCR* are the candidate ones.

The 2ⁿᵈ Stage. For all candidate doctors in the same department, calculate the final centroid (C2) axis of them by Eq.(3) (see the shadow circle region in Fig. 3(c)), where C2 is a final center and *r* is a radius.

$$C2.x = \frac{1}{|Can|} \sum_{i=1}^{|Can|} d_i.Loc.x \; , \quad C2.y = \frac{1}{|Can|} \sum_{i=1}^{|Can|} d_i.Loc.y \tag{3}$$

$$r = \text{argMax} \left\{ \sum_{i=1}^{|Can|} (Dis(C2, d_i.Loc)) \right\} \tag{4}$$

where $|Can|$ means the total number of the candidate doctors in the department *Dep*.

Algorithm 1. Optimal learning-based data placement algorithm
Input: Ω: the image set, all doctors;
Output: the optimal data placement;
1.   **for** each department in a hospital **do**
2.   **for** all doctors d_i in the same department **do**
3.   calculate their initial centroid(C1) according to Eq. (2);
4.   **end for**
5.   **for** each doctor d_i in the same department **do**
6.   calculate the distance between C1 and d_i
7.   **if** the distance is less than a threshold value(ε) **then**
8.   add d_i as a candidate element   // *represented by blue points in Fig. 3(b)*
9.   **end if**
10.   **for** each candidate doctor d_i in the same department **do**
11.   calculate their final centroid(C2) according to Eq. (3);
12.   **end for**
13.   the data server of the department can be placed at the position of C2.
14. **end for**

4.2   Multi-resolution-Based Adaptive Data Transfer Scheme

As mentioned before, since the pixel resolution of a medical image is usually very high (e.g., 2040*2040), the data size of the image is large accordingly. It is not trivial to transfer an image with big size to the destination nodes especially in a wireless network.

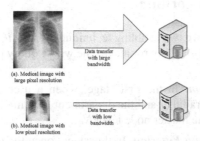

(a). Medical image with
large pixel resolution

Data transfer
with large
bandwidth

(b). Medical image with
low pixel resolution

Data transfer
with low
bandwidth

Fig. 4. Bandwidth-conscious adaptive data transfer scheme

As shown in Fig. 4, the basic idea of our proposed enabling technique is that for a same medical image, the image with different pixel resolutions can be transferred according to the variance of the network bandwidth. Specifically, for high network bandwidth, the medical image with high-resolution for which doctors can get a high-quality image can be transferred in a reasonable short period of time. On the contrary, in order to get a short response time, the low resolution version of the same image is sent to the destination node in a network with low bandwidth. Therefore, the objective of our method is to get a tradeoff between the quality of a medical image and the transferring time under different resolution and current network bandwidth.

The resolution of the i-th image is defined as $I_{RES}(i) \in [x_{LOW}, x_{UPPER}]$, where x_{LOW} and x_{UPPER} mean the lower bound and upper bound resolution of the i-th image. Similarly, the bandwidth of the j-th edge is defined as $E_j \in [y_{LOW}, y_{UPPER}]$, where y_{LOW} and y_{UPPER} are the lower bound and upper bound bandwidth of the j-th edge. So the optimal image resolution under the current network bandwidth can be derived as follows:

$$I_{RES}(i) = x_{LOW} + \frac{i|x_{UPPER} - x_{LOW}|}{\Delta}, \; if \; E_j \in \left[y_{LOW} + \frac{(i-1)|y_{UPPER} - y_{LOW}|}{\Delta}, y_{LOW} + \frac{i|y_{UPPER} - y_{LOW}|}{\Delta} \right] \quad (5)$$

where $i \in [1, \Delta]$ and Δ is a granularity value.

As $E_j \in \left[y_{LOW} + \frac{(i-1)|y_{UPPER} - y_{LOW}|}{\Delta}, y_{LOW} + \frac{i|y_{UPPER} - y_{LOW}|}{\Delta} \right]$, so we have $i \in \left[\frac{(E_j - y_{LOW})\Delta}{|y_{UPPER} - y_{LOW}|}, \frac{(E_j - y_{LOW})\Delta}{|y_{UPPER} - y_{LOW}|} + 1 \right]$.

Since i should be an integer value, so $i = \left\| \frac{(E_j - y_{LOW})\Delta}{|y_{UPPER} - y_{LOW}|} + 1 \right\|$, where $\|\bullet\|$ is the integer value of $\bullet$.

Therefore, the pixel resolution of i-th image can be rewritten by Eq. (6).

$$I_{RES}(i) = x_{\text{LOW}} + \left\| \frac{(E_j - y_{\text{LOW}})\Delta}{|y_{\text{UPPER}} - y_{\text{LOW}}|} + 1 \right\| \times \frac{|y_{\text{UPPER}} - y_{\text{LOW}}|}{\Delta} \tag{6}$$

5   The MiMiC Algorithm

On the support of the above two enabling techniques, a *MiMiC* query can be efficiently conducted in the mobile cloud environment. The whole query process can be composed of three stages:

(1). Query submission. In the first stage, when a user submits a query request (namely query image I_q, radius r and the department information) from the query node N_q, the query is sent to the master node level $N_{m,}$.

(2). Global Image Data Filtering. In this stage, once the query request is received by the corresponding master node N_m^j, then the irrelevant images in the N_m^j are filtered quickly by using the iDistance [8] index, thereby the transfer cost from the master node level to the slave one can be reduced significantly.

Specifically, in this master node N_m^j, an input buffer called *IB* is created for caching the images in $\Omega(j)$, where $\Omega(j)$ refers to the images in N_m^j and $j \in [1, \alpha]$. Meanwhile, an output buffer *OB* which is used to store the candidate image set $\Omega'(j)$ is also created. Once the data size of the candidate images in *OB* reaches the size of a package, the candidate images are transferred to the slave node through the package-based data transfer mode.

Algorithm 2. *IFilter*(I_q, r)
Input: $\Omega(j)$: the sub image set in the j-th master node,
        $\Theta(I_q, r)$: the query sphere,
Output: the candidate image set $\Omega'(j)$
1. $\Omega'(j) \leftarrow \Phi$;   /* initialization */
2. **for** the medical images($\Omega(j)$) in the j-th master node **do**
3.   **for** $k=1$ to T **do**   /* for each cluster in the j-th master node */
4. $\Omega'(j) \leftarrow \Omega'(j) \cup iDistance(I_q, r, k)$;
5. $\Omega'(j)$ is cached in the output buffer *OB*;
6.   **end for**
7. **end for**

In Algorithm 2, the routine *iDistance*(I_q, r, k) returns the candidate image set in the j-th cluster (cf. [8]).

(3). Data Refinement. In the final stage, the distances between the candidate images and the query image I_q are computed in the corresponding slave node. If the distance is less than or equal to r, then the candidate image is sent to the query node N_q in the package-based manner. Specifically, in the j-th slave node N_s^j, we need to set an input buffer *IB* and the memory M for the candidate image set $\Omega'(j)$. Additionally, an output buffer *OB* is set, which is used to store the answer images temporarily. If the data size of the answer image set in *OB* equals to the package size, then the answer images are sent to the query node N_q in the package-based transfer manner.

Algorithm 3. *Refine*($\Omega'(j)$, I_q, r)
Input: the candidate images($\Omega'(j)$) from the j-th master node,
    a query image I_q and a radius r
Output: the answer image set $\Omega''(j)$ from the j-th slave node
1. $\Omega''(j) \leftarrow \Phi$;
2. **for** the candidate images($\Omega'(j)$) in the j-th slave node **do**
3.   **if** $d(I_i, I_q) \leq r$ **and** $I_i \in \Omega'(j)$ **then**
4. $\Omega''(j) \leftarrow \Omega''(j) \cup I_i$;
5.   **end if**
6. $\Omega''(j)$ is cached in the output buffer OB;
7. **end for**

Algorithm 4 shows the detailed steps of our proposed *MiMiC* algorithm.

Algorithm 4.   MiMiCSearch(I_q, r, j)
Input: a query image: I_q, r, department ID: j
Output: the query result S
1. $\Omega'(j) \leftarrow \Phi$, $\Omega''(j) \leftarrow \Phi$;   /* *initialization*   */
2. a query request with the department information is submitted to the corresponding master
    node N_m^j;
3. $\Omega'(j) \leftarrow$ *IFilter*(I_q, r);
4. the candidate image set $\Omega'(j)$ is sent to the corresponding slave node;
5. $\Omega''(j) \leftarrow$ *Refine*($\Omega'(j)$, I_q, r);
6. the answer image set $\Omega''(j)$ is sent to the query node N_q;

6   Experimental Results

To verify the efficiency of the proposed *MiMiC* method, we conduct simulation experiments to demonstrate the query performance. The retrieval system is run on Android platform [16] and the backend system is simulated by the Amazon EC2 [17]. For the index part of the system, iDistance [8] is adopted to support quick filtering of medical images deployed at the master node level. The medical image dataset we used is download from *Medical image archive* [15], which includes 100,000 64-D color histogram features, the value range of each dimension is between 0 and 1.

6.1   Effect of Adaptive Data Transfer Scheme

In the first experiment, we study the effect of the adaptive(i.e., multi-resolution- based) image data transfer scheme on the performance of the *MiMiC* query processing. Method 1 does not adopt the adaptive data transfer algorithm and method 2 uses it. Fig. 5 shows when r is fixed and the bandwidth is relatively stable, the total response time using the method 2 is superior to that of method 1. Meanwhile, with the condition that the bandwidth is stable and r is increasing gradually, the performance gap becomes larger since the data size of the candidate images to be transferred is increasing so rapidly that the images can not be sent to the destination nodes quickly.

6.2   Effect of Data Size

This experiment studies the effect of data size on the performance of the *MiMiC* query processing.

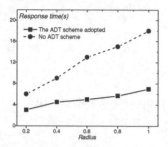

Fig. 5. Effect of adaptive data transfer scheme

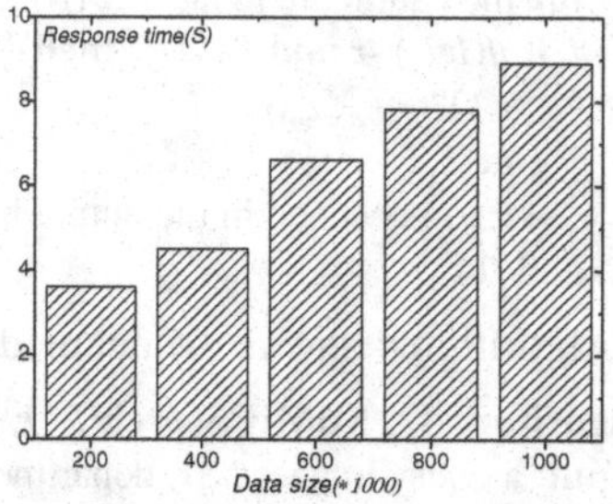

Fig. 6. Effect of data size

In Fig. 6, with the increase of data size, the response time keeps increasing. When the data size is larger than 60,000, the increasing tendency becomes slowly. This is because for the small or medium-scale queries based on *MC*, their network transfer costs are always larger than the distance computation cost in the slave nodes. The total response time can be mainly dominated by the transfer cost. Therefore it is more suitable and effective for the large high-dimensional dataset.

6.3   Effect of Radius

This experiment tests the effect of radius on the query performance. Suppose that the data size and the number of master(slave) nodes are fixed, when r increases from 0.2 to 1, it is clear that the query response time is gradually increasing. This is because with the increase of r, the search region in the high-dimensional spaces becomes larger and larger which leads to the fact that the number of candidate images is increasing accordingly.

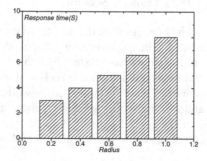

Fig. 7. Effect of radius

7   Conclusions

In this paper, we have presented a mobile-cloud-based similar(*MiMiC*) query processing, which specifically caters for the different bandwidth of nodes in the mobile cloud. Two enabling techniques, namely, *learning-based adaptive data placement scheme*, and *multi-resolution-based adaptive data transfer scheme* are proposed to reduce the communication cost. The experimental studies indicate that the proposed *MiMiC* method is more suitable for the large-scale medical image retrieval in minimizing the network communication cost and maximizing the parallelism in I/O and CPU.

References

[1] http://rvl2.ecn.purdue.edu/~cbirdev/www/CBIRmain.html
[2] http://irma-project.org/
[3] Ghemawat, S., Gobioff, H., Leung, S.-T.: The Google File System. In: 19th ACM Symposium on Operating Systems Principles, Lake George, NY (October 2003)
[4] DeCandia, G., Hastorun, D., Jampani, M., et al.: Dynamo: Amazon's Highly Available Key- value Store. In: SOSP 2007 (2007)
[5] Cooper, B.F., Ramakrishnan, R., Srivastava, U., et al.: PNUTS: Yahoo!'s hosted data serving platform. PVLDB 1(2), 1277–1288 (2008)
[6] Dean, J., Ghemawat, S.: MapReduce: Simplified Data Processing on Large Clusters. In: OSDI 2004: Sixth Symposium on Operating System Design and Implementation, San Francisco, CA (December 2004)
[7] Böhm, C., Berchtold, S., Keim, D.: Searching in High-dimensional Spaces: Index Structures for Improving the Performance of Multimedia Databases. ACM Computing Surveys 33(3) (2001)
[8] Jagadish, H.V., Ooi, B.C., Tan, K.L., Yu, C., Zhang, R.: iDistance: An Adaptive B^+-tree Based Indexing Method for Nearest Neighbor Search. TODS 30(2), 364–397 (2005)
[9] Berchtold, S., Bohm, C., Braunmuller, B., Keim, D.A., et al.: Fast Parallel Similarity Search in Multimedia Databases. In: SIGMOD, pp. 1–12
[10] Papadopoulos, A.N., Manolopoulos, Y.: Similarity query processing using disk arrays. In: SIGMOD 1998, pp. 225–236 (1998)
[11] Ratnasamy, S., Francis, P., Handley, M., Karp, R., Shenker, S.: A scalable content-addressable network. In: SIGCOM, pp. 161–172 (2001)
[12] Schmidt, C., Parashar, M.: Flexible information discovery in decentralized distributed systems. In: HPDC-12 (2003)
[13] Sahin, O.D., Gupta, A., Agrawal, D., El Abbadi, A.: A peer-to-peer framework for caching range queries. In: ICDE   (2004)
[14] Kalnis, P., Ng, W.S., Ooi, B.C., Tan, K.L.: Answering Similarity Queries in Peer-to-Peer Networks. Information Systems (2006)
[15] Medical image databases (2002),
    http://www.ece.ncsu.edu/imaging/Archives/ImageDataBase/Medical/index.html
[16] The Android platform (2010),
    http://code.google.com/intl/zh-CN/android/
[17] The Amazon EC2 (2009), http://aws.amazon.com/ec2/

Efficient Algorithms for Constrained Subspace Skyline Query in Structured Peer-to-Peer Systems

Khaled M. Banafaa and Ruixuan Li

School of Computer Science and Technology,
Huazhong University of Science and Technology,
Wuhan, Hubei 430074, P.R. China
kbanafaa@smail.hust.edu.cn, rxli@hust.edu.cn

Abstract. To avoid complex P2P architectures, some previous research studies on skyline queries have converted multi-dimensional data into a single index. Their indexing, however, does not solve constrained subspace queries efficiently. In this paper, we have devised algorithms and techniques to solve *constrained subspace skyline queries* efficiently in *a structured peer-to-peer architecture* using a single index. Dataspace is horizontally partitioned; and peers are given Z-order addresses. Each subspace query traverses peers using the *subspace Z-order filling curve*. Such partitioning and traversal approaches allow parallelism for incomparable data as well as the use of some techniques to reduce the data traveled in the network. The order of traversal also preserves progressiveness. Our experiments, applied on Chord [1], have shown a reduction in the number of traversed peers and an efficient usage of bandwidth.

Keywords: Skyline query, peer-to-peer system, subspace skyline, constrained.

1 Introduction

Due to the huge available distributed data, advanced queries, such as skyline queries, identify some interesting data objects for the users. Peer-to-peer (P2P) networks have also become very popular for storing, sharing and querying data. Due to some application requirements for multi-dimensional data, P2P had to be adapted to host such kind of data. To cope with this requirements, complex network overlay architectures have been suggested. To maintain simple P2P architectures, researchers have looked for ways to convert multi-dimensional data to a single dimensional index (1D) for different applications. Skyline operator [2] has also attracted considerable attention in the database research community recently. A user may be interested in purchasing a computer on the web. Agents (peers) may have different specifications with different prices. As an example of CPU performance versus price of computers is shown Fig. 1. Other attributes will also affect the price (eg. HDD and RAM). A purchaser will definitely not be interested in a computer while there is a same or a better one with a better or

H. Gao et al. (Eds.): WAIM 2012, LNCS 7418, pp. 334–345, 2012.

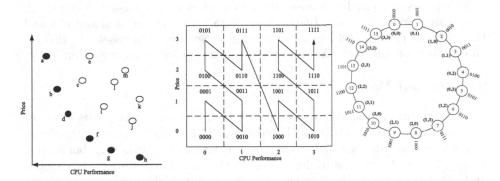

Fig. 1. Skyline   **Fig. 2.** Space partitioning   **Fig. 3.** Chord: Z-order

the same price. Purchasers may, however, have different preferences or requirements. A purchaser may be interested in a high capacity HDD; another may be interested in CPU performance. Answers to such preferences are called skyline [2] and [3]. Black filled points are called skyline in Fig. 1.

A purchaser may have ranges or constraints on the attributes. Price ranges and RAM sizes are some examples of constraints. Researches (eg. DSL [4], Sky-Plan [5] and PaDSkyline [6]) have studied such constraints. These approaches, however, requires the query node to contact all nodes for their minimum bounding rectangles (MBRs) before building a plan. Other studies (e.g. SUBSKY [7] and Skypeer [8]) considered subspace skylines. Those studies, however, do not consider constraints due to their subspace pre-processing.

Our motivation is to efficiently process subspace queries [8] with constraints [5]. In high dimensional databases, a user may be interested only in some attributes with some constraints. Considering all attributes might return a huge number of data. A purchaser may be interested in a price within a range ($500-$800) and a minimum CPU performance (1.5GHz). They may not be interested in other attributes (eg. HDD, memory, video cards, ... etc). Such queries have been considered in centralized systems [3] using R-trees. Those methods are, however, not applicable in distributed systems for their requirements of having all data in one place.

In this paper, we aim at solving constrained subspace skyline queries in structured P2P systems using a single index. By partitioning the dataspace horizontally, Z-address scheme is used to address partitions as well as peers. Depending on query subspace and constraints, peers are traversed. Our contributions are:

- By Z-order addressing a horizontally partitioned data space on a simple, stable overlay structure such as Chord [1], progressive algorithms for *constrained subspace query* traversal are suggested. They increase the pruning ability.
- Incomparable partitions are exploited by parallelizing our algorithms.
- Experiments have shown that such approaches resulted in reduction in traversed peers and bandwidth usage while preserving progressiveness.

This paper is organized as follows. Related work is first discussed in Section 2. The main part of our work (partitioning data space, traversal algorithms and the used techniques) is discussed in Section 3. Section 4 discusses experiments and findings of this work. We end up our paper by conclusion in Section 5.

2   Related Work

Skyline was first introduced to database by Börzsönyi in [2]. All points are compared using a window in Block Nested Loop(BNL). BBS[9] is an optimal and progressive algorithm for skyline queries. It is based on nearest neighbor search. It used a heap to gets rid of duplicates. In [10], assuming integers, RZ-regions and ZB-tree are used. The above algorithms are not efficient for distributed and P2P systems for their centralized requirements.

In distributed systems, such as [11], data are distributed vertically. A round-robin is used to get the traditional skyline on the presorted attributes. Constrained queries are neither supported for fullspace nor for subspace. In DSL [4], data space is partitioned horizontally. It does not, however, support subspace and constrained subspace skyline. SkyFrame uses greedy and relaxed skyline search on a balanced tree structure for peer-to-peer networks (Baton) [12]. Even though it supports constrained skyline, it does not support subspace and constrained subspace queries. In PaDSkyline [6], the querying peer collects MBRs from other peers. A plan is made for incomparable peers to work in parallel. In SkyPlan [5], using weighted edges, plans are also mapped. The maximum spanning trees are used to maximize the pruning ability. In FDS [13], iterations and feedback are carried out between the coordinator and the other nodes. The above studies [5], [6], and [13] require the querying peer to contact all nodes. Some algorithms [14] have converted multi-dimensional data into a single-data index and adapted it into P2P. However, they support neither constrained queries nor subspace queries.

Subspaces have also been studied in both centralized and distributed systems. Centralized algorithms [7], [15], [16], and [17] and use different approaches. Sky-Cube [15], [16] precomputes all possible subspaces skylines exploiting various sharing strategies using bottom-up and top-down methods. SUBSKY [7] builds B+ tree by converting the multi-dimensional data into one-dimensional anchors. Authors in[17] used materialization and proposed the maximal space index to answer subspaces in a high dimensional data. Due to their pre-computation, updating any point may result in rebuilding the solutions. They are centralized and do not support constrained subspace queries.

Skypeer [8] and DCM [18] are meant for distributed systems. Skypeer uses *extended skylines* on a super-peer architecture. A querying peer submits its subspace query to its super-peer. The super-peer, then, contacts the other super-peers for subspace skyline. Skypeer is nonprogressive and does not support constrained subspace query. In DCM, the results of subspaces queries are cached in peers and indexed using a distributed cache index (DCI) on Baton or Chord. Subspace queries are then forwarded to cached skylines using the DCI. Even though it does not consider constrained subspace skylines, this work is orthogonal to our work for unconstrained subspaces.

Even though work in [19] (we call Chordsky) supports constrained full-dimension skyline in a Chord structure. ChordSky's usage of sum for the the monotonic functions makes it inefficient for subspace skyline queries. This is due to the disturbance of the monotonic order when a subspace is considered. The authors' objectives, however, were having a simple stable structure like Chord to answer constrained skyline queries.

We have the same objectives as ChordSky but for a more general skyline query (i.e. constrained subspace skyline). Thus, we consider ChordSky as a baseline for our work. We first use it with no modification (i.e. for constrained full-dimension skyline). We, then, modified it to support subspace. Our work has shown to be more efficient. Even though ChordSky is progressive for full-dimension queries, it is not for subspace skyline. Our work is, however, progressive for both constrained full-space as well as constrained subspace skyline. A Z-order structure [10] inspired our work. Our system uses the idea of Z-order on P2P architecture. It uses a one-dimensional index. We have applied our work on Chord [1] but it can also be applied on other architecture like Baton.

To the best of our knowledge, constrained subspace skyline has not been considered in distributed and P2P systems. Our aim is to minimize the visited peers and data transferred in the network while preserving progressiveness.

3   Constrained Subspace Skyline

In this section, we first give some formalization to constrained subspace skyline in Section 3.1. Data space partitioning and traversal techniques and algorithms are discussed in Section 3.2 and Section 3.3 respectively. Parallelism of our algorithms and load balancing are explored in Section 3.4.

3.1   Preliminaries

Without loss of generality, we assume minimum values of attributes are preferred (e.g. cheaper is preferred to expensive, near is preferred to far, etc). For maximum values preferences, the inverse of the values can be used. Let $S = \{d_1, d_2, ..., d_d\}$ be a d-dimensional space and PS be a set of points in S. A point $p \in PS$ can be represented as $p = \{p_1, p_2, ..., p_d\}$ where every p_i is a value on dimension d_i. Each non-empty subset S' of S ($S' \subseteq S$) is called a subspace. A point $p \in PS$ is said to dominate another point $q \in PS$ on subspace S' (denoted as $p \prec_{S'} q$) if (1) on every dimension $d_i \in S', p_i \leq q_i$; and (2) on at least one dimension $d_j \in S', p_j < q_j$. The skyline of a space $S' \subseteq S$ is a set $PS' \subseteq PS$ of so-called skyline points which are not dominated by any other point of space S'. That is, $PS' = \{p \in PS| \nexists q \in PS : q \prec_{S'} p\}$.

Let $C = \{c_1, c_2, ..., c_k\}$ be a set of range constraints on a subspace $S' = \{d'_1, d'_2, ..., d'_k\}$ where $k \leq d$. Each c_i is expressed by $[c_{i,min}, c_{i,max}]$, where $c_{i,min} \leq c_{i,max}$, representing the min and max value of d'_i. A constrained subspace skyline of a space $S' \subseteq S$ refers to the set of points $PS'_c = \{p \in PS_c| \nexists q \in PS_c : q \prec_{S'} p\}$, where $PS_c \subseteq PS$ and $PS_c = \{p \in PS| \forall d_i \in S' : c_{i,min} \leq p_i \leq c_{i,max}\}$.

3.2   Partitioning and Assignment

We use a shared-nothing architecture (SN) where the data space is horizontally partitioned; and each peer is assigned a partition. Such architecture has also been used in other previous works like [19].

As in [14], the range of each attribute is assumed to be normalized into [0,1]. Each dimension i (i.e. $1 \leq i \leq d$ of the data space is divided into k_i equal partitions. Each partition is assigned an integer number within $[0, k_i - 1]$ in an ascending order starting from 0. The whole space is, thus, divided into a grid of $\prod k_i$ cells. The lower left cell is $Cell(0,0,0,0)$ while the upper right cell is $Cell(k_0 - 1, k_1 - 1, k_2 - 1, ..., k_d - 1)$ as shown if Fig. 2.

A cells' Z-addresses are obtained by interleaving their dimensions' values. Peers are also assigned a Z-order address using the same grid above. In Chord, each peer is assigned the next Z-order address starting with '0' as in Fig. 3.

A data point obtains its value for each dimension using Equation 1. The Z-address of a peer responsible of a point is found by interleaving these values.

$$IntValue(d_i) = \lfloor p(d_i) * k \rfloor \tag{1}$$

3.3   Skyline Query Traversal

Fig. 2 and Fig. 3 demonstrate the previous running (computer) example of a 2-dimension data space where $k_i = 4$.

Since dataspace has been partitioned and assigned to peers, query traversal needs exploit such partitioning. Before explaining our traversal and prunability, some definitions are introduced to show the domination relation between cells and points using lower left point (LLP) and upper right point(URP):

Definition 1. *A point p completely(partially) dominates a cell α if p dominates α's LLP(URP).*

Definition 2. *Cell Domination. A cell α completely or partially dominates another cell β if α's URP or α's LLP dominates β's LLP respectively.*

Property 1. Monotonic Ordering. Cells ordered by non-descending Z-addresses are monotonic such that cells are always placed before their (completely and partially) dominated cells.

Constrained Fullspace Skyline Queries. For a constrained fullspace skyline queries, the querying peer sends its query to the peer responsible of the minimum constraints C_{min}. By Property 1, a traversal is straight forward. Each peer calculates its and previous peers' constrained skyline and sends it to the next unpruned peer. An unpruned peer is a peer within the constraints and is not completely dominated by any discovered constrained skyline point. Chord nodes can also prune empty nodes by maintaining an empty nodes list in each node.

Algorithm_1-(Fullspace)

1: Input:
2: RS: Received Skyline
3: Output:
4: DS: Discovered Skyline
5: BEGIN
6: LS: Constrained Local
    Skyline Points
7: $DS = \phi$
8: **for all** $P \in LS$ **do**
9:     **if** $\not\exists Q \in RS : Q \prec P$ **then**
10:         $DS = DS \cup P$;
11:     **end if**
12: **end for**
13: Send DS to querying
    peer as final skyline points
14: $RS = RS \cup DS$
15: Send RS to next
    unpruned peer
16: END

Algorithm_2-(CSSA)

1: Input:
2: RFS: Rcvd Final Sky Pts
3: RGS: Rcvd Group Sky Pts
4: Output:
5: DS: Discovered Skyline
6: BEGIN
7: LS: Constrained Local Sky Pts
8: $LS = \text{findSkyline}(LS \cup RGS)$
9: $DS = \phi$
10: **for all** $P \in LS$ **do**
11:     **if** $\not\exists Q \in RFS : Q \prec P$ **then**
12:         $DS = DS \cup P$;
13:     **end if**
14: **end for**
15: **if** (Last-peer-of-subspace-group) **then**
16:     Send DS to query peer as a final sky
17:     $RFS = RFS \cup DS$
18:     Send RFS to first unpruned peer in SG
19: **else**
20:     $RGS = DS$
21:     Send RFS and RGS to next peer in SG
22: **end if**
23: END

Fig. 4. Constrained full space vs. constrained subspace skyline algorithms

Constrained Subspace Skyline Queries (CSSA). A more general skyline query type of the above query is *constrained subspace skyline queries* . Some issues are, therefore, reconsidered. First, the starting and ending peers need be calculated. Second, some peers partially dominate each other with respect to the query subspace. This disturbs progressiveness. A way to preserve progressiveness as much as possible needs be used. Third, in each step, the next peers needs be clear and deterministic. Last, peer's pruning must be reconfigured for subspaces.

Starting and ending peers: To get their Z-addresses, the minimum and maximum values are placed in Equation 1 for the missing dimensions respectively.

Peers partial domination: Peers partially dominate each other with respect to the queried subspace if they have the same subspace Z-address. Thus, grouping those peers with the same subspace Z-address value creates subspace-groups.

Definition 3. *A subspace-group (SG) is all peers with the same IntValues (Equation 1) in all dimensions of the subspace.*

Peers in a subspace-group partially dominate each other. Thus the data transferred within the subspace-group may have false skyline points. Once all members of the subspace-group are visited, the final skyline points of a subspace-group are reported to the querying peer as a final skyline points. Thus progressiveness can be preserved between subspace-groups. From Lemma 1, the traversal between subspace-groups is deterministic and progressive.

Lemma 1. *Subspace Group Traversal. Traversing the subspace-groups in a non-descending order of their subspace represented by peers' subspace Z-address ensures traversing a subspace-group before their dominated subspace-groups.*

Proof. Suppose a subspace-group β with a subspace address ω comes after subspace-group α with a subspace address v. Let $\beta \prec \alpha$. Therefore, ω has less than or equal values to α's values in all considered subspace dimensions. Thus $\omega < v$ which contradicts our assumption. This means β can not dominate α.

Subspace pruning: A subspace-group is pruned if it is dominated by any discovered point with respect to the query's subspace.

Algorithm 2 in Fig. 4 is used for computing constrained subspace skyline. The traversal order depends on the bits of dimensions in favor. Using the subspace bits in favor as a prefix, the list of peers can be sorted. Thus traversal order takes the sorted peers order. For example, in Fig. 2 and Fig. 3, the traversal order for fullspace skyline queries is {0, 1, 2, 3, 4, 5, 6, 7, 8, 9, 10, 11, 12, 13, 14, 15} because all bits are considered. For the CPU performance attribute subspace only or price subspace query traversals, the order would be {0, 1, 4, 5, 2, 3, 6, 7, 8, 9, 12, 13, 10, 11, 14, 15} and {0, 2, 8, 10, 1, 3, 9, 11, 4, 6, 12, 14, 5, 7, 13, 15} respectively. As seen in Fig. 2, peers {0,1,4,5} are in column 1. Their CPU performance attribute value (IntValue =0). while peers {2,3,6,7} are in the second column with their CPU performance attribute value (IntValue =1). Column 1 precedes column 2 because it partially dominates column 2. The same is done for the other groups. Notice that for price attribute, rows are taken as subspace groups. The peers within a subspace group can be ordered in any order because they partially dominated each other. We, however, used order of the fullspace within the subspace groups. Once a query is triggered, the starting and ending nodes' Z-addresses are calculated using Equation 1. The query traverses nodes using the order determined by the subspace of the query between the starting node and the ending node excluding pruned peers. Each node can calculate its next peer to send the query and its results to.

3.4   Parallelization of Our Approach

In P2P systems, total parallelism can be achieved by having all peers involved in the query work concurrently. But it results in an increase in transferred data and it will disable prunability and progressiveness. We, however, exploit the incomparable cells' features in skyline. The transferred data reduction is also explored. One-dimension neighbors are used to find the next incomparable peers.

Definition 4. *One-dimension Neighbor. A one-dimension neighbor to a cell α is any cell β that can be converted to α by subtracting 1 from only one dimension.*

Property 2. All one-dimension neighbors of a cell are incomparable.

Constrained Subspace Skyline Parallel Algorithms (CSSPA). Using Property 2, a peer sends the query to all unpruned one-dimension neighbors. This can be achieved by having the query traverses from a peer to all peers with Z-values higher by one in one dimension and keeping the other values of the other dimensions. An example of fullspace queries in Fig. 2, a node 0 sends the query to its one-dimension neighbor (i.e 1 and 2). The same is done for subspaces queries. The difference is that dimensions of the query's subspace are only considered for the one-dimension neighbors. For example, for the CPU performance attribute subspace query, node 0 sends to node 2 only because it is the only one-dimension neighbor to node 0. Node 1 is in the same subspace group of node 0. The one-dimension neighbors here are subspace-groups. All peers within a group can work concurrently for total concurrency. But as mentioned earlier, it resulted in more data to transfer. A serial traversal can be used for nodes within a subspace group. The first peer of each group is determined as explained above.

Constrained Subspace Skyline Parallel Algorithms with Data Reduction (CSSPA-DR). Parallelism of the previous section implies that queries travel from lower values in each dimension to a higher value in that dimension. The skyline points received from previous peer are used to prune points. Since the parallel traversal comes from low values to higher values in a dimension, that dimension can be excluded and only the subspace skyline of the rest of queried subspace can be sent. Thus, the points sent (PS_i) through dimension D_i are PS_i = subspace-skyline-of(Query-dimensions - D_i). Thus, some of the points may only be sent. In our running example, if CPU performance attribute is used, a maximum of one point is traveled between the subspace groups.

Load Balancing. Load imbalance can be introduced by query imbalance as well as data in each cell [4]. To overcome this problem, we adopt the method in [4]. We use probing and replication. Each peer randomly chooses m points in the d-dimensional space. Each peer responsible for a point is probed for its query load balance. By sampling replies, a peer whose load exceeds a threshold will make a copy of its contents to a peer with minimum load. Then, queries arrived to such peer are distributed among the replicas in a round-robin fashion.

4   Performance Evaluation

The work in [19], we called ChordSky, is used as a baseline. Our algorithms (CSSA, CSSPA and CSSPA-DR) as well as the modified ChordSky were built using Peersim 1.0.5 simulator on Intel Pentium 4 CPU 2.00GHz, 512M memory. A uniform data with cardinality of 1 million and network sizes of 100-4000 nodes. Different numbers of dimensions are used {2,3,4,5,6,7}. Subspace queries are also randomly chosen with random dimensions. Extensive skyline queries are randomly produced and results were reported as shown later. Our aim is at minimizing the visited peers and transmitted data while preserving progressiveness.

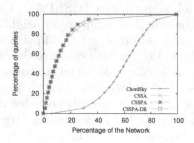

Fig. 5. Queries vs. traversed peers

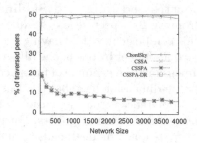

Fig. 6. Percentage of traversed peers

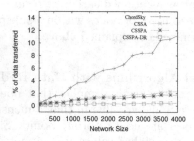

Fig. 7. Transferred data

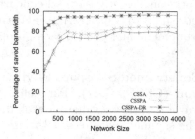

Fig. 8. Percentage of saved data

4.1 Accessed Nodes

Fig. 5 demonstrates the relationship between the percentage of queries and the needed traversed peers to answer queries. For the baseline algorithms, the slope is around 1.7 as opposed to our new algorithms whose slope are around 5.3. The baseline algorithm's traversed peers size increases by only 20% the answered queries percentage increase while it is 60% for the new algorithms. Thus, the new algorithms results in fewer number of traversed peers to answer the same percentage of queries. For example, less than 20% of peers can answer upto 80% queris in the new algorithms while the modified ChordSky (i.e. baseline) needs around 80% peers for same number of queries. In general, the reduction of traversed peers by the new algorithms(i.e. CSSA, CSSPA, and CSSPA-DR) as compared to ChordSky is between 40% to 50% of the network size to answer between 20% to 90% of the queries. This big reduction is due to pruning ability obtained by our partitioning and traversal algorithms. Comparing our new algorithms (i.e. CSSA, CSSPA, and CSSPA-DR) with each other, they have similar results to each other because they are using the same pruning method.

Fig. 6 shows traversed peers percentage average for different network sizes. The average traversed peers keeps its around 50% percentage with variant network sizes for ChordSky. For different sizes of the network, a slight decrease in the traverse peers' percentage is shown in our new approaches as the size increases. It decreases slowly to reach around 7%. Our approach's prunability power, thus, increases as network size increases.

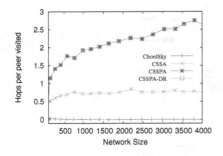

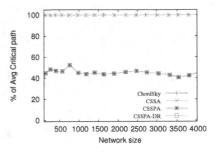

Fig. 9. Lookup hops per traversed peer   **Fig. 10.** Avg. critical line

4.2   Bandwidth Consumption

Fig. 7 shows transferred data for the algorithms. The pruning power and the reduction in the visited peers resulted in a reduction in transferred data points. As the size of the network increases, the traveled data also increases. Due to lack of peer prunability in ChordSky, more data are transferred. Our parallel algorithms show less traveled data than our serial approach because it does not exploit incomparable peers. Thus, it sends irrelevant data to incomparable peers. As a result of the increase number of traversed peers, which due to the increase in network size, a slight increase in the transferred data is noted.

Fig. 8 shows the huge saving for our approaches as compared to ChordSky. Up to 80% of the transferred data in ChordSky is saved. In CSSPA-DR, more than 90% is saved due the new data reduction techniques . As for CSSA and CSSPA, upto 80% are saved. Because CSSPA shows better saving than CSSA.

4.3   Expenses

ChordSky's unprunablity results in no cost. But the new algorithms' pruning ability may require a target lookup. A target in Chord may need up to $O = log_2 n$ hops, but it is not the case for our Z-order distribution. Queries are usually traversed between neighboring due to the addressing scheme. Jumps could mean pruning. Fig. 9 shows that the average number of hops is less than three hops per traversed peer. The small cost in the figure encourages the usage of those new algorithms. The cost of parallel algorithms is more than that of serial ones since a one-dimension neighbor peer could be looked by different peers.

4.4   A Bird's-eye View of the Algorithms' Results

Fig. 11 summarizes our findings. The percentage of visited peers to the overall network size is the highest when the baseline is used. Using the same new pruning techniques in all new algorithms is reflected in the same traversed peers for the new algorithms. The percentage of the transferred data to the overall data is low for our algorithms. It is minimum when the data reduction technique is used. ChordSky's visited peers seems to be strongly directly related to the

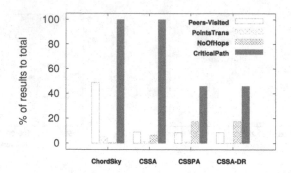

Fig. 11. A bird's eye view on adv. and disadv of each algorithm

network size. ChordSky's serial traversal is reflected in the minimum lookup cost. Parallelism increases cost because the peer is reached by different previous peers. Its bandwidth usage, however, is the minimum. By assuming that n is the number of traversed peers, both the baseline and the CSSA are completed only after n steps. When parallelisms are used, only 40% of the n steps are needed.

5   Conclusion

This paper addresses the *constrained subspace skyline* queries that have not been addressed in structured P2P systems. By converting the multi-dimensional data into a single index using a Z-order filling space, simpler P2P structures are used for a more general skyline queries (i.e. *constrained subspace skyline* queries). By a horizontal partitioning of the dataspace and assigning them Z-addresses, serial and parallel algorithms for such queries traversal are designed. They exploit incomparable partitions featured in skyline computations. Sending only necessary points to each next unpruned *one-dimension neighbor* and exploiting feature of this neighboring relation, a reduction in bandwidth usage is achieved. The efficiencies of the algorithms are shown in progressiveness and reduction in both the bandwidth consumption and traversed peers. New efficient approaches for such queries need to be found for unstructured P2P. Streams and partial order attributes should also be considered in our future work.

Acknowledgments. This research is partially supported by National Natural Science Foundation of China under grants 61173170 and 60873225, Innovation Fund of Huazhong University of Science and Technology under grants 2011TS135 and 2010MS068, and CCF Opening Project of Chinese Information Processing.

References

1. Stoica, I., Morris, R., Karger, D.R., Kaashoek, M.F., Balakrishnan, H.: Chord: A scalable peer-to-peer lookup service for internet applications. In: SIGCOMM, pp. 149–160 (2001)

2. Börzsönyi, S., Kossmann, D., Stocker, K.: The skyline operator. In: 17th International Conference on Data Engineering, ICDE 2001, pp. 421–432. IEEE, Washington (2001)
3. Dellis, E., Vlachou, A., Vladimirskiy, I., Seeger, B., Theodoridis, Y.: Constrained subspace skyline computation. In: CIKM, pp. 415–424 (2006)
4. Wu, P., Zhang, C., Feng, Y., Zhao, B.Y., Agrawal, D., El Abbadi, A.: Parallelizing Skyline Queries for Scalable Distribution. In: Ioannidis, Y., Scholl, M.H., Schmidt, J.W., Matthes, F., Hatzopoulos, M., Böhm, K., Kemper, A., Grust, T., Böhm, C. (eds.) EDBT 2006. LNCS, vol. 3896, pp. 112–130. Springer, Heidelberg (2006)
5. Rocha-Junior, J., Vlachou, A., Doulkeridis, C., Nørvåg, K.: Efficient execution plans for distributed skyline query processing. In: Proceedings of the 14th International Conference on Extending Database Technology, pp. 271–282. ACM (2011)
6. Chen, L., Cui, B., Lu, H.: Constrained skyline query processing against distributed data sites. IEEE Trans. Knowl. Data Eng. 23(2), 204–217 (2011)
7. Tao, Y., Xiao, X., Pei, J.: Subsky: Efficient computation of skylines in subspaces. In: ICDE, vol. 65 (2006)
8. Vlachou, A., Doulkeridis, C., Kotidis, Y., Vazirgiannis, M.: SKYPEER: Efficient subspace skyline computation over distributed data. In: ICDE, pp. 416–425. IEEE (2007)
9. Papadias, D., Tao, Y., Fu, G., Seeger, B.: Progressive skyline computation in database systems. ACM Transactions on Database Systems 30(1), 41–82
10. Lee, K.C., Lee, W.C., Zheng, B., Li, H., Tian, Y.: Z-SKY: an efficient skyline query processing framework based on Z-order. VLDB Journal: Very Large Data Bases 19(3), 333–362 (2010)
11. Balke, W.-T., Güntzer, U., Zheng, J.X.: Efficient Distributed Skylining for Web Information Systems. In: Bertino, E., Christodoulakis, S., Plexousakis, D., Christophides, V., Koubarakis, M., Böhm, K. (eds.) EDBT 2004. LNCS, vol. 2992, pp. 256–273. Springer, Heidelberg (2004)
12. Wang, S., Vu, Q.H., Ooi, B.C., Tung, A.K.H., Xu, L.: Skyframe: a framework for skyline query processing in peer-to-peer systems. VLDB J. 18(1), 345–362 (2009)
13. Zhu, L., Tao, Y., Zhou, S.: Distributed skyline retrieval with low bandwidth consumption. IEEE Trans. Knowl. Data Eng. 21(3), 384–400 (2009)
14. Cui, B., Chen, L., Xu, L., Lu, H., Song, G., Xu, Q.: Efficient skyline computation in structured peer-to-peer systems. IEEE Trans. Knowl. Data Eng. 21(7), 1059–1072 (2009)
15. Pei, J., Jin, W., Ester, M., Tao, Y.: Catching the best views of skyline: A semantic approach based on decisive subspaces. In: Proceedings of the 31st International Conference on Very large Data Bases, pp. 253–264. VLDB Endowment (2005)
16. Yuan, Y., Lin, X., Liu, Q., Wang, W., Yu, J., Zhang, Q.: Efficient computation of the skyline cube. In: Proceedings of the 31st International Conference on Very Large Data Bases, pp. 241–252. VLDB Endowment (2005)
17. Jin, W., Tung, A., Ester, M., Han, J.: On efficient processing of subspace skyline queries on high dimensional data. In: 19th International Conference on Scientific and Statistical Database Management, SSBDM 2007, pp. 12–12. IEEE (2007)
18. Chen, L., Cui, B., Xu, L., Shen, H.T.: Distributed Cache Indexing for Efficient Subspace Skyline Computation in P2P Networks. In: Kitagawa, H., Ishikawa, Y., Li, Q., Watanabe, C. (eds.) DASFAA 2010. LNCS, vol. 5981, pp. 3–18. Springer, Heidelberg (2010)
19. Zhu, L., Zhou, S., Guan, J.: Efficient skyline retrieval on peer-to-peer networks. In: FGCN, pp. 309–314. IEEE (2007)

Processing All k-Nearest Neighbor Queries in Hadoop

Takuya Yokoyama[1], Yoshiharu Ishikawa[2,1,3], and Yu Suzuki[2]

[1] Graduate School of Information Science, Nagoya University, Japan
[2] Information Technology Center, Nagoya University, Japan
[3] National Institute of Informatics, Japan
{yokoyama,suzuki}@db.itc.nagoya-u.ac.jp, y-ishikawa@nagoya-u.jp

Abstract. A k-nearest neighbor (k-NN) query, which retrieves nearest k points from a database is one of the fundamental query types in spatial databases. An *all k-nearest neighbor query* (AkNN query), a variation of a k-NN query, determines the k-nearest neighbors for each point in the dataset in a query process. In this paper, we propose a method for processing AkNN queries in *Hadoop*. We decompose the given space into cells and execute a query using the MapReduce framework in a distributed and parallel manner. Using the distribution statistics of the target data points, our method can process given queries efficiently.

1 Introduction

An *all k-nearest neighbor query* (an *AkNN query*) is a variation of a k-nearest neighbor query and determines the k-NNs for each point in the given dataset in one query process. Although efficient algorithms for AkNN queries are available for centralized databases [2,4,11], we need to consider to support distributed environments where the target data is managed in multiple servers in a distributed way. Especially, *MapReduce*, which is a fundamental framework for processing large-scaled data in distributed and parallel environments, is promising for enabling scalable data processing. In our work, we focus on the use of *Hadoop* [5] since it is quite popular software for MapReduce-based data processing.

In this paper, we propose a method for efficiently processing AkNN queries in Hadoop. The basic idea is to decompose the target space into smaller cells. At the first phase, we scan the entire dataset and get the summary of the point distribution. According to the information, we determine an appropriate cell decomposition. Then we determine k-NN objects for each data points by considering the maximal range in which possible k-NN objects are located.

2 Related Work

In this work, we assume the use of the distributed and parallel computing framework *Hadoop* [5,9]. *MapReduce* [3] is the foundation of Hadoop data processing.

H. Gao et al. (Eds.): WAIM 2012, LNCS 7418, pp. 346–351, 2012.

Due to the page length limitation, we omit these details. If you are interested in the framework, please refer to textbooks like [9].

An AkNN query is regarded as a kind of a *self-join* query. Join processing in the MapReduce framework has been studied intensively recent years [1,6], but generally speaking, MapReduce only supports equi-joins; development of query processing methods for non-equi joins is one of the interesting topics on the MapReduce technology [8]. For AkNN queries, there are proposals that use R-trees and space-filling curves [2,4,11], but they are limited for the use in a centralized environment. For processing AkNN queries in Hadoop in an efficient manner, we need to develop a query processing method that effectively uses the MapReduce framework.

3   Cell Decomposition and Merging

3.1   Basic Idea

We now describe the basics of our AkNN query processing method. We consider two dimensional points with x and y axes. Basically, we decompose the target space into $2^n \times 2^n$ small cells. The constant n determines the granularity of the decomposition. Since the k-nearest neighbor points for a data point is usually located in the nearby area of the point, we can expect that most of the k-NN objects are found in the nearby cells. A simple idea is to classify data points into the corresponding cells and compute candidate k-NN points for each point. The process can be parallelized easily and is suited to the MapReduce framework.

However, we may not be able to determine k-NN points at one step; we need to perform an additional step for such a case. Data points in other nearby cells may belong to the k-nearest neighbors. To illustrate this problem, consider Fig. 1, where we are processing an AkNN query for $k = 2$. We can find 2-NN points for A by only investigating the inside of cell 1 since the circle centered at A and tightly covers 2-NN objects (we call such a circle a *boundary circle*) does not overlap the boundary of cell 0. In contrast, the boundary circle for B overlaps with cells 1, 2, and 3. In this case, there is a possibility that we can find 2-NN objects in the three cells. Therefore, an additional investigation is necessary.

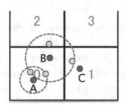

Fig. 1. Cell-based k-NN processing ($k = 2$)

The idea is simple but there is a problem; we may not be able to draw the boundary circle for a point. Consider point C in Fig. 1. For this point, there is only one (less than k) point in cell 1. Thus, we cannot draw the boundary circle. We solve the problem in the following subsection.

3.2   Merging Cells Using Data Distribution Information

We solve the problem described above by prohibiting the situation that there are not enough points in each cell. We first check the number of points within each cell. If we find a cell with less number of points, we merge the cell with the neighboring cells to assure that the number of points in the merged cell is greater than or equal to k. After that, we can draw the boundary circle.

The outline of the idea is illustrated in Fig. 2, where 4×4 decomposition is performed. At the first step, we count the number of points in each cell. Then, we merge the cells with less number of objects with the neighboring cells. In our method, we employ the hierarchical space decomposition used in quadtrees [7]. When we perform merging, we merge four neighboring cells which correspond to the same node at the parent level.

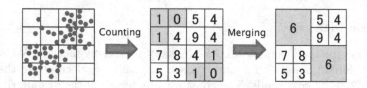

Fig. 2. Cell merging using distribution information

The problem of this approach is that we need to perform an additional counting phase before the nearest neighbor computation. However, it can simplify the following steps. The distribution information is useful in other ways. If we can know there is no points in a cell beforehand, we do not need to consider the cell in the following processes. As shown in the experiments, the cost of the counting is relatively small compared to the total cost.

4   Illustrative Example

The query process consists of four MapReduce steps:

1. MapReduce1: Data distribution information is obtained and cell merging is performed.
2. MapReduce2: We collect input records for each cell and then compute candidate k-NN points for each point in the cell region.
3. MapReduce3: It updates k-NN points for each point. We use the idea described in Section 3 that uses the notion of a boundary circle.
4. MapReduce4: Integrating multiple k-NN lists for each point and get the resulting k-NN list.

Due to the page length limitation, we only show examples. Please refer to [10] for the detail of the algorithms,

We give an example of MapReduce1 to 3 steps that finds AkNN points. Figure 3 shows the distribution of points and we focus on points A, B, and C. The three points are the representative example patterns:

- A (the bounding circle does not overlap with other cells): k-NN points are determined at MapReduce2 step.
- B (the bounding circle overlaps with one cell): We can determine k-NN points at MapReduce3 step by investigating additional cell 2.
- C (the bounding circle overlaps with multiple cells): We investigate additional cells 1, 2, and 3 at MapReduce3 step. Then we integrate their results at MapReduce4 step and determine the k-NN points.

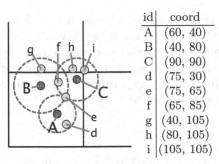

id	coord
A	(60, 40)
B	(40, 80)
C	(90, 90)
d	(75, 30)
e	(75, 65)
f	(65, 85)
g	(40, 105)
h	(80, 105)
i	(105, 105)

Fig. 3. k-NN example ($k = 2$)

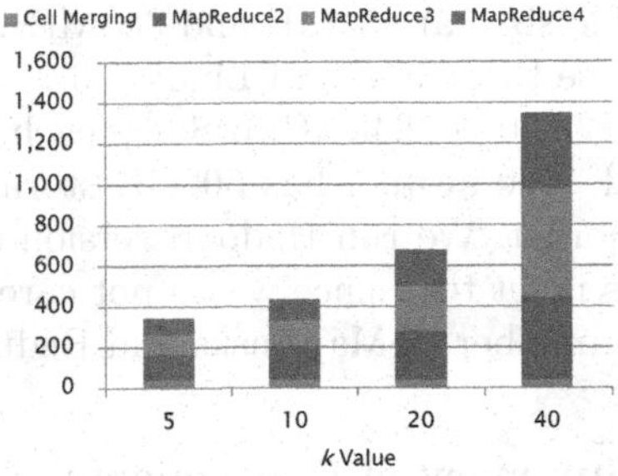

Fig. 4. Execution time for different k values (10M, $n = 8$)

Figure 5 illustrates the execution steps of the entire MapReduce steps. We can see that the k-NN lists for points A, B, and C are incrementally updated and finally fixed.

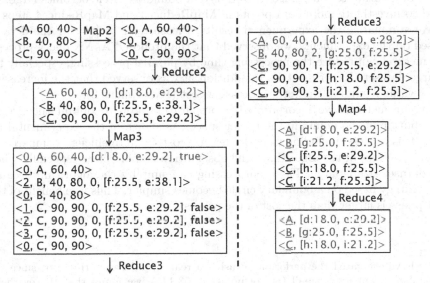

Fig. 5. Processing AkNN query on example dataset

5   Experiments

We have implemented the proposed method. In this section, we evaluate the performance of the MapReduce program running in a Hadoop environment.

5.1   Datasets and Experimental Environment

The experiments are performed using two synthetic datasets: the datasets 1M and 10M consist of 1,000,000 and 10,000,000 points in the target space, respectively. Their file sizes are 34MB and 350MB[1].

We use three nodes of Linux 3.0.0-14-server (Ubuntu 11.10) with Intel Xeon CPU (E5620 @ 2.40GHz). Since each CPU has 4×2 cores, we have 24 cores in total. The system has 500GB storage and the servers are connected by 1G bit Ethernet. We run Hadoop version 0.20.203.0 in the system. The number of replicas is set to 1 since we do not care failures in this experiment. The number of max number of Map tasks and Reduce tasks are set to 8.

5.2   Summary of Experiments

As an example, Fig. 4 shows an experimental result for Experiment 2, where the granularity parameter is set to $n = 8$, the number of Reduce tasks is 24, and the dataset is 10M. As the figure illustrates, the processing time increases as the k value increases. Especially, the increases of the cost for MapReduce3 and MapReduce4 are large. The reason is that a large k value results in a large size of each intermediate record, and it results in the increase of data processing time. In addition to that, since the radius of a boundary circle becomes large, we need to investigate more data points in MapReduce3 and MapReduce4 steps.

We summarize the experimental results. Please refer to [10] for the details. Based on the experiments, we observed that the proposed AkNN query processing method can reduce the processing time by parallel processing, especially for a large dataset (Experiment 1). In addition, it was observed that the increase of the k value results in the total processing cost (Experiment 2).

In our method, we incorporated a preparation step to obtain the overall distribution of the points in the target space. As shown in the experimental results, this process, including the cell merging cost, is quite efficient compared to other processes of the algorithm. The observation is more clear especially when the dataset size is large. The preprocessing can simplify the algorithm because the strategy based on a boundary circle becomes simpler. Thus, the benefit of the first phase is larger than the cost of the process.

---

[1] We have evaluated the performance using a real map dataset. However, since the number of entries is small (no. of points = 53,145), we found that the overhead dominates the total cost and there is no merit to use Hadoop. Therefore, we used the synthetic dataset here to illustrate the scalability of the method.

6   Conclusions

In this paper, we have proposed an AkNN query processing method in the MapReduce framework. By using cell decomposition, the method adapts the distributed and parallel query framework of MapReduce. Since k-NN points may be located outside of the target cell, we may need additional steps. We solved the problem by considering a boundary circle for the target point. In addition, to simplify the algorithm, we proposed to collect distribution statistics beforehand and the statistics is used for cell merging. In the experiments, we have investigated the behaviors of the algorithm for different parallelization parameters and different k values.

The future work includes how to estimate the appropriate number of cell decomposition granularity (n) by using statistics of the underlying data. In addition, the number of parallel processes is also an important tuning factor. As shown in the experiments, too many parallel processes may result in the increase of the total processing cost due to the overhead. If we can estimate these parameters accurately and adaptively, we would be able to achieve nearly optimal processing for any environments.

Acknowledgments. This research is partly supported by the Grant-in-Aid for Scientific Research (22300034) and DIAS (Data Integration & Analysis System) Program, Japan.

References

1. Afrati, F.N., Ullman, J.D.: Optimizing joins in a map-reduce environment. In: Proc. EDBT, pp. 99–110 (2010)
2. Chen, Y., Patel, J.M.: Efficient evaluation of all-nearest-neighbor queries. In: Proc. ICDE 2007, pp. 1056–1065 (2007)
3. Dean, J., Ghemawat, S.: MapReduce: Simplified data processing on large clusters. In: OSDI, pp. 137–150 (2004)
4. Emrich, T., Graf, F., Kriegel, H.-P., Schubert, M., Thoma, M.: Optimizing All-Nearest-Neighbor Queries with Trigonometric Pruning. In: Gertz, M., Ludäscher, B. (eds.) SSDBM 2010. LNCS, vol. 6187, pp. 501–518. Springer, Heidelberg (2010)
5. The apache software foundation: Hadoop homepage, http://hadoop.apache.org/
6. Jiang, D., Tung, A.K.H., Chen, G.: MAP-JOIN-REDUCE: Toward scalable and efficient data analysis on large clusters. IEEE TKDE 23(9), 1299–1311 (2011)
7. Samet, H.: The quadtree and related hierarchical data structures. ACM Computing Surveys 16(2), 187–260 (1984)
8. Vernica, R., Carey, M.J., Li, C.: Efficient parallel set-similarity joins using MapReduce. In: Proc. SIGMOD, pp. 495–506 (2010)
9. White, T.: Hadoop: The Definitive Guide. O'Reilly (2009)
10. Yokoyama, T., Ishikawa, Y., Suzuki, Y.: Processing all k-nearest neighbor queries in hadoop (long version) (2012), http://www.db.itc.nagoya-u.ac.jp/papers/2012-waim-long.pdf
11. Zhang, J., Mamoulis, N., Papadias, D., Tao, Y.: All-nearest-neighbors queries in spatial databases. In: Proc. SSDBM, pp. 297–306 (2004)

RGH: An Efficient RSU-Aided Group-Based Hierarchical Privacy Enhancement Protocol for VANETs

Tao Yang[1,2,3], Lingbo Kong[4], Liangwen Yu[1,2,3], Jianbin Hu[1,2,3,*], and Zhong Chen[1,2,3]

[1] MoE Key Lab of High Confidence Software Technologies,
Peking University, 100871 Beijing, China
[2] MoE Key Lab of Computer Networks and Information Security,
Peking University, 100871 Beijing, China
[3] School of Electronics Engineering and Computer Science,
Peking University, 100871 Beijing, China
[4] School of Software Engineering, Beijing Jiaotong University, 100044 Beijing, China
{ytao,yulw,hujb,chen}@infosec.pku.edu.cn

Abstract. VANETs are gaining significant prominence from both academia and industry in recent years. In this paper, we introduce an Efficient RSU-aided Group-based Hierarchical Privacy Enhancement Protocol for VANETs. The protocol exploits the Road-Side-Unit (RSU) to form a temporary group based on the vehicles around it. As the group leader, the RSU takes charge of the verification of the member car through the Revocation List from the DoT (Department of Transportation). The RSU also generates the session key and determines the configuration for the group. The RSU handles the message from the group member, and then sends it to other group members. RSUs have hierarchical security because of the real environment secure threats. RSUs are divided into High-level RSU (HRSU) and Low-level RSU (LRSU). HRSU inserts a message-related trace entry into the trace log while LRSU attaches trace Tag on the message. HRSU has to submit the trace log to the Department of Audit (DoA) by secure way periodically. If required, DoA can trace out the disputed message's real signer with the cooperation of the DoT. Comparison with other existing classic schemes in the literature has been performed to show the efficiency and applicability of our scheme.

Keywords: vehicular ad-hoc networks, privacy enhancement, message authentication, traceability.

1 Introduction

VANETs (vehicular ad-hoc networks) are gaining significant prominence from both academia and industry in recent years. VANETs consist of three fundamental components namely RSU (Road-Side Unit), OBU (On-Board Unit), and an appropriate network (such as Dedicated Short Range Communications, denotes DSRC, TCP/IP, etc) to coordinate with the whole system. VANETs aim to enhance the safety and

---

* Corresponding author.

H. Gao et al. (Eds.): WAIM 2012, LNCS 7418, pp. 352–362, 2012.

efficiency of road traffic through the V2I (Vehicle to Infrastructure) and V2V (Vehicle to Vehicle) communication. Due to its various applications and potential tremendous benefits it would offer for future users, VANETs are receiving significant prominence from academia and industry in recent years.

VANETs enable useful Intelligent Transportation System functionalities such as cooperative driving and probe vehicle data that increase vehicular safety and reduce traffic congestion. Served by VANETs, people can enjoy a safer and easier environment on the road. However, it also faces serious security threats about privacy preservation, because of its high privacy sensitivity of drivers, its huge scale of vehicle number, its variable node velocity, and its openness. The main privacy concern is the vehicle identity disclosure and tracing, which could be exploited by the criminals for some evil motivation. To address the privacy issues, the privacy-preserving communication scheme should be introduced in VANETs. Besides the basic vehicle identity privacy-protecting objectives (such as anonymity, integrity, authentication, non-repudiation, etc.), a good privacy preserving scheme for VANETs must have low computational/communication complexity, low communication/storage load, as well as an efficient accountable/traceable mechanism.

Outline: The rest of this paper is organized as follows. Section 2 introduces the preliminaries such as system model and security objectives. Section 3 describes the design rationale and details of our scheme. Then, the security and efficiency analysis results are discussed in Section 4 and Section 5 respectively. We will present the related work in Section 6. Finally, we will give conclusion in Section 7.

2   Preliminaries

2.1   System Model

We consider our VANETs system model as shown in Fig. 1.

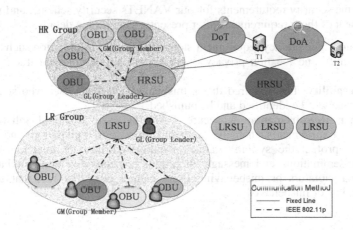

Fig. 1. The system model

1) **DoT**(Department of Transportation): The DoT is in charge of the registration of all RSUs and OBUs each vehicle is equipped with. The DoT can help DoA trace the real identity of a safety message sender by checking the trace table T_1 in local. To the end, the DoT is assumed powered with sufficient computation and storage capability, and the DoT cannot be compromised and is fully trusted by all parties in the system.

2) **DoA**(Department of Audit): The DoT is in charge of the trace task for the real message signer if the message has been disputed. DoA can finish a trace process by match trace table T_2 with the cooperation of DoT. T_2 is accumulated by merging the submitted trace log from RSUs periodically.

3) **RSU**(Road-Side Unit): The RSUs are subordinated by the DoT/DoA, which act as the infrastructure of the VANETs and hold storage units for storing information coming from the DoT/OBUs. RSUs connect to DoT /DoA by a wired network and communicate with OBUs through DSRC. As a distributed unit is deployed on the roadside, an RSU has a risk to be compromised. RSUs cannot generate signatures on behalf of either the OBU or the DoT/DoA. RSUs have hierarchical security because of the real environment secure threats. RSUs are divided into HRSU and LRSU, and are deployed in 1-HRSU-n-LRSU mode. HRSU inserts a message-related trace entry into the trace log while LRSU attaches trace Tag on the message. HRSU has to submit the trace log to DoA by secure way periodically.

4) **OBU**(On-Board Unit):: Vehicles equipped with OBUs mainly communicate with each other to share local traffic information and improve the driving experience. If necessary, OBUs can also communicate with RSU to exchange some information. A vehicle needs to be registered to the DoT with its public system parameters and corresponding private key before it joins the VANETs. The use of secret information such as private keys generates the need for a tamper-proof device (TPD) in each vehicle. Similar to previous work we assume that access to TPD is restricted to authorized parties.

2.2   Secure Objectives

We define the security requirements for our VANETs security scheme, and will show the fulfillment of these requirements after presenting the design details.

1) **Privacy:** The privacy requirement states that private information such as vehicle owner's identity and location privacy is preserved against unlawful tracing and use profiling.

2) **Traceability:** It is required that a misbehaving vehicle user who send out the bogus message will be identified and be punished.

3) **Other Requirements:** A secure VANETs system should satisfy several fundamental requirements, namely, authentication, non-repudiation and message integrity, to protect the system against unauthorized-message injection, denial of message disseminations and message alteration, respectively. Non-repudiation also requires that violators or misbehaving users cannot deny the fact that they have violated the law or misbehaved.

3   RGH Protocol

Table 1. Notation Table

Notation	Description	Notation	Description
RSU_j	j^{th} RSU	G_j	RSU_j's group key
V_i	i^{th} Vehicle	$H(.)$	Hash function $H : \{0,1\}^* \to \mathbb{Z}_q^*$
RID_i	V_i's true ID	$Sign(.)$	Signature function
PID_i	V_i's pseudo ID	$Enc(.)$	symmetric encryption function
s	DoT's master key	$a\|\|b$	String concatenation of a and b
P_{pub}	DoT's public key	T_1	The trace table in DoT
$K_{RSU_j}^-, K_{RSU_j}^+$	RSU_j's secret and public key	T_2	The trace table in DoA
K_i^-, K_i^+	V_i's secret and public key	T_{2j}	The trace table in RSU_j
DoA^-, DoA^+	DoA's secret and public key		

3.1   System Initialization

In the system initialization, the DoT initializes the system parameters in offline manner, and registers vehicles and *RSUs*.

1) **Setup:** Given the security parameter k, the bilinear parameters $(q, G_1, G_2, \hat{e}, P, Q)$ are first chosen, where $|q| = k$. Then DoT randomly selects a master secret $s \in_R \mathbb{Z}_q^*$, and computes the corresponding public key $P_{pub} = sP$. In addition, the DoT chooses a secure symmetric encryption algorithm $Enc()$, a cryprographic hash function $H : \{0,1\}^* \to \mathbb{Z}_q^*$, and a secure signature algorithm $sign()$. In the end, DoT sets the public system parameters as $(q, G_1, G_2, \hat{e}, P, P_{pub}, H, HMAC, Enc, Sign)$.

2) **DoA-KeyGen:** DoT randomly select $K_{DoA}^- \in_R \mathbb{Z}_q^*$ for DoA, and computes the corresponding public key $DoA^+ = DoA^- \cdot P$. DoT sends the pair (DoA^+, DoA^-) to DoA by secure channel, and sends DoA^+ to all the *RSUs* by secure channel.

3) **RSU-KeyGen:** DoT randomly select $K_{RSU}^- \in_R \mathbb{Z}_q^*$ for every RSU, and computes the corresponding public key $K_{RSU}^+ = K_{RSU}^- \cdot P$. DoT sends the pair (K_{RSU}^+, K_{RSU}^-) to the target *RSU* by secure channel.

4) **OBU-KeyGen:** Suppose every vehicle has a unique real identity, denote RID. Vehicle V_i first randomly chooses $K_i^- \in_R \mathbb{Z}_q^*$ as its secret key and computes the corresponding public key $K_i^+ = K_i^- \cdot P$. Then V_i submits a four-tuple (K_i^+, RID_i, a_i, b_i) to DoT, where $a_i = H(t_i P \| RID_i), b_i = (t_i - K_i^- \cdot a_i), t_i \in_R \mathbb{Z}_q^*$

DoT verifies the tuple's validity by judge $a_i = H((b_i P + (K_i^+)^{a_i}) \parallel RID_i)$. If pass, DoT assigns a PID for V_i, issues a anonymous certification binding K_i^+ to V_i, and then store the entry $[K_i^+, RID_i, PID_i]$ in the trace table T_1.

3.2   Group Construction

The group construction includes the following 3 procedures:

1) **RSU-Hello:** *RSU* broadcasts Hello-Message periodically to show its existence for the near vehicle. Hello-Message is $[K_{RSU}^+]$.

2) **OBU-Join-Req:** After received the Hello-Message from RSU_j, Vi sends back a Join-Message which is $[K_i^+ \parallel T]$'s encryption version by a session key between V_i and RSU_j. The OBU-Join-Req Algorithm is shown as following.

Table 2. OBU-Join-Req Algorithm

Algorithm 1: OBU-Join-Req Algorithm
1. V_i randomly selects a secret $r \in_R Z_q^*$;
2. V_i computes a session key $\phi = r(K_{RSUj}^+)$;
3. V_i computes a hint $\psi = rP$;
4.   Let $M = [K_i^+ \parallel T]$ where T denotes Timestamp;
5.   Let $M' = [\psi, Enc_\phi(M)]$, and send it to RSU_j.

3) **RSU-Confirm:** After received the Join-Message from V_i, RSU_j sends a Confirm-Message include the group key G_j and other configuration back. The RSU-Confirm Algorithm is shown as following.

Table 3. RSU-Confirm Algorithm

Algorithm 2: RSU-Confirm Algorithm
1. RSU_j computes the session key $\phi' = \psi K_{RSU}^-$ where hint ψ comes from the M' in Algorithm 1;
2. RSU_j uses ϕ' to decrypt the $Enc_\phi(M)$ part and gets T and K_i^+ ;
3. RSU_j checks the validity of timestamp T;
4.   If valid, RSU_j checks the Revoke List from DoT for the validity of K_i^+ ;
5.   If valid, RSU_j randomly select a group session key $G_j \in_R Z_q^*$ (if G_j has exist, skip this step);
6.   Let $M'' = [Enc_{\phi'}(\omega \parallel G_j \parallel T)]$ where ω means the configuration description of the group, and send it to V_i.

3.3   Message Sending

The format of the safety messages sent by the *OBU* is defined in Table 2, which consists of five fields: message ID, payload, timestamp, *RSU$_j$*'s public key and signature by the *OBU*. The message ID defines the message type, and the payload field may include the information on the vehicle's position, direction, speed, traffic events, event time, and so on. A timestamp is used to prevent the message replay attack. The next field is K_{RSUj}^+, the public key of *RSU* which is the group leader. The first four fields are signed by the vehicle V_i, by which the "signature" field can be derived. Table IV specifies the suggested length for each field.

Table 4. Message Format for OBU

Message ID	Payload	Timestamp	RSU$_j$'s Public Key	V$_i$'s Signature
2 bytes	100 bytes	4 bytes	21 bytes	20 bytes

To endorse a message M, V_i generates a signature on the message, and then encrypts and sends it to RSU_j. The OBU-Message algorithm is shown in as Table 5.

Table 5. OBU-Message Algorithm

Algorithm 3: OBU-Message Algorithm

1. For   a   message   content   DATA, V_i computes
    $\sigma = Sign_{K_i^-}(ID \parallel DATA \parallel K_{RSUj}^+ \parallel T)$;

2. Let $\bar{M} = Enc_\phi([ID \parallel DATA \parallel T \parallel K_{RSUj}^+ \parallel \sigma])$, and send it to RSU_j.

3.4   Message Transfering

After receiving a valid signature from the vehicles, RSU_j anonymizes the message and broadcasts the anonymous message to the group member. According to the secure level of different *RSUs*, the RSU-Transfer algorithms are divided into the following two different algorithm.

Table 6. RSU-Transfer Algorithm

Algorithm 4: HRSU-Transfer Algorithm

1. $HRSU_j$ uses   the   session   key ϕ' to   decrypt
    $\bar{M} = Enc_\phi([ID \parallel DATA \parallel T \parallel K_{RSUj}^+ \parallel \sigma])$ (ref Algorithm 3);

2. $HRSU_j$ checks the validity of timestamp T;

3. If valid, $HRSU_j$ checks σ 's validity using K_i^+ got from Algorithm 2;

4. If valid, $HRSU_j$ inserts an entry (H(m), σ) into local trace table T_{2j} (this table would be submitted to DoA's trace table T_2 periodically through secure channel such as SSL);

5. $HRSU_j$ uses the group session key G_j to HMAC the anonymous message
    $\tilde{M} = [M \parallel HMAC_{G_j}(M)]$ where $M = [ID \parallel DATA \parallel T \parallel K_{RSUj}^+]$;

6. $HRSU_j$ broadcasts $\tilde{M}$ to all the members of the group.

1) **HRSU-Trans:** *HRSUs* have good secure conditions and are controlled strongly by DoA. The transfer algorithm of *HRSU* is shown as Table 6.

2) **LRSU-Trans:** *LRSUs* have weaker secure conditions than *HRSU*, and the logs in them are not secure enough. The transfer algorithm of *LRSU* is shown as Table 7.

Table 7. LRSU-Transfer Algorithm

Algorithm 5: LRSU-Transfer Algorithm
1.   *LRSU$_j$*   uses   the   session   key ϕ' to   decrypt $\overline{M} = Enc_\phi([ID \parallel DATA \parallel T \parallel K^+_{RSUj} \parallel \sigma])$ (ref Algorithm 3);
2.   *LRSU$_j$* checks the validity of timestamp T;
3.   If valid, *LRSU$_j$* checks σ 's validity using K^+_i got from Algorithm 2;
4.   If valid, *LRSU$_j$* computes Tag=*Sign*$_{DoA+}$(σ);
5.   *LRSU$_j$* uses the group session key G_j to HMAC the anonymous message $\tilde{M} = [M \parallel HMAC_{G_j} (M) \parallel Tag]$ where $M = [ID \parallel DATA \parallel T \parallel K^+_{RSUj}]$;
6.   *LRSU$_j$* broadcasts $\tilde{M}$ to all the members of the group.

3.5   Message Verification

After receiving the anonymous message $\tilde{M}$ from the group leader *RSU$_j$*, the group member vehicles use group session key G_j to HMAC the $\tilde{M}$'s $[ID \parallel DATA \parallel T \parallel K^+_{RSUj}]$ part and to check the validity of the message.

3.6   Trace

If a message is found to be fraudulent, a tracing operation is started to determine the real identity of the signature originator. In detail, the trace process include the following two conditions:

1) If $\tilde{M} = [ID \parallel DATA \parallel T \parallel K^+_{RSUj} \parallel HMAC_{G_j} (ID \parallel DATA \parallel T \parallel K^+_{RSUj}) \parallel Tag]$,   the DoA first uses its private key *DoA$^-$* to decrypt the Tag section of the message to get signature σ . Then DoA uses σ to locate the corresponding K^+_i . According to the DoA's demand, the DoT check table T_1 to retrieve the real identity RID of the K^+_i and returns it to the DoA.

2) If $\tilde{M} = [ID \parallel DATA \parallel T \parallel K^+_{RSUj} \parallel HMAC_{G_j} (ID \parallel DATA \parallel T \parallel K^+_{RSUj})]$, the DoA first finds the *RSU* by extracting the *RSU*'s public key from the message. Then DoA checks table T_2 to locate the corresponding K^+_i. The rest of the trace process is the same as above.

4   Security Analysis

4.1   Privacy

Privacy is achieved by the pseudo identity PID which conceals the real identity RID such that peer vehicles and *RSUs* cannot identify the sender of a specific message while are still able to authenticate the sender.

4.2   Traceability

Given the disputed signature, the lookup tables T_1 in DoT and T_2 in DoA enable the eventual tracing misbehaving vehicles. The tracking procedure executed by corporation between DoA and DoT guarantees the traceability.

4.3   Authentication, Non-Repudiation and Integrity

Authentication, non-repudiation, and integrity are guaranteed by digital signatures σ which bound the message to a PID and consequently the corresponding identity. The message integrity can be protected by utilizing HMAC both in HRSU-Transfer and LRSU-Transfer Algorithm.

5   Performance Analysis

We carry out performance analysis in this section in terms of storage and communication performance for our scheme, and the reference protocols are HAB[1,2], GSB[3] and ECPP[4].

5.1   Storage

The storage requirements on DoT/DoA/RSU are not stringent since these entities are resource-abundant in nature. We are mainly concerned with the storage cost in vehicles *OBU* and list the contrast table as following (α means the RL's entry number and β means the anonymous certification number from each *RSU* which depends on the density of *RSUs*.):

Table 8. OBU storage cost contrast

Scheme	Cost (Bytes)
HAB	$500*43800+21*\alpha$
GSB	$126+43*\alpha$
ECPP	$87*\beta$
Our	42

From above table, the storage advantage of our scheme is obvious.

5.2   Communication

This section compares the communication overheads of the protocols studied. We assume that all protocols generate a timestamp to prevent replay attacks so we exclude the length of the timestamp in this analysis.

In HAB, each message generates yields 181 bytes as the additional overhead due to cryptographic operations. In GSB, each message generates 197 bytes as the additional overhead. For ECPP protocols, the additional communication overhead is 42+147=189 bytes, where the first term represents the signature's length, the second term represents the length of the anonymous key and its corresponding certificate. For RGH, the additional communication overhead is 2+21+20=43 bytes.

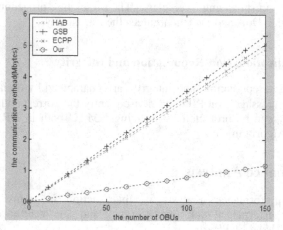

Fig. 2. The commucation overhead contrast

Fig. 2 shows the relationship between the overall communication overhead in 1 minute and the traffic load within a vehicle. Obviously, as the number of messages increases, the transmission overhead increases linearly. Clearly, we can observe that our protocol has much lower communication overhead than the other protocols.

6   Related Work

Security and privacy in VANETs raise many challenging research issues and there are many literatures related to the security and privacy in VANETs. We only review the most related and classic schemes focused on vehicle identity relative privacy-preserving.

Raya et al . [1,2] investigated the privacy issue and   proposed a landmark pseudonym-based protocol (huge anonymous public keys based, denote HAB) employing the public key infrastructure (PKI). A huge number of private keys (about 43,800) and their corresponding anonymous certificates are preload into each OBU by authorized department. To sign each launched message, a vehicle randomly selects one of its anonymous certificates and uses its corresponding private key. The other vehicles use the public key of the sender enclosed with the anonymous certificate to authenticate the source of the message. These anonymous certificates are generated by employing the pseudo-identity of the vehicles, instead of taking any real identity information of the drivers. Each certificate has a short life time to meet the privacy requirement. Although HAB protocol can effectively meet the conditional privacy requirement, it is inefficient, heavy-cost, and may become a scalability bottleneck. And more, the certificates database to be searched by the tracer in order to match a compromised certificate to its owner's identity is huge. To address the revoking issue, the authors later proposed three credential revocation protocols tailored for VANETs, namely RTPD, RC^2RL, and DRP [7]. All the three protocols seem to work well under conventional PKI. However, the authors also proposed to use frequently updated anonymous public keys to fulfill users' requirement on identity and location privacy. However, if HAB is used in conjunction with RC^2RL and DRP, the CRL will become huge in size, rendering the revocation protocols highly inefficient.

Lin et al. proposed the GSB[3] protocol, based on the group signature[8]. With GSB, each vehicle stores only a private key and a group public key. Messages are signed using the group signature scheme without revealing any identity information to the public. Thus privacy is preserved while DoA is able to track the identity of a sender. However, the time for safety message verification grows linearly with the number of revoked vehicles in the revocation list in the entire network. Hence, each vehicle has to spend additional time on safety message verification. Furthermore, when the number of revoked vehicles in the revocation list is larger than some threshold, it requires every remaining vehicle to calculate a new private key and group public key based on the exhaustive list of revoked vehicles whenever a vehicle is revoked. Lin et al. do not explore solutions to effectively updated the system parameters for the participating to vehicles in a timely, reliable and scalable fashion. This issue is not explored and represents an important obstacle to the success of this scheme.

Lu et al. [4] introduced an efficient conditional privacy preservation protocol (ECPP) based on generating on-the-fly short-lived anonymous keys for the communication between vehicles and RSUs. ECPP used RSUs as the source of certificates. In such an approach, RSUs (as opposed to OBUs) check the group signature to verify if the sender has been revoked and record values to allow tracing. OBUs then use a RSU provided certificate to achieve authenticity and short-term linkability. However, ECPP is vulnerable to Sybil attacks and requires an unreasonable amount of computation for RSUs (i.e., linear in the size of the revocation information for every certificate request).

7   Conclusion

In this paper, a lightweight RSU-aided group-based Hierarchical privacy-protecting protocol for VANETs has been proposed. The communication group is formed based on RSU and can provide anonymous communication between the group member vehicles and inter-group vehicles. For the better security, RSUs are divided into HRSU and LRSU, and are deployed in 1-HRSU-n-LRSU mode. HRSU inserts a message-related trace entry into the trace log while LRSU attaches trace Tag on the message. HRSU has to submit the trace log to DoA by secure way periodically. Furthermore, the scheme can preserve vehicle privacy, and simultaneously provide traceability. The protocol is characterized as low-complexity, low-load, efficient and accountable. Analysis about the security and the performance shows it can match the objectives well.

Acknowledgments. This work was supported in part by the NSFC under grant No. 61170263 and No. 61003230.

References

1. Raya, M., Hubaux, J.: The security of vehicular ad hoc networks. In: Proceedings of the 3rd ACM Workshop on Security of Ad Hoc and Sensor Networks, pp. 11–21. ACM, Alexandria (2005)
2. Raya, M., Hubaux, J.: Securing vehicular ad hoc networks. Journal of Computer Security, Special Issue on Security of Ad Hoc and Sensor Networks 15, 39–68 (2007)
3. Lin, X., Sun, X., Ho, P., Shen, X.: GSIS: A Secure and Privacy-Preserving Protocol for Vehicular Communications. IEEE Transactions on Vehicular Technology 56, 3442–3456 (2007)
4. Lu, R., Lin, X., Zhu, H., Ho, P., Shen, X.: ECPP: Efficient Conditional Privacy Preservation Protocol for Secure Vehicular Communications. In: Proceedings of the 27th IEEE International Conference on Computer Communications, INFOCOM 2008, pp. 1229–1237. IEEE (2008)
5. Wasef, A., Shen, X.: Efficient Group Signature Scheme Supporting Batch Verification for Securing Vehicular Networks. In: Proceedings of IEEE International Conference on Communications, ICC 2010, pp. 1–5. IEEE (2010)
6. Calandriello, G., Papadimitratos, P., Hubaux, J., Lioy, A.: Efficient and robust pseudonymous authentication in VANET. In: Proceedings of the Fourth ACM International Workshop on Vehicular Ad Hoc Network, VANET 2007, pp. 19–28. ACM, Montreal (2007)
7. Raya, M., Papadimitratos, P., Aad, I., Jungels, D., Hubaux, J.P.: Eviction of Misbehaving and Faulty Nodes in Vehicular Networks. IEEE Journal on Selected Areas in Communications 25, 1557–1568 (2007)
8. Chaum, D., van Heyst, E.: Group Signatures. In: Davies, D.W. (ed.) EUROCRYPT 1991. LNCS, vol. 547, pp. 257–265. Springer, Heidelberg (1991)

Locating Encrypted Data Precisely without Leaking Their Distribution

Liqing Huang[1,2] and Yi Tang[1,2,*]

[1] School of Mathematics and Information Science
Guangzhou University, Guangzhou 510006, China
[2] Key Laboratory of Mathematics and Interdisciplinary Sciences of Guangdong
Higher Education Institutes
Guangzhou University, Guangzhou 510006, China
ytang@gzhu.edu.cn

Abstract. Data encryption is a popular solution to ensure the privacy of
the data in outsourced databases. A typical strategy is to store sensitive
data encrypted and map those original values into bucket tags for querying on encrypted data. To achieve computations over encrypted data,
the homomorphic encryption (HE) methods are proposed. However, performing those computations needs locating data precisely. Existing test-
over-encrypted-data methods cannot prevent a curious service provider
doing in the same way and causing the leaks of original data distribution. In this paper, we propose a method, named Splitting-Duplicating,
to support encrypted data locating precisely by introducing an auxiliary
value tag. To protect the privacy of original data distribution, we limit
the frequencies of different tag values in a given range. We use an entropy based metric to measure the degree of privacy protected. We have
conducted some experiments to validate our proposed method.

1 Introduction

Outsourcing data to a datacenter is a typical type of computing paradigm in the
cloud. To ensure the privacy of the sensitive data, storing those data encrypted
on remote servers becomes a common view. However, encryption also destroys
the natural structure of data and introduces many challenges in running database
applications.

One of the main challenges is how to execute queries over encrypted data. Developing new ciphers for keyword searching on encrypted data seems a comprehensive solution for this issue. However, either the symmetric encryption scheme
[6] or the asymmetric encryption scheme [7] cannot prevent the curious service
provider locating the positions with the same method. The DAS (Database as a
Service) model [4] addresses a bucketization method. According to this model,
the data is encrypted by a traditional block cipher and the domain of each
sensitive attribute is divided into a set of *bucket*s which is labeled by bucket

---

* Corresponding author.

H. Gao et al. (Eds.): WAIM 2012, LNCS 7418, pp. 363–374, 2012.
© Springer-Verlag Berlin Heidelberg 2012

tags. Thus some values can be mapped into a single bucket and the tuples with the same bucket tag may have different original values. This scheme obviously introduces false answers when retrieving data from remote servers.

Another challenge is on the computations over encrypted data. Since homomorphic encryption (HE) enables an equivalent relation exists between one operation performed on the plaintext and another operation on the ciphertext, it is considered as an effective solution to this issue. According to the relations supported, the homomorphic encryption methods can be partially [2] or fully [3]. The introduced nonce in both methods makes it difficult to distinguish two HE-encrypted values even if they are the same in plaintext. The HE methods require locating the encrypted tuples precisely before performing any computation.

Locate encrypted tuples implies execute comparison operations over encrypted data on server without decryption. As mentioned above, if the comparison results could be distinguished on server, the curious service provider could also manipulate in the same way to obtain information about sensitive values. The CryptDB scheme [8] defines layers of encryption for different types of database queries. For executing a specific query, layers of encryption can be removed by decrypting to an appropriate layer and the tuple locating is based on the ciphertext comparing. This may lead many sensitive values be stored to the lowest level of data confidentiality provided by the weakest encryption scheme, and leak sensitive details such as the relations between sensitive values and the distribution of values. Furthermore, the introduced nonce in this scheme makes the ciphertext comparing become complicated even if a proxy at client side maintains the relations between the plaintext and ciphertext. ·

It is necessary to introduce the HE methods in encrypting outsourced data. In this paper, we aim at providing a simple method for locating HE-encrypted data precisely. Compared with the matching methods of an encrypted string, we introduce the notion of value tags and perform the comparisons based on the value tags. A sensitive value being encrypted is mapped into an appropriate value called value tag. It implies that two encrypted tuples with a same value tag will have a same sensitive value.

The contributions of this paper can be enumerated as follows.

1. We introduce an auxiliary attribute for precisely locating HE-encrypted values. The original sensitive values are mapped into some artificial value tags. To prevent value tags disclosing the sensitive data distribution, the same sensitive values are actually mapped to a set of value tags.
2. We propose a method, named *Splitting-Duplicating*, to construct the value tag mapping. The goal of this method is to keep the frequencies of each tag varying in a limited range. The original value with higher frequency will be mapped into some different tags via a *Splitting* process. For the one with lower frequency, we propose a *Duplicating* process to create some noise tuples to make the frequency of the associated tag fall in the given range.
3. We develop an entropy-based metric to measure the protected degree of data distribution via our proposed *Splitting-Duplicating* method.

The rest of this paper is structured as follows. In Section 2, we give our motivation. In Section 3, we formulate the proposed Splitting-Duplicating method. In Section 4, conduct some range query experiments to validate our proposed method. And finally, the conclusion is drawn in Section 5.

2   Motivation

We assume the database is outsourced to a database service provider and the provider is curious but honest.

Considering the relation, $R = (A_1, A_2, ..., A_n)$, with n sensitive attributes. Its corresponding encrypted relation on server can be described as $R_1^s = (A_{1_e}^s, A_{2_e}^s, ..., A_{n_e}^s)$ where the attribute $A_{i_e}^s$ denotes the encrypted version of A_i. Since not all the attribute value need to be modified during the database running, the attributes can be simplified into two types, the modifiable attributes and the non-modifiable ones.

Without loss of generality, assume the attribute A_n is modifiable and the others are non-modifiable in relation R. Considering the encryption features in traditional ciphers and the computation feature in HE methods, we redescribe R_1^s as $R_2^s = (enc\_other\_attr, A_{n_{HE}}^s)$ where enc_other_attr denotes the encrypted version of $n-1$ tuple $\langle A_1, A_2, ..., A_{n-1}\rangle$, $A_{n_{HE}}^s$ denotes the HE-encrypted A_n attribute.

However, the data in a database is not only for storage but also for queries and computations. The introduced nonce in HE methods makes it difficult to perform equality tests on server. To support the encrypted tuple locating, we introduce *value tag*. A value tag is used to identify an attribute value. If two value tags of an attribute are equal, the corresponding two attribute values are also equal. Therefore, the encrypted relation stored on server can be $R^s = (enc\_other\_attr, A_{n_{HE}}^s, A_1^s, A_2^s, ..., A_{1-1}^s, A_n^s)$ where A_i^s denotes the value tag of attribute A_i.

Table 1. A table in original (left) and HE-encrypted (right)

#tuple	name	salary	enc_other_attr	$salary_{HE}^s$	$name^s$	salary_tag
1	Alice	22	1100010...	0110011...	x_1	z_1
2	Bob	15	0101010...	1010110...	x_2	z_2
3	Cindy	20	0011101...	1001010...	x_3	z_3
4	Donald	20	0101011...	0010101...	x_4	z_3
5	Eva	25	1010110...	1110001...	x_5	z_4

Table 1 shows an example for a relation in original and on server. When executing an SQL operation over the encrypted data, a translator at client will translate the plain operation in R into the server-side operation in R^s.

For example, the SQL Update operation:

Table 2. An HE-encrypted table after Splitting (right)

#tuple	name	salary	enc_other_attr	$salary^s_{HE}$	$name^s$	salary_tag
1	Alice	22	1100010...	0110011...	x_1	z_1
2	Bob	15	0101010...	1010110...	x_2	z_2
3	Cindy	20	0011101...	1001010...	x_3	z_{31}
4	Donald	20	0101011...	0010101...	x_4	z_{32}
5	Eva	25	1010110...	1110001...	x_5	z_4

Table 3. An HE-encrypted table after Duplicating (right)

#tuple	name	salary	enc_other_attr	$salary^s_{HE}$	$name^s$	salary_tag
1	Alice	22	1100010...	0110011...	x_1	z_1
2	Bob	15	0101010...	1010110...	x_2	z_2
3	Cindy	20	0011101...	1001010...	x_3	z_3
4	Donald	20	0101011...	0010101...	x_4	z_3
5	Eva	25	1010110...	1110001...	x_5	z_4
6	-	-	1001010...	0111001...	x_6	z_1
7	-	-	1101110...	1101110...	x_7	z_2
8	-	-	0110100...	1101001...	x_8	z_4

update R **SET** $R.salary = R.salary + 2$ **where** $R.salary \leq 20$
can be translated into the following on server:

update R^s **SET** $R^s.salary^s_{HE} = R^s.salary^s_{HE} + 2_{HE}$ **where** $R^s.salary\_tag = z_2$ **or** z_3

where 2_{HE} is the HE-encrypted result of 2.

It is easy to distinguish the two Update operations. The server-side operation can be viewed as the encrypted version of the client-side operation. The translation, besides changes the plain value modification into an encrypted one, changes the where condition into an OR-Expression whose length is 2..

We also note that the distribution of salary_tag is the same as the attribute salary in plaintext. The link attack could immediately be proceeded to this encrypted relation [1]. A simple strategy against the link attack is to destroy the consistency between the two distributions.

In this paper, we consider to change the distribution of the artificial tag values. We intend to limit the frequencies of value tags in a given range and make the distribution as uniform as possible. Two approaches can be adopted to reach the target.

1. *Splitting.* Adding new *tag* values to the table. For example (as in Table 2), we can define the salary_tag of item *Cindy* and item *Donald* as z_{31} and z_{32}, respectively. This method cuts a higher frequency into some lower pieces.

2. *Duplicating.* Adding new *noisy* items to the table. For example (as in Table 3), we can simply duplicate the items of *Alice*, *Bob*, and *Eva* into the

table and make all the frequencies of *salary_tag* as the same. This method increases the frequency of a value tag with lower original frequency.

3   The Splitting-Duplicating Method

3.1   The Value Tag Entropy

We follow the basic notations described previously. The tag construction for traditional encrypted attributes are referred in [4], we only need to consider how to construct the value tags for the HE-encrypted attributes. Without causing confusion, we suppose the attribute A is the HE-encrypted attribute.

Definition 1. *Given an attribute A with domain D_A and a value $v \in D_A$, a value tag mapping is the mapping, $VTmap : D_A \to Integer$, and the value tag for v is the integer $VTmap(v)$, denoted by vt.*

Some typical cryptology techniques, such as the block ciphers and the keyed hash functions, can be used to define the mapping $VTmap$.

Definition 2. *An HE-related tuple is the tuple $\langle v, vt, n_{vt}, n_a \rangle$ where v is the original value, vt is the corresponding value tag stored on server, n_{vt} is the frequency of vt, and n_a is the additional frequency increment for vt.*

The HE-related tuples are stored on clients for translating the SQL operations. Initializing a set of HE-related tuples is straightforward. The following procedure demonstrates the sketch for initializing the HE-related tuple set T with size N, where the HE-related tuple set is constructed as a *RTLtup* structure.

```
procedure Initializing(RLTtup T)
   for i = 1 to N
      v_a ← R.A
      if {⟨v, vt, n_vt, n_a⟩|⟨v, vt, n_vt, n_a⟩ ∈ T ∧ v = v_a} ≠ φ
         n_vt ← n_vt + 1
      else
         vt ← VTmap(v_a)
         T = T ∪ {⟨v_a, vt, 1, 0⟩}
      endif
   endfor
endprocedure
```

Given the value tag set $VT = \{vt | \langle v, vt, n_{vt}, n_a \rangle \in T\}$ and $n = \sum_{vt} n_{vt}$, the following formula can be used to compute the entropy of VT:

$$H(VT) = -\sum_{vt} \frac{n_{vt}}{n} \cdot \log \frac{n_{vt}}{n} \tag{1}$$

We use the above formula to measure the degree of original data distribution protected in a given value tag set. It is obviously that the $H(VT)$ reaches its maximize value $\log_2 |VT|$ when the frequencies of each value tag are the same.

Intuitively, if the frequencies of each value tag are nearly the same, an adversary need more efforts to guess the whole value distribution of sensitive data on the database server, even if he knows the probability distribution of sensitive data and a few exact values of sensitive data. This intuition is consistent with the notion of the value tag set entropy. It is noted that the information entropy is a measure of unpredictability. The larger the entropy, the more privacy protected in a value tag set. It implies that the approximative frequencies of each value tag may introduce more power against the link attacks.

3.2   Limiting the Frequencies of Value Tags

Let t_{low}, t_{high} be two threshold values where $t_{low} < t_{high}$. We try to limit the frequencies of each value tag in a range $(t_{low}, t_{high}]$.

Definition 3. *The HE-related tuple set $HFset = \{\langle v, vt, n_{vt}, n_a \rangle | \langle v, vt, n_{vt}, n_a \rangle \in T \wedge n_{vt} \geq t_{high}\}$ is called as the high frequency set.*

Definition 4. *The HE-related tuple set $LFset = \{\langle v, vt, n_{vt}, n_a \rangle | \langle v, vt, n_{vt}, n_a \rangle \in T \wedge n_{vt} < t_{low}\}$ is called as the low frequency set.*

In general, when an HE-related tuple set is initialized, both $HFset$ and $LFset$ are not empty. Our goal is to reconstruct the HE-related tuple set T with emptied $HFset$ and $LFset$. We will adopt two strategies, *Spiltting* and *Duplicating*, to empty $HFset$ and $LFset$, respectively.

We first give a method for decomposing n into a sequence: $sn_1, sn_2, ...sn_k$ s.t. $\forall i : sn_i \in (t_{low}, t_{high}]$ and $\sum_i sn_i = n$. We only consider two scenarios, the case of $(0, t]$ and the case of $(t, 2t]$.

The decomposition in case $(0, t]$ is trivial and we only need iteratively pick an integer from $(0, t]$. For the case of $(t, 2t]$, the procedure can be showed as follows.

```
procedure Decomposing(int n, t, int[ ] sn)
    tn ← n, i ← 1
    while tn > 2t
        random select r ∈ (t, 2t]
        tn ← tn − r, sn[i] ← r, i ← i + 1
    endwhile
    if tn ∈ (t, 2t]
        sn[i] ← tn
    else if tn ≤ t
        sn_min = min{sn[1], sn[2], ..., sn[i − 1]}
        if sn_min + tn ∈ (t, 2t]
            sn_min ← sn_min + tn
        else
            sn_max = max{sn[1], sn[2], ..., sn[i − 1]}
            random select r ∈ (t, sn_max + tn − t]
            sn_max ← sn_max + tn − r, sn[i] ← r
        endif
```

endif
endprocedure

Theorem 1. *The procedure of Decomposing can decompose the integer n (n ¿ $2t$) into a sequence of integers that are in $(t, 2t]$.*

Definition 5. *For a tuple $t = \langle v, vt, n_{vt}, n_a \rangle \in T$, we say t is split into two tuples $t_1 = \langle v, vt_1, n_{vt_1}, n_{a_1} \rangle$, $t_2 \langle v, vt_2, n_{vt_2}, n_{a_2} \rangle$ iff the original value v (whose value tag is vt) are mapped into tags vt_1 and vt_2, and $n_{vt_1} \neq 0, n_{vt_2} \neq 0$.*

We denote VT_s as the new value tag set after a tuple is split.

Theorem 2. *If a tuple $t \in T$ is split into two tuples t_1, t_2, we have $H(VT_s) > H(VT)$.*

This theorem indicates that when a tuple is split into two parts, the entropy of value tag set becomes larger.

3.3   The *Splitting* Procedure and *Duplicating* Procedure

The *Splitting* procedure and the *Duplicating* procedure are simple and directly. We sketch them as following.

```
procedure Splitting(RLTtup T, int t_low, t_high)
   while HFset(T) ≠ φ
      select ⟨v, vt, n_vt, 0⟩ ∈ HFset(T)
      T ← T − {⟨v, vt, n_vt, 0⟩}
      decompose n_vt into a sequence: k_1, k_2, ..., k_m
         where k_i ∈ (t_low, t_high] and ∑_i k_i = n
      for each k_i
         construct a value tag vt_i
         pick k_i tuples whose value tag is vt in R^s
         replace these k_i tuples's value tag with vt_i, repectively
         T ← T ∪ {⟨v, vt_i, k_i, 0⟩}
      endfor
   endwhile
endprocedure
```

In the *Splitting* procedure, the HE-related tuple set T is scanned to check whether or not the $HFset(T)$ is empty. If the element $\langle v, vt, n_{vt}, 0 \rangle$ is in $HFset(T)$, we will decompose n_{vt} into a sequence of integer $\{k_i\}$ such that $\forall k_i : k_i \in (t_{low}, t_{high}]$ and $\sum_i k_i = n$. It implies that we have split n_{vt} tuples into a set of tuple set whose size is k_i, respectively. For the tuple set with size k_i, we generate a new value tag vt_i for those k_i tuples and append the tuple $\langle v, vt_i, k_i, 0 \rangle$ into value tag set T. The *Splitting* procedure empties the subset $HFset(T)$.

```
procedure Duplicating(RLTtup T, int t_low, t_high)
   while LFset ≠ φ
      select ⟨v, vt, n_vt, n_a⟩ ∈ LFset
      T ← T − {⟨v, vt, n_vt, n_a⟩}
      generate k' : t_low − n_vt ≤ k' < t_high − n_vt
      for  i = 1  to  k'
         ev ← HE(v)
         construct other R^s attribute values: enc_others, v_1, v_2, ..., v_{n−1}
         store ⟨enc_others, ev, v_1, v_2, ..., v_{n−1}, vt⟩ on server
      endfor
      T ← T ∪ {⟨v, vt, n_vt + k', n_a + k'⟩}
   endwhile
endprocedure
```

In the *Duplicating* procedure, the iterative check targets at $LFset(T)$. If the element $⟨v, vt, n_{vt}, n_a⟩$ is in $LFset(T)$, some random *noise* tuples with the same value tag vt will be created and stored into the encrypted database on server. This makes the frequency of vt in R^s increase to the range $(t_{low}, t_{high}]$. The *Duplicating* procedure empties the subset $LFset(T)$ but also introduces some noise tuples.

4   Experiments and Discussion

4.1   Measuring SQL Operation Time Cost Based on OR-Expression

After performing *Splitting-Duplicating* over the encrypted data, the where condition of a plain SQL operation will be translated into an OR-expression. The number of operands in an OR-expression is called as the length of this OR-expression. Different lengths of OR-expression may lead to different SQL operation response time. We run a set of experiments to confirm this intuition.

These experiments are on an HP mini-210 with Intel Atom N450 1.66GHz and 1GB memory. We create 5 tables, each is with 20k integers, under Windows XP sp3 and MS SQL2000. The integers is evenly distributed in intervals $[1, 1k], [1, 2k], [1, 5k], [1, 10k]$, and $[1, 20k]$, respectively.

We perform the query, **select** * *from* T **where** *OR-expression*, and regulate the *OR-expression* in order to retrieve 20 tuples on each table. We randomly generate the *OR-expression*, execute the query, and record the response time. The average time in 1,000 tests is shown as in Fig. 1

Fig. 1 demonstrates that as the length of *OR-expression* increases, the query retrieved the same number of tuples needs more response times. Based on this observation, we use the length of *OR-expression* to evaluate the SQL operation time cost for our proposed *Splitting-Duplicating* method.

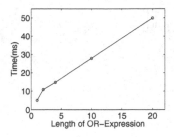

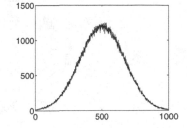

Fig. 1. Time Cost on OR-expression   **Fig. 2.** The Original Data Distribution

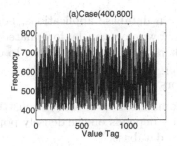

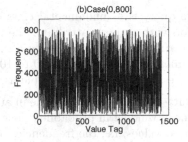

Fig. 3. The Value Tag Distributions after Splitting-Duplicating

4.2 Performing *Splitting-Duplicating* on Synthesized Dataset

The next experiments are on synthesized data. We assume the sensitive data are on normal distribution and generate a test dataset $data_{500k}$ with 500,000 integers in the interval $[1, 999]$. These $500k$ integers are viewed as the attribute values needed to be HE-encrypted, and we adopt our proposed method to construct corresponding value tags.

The original value distribution in $data_{500k}$ is demonstrated in Fig.2. Fig. 3 demonstrates two instances of value tag distribution after performing the *Splitting-Duplicating* procedure. Fig. 3(a) is for the case of limiting the frequencies in $(400, 800]$ and (b) is for limiting in $(0, 800]$. The tag values in both instances demonstrate completely different distribution comparing to the original value distribution.

Fig. 4 demonstrates how the entropy changes with the parameter t of the two scenarios, $(t, 2t]$ and $(0, t]$. We find that as t increases, the value of entropy decreases in both cases. It implies that smaller ts will bring larger entropy because of the larger state variable space. This means the smaller ts will provide more degrees of the capability of privacy protected.

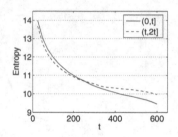

Fig. 4. The Entropy for the Scenario (0, t] and the Scenario (t, 2t]

4.3   Range Queries

Range query is a fundamental class of queries in database applications. For a relation with a single attribute, the range query window can be formulated as $[x, x + size]$, denoted by $querywin(x, size)$, where $size$ is the range size. In our experiments, we set $size$ from 0 to 100 with step size 1.

There are two kinds of possible user behaviors in range queries [5].

- Data-based query. Each value in attribute domain can be chosen to construct query window in probability. The probability depends on the densely populated values, i.e, the frequencies of values in database.
- Value-based query. Each value in attribute domain is equally chosen to construct query window. This query demonstrates the case where no user preference is known previously.

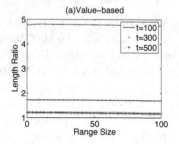
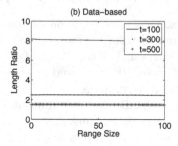

Fig. 5. Range Queries for the Scenario (0,t]

Thus, the range query windows, $querywin(x, size)$, can be constructed by different x selection methods. The data-based query window is constructed by randomly choosing x from $data_{500k}$ while the value-based query window is constructed by randomly choosing x from $[1, 999]$.

For each range size $size$, we respectively generate 1,000 random queries on the two user behaviors, and apply these queries on the constructed value tag set. We evaluate the experiment results on the following items.

- The OR-expression length ratio. It reflects the average query response time cost comparing to the original plain queries.
- The proportion of noise tuple cases. It is the average proportion of the query responses included noise tuples in each query and reflects the possibility of refilter the retrieved tuples.
- The number of noise tuples. It defines average number of noise tuples introduces in queries and demonstrates the extra network payloads.

Fig.5, Fig.6, and Fig.7 demonstrate three sets of experiment results.

As shown in Fig.5(a), Fig.5(b), Fig.6(a), and Fig.7(a), we find that the OR-expression length ratio values almost keep invariable when varying the query window sizes in our conducted experiments. The larger the frequency limitation parameter t, the smaller the OR-expression length ratio. The length ratios introduced in two query behaviors for the Scenario $(t, 2t]$ are almost the same, while for the Scenario $(0, t]$, the value-based queries can be responded more efficiently.

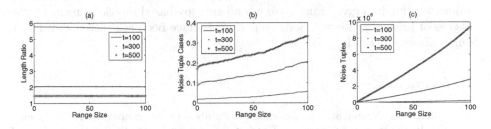

Fig. 6. Data-based Range Queries for the Scenario $(t, 2t]$

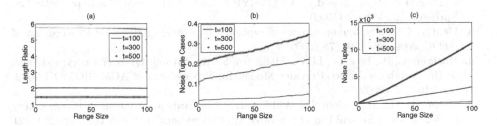

Fig. 7. Value-based Range Queries for the Scenario $(t, 2t]$

There are not noisy tuples in query results because of no needing *Duplicating* procedure in the scenario $(0, t]$. For the scenario $(t, 2t]$, the two query behaviors demonstrate similar characteristics in noise tuple appearance. As demonstrated in Fig.6(b), Fig.7(b), the proportions of noise tuple cases are almost the same in both behaviors, and the proportion values increase slightly when the query range increases. Both two figures show that the larger frequency limitation parameter t leads to larger proportions of noise tuple cases. For the average number of noise tuples, as shown in Fig.6(c) and Fig.7(c), the value-based queries may receive less noise tuples obviously.

4.4   Trade-Off Between the Privacy and the Performance

According to the results of conducted experiments, we can define a trade-off strategy between the privacy and the performance. On one hand, as shown in Fig.4, the smaller the frequency limitation parameter t, the larger entropy the system has. On the other hand, as shown in Fig.5, Fig.6, and Fig.7, the smaller t will introduce more response cost. Therefore, given a degree of privacy protected, we can define an entropy-based method to find a proper parameter t with minimum query response time cost.

5   Conclusion

We have proposed *Splitting-Duplicating* method to support locating HE-encrypted data precisely by introducing an auxiliary value tag. To protect the privacy of original data distribution, we limit the frequencies of different tag values varying in a given range. We use an entropy based metric to measure the degree of privacy protected. Some experiments have been conducted to validate our proposed method.

References

1. Damiani, E., Vimercati, S., Jajodia, S., Paraboschi, S., Samarati, P.: Balancing Confidentiality and Efficiency in Untrusted Relational DBMSs. In: Proceedings of ACM CCS 2003, pp. 93–102 (2003)
2. Paillier, P.: Public-Key Cryptosystems Based on Composite Degree Residuosity Classes. In: Stern, J. (ed.) EUROCRYPT 1999. LNCS, vol. 1592, pp. 223–238. Springer, Heidelberg (1999)
3. Gentry, C.: Fully homomorphic encryption using ideal lattices. In: Proceedings of STOC 2009, pp. 169–178 (2009)
4. Hacigumus, H., Iyer, B., Li, C., Mehrotra, S.: Executing SQL over Encrypted Data in the Database-Service-Provider Model. In: Proceedings of ACM SIGMOD 2002, pp. 216–227 (2002)
5. Pagel, B., Six, H., Toben, H., Widmayer, P.: Towards an Analysis of Range Query Performance in Spatial Data Structures. In: Proceedings of PODS 1993, pp. 214–221 (1993)
6. Song, D., Wagner, D., Perrig, A.: Practical techniques for searches on encrypted data. In: Proceedings of IEEE S&P 2000, pp. 44–55 (2000)
7. Yang, G., Tan, C.H., Huang, Q., Wong, D.S.: Probabilistic Public Key Encryption with Equality Test. In: Pieprzyk, J. (ed.) CT-RSA 2010. LNCS, vol. 5985, pp. 119–131. Springer, Heidelberg (2010)
8. Popa, R., Redfield, C., Zeldovich, N., Balakrishnan, H.: CryptDB: Protecting Confidentiality with Encrypted Query Processing. In: Proceedings of SOSP 2011, pp. 85–100 (2011)

LB-Logging: A Highly Efficient Recovery Technique for Flash-Based Database

Zeping Lu, Xiaoying Qi, Wei Cao, and Xiaofeng Meng

Renmin University of China, Beijing, China
{zplu,87qixiaoying,caowei,xfmeng}@ruc.edu.cn
http://idke.ruc.edu.cn/

Abstract. Nowadays, due to users' increasing requirements of fast and reliable data management for mobile applications, major electronic device vendors use embedded DBMS on their mobile devices such as MP3 players, mobile phones, digital cameras and PDAs. However, in embedded database, data logging is the bottleneck against fast response time. There has been a lot of work on minimizing logging overhead to provide the best online performance to database workloads. However, to the best of our knowledge, there is still no recovery method taken into consideration. In this paper, we propose a novel logging method called LB-logging to support high efficiency in recovery of crashed databases. LB-logging is based on list structures instead of sequential structures in traditional databases. In addition, by making use of the history data versions which are naturally located in flash memory due to the out-of-place update, we take the full advantage of high I/O performance of flash memory to accelerate our recovery algorithm. Experimental results on Oracle Berkeley DB show that our LB-Logging method significantly outperforms the traditional recovery by 2X-15X, and other logging methods for SSD by 1.5X-6X.

1 Introduction

Flash memory is a new kind of data storage media. Different from Hard Drive Disk (HDD), it has a lot of attractive characteristics such as fast access speed, shock resistance, low power consumption, smaller size, lighter weight and less noise. Flash memory is widely used in a large number of electronic devices. The latest mobile phones, digital cameras, DV recorders, MP4 players, and other electronic handheld devices use flash memory as the main data storage devices. During the past years, the capacity of flash memory doubles every year, which is faster than Moore's law, and flash chip of 1TB has been reported available in market [1]. As the capacity increases and price drops dramatically during the past years, flash memory (instead of magnetic disk) has been considered as the main storage media in the next generation of storage devices.

In database systems, transaction processing like instantaneous queries and updates incurs frequent random IO. Additionally, in data warehousing, multiple streams of queries, insertions and deletions, also need large number of random

H. Gao et al. (Eds.): WAIM 2012, LNCS 7418, pp. 375–386, 2012.

IO operations. While, in current disk based database (or warehousing) systems, the major bottleneck is the IO disk performance. As we know, random IO is the drawback of hard disks. It is because that with the mechanical moving arms, a hard disk can only produce very limited number of IO operations per second. To overcome the shortcoming, flash memory would be the perfect alternative to traditional magnetic disks.

The access characteristics of flash memory are different from those of magnetic disks. Since traditional databases were designed to utilize disk features, thus we can't take full advantage of high I/O performance of flash memory if we transfer traditional database system onto flash memory without any modification. The rationale is the write granularity and erase-before-rewrite limit of NAND flash memory. Particularly, in NAND flash memory the write granularity is a page. And we cannot overwrite the same address unless we erase the whole block containing that page.

Logging is an important component of DBMS [2, 3]. It aims at maintaining the ACID property of transactions. However, erase-before-rewrite and page write characteristics lead to lower performance for transaction processing when the underlying hardware changes from traditional magnetic disks to flash memory. As for logging and recovery, the situation becomes more serious because of the out-of-place update model which leads to high cost with large quantity of minor random writes during the course of recovery. Therefore, it is necessary to design a new logging technique for flash-based DBMS [4, 5, 6].

In this paper we analyze the logging design problems in flash memory based databases and propose a new solution called LB-Logging. It makes use of the history versions of data which is naturally disposed in flash memory due to out-of-place updates. Furthermore, LB-Logging uses list structures instead of sequential structures in the traditional databases to store log records. We summarize our contributions as follows.

- A novel logging method called LB-Logging is proposed for the first time to overcome the limitation of traditional logging on flash-based databases.
- By making use of list structures instead of sequential structures to store log records, LB-Logging utilizes the high performance of random reads of flash memory to greatly shorten the recovery time.
- Additionally, to effectively reduce the log redundancy and improve the space utilization of databases, LB-Logging records data addresses instead of data values in logs.
- Results of empirical studies with implementations on a real database system of the proposed LB-Logging algorithm demonstrate that LB-Logging reduces the recovery time effectively.

The rest of this paper is organized as follows: In Section 2, we give the related work and compare our approach with them. Section 3 discusses the characteristics of flash memory and their impact on traditional disk-based databases. Section 4 and section 5 introduces the basic concepts and the design of the LB-Logging schema. Experimental results are given in Section 6 and we conclude in Section 7.

2   Related Work

It's not been long since flash memory began to be used as data storage media for computers. There is not much work to solve the problem of logging in flash-based DBMS [7-10].

Some researchers tried to change the storage of data files and use logs to record the updates instead of in-place-updates. IPL [11] is an influential work among them. In IPL design principles, each block,the erase unit on flash memory, is divided into two segments, data pages and log region [12, 13]. All the update operations are transformed into logs in main memory buffer firstly. Later the logs are flushed out to log sectors of the erase unit allocated for the corresponding data pages. If data pages fetched from the same erase unit get updated often, the erase unit may run out of free log sectors. It is when merging data pages and their log sectors is triggered by the IPL storage manager. This storage model could provide indirect recovery through the logs stored in database. But it requires a lot of modifications of traditional DBMS, so it can not be easily added to an existing DBMS.

FlashLogging [14] is trying to exploit multiple flash drives for synchronous logging. As USB flash drive is a good match for the task of synchronous logging because of its unique characteristics compared to other types of flash devices. FlashLogging designed an unconventional array organization to effectively manage these dispersed synchronous logging stored in different USB devices. Similarly, Lee also tried to combine USB flash drives and magnetic disks as a heterogeneous storage for better performance. However, as the price of SSD continues to decline, the advantage of USB device's price is gradually disappearing. FlashLogging is not a convenient model to build.

3   Motivation

In this section, we will present the problems caused in logging and recovery when transferring traditional databases to flash memory without any modification. And first we will describe the characteristics of flash memory to help us understand the problems.

3.1   Flash Memory

The characteristics of flash memory are quite different from those of magnetic hard disks. In flash memory, data is stored in an array of flash blocks. Each block spans 32-64 pages, where a page is the smallest unit of read and write operations. The read operations of flash memory are very fast compared to that of magnetic disk drive. Moreover, unlike disks, random read operations are as fast as sequential read operations as there is no mechanical head movement. The major drawback of the flash memory is that it does not allow in-place updates. Page write operations in a flash memory must be preceded by an erase operation and within a block, pages need be to written sequentially. The typical access latencies for read, write, and erase operations are 25 microseconds, 200 microseconds, and 1500 microseconds, respectively.

3.2 Problem Definition

Log-based recovery techniques are widely used in traditional databases. Different protocols determine different designs of log storage format, buffer management, checkpoint and recovery mechanisms. Take undo logs as an example. When transaction T updates element X whose original value is v, undo log will generate a log record like $\langle T,X,v \rangle$ in the DRAM. And finally the log will be flushed into disk. When transaction T rolls back, we have to re-write X to its original value v.

Table 1. Undo Log on Flash-based Database

(a) The Original Table

	value	flag
A	v_1	1
B	v_b	1
C	v_c	1
...	...	...

(b) After A Updated

	value	flag
A	v_1	0
B	v_b	1
C	v_c	1
A	v_2	1
...	...	...

(c) After A Rolled Back

	value	flag
A	v_1	0
B	v_b	1
C	v_c	1
A	v_2	0
A	v_1	1
...	...	...

Here, we replay this process on flash memory. The original table is as shown in Tab. 1(a). When a transaction updates A from v1 to v2, it is necessary to insert a new record of A, as shown in Tab. 1(b) in the last line. And if T has to roll back sometimes later, we must re-write A's original value v1 again because of out-place update, as shown in Tab. 1(c). We can see that the last record is the same as the first record. In other words, the last one is actually redundant in this situation. We can infer that there may be large amounts of data's history versions. In fact, the recovery process doesn't have to write the data which has already existed. It is not only a waste of space, but also a waste of time.

As we discussed earlier, every write operation occupies at least one page (typically 2KB) regardless of the size of the data. But generally speaking, the size of the rolled back element may be less than 2KB. It brings extra space wasted. Apart from that, the additional writes may bring some unnecessary erase operations with time cost even greater. Therefore, to avoid rewrite operations, a new design of logging and recovery method is needed for flash-based DBMS.

4   LB-Logging Approach

In this section, we present the basic concepts of LB-Logging approach for flash-based databases that we propose to address the problems of the conventional logging designs for disk-based databases. We will introduce LB-Logging's log structures, recovery process and check-point strategy to give a detailed explaining of its principles and advantages.

4.1 Logging Algorithm

From the previous analysis, it can be found that in flash memory, there is no need to use explicit rollback operation to re-write the original data elements in recovery as is done in disk-based databases. Taking into account that the history versions of data exist, we can make full use of it for rollback and recovery. It would speed up the process without executing write operations which are more expensive. LB-Logging uses list structures to store log records. In this sense, LB-logging is a redo logging. It maintains a chain across different log records for the sequence of data operations logged in each transaction and a list of different versions of each data element in one log record. So during the recovery process, we can search all the log records of every operation in each transaction.

Table 2. Log File Structure of LB-logging

T_Id	Element	Pre_Element	Address_List
T_1	X	Begin	P(X1) →P(X2) →P(X3)
T_1	Y	P(X)	P(Y1) →P(Y2)
T_2	A	Begin	P(A1)
T_1	Z	P(Y)	P(Z1) →P(Z2) →NULL
T_2	B	P(A)	NULL →P(B1)
T_1	Commit	P(Z)	NULL
T_2	Rollback	P(B)	NULL
...	...	...	...

Information stored in each log record includes transaction ID (*T_Id*), modified data item name (*Element*), former element updated by this transaction (*Pre_Element*), and the list of all versions of current data item (*Address_List*). The list is arranged by the order of operation time. The log file structure is shown in Table 3.

When a transaction starts, LB-Logging creates a log record, but wait until the first database changed by the transaction, it will insert a log record for the first updated element with *Pre_Element* Field marked with *Begin*. As the transaction updates the data subsequently, we insert new log records and the list of addresses of data elements needed to be maintained. For insert and delete operations, several identifiers are used to distinguish them in *Address_List*. For example, if an element's history list ends with a *NULL* address, that means this element is deleted from the database. Similarly, if the first address of an element's history version is *NULL*, that means this element is inserted by this transaction. If the transaction is committed or rolled back, we will insert a log record for commit or rollback operation. The *Pre_Element* field stores the former updated item and the *Address_List* field is set to empty. Thus, using this structure of log records, all the operations to the database can be recorded completely.

We follow WAL rule to decide when to flush log records, which means as long as the local database is modified, there must be associated log records. But the opposite is not true. In other words, there may be cases that when system crashes, changes are only logged in stable storage, but these changes may not be propagated to the database. When we need to execute a recovery operation, according to the database redo log files, the above mechanisms can ensure the consistency and integrity of the database.

Algorithm 1. Recovery algorithm of LB-Logging

```
procedure REDOLOGFILE(file logFile)
    BOOL flag = true;
    logRecord current;
    current = getFirstLogRecord(logFile);
    while flag do
        if current != checkpointLogRecord then
            if current == insertLogRecord then
                delete data;
            end if
            if current == deleteLogRecord then
                insert data;
            end if
            if current == updateLogRecord then
                reupdate data;
            end if
        else
            flag = false;
        end if
    end while
    return 1;
end procedure
```

4.2   Recovery Process

In LB-Logging, what we mainly do in recovery is redo the database operations in committed transactions according to the redo log files. Recovery manager finds all the committed transaction log records from the list structure of log files, and redo them one by one.

When a transaction which has not been submitted is rolled back, we just need to delete the log records simply. Because the corresponding logs are still in memory, and the data is not flushed out to the external storage media. However, if the application has a lot of long transactions, there may be some log records which haven't been committed. These log records may take up a large proportion of the memory space. Part of the logs may be forced to flush out of the memory. This is acceptable. Because we can determine that the transaction's changes do not reach the flash memory database since the transaction has not been committed. So we just need to insert a roll back log to ensure that the operations has no effect on the database.

If the system crashes, it needs to recover immediately. We have to read the log file from secondary storage media to memory. Typically, if a system crashes when writing logs. There are usually some uncommitted log records at the end of the log file. We do not do anything about these operations. Since LB-logging ensures that as long as there is no commit log records, the changes have not been flushed to the database yet. For the transactions that have been submitted, we need to read the log records and redo them one by one. The detailed procedure is shown in Algorithm 1.

5   Discussion

In this section, we present some improvement of LB-Logging approach that we propose to overcome the problems of basic LB-Logging approach to provide an much better performance.

5.1   Checkpoint Policy

We can notice that log file is frequently updated. With the updated data being propagated to storage, a large number of log records becomes useless. Under normal cases, transaction rollback rate is usually not too high. Therefore, if there are a large number of useless log records, the length of log file will be unnecessarily increased, thus taking too much flash storage space. Aside from that, we need to read logs during recovery, the long log file will harm the efficiency of recovery. This inspires us to establish check points to avoid an overly long log file that would affect the overall performance of the system.

We take a simple checkpoint policy for flash-based DBMS. We only transfer the log records that are still valid. In other words, when set up one check point, we will do as follows. Firstly, find a clean block. Then, check the validity of each log record one by one. Finally, select the records which are still valid and write them to the new free block. After all the log records of the old log block are scanned, the original records on the log block are no longer useful to us. Therefore, we can erase the old block. The actual amount of log records that need to be transferred is quite small. So the transfer cost is acceptable. The specific steps of this transfer operation are shown in Algorithm 2.

Algorithm 2. Checkpoint algorithm of LB-Logging

```
procedure TRANSFER(file logFile)
    bool flag = true;
    block newBlock = new block();
    logRecord current;
    current = getFirstLogRecord(logFile);
    while flag do
        if current is still valid then
            Copy current to newBlock;
        end if
        current = getNextLogRecord(logFile);
        if newBlock is full then
            Get another free block;
        end if
        if current==NULL then
            flag = false;
        end if
    end while
    Erase the old log file;
end procedure
```

The value of checkpoint interval requires to be examined. Long interval will bring to large log file. And short checkpoint interval may shorten the lifetime of flash memory and affect overall performance. So the checkpoint interval can be variably set according to applications' characteristics. Here we assume that features of the application is very clear. And database administrator can choose to do checkpoints off line or in non-peak time to minimize impact on the performance of the database application.

5.2   Heterogeneous Storage

The main operations of log files include logging, recycling the invalid log records and reading log records to recover the system. So the most frequent operations for log files are small random write and erase operations. As we introduced before, small sized write operations are the biggest limitation of flash memory whose performance may be worse than those of magnetic disks. The other critical technical constraints of flash memory is limited erase cycles. Thus too many unnecessary erase operations will greatly shorten the service life of flash memory. Therefore, log file is not suitable for flash memory.

Current databases generally support log files and data files to be stored separately. Here we use hybrid storage systems to store different types of database objects. As shown in Fig. 1, data records are stored in flash disks, while log records are stored in magnetic disk. In this design, we can restore the system without increasing the complexity of the algorithm and save the space of flash memory occupied by log records. Thus it improves the flash space utilization,

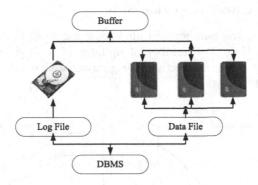

Fig. 1. Heterogeneous Storage of LB-Logging

reduces the cost to build the database system, and enhances the overall performance of database system.

6   Performance Evaluations

In this section, we present real system experimental evaluations of LB-Logging. We first describe the experimental setup in section 6.1. Then we present experimental results with different update transaction size and different update times for each data in section 6.2 and section6.3, respectively.

6.1   Experimental Setup

We examine the performance of LB-logging compared with ARIES and HV-Logging[15]. The latter one only makes use of the history versions of data without linked structures. We implement both our approaches and the comparable methods in a real database called Oracle Berkeley DB[16]. And our experiments run on two platforms which are exactly identical except that one is equipped with an HDD and the other with an SSD. We used two HP Compaq 6000 Pro MT PC. Each machine is equipped with an Intel(R) Core(TM) 2 Quad Q8400 @ 2.66GHz 2.67GHz CPU, 4GB DRAM, running Windows 7 Professional. The SSD we use is Intel SSDSA2MH080G1GC 80G, and the HDD we use is 250G 7200rpm ST3250310AS with 8MB cache.

Our experimental process is as follows. A transaction starts doing all the required updates. Before the transaction's commission, we roll back the transaction. The performance is measured by recovery time which we carefully records. However, when we try to record the elapsed time of recovery process, there are some little differences between multiple runs for one workload. Here we use the average value. Through the description of our algorithm above, we can find that two key factors are important to the system. They are update transaction size and average update times for each data. So we varied the two critical parameters in our experiments to see how the performance changes.

6.2   Varying Update Transaction Size

In recovery process, the number of rolled back log records directly influences the recovery time. Here, the number of updates of a single transaction is an important parameter. In this section, we guarantee that all data individually have 2 times update in average.

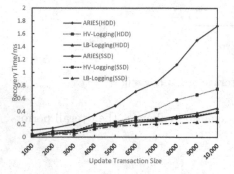 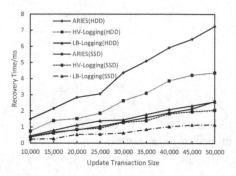

Fig. 2. Recovery time with different update transaction size (small amount)

Fig. 3. Recovery time with different update transaction size (big amount)

Fig. 2 and Fig. 3 show our experimental results. Fig. 2 describes the result when the update transaction size is small. Fig. 3 shows the result when increasing update transactions' size. From the experimental result, we can find that the advantage of SSD over HDD is obvious. Whether it is the traditional logging method, or the improved logging method for flash memory, the recovery time on SSD is much less than that on disk. We also observe that whether on SSD or HDD, the recovery efficiency is much higher for LB-Logging. Compared with HV-Logging, LB-Logging costs only 70% recovery time. With the increasing amount of data, the advantage of LB-Logging is even more obvious. This could fully reflect the superiority of our algorithm.

6.3   Varying Update Times for Each Data

In this section, we discuss how the variation of update times for each data reflects the recovery performance. As previously described, in order to employ flash memory's high-speed random read advantage, LB-Logging uses a kind of linked structure for log records. In the linked structure, the list length is a key factor. In our design, the length of the list depends on how often each data gets updated.

Fig. 4 shows the experimental results. When the updates are less frequent, it is difficult to reflect the superiority of our design. With the increasing times of updates per data, LB-Logging's advantage is highlighted increasingly. We find that when the update frequency is increased to 8 times, disk-based recovery time is about 15 times longer than that of LB-Logging. These results sufficiently justify the LB-Logging's superiority over other logging schemes.

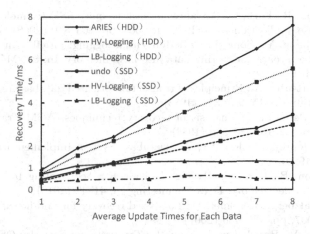

Fig. 4. Recovery time with different update times for each data

7   Conclusion

LB-Logging makes use of the data's history versions which naturally exists in flash-based database for logging and exploits a kind of list structure as the replace of sequential structure to store log records to provide efficient recovery. Through periodic checkpoints mechanism, LB-Logging reduces the length of the log file by removing invalid log records and saves the space for flash memory. Using hybrid storage system, LB-Logging stores the log records in both SSD and HDD separately to improve the recovery performance on flash-based database. So the proposed algorithm LB-Logging provides stronger reliability, faster recovery, smaller space consumption for log files.

The results show that the recovery time of traditional logging algorithm is 15 times longer than LB-Logging in the best conditions. And the recovery time of traditional logging algorithm on SSD is 7 times longer than LB-Logging. The recovery time of optimized algorithm for flash memory is 5 times longer than LB-Logging. This fully demonstrates the superiority of LB-Logging.

Acknowledgements. This research was partially supported by the grants from the Natural Science Foundation of China (No, 60833005, 91024032, 91124001, 61070055,); the Research Funds of Renmin University of China (No: 11XNL010, 10XNI018); National Science and Technology Major Project (No: 2010ZX01042-002-003).

References

1. Jim, G.: Tape is dead disk is tape flash is disk RAM locality is king. In: Pacific Grove: Microsoft, Gong Show Presentation at Third Biennial Conference on Innovative Data Systems Research, vol. 1 (2007)

2. Lee, S., Moon, B., Park, C.: Advances in flash memory SSD technology for enterprise database applications. In: SIGMOD 2009, pp. 863–870 (2009)
3. Kim, Y., Whang, Y., Song, I.: Page-differential logging: an efficient and DBMS-independent approach for storing data into flash memory. In: SIGMOG 2010, pp. 363–374 (2010)
4. Haerder, T., Reuter, A.: Principles of transaction-oriented database recovery. ACM Computing Surveys 15, 287–317 (1983)
5. Reuter, A.: Performance analysis of recovery techniques. ACM Transactions on Database Systems 15, 526–559 (1984)
6. Hector, G., Jeffrey, D., Jennifer, W.: Database System Implementation. Prentice Hall, USA (1999)
7. Lee, S., Moon, B., Park, C., Hwang, J., Kim, K.: Accelerating In-Page Logging with Non-Volatile Memory. Data Engineering 33, 41–47 (2010)
8. Wang, R., Salzberg, B., Lomet, D.: Log-based recovery for middleware servers. In: SIGMOD 2007, pp. 425–436 (2007)
9. Prabhakaran, V., Rodeheffer, T., Zhou, L.: Transactional flash. In: OSDI 2008, pp. 147–160 (2008)
10. On, S.T., Xu, J., Choi, B., Hu, H., He, B.: Flag Commit: Supporting Efficient Transaction Recovery in Flash-based DBMSs. TKDE 99, 1–1 (2011)
11. Lee, S., Moon, B.: Design of flash-based DBMS: an in-page logging approach. In: SIGMOD 2007, pp. 55–66 (2007)
12. Nath, S., Kansal, A.: FlashDB: dynamic self-tuning database for nand flash. In: IPSN, pp. 410–419 (2007)
13. Elnozahy, E., Alvisi, L., Wang, Y., Johnson, D.: A survey of rollback-recovery protocols in message-passing systems. ACM Computer Survey 34(3), 375–408 (2002)
14. Chen, S.: FlashLogging: exploiting flash devices for synchronous logging performance. In: SIGMOD 2009, pp. 73–86 (2009)
15. Lu, Z., Meng, X., Zhou, D.: HV-Recovery: A High Efficient Recovery Techniques for Flash-Based Database. Chinese Journal of Computers 12, 2258–2266 (2010)
16. Oracle Berkeley DB, http://www.oracle.com/technetwork/database/berkeleydb/overview/index.html

An Under-Sampling Approach to Imbalanced Automatic Keyphrase Extraction

Weijian Ni, Tong Liu, and Qingtian Zeng*

Shandong University of Science and Technology
Qingdao, Shandong Province, 266510 P.R. China
niweijian@gmail.com, liu_tongtong@foxmail.com,
qtzeng@163.com

Abstract. The task of automatic keyphrase extraction is usually formalized as a supervised learning problem and various learning algorithms have been utilized. However, most of the existing approaches make the assumption that the samples are uniformly distributed between positive (keyphrase) and negative (non-keyphrase) classes which may not be hold in real keyphrase extraction settings. In this paper, we investigate the problem of supervised keyphrase extraction considering a more common case where the candidate phrases are highly imbalanced distributed between classes. Motivated by the observation that the saliency of a candidate phrase can be described from the perspectives of both morphology and occurrence, a multi-view under-sampling approach, named co-sampling, is proposed. In co-sampling, two classifiers are learned separately using two disjoint sets of features and the redundant candidate phrases reliably predicted by one classifier is removed from the training set of the peer classifier. Through the iterative and interactive under-sampling process, useless samples are continuously identified and removed while the performance of the classifier is boosted. Experimental results show that co-sampling outperforms several existing under-sampling approaches on the keyphrase exaction dataset.

Keywords: Keyphrase Extraction, Imbalanced Classification, Under-sampling, Multi-view Learning.

1 Introduction

Keyphrases in a document are often regarded as a high-level summary of the document. It not only helps the readers quickly capture the main topics of a document, but also plays an essential role in a variety of natural language processing tasks such as digital library [1], document retrieval [2] and content based advertisement [3]. Since only a minority of documents have manually assigned keyphrases, there is great need to extract keyphrases from documents automatically. Recently, several automatic keyphrase extraction approaches have been proposed, most of them leveraging supervised learning techniques [4] [5]. In these

---

* Corresponding author.

H. Gao et al. (Eds.): WAIM 2012, LNCS 7418, pp. 387–398, 2012.
© Springer-Verlag Berlin Heidelberg 2012

approaches, the task of automatic keyphrase extraction is basically formalized as a binary classification problem where a set of documents with manually labeled keyphrases are used as training set and a classifier is learned to distinguish keyphrases from all the candidate phrases in a given document.

One of the general assumptions made by traditional supervised learning algorithms is the balance between distributions of different classes. However, as we have observed, the assumption may not be hold in the real settings of automatic keyphrase extraction. The more common case is that the number of keyphrases (positive samples) is much fewer than that of non-keyphrases (negative samples) appearing in the same document. Taking a document of length n as an example, the number of possible phrases of lengths between p to q ($p \leq q$) would amount to $O((q - p + 1) \cdot n)$, while the number of keyphrases in a document is often less than ten. It has since been proven that the effectiveness of most traditional learning algorithm would be compromised by the imbalanced class distribution [6], we argue that the performance of automatic keyphrase extraction could be promoted by exploring the characteristics of imbalanced class distribution explicitly. However, to the best of our knowledge, the issue of class-imbalance in automatic keyphrase extraction has not been well studied in the literature.

In this paper, we adopt under-sampling mechanism to deal with the imbalanced data in supervised automatic keyphrase extraction. Particularly, the useless samples in training set are removed and thus traditional supervised learning approaches could be utilized more efficiently and effectively.

One of the reasons accounts for the successes of supervised keyphrase extraction approaches is that the saliency of each candidate phrase could be described using various features. In general, these features are calculated based on either the morphology (e.g., *phrase length*, *part-of-speech tag*) or the occurrence (e.g., *first occurrence*, *PageRank value*) of candidates. In another word, the dataset in supervised keyphrase extraction are comprised of two views. Inspired by the advantageous of multi-view learning approaches, we propose a novel under-sampling approach, named **co-sampling**, which aims to exploit the multiple views in the task of keyphrase extraction to remove the useless samples.

The rest of the paper is organized as follows. After a brief overview of related work in Section 2, we present the details of co-sampling algorithm in Section 3. Section 4 reports experimental results on a keyphrase extraction dataset. Finally, we conclude the paper and discuss the further work of co-sampling in Section 5.

2   Related Work

2.1   Keyphrase Extraction

Generally, automatic keyphrase extraction approaches can be categorized into two types: supervised and unsupervised.

In most supervised approaches, the task of keyphrase extraction is formulated as a classification problem. The accuracy of extracting results relies heavily on the features describing the saliency of candidate phrases. TF×IDF and first occurrence of candidate phrase are the two features used in early work [4]. Besides, the features such as part-of-speech tag pattern, length and frequency of candidate phrase have shown their benefits to recognize keyphrases in a given document [8]. Recently, much work has been conducted on extracting keyphrases from particular types of documents including scientific literatures [9], social snippets [10], web pages [11] and etc. One of the keys in the work is to calculate domain-specific features that capture the special salient characteristics of keyphrase in the particular type of documents. For example, section occurrences and acronym status play a critical role in finding keyphrases in scientific publications [9].

The basic idea of most unsupervised approaches is to leverage graph-based ranking techniques like PageRank [12] and HITS [13] to give a rank of all the candidate phrases. In general, the ranking scores are computed via random walk over co-occurrence graph of a given document. Recent extensions of unsupervised approaches mainly focus on building multiple co-occurrence graphs to reflect the characteristics of various keyphrase extraction settings. Wan et al. [14] built a global affinity graph on documents within a cluster in order to make use of mutual influences between documents. Liu et al. [15] took the semantic topics of document into account and built co-occurrence graph with respect to each topics.

2.2   Imbalanced Classification

Imbalanced class distribution is a common phenomenon in many real machine learning applications. With imbalanced data, the classifiers can be easily overwhelmed by the majority class and thus ignore the minority but valuable one.

A straightforward but effective way to handle the imbalanced data is to re-balance the class distribution through sampling techniques, including removing a subset of samples from the majority class and inserting additional artificial samples in the minority class, which are referred to as under-sampling and over-sampling, respectively. EasyEnsemble and BalanceCascade [16] are the two typical under-sampling approaches, which learn an ensemble of classifiers to select and remove the useless majority samples. SMOTE is an example of over-sampling approaches [17]. The algorithm randomly selects a point along the line joining a minority sample and one of its k nearest neighbors as a synthetic sample and adds it into the minority class. For the recent extensions of SMOTE, see [18], [19].

Cost-sensitive learning is another solutions to the problem of imbalanced classification. The basic idea is to define a cost matrix to quantify the penalties of mis-classifying samples from one class to another. For most traditional learning algorithms, the cost-sensitive versions have been proposed for imbalanced classification. For example, Yang et al. proposed three cost-sensitive boosting algorithms named AdaC1, AdaC2 and AdaC3 [20], Zhou et al. studied empirically the effect of sampling and threshold-moving strategy in training cost-sensitive neural networks [21].

Different from most existing work focusing on either sampling technique or cost-sensitive learning, there have been several proposed imbalance classification approaches combining the above two types of mechanisms [22].

3   Under-Sampling for Keyphrase Extraction

3.1   Problem Formulations and Algorithm Sketch

Let $\mathcal{X} \subseteq \mathbf{R}^d$ denote the input feature space of all possible candidate phrases and $\mathcal{Y} = \{+1, -1\}$ denote the output space. Because of the two-view characteristic of keyphrase extraction, the input space $\mathcal{X}$ can be written as $\mathcal{X} = \mathcal{X}_1 \times \mathcal{X}_2$, where $\mathcal{X}_1$ and $\mathcal{X}_2$ correspond to the morphology and the occurrence view of candidate phrases, respectively. That is, the feature vector of each candidate phrase $\mathbf{x}$ can be denoted as $\mathbf{x} = (\mathbf{x}^1, \mathbf{x}^2)$.

Given a set of training samples $S = P \cup N$ where $P = \{((\mathbf{x}_i^1, \mathbf{x}_i^2), +1) \mid i = 1, \cdots, m\}$ and $N = \{((\mathbf{x}_i^1, \mathbf{x}_i^2), -1) \mid i = 1, \cdots, n\}$ (in imbalanced classification problems, $m \ll n$), the goal of co-sampling is to boost the performances of the classifiers through excluding redundant negative samples from training process.

As a co-training [7] style algorithm, co-sampling works in an iterative manner as shown in Table 1. During the iterations, redundant negative samples are removed from the training set of a classifier according to the predicted results of the classifier learned on another view. Figure 1 gives an illustration of the procedure of co-sampling. The iterative process stops when one of the following criteria is met:

1. The number of iterations exceeds a predefined maximum number.
2. The confidences of the predictions of either classifier on any negative samples is below a predefined threshold, i.e., no reliably predicted negative samples can be found.
3. The redundancies of any reliably predicted negative samples is below a predefined threshold, i.e., no redundant samples can be found.

At the end of co-sampling, the two classifiers f_1 and f_2 learned on the under-sampled training sets are combined to give prediction for a new sample $\mathbf{x} = (\mathbf{x}^1, \mathbf{x}^2)$. In particular, the final output is calculated as follows:

$$f(\mathbf{x}) = \begin{cases} f_1(\mathbf{x}^1), & |P(+1|f_1(\mathbf{x}^1)) - 0.5| > |P(+1|f_2(\mathbf{x}^2)) - 0.5| \\ f_2(\mathbf{x}^2), & \text{Otherwise} \end{cases}$$

where $P(+1|f_i(\mathbf{x}^i))$ is the posterior probability of $\mathbf{x}$ to be a positive sample based on the prediction result $f_i(\mathbf{x}^i)$ $(i = 1, 2)$.

During the iterations, we employ an online learning algorithm, i.e., *Perceptron with uneven margins* [23], to learn the classifiers on each view for its easy implementation, efficient training and theoretical soundness.

The key issue of co-sampling is to select the appropriate negative samples to be removed. To address the problem, we adopt a two-stage method. The first stage is to estimate the confidence of either classifier's predictions on the negative samples and take these reliably predicted ones as the candidates for further removal. The second stage is to quantify the redundancy of each candidate samples and remove these most redundant ones. The above two stages will be introduced in Section 3.2 and 3.3 detailedly, respectively.

Table 1. The co-sampling algorithm

Co-sampling

Input:
  a set of positive samples represented by two views: P^1 and P^2
  a set of negative samples represented by two views: N^1 and N^2
1:**repeat**
2: **for** $i = 1$ **to** 2
3:    Learn classifier f_i on $P^i \cup N^i$ and get the predicted
      results of f_i, through performing 10-fold cross validation.
4:    Estimate the confidences of predicted results of f_i.
5:    Select redundant samples N'_i from these reliably predicted
      negative samples of f_i.
6:    $N_{3-i} \leftarrow N_{3-i} - N'_i$
7: **end for**
8:**until** some stopping criterion is met
9:$f(\mathbf{x}) = f_1(\mathbf{x}^1) \oplus f_2(\mathbf{x}^2)$
Output: $f(\mathbf{x})$

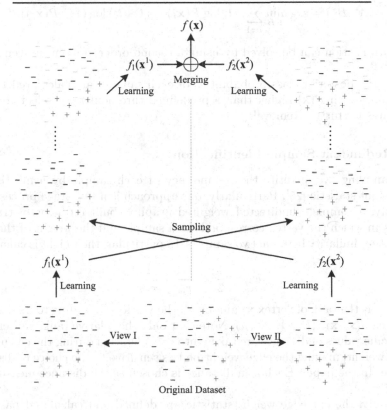

Fig. 1. Illustration of the procedure of co-sampling. The plus and minus signs denote the positive and negative samples, respectively; the gray markers denote the samples have been removed by the end of each iteration.

3.2   Prediction Confidence Estimation

Intuitively, the sample with high posterior class probability $P(y = -1|f(\mathbf{x}))$ can be viewed as the reliably predicted negative sample. We thus leverage posterior class probability to estimate the confidence of either classifier's prediction on each negative sample. As the perceptron algorithm outputs a linear prediction function $f(\mathbf{x}) = \langle \mathbf{w}, \mathbf{x} \rangle + b$, we need to map the real valued outputs to probabilities.

Following [24], the posterior class probability of each predicted negative sample is derived through fitting a sigmoid function:

$$\bar{P}(\mathbf{x}) \triangleq P(y = -1|f(\mathbf{x})) = \frac{1}{1 + \exp(A \cdot f(\mathbf{x}) + B)} \tag{1}$$

To find the best parameters A and B, we generate a set of training samples $T = \{(f(\mathbf{x}_i), t_i) \,|\, t_i = \frac{1-y_i}{2}, \, i = 1, \cdots, n\}$ and minimize the Kullback-Leibler divergence between $\bar{P}$ and its empirical approximation derived from T, i.e.,

$$(A^*, B^*) = \arg\min_{A,B} \sum_{i=1}^{n} -t_i \log \bar{P}(\mathbf{x}_i) - (1 - t_i) \log (1 - \bar{P}(\mathbf{x}_i))$$

This optimization can be solved by using a second order gradient descent algorithm [24].

After the posterior class probabilities of each samples are calculated by (1), the samples with $\bar{P}(\mathbf{x})$ higher than a predefined threshold τ are selected as the candidates for further removal.

3.3   Redundant Sample Identification

In co-sampling, we quantify the redundancy of each sample by using the cut edge weight statistic [25]. Particularly, our approach is a two-stage process.

Firstly, a complete undirected weighted graph is built on a set of training samples in which the vertex corresponds to a sample and the weight of the edge reflects the similarity between two samples. In particular, the weight is calculated as:

$$w_{ij} = \frac{1}{2} \times \left(\frac{1}{r_{ij}} + \frac{1}{r_{ji}} \right)$$

where r_{ij} is the rank of vertex $\mathbf{x}_j$ among all the vertices according to its distance to $\mathbf{x}_i$, i.e., $d(\mathbf{x}_i, \mathbf{x}_k) \, (k = 1, \cdots, n)$. Note that only the relative ranks rather than the specific values of distance are taken into consideration during the calculation of edge weight, making the edge weight be less sensitive to the choice of distance measure. In the paper, Euclidean distance is chosen as the distance measure for simplicity.

Secondly, the cut edge weight statistic are defined and calculated based on the graph. Intuitively, a sample is supposed to be less informative for training if it appears to be easily distinguished from the samples of other classes. As each sample is represented as a vertex in graph, this implies that a redundant sample would be the one whose sum of the weights of the edges linking to the vertices of the same class is significantly larger than that of the edges linking to the vertices of different classes. The edges of the latter type are often referred to as cut edges. In order to identify the redundant samples, we define a null hypothesis H_0 as that the classes of the vertices of the graph are drawn independently from the probability distribution $P(Y = k)$ $(k = -1, +1)$. Usually, $P(Y = k)$ is estimated empirically as the proportion of the class k in the training set. Then, under H_0, the cut edge statistic of a sample $(\mathbf{x}_i, y_i)$ is defined as:

$$J_i = \sum_{j=1}^{n} w_{ij} I_{ij}$$

where n is the number of training samples and I_{ij} is an independent and identically distributed random variables drawn from a Bernoulli distribution, i.e.,

$$I_{ij} = \begin{cases} 1, & \text{if } y_i \neq y_j \\ 0, & \text{otherwise} \end{cases}$$

Accordingly, $P(I_{ij} = 1) = 1 - P(Y = y_i)$.

According to the de Moivre-Laplace theorem, we can derive that the distribution of the cut edge statistic J_i is approximately a normal distribution with mean $\mu_{i|H_0}$ and variance $\sigma^2_{i|H_0}$ if n is large enough:

$$\mu_{i|H_0} = (1 - P(Y = y_i)) \sum_{i=1}^{n} w_{ij}$$

$$\sigma^2_{i|H_0} = P(Y = y_i)(1 - P(Y = y_i)) \sum_{i=1}^{n} w_{ij}^2$$

In another words,

$$\bar{J}_i \triangleq \frac{J_i - \mu_{i|H_0}}{\sigma_{i|H_0}} \sim \mathcal{N}(0, 1)$$

As for a negative sample, it would be supposed to be redundant if its value of the cut edge statistic is significantly smaller than the expected under H_0. Consequently, the redundancy of a negative sample $\mathbf{x}_i$ is quantified by using the right unilateral p-value of $\bar{J}_i$, i.e.,

$$r(\mathbf{x}_i) = \frac{1}{\sqrt{2\pi}} \int_{\bar{J}_i}^{+\infty} e^{-\frac{1}{2}t^2}\, dt \qquad (2)$$

Finally, the candidate negative samples selected from Section 3.2 with $r(\mathbf{x})$ higher than a predefined threshold θ will be excluded from the next co-sampling iterations of the peer classifier.

3.4   Two Views for Keyphrase Extraction

In the section, we describe the partition of feature set to generate the two-view representations of the task of supervised keyphrase extraction.

In general, each candidate phrase has views of morphology and occurrence. For example, *ending with a noun* is a feature in the morphology view of candidate phrases while *appearing in title* is a feature in the occurrence view. Basically, the co-training style algorithms require that each view is sufficient for learning a strong classifier and the views are conditionally independent to each other given the classes. However, the conditions are so strong that it hardly holds in most real applications. Therefore, we partition the features describing the saliency of candidate phrases into two disjoint sets according to whether the feature carries either morphology or occurrence information about the phrase, together with practical concerns about the above sufficient and redundant conditions.

Particularly, the morphology view consists of the following features:

1. *Length.* That is, the number of words in a candidate phrase.
2. *Part of speech tags.* The type of features include a number of binary features indicating whether the phrase starts with/ends with/contains a noun/ adjective/verb. Besides, the part-of-speech tag sequence of the candidate is used as a feature.
3. *TFIDF.* TF, IDF and TF×IDF of every words in the candidate phrase are calculated and the average/minimum/maximum of each set of values are used as the features.
4. *suffix sequence.* The suffices like *-ment* and *-ion* of each word are extracted, then the sequence of the suffixes of the candidate is used as a feature.
5. *Acronym form.* Whether the candidate is an acronym may be a good indicator of keyphrase. In order to identify the acronyms, we employ the approach proposed in [26].

The occurrence view consists of the following features:

1. *First occurrence.* The feature is calculated as the number of words between the start of the document and the first appearance of the candidate, normalized by the document length.
2. *Occurrence among sections.* The type of features include a number of binary features indicating whether the candidates appear in a specific logical section. As for scientific literatures, the sections include: *Title, Abstract, Introduction/Motivation, Related Work/Background, Approaches, Experiments/ Evaluation/Applications, Conclusion, References* and etc.
3. *PageRank values.* As in [12], the PageRank values of every words in the phrase are calculated and the average/minimum/maximum of the values are used as features.

4   Experiments

4.1   Dataset

To avoid manually annotation of keyphrases which is often laborious and erroneous, we constructed an evaluation dataset using research articles with author provided keyphrases. Specifically, we collected the full-text papers published in the proceedings of two conferences, named ACM SIGIR and SIGKDD, from 2006 to 2010. After removing the papers without author provided keyphrases, there are totally 3461 keyphrases appear in 997 papers in our evaluation dataset. For each paper, tokenization, pos tagging, stemming and chunking were performed using NLTK (Natural Language Toolkit)[1]. We observed that the keyphrases make up only 0.31% of the total phrases in the dataset, which practically confirms that there exists the problem of extreme class-imbalance in the task of supervised keyphrase extraction.

4.2   Baselines

Since co-sampling is a multi-view supervised keyphrase extraction approach, several supervised approaches that make use of single view consisting of all the features referred in Section 3.4 are used as the baselines.

The first baseline is the supervised keyphrase extraction approach that learns the classifier using SVM on the original dataset without sampling. One of the existing under-sampling approaches, named EasyEnsemble [16], are taken as the second baseline. Besides, we also compare co-sampling with random under-sampling approach. In the random under-sampling approaches, a balanced training set is generated by sampling a subset of negative samples randomly according to a class-imbalance ratio defined as:

$$\alpha = \frac{|N'|}{|P|}$$

where $|N'|$ and $|P|$ are the size of sampled negative set and original positive set, respectively. The class-imbalance ratio is set to 50, 20, 10, 5 and 1 experimentally. SVM is employed to learn the classifiers on the sampled training set.

4.3   Evaluation Measures

The traditional metrics namely *Precision, Recall* and *F1-score*, are use to evaluate our approach and all the baselines.

4.4   Experimental Results

For co-sampling, the parameters τ and θ are tuned using an independent validation set randomly sampled from the total dataset and the result with highest F1-score are reported. For the baseline EasyEnsemble, the parameter T (the number of subsets to be sampled from negative set) is set to 4 as in [16]. For all the baselines, the learner, i.e. SVM, is implemented using SVM^{light}[2] toolkit. As for the slack parameter C in SVM, the default values given by SVM^{light} are used.

---

[1] www.nltk.org
[2] svmlight.joachims.org

Table 2. Performances of co-sampling and baselines

Methods	Precision	Recall	F1-score
Co-sampling	**0.2797**	**0.4803**	**0.3535**
EasyEnsemble	0.2644	0.3030	0.2824
Random($\alpha = 1$)	0.2101	0.2242	0.2169
Random($\alpha = 5$)	0.2270	0.2360	0.2314
Random($\alpha = 10$)	0.2081	0.2163	0.2121
Random($\alpha = 20$)	0.2510	0.1769	0.2076
Random($\alpha = 50$)	0.2514	0.1375	0.1778
SVM	NIL	0.0000	NIL

Comparison of Keyphrase Extraction Accuracies. Table 2 shows the performances of co-sampling and baselines which are averaged over 10 fold cross validation. We can see that co-sampling outperforms all the baselines in terms of Precision, Recall and F1-score. We conducted paired t-test over the results of the 10 folds and found the improvements of co-sampling over all the baselines (except the improvement over EasyEnsemble in terms of Precision) are significant at the 0.01 level. Moreover, the baseline employing SVM on the original dataset without sampling is failed to give a nontrivial classifier because all the candidate phrases are classified as non-kephrase, which gives an example of class-imbalance problem on traditional learning algorithms.

Training Performance versus Number of Iterations. We investigated how the performances of each classifiers learned on different views varies while the training set is under-sampled continuously during the iteration process. Particularly, we measured the training accuracy of each classifier by calculating F1-score of the predicted results obtained through performing 10-fold cross validation as

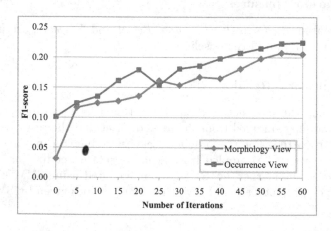

Fig. 2. Training Performance versus Number of Iterations

in Table 1. Figure 2 gives the plots of training accuracies versus number of iterations. We can see that the classifier learned on either the morphology or the occurrence views can be boosted into a "stronger" classifier during the co-sampling process. This gives an evidence that the two classifiers are capable of mutually reinforcing through removing the reliably predicted redundant samples for each other.

5   Conclusion and Future Work

In this paper, we have argued that it is more essential to formalize the task of supervised keyphrase extraction as an imbalanced classification problem and have proposed a novel under-sampling approach, named co-sampling, to tackle the class-imbalance problem. Co-sampling is by nature a multi-view learning algorithm in which the keyphrase extraction dataset is partitioned into the morphology and the occurrence views. Experimental results on a keyphrase extraction dataset verified the advantages of co-sampling.

In the future, we will evaluate co-sampling experimentally by using more imbalanced multi-view datasets from various domains. Furthermore, it is necessary to give an in-depth theoretical analysis of co-sampling as has been done for co-training style algorithms.

Acknowledgments. This paper is supported partly by Chinese National Natural Science Foundation (61170079); Shandong Province Higher Educational Science and Technology Program (J12LN45); Key Research Program of Statistics Science of Shandong Province (KT11017); Research Project of "SDUST Spring Bud" (2010AZZ179); Sci. & Tech. Development Fund of Shandong Province (2010GSF10811); Sci. & Tech. Development Fund of Qingdao(10-3-3-32-nsh); Excellent Young Scientist Foundation of Shandong Province (BS2010DX009 and 2010KYJQ101); China Postdoctoral Science Foundation (2011M501155).

References

1. Song, M., Song, I.Y., Allen, R.B., Obradovic, Z.: Keyphrase extraction-based query expansion in digital libraries. In: Proceedings of the 6th ACM/IEEE-CS JCDL, pp. 202–209 (2006)
2. Lehtonen, M., Doucet, A.: Enhancing Keyword Search with a Keyphrase Index. In: Geva, S., Kamps, J., Trotman, A. (eds.) INEX 2008. LNCS, vol. 5631, pp. 65–70. Springer, Heidelberg (2009)
3. Wu, X., Bolivar, A.: Keyword extraction for contextual advertisement. In: Proceedings of the 17th WWW, pp. 1195–1196 (2008)
4. Witten, I.H., Paynter, G.W., Frank, E., Gutwin, C., Nevill-Manning, C.G.: KEA: Practical Automatic Keyphrase Extraction. In: Proceedings of the 4th ACDL, pp. 254–255 (1999)
5. Turney, P.D.: Learning Algorithms for Keyphrase Extraction. Information Retrieval 2, 303–336 (2000)
6. Weiss, G.M., Provost, F.: The Effect of Class Distribution on Classifier Learning: An Empirical Study. Technical Report, Department of Computer Science, Rutgers University (2001)

7. Blum, A., Mitchell, T.: Combining labeled and unlabeled data with co-training. In: Proceedings of the 11th COLT, pp. 92–100 (1998)
8. Turney, P.D.: Learning Algorithms for Keyphrase Extraction. Information Retrieval 2, 303–336 (2000)
9. Nguyen, T.D., Kan, M.-Y.: Keyphrase Extraction in Scientific Publications. In: Goh, D.H.-L., Cao, T.H., Sølvberg, I.T., Rasmussen, E. (eds.) ICADL 2007. LNCS, vol. 4822, pp. 317–326. Springer, Heidelberg (2007)
10. Li, Z., Zhou, D., Juan, Y., Han, J.: Keyword Extraction for Social Snippets. In: Proceedings of the 19th WWW, pp. 1143–1144 (2010)
11. Yih, W., Goodman, J., Carvalho, V.R.: Finding Advertising Keywords on Web Pages. In: Proceedings of the 15th WWW, pp. 213–222 (2006)
12. Mihalcea, R., Tarau, P.: TextRank: Bringing Order into Texts. In: Proceedings of the 1st EMNLP, pp. 404–411 (2004)
13. Litvak, M., Last, M.: Graph-Based Keyword Extraction for Single-Document Summarization. In: Proceedings of the Workshop on Multi-source Multilingual Information Extraction and Summarization, pp. 17–24 (2008)
14. Wan, X., Xiao, J.: CollabRank: Towards a Collaborative Approach to Single-Document Keyphrase Extraction. In: Proceedings of the 22nd COLING, pp. 969–976 (2008)
15. Liu, Z., Huang, W., Zheng, Y., Sun, M.: Automatic Keyphrase Extraction via Topic Decomposition. In: Proceedings of the 7th EMNLP, pp. 366–376 (2010)
16. Liu, X., Wu, J., Zhou, Z.: Exploratory Under-Sampling for Class-Imbalance Learning. IEEE Transactions on Systems, Man, and Cybernetics, Part B 39, 539–550 (2009)
17. Chawla, N.V., Bowyer, K.W., Hall, L.O., Kegelmeyer, W.P.: SMOTE: Synthetic Minority Over-Sampling Technique. Journal of Artificial Intelligence Research 6, 321–357 (2002)
18. Han, H., Wang, W.-Y., Mao, B.-H.: Borderline-SMOTE: A New Over-Sampling Method in Imbalanced Data Sets Learning. In: Huang, D.-S., Zhang, X.-P., Huang, G.-B. (eds.) ICIC 2005, Part I. LNCS, vol. 3644, pp. 878–887. Springer, Heidelberg (2005)
19. Fan, X., Tang, K., Weise, T.: Margin-Based Over-Sampling Method for Learning from Imbalanced Datasets. In: Huang, J.Z., Cao, L., Srivastava, J. (eds.) PAKDD 2011, Part II. LNCS, vol. 6635, pp. 309–320. Springer, Heidelberg (2011)
20. Sun, Y., Kamel, M.S., Wong, A.K.C., Wang, Y.: Cost-sensitive boosting for classification of imbalanced data. Pattern Recognition 40, 3358–3378 (2007)
21. Zhou, Z., Liu, X.: Training Cost-Sensitive Neural Networks with Methods Addressing the Class Imbalance Problem. IEEE Transactions on Knowledge and Data Engineering 18, 63–77 (2006)
22. Nguyen, T., Zeno, G., Lars, S.: Cost-Sensitive Learning Methods for Imbalanced Data. In: Proceedings of the 2010 IJCNN, pp. 1–8 (2010)
23. Li, Y., Zaragoza, H., Herbrich, R., Shawe-Taylor, J., Kandola, J.: The perceptron algorithm with uneven margins. In: Proceedings of the 19th ICML, pp. 379–386 (2002)
24. Platt, J.: Probabilistic outputs for support vector machines and comparison to regularized likelihood methods. In: Advances in Large Margin Classifiers, pp. 61–74 (2000)
25. Muhlenbach, F., Lallich, S., Zighed, D.A.: Identifying and Handling Mislabelled Instances. Journal of Intelligent Information Systems 22, 89–109 (2004)
26. Ni, W., Huang, Y.: Extracting and Organizing Acronyms based on Ranking. In: Proceedings of the 7th WCICA, pp. 4542–4547 (2008)

A Tourist Itinerary Planning Approach
Based on Ant Colony Algorithm

Lei Yang, Richong Zhang, Hailong Sun, Xiaohui Guo, and Jinpeng Huai

Schoole of Computer Science and Engineering, Beihang University,
Beijing, 100191 China
{yanglei,zhangrc,sunhl,guoxh,huaijp}@act.buaa.edu.cn

Abstract. Many itinerary planning applications have been developed to assist travelers making decisions. Existing approaches model this problem as finding the shortest path of tourism resources and generate either inter-city or intra-city visiting plans. Instead of following the conventional route, we propose an approach to automatically generate tourist itineraries by comprehensively considering transportations, lodgings and POIs between/inside each destination. In addition, due to the NP-complete nature of the itinerary planning, we develop an approach based on the ant colony optimization (ACO) algorithm to solve the tourist planning problem. To show the versatility of the proposed approach, we design an itinerary planning system which arranges all available tourism resources collected from the web, such as POIs, hotels, and transportations. Our experimental result confirms that the proposed algorithm generates high utility itineraries with both effectiveness and efficiency.

Keywords: itinerary planning, ant colony algorithm, tourist guide.

1 Introduction

The online travel agency, or e-Tourism website, is more and more prevalent in the web, such as expedia.com, priceline.com, hotels.com, booking.com, etc. By taking the advantages of the increasing availability of such rich choices, users enjoy the convenience and lower price than ever before. Meanwhile, with such rich information, it is difficult for travelers to compare tourism resources and to determine itineraries in a short time. Therefore, the itinerary planning system is necessary to help travelers making tourism decisions.

Traditional recommender systems [1,3] have been applied to recommend eligible POIs for tourists. However, these approaches could not generate sequences and routes between POIs. Existing itinerary planning approaches [2,4,5,6] solely consider arranging tourism resources in a single destination. When travelers are going to make a tourism decision for several destinations, such systems are unable to provide a multi-city itinerary with transportation and lodgings between cities. Furthermore, many travel websites, such as booking.com and so on, which are static or interactive systems, highly require users involving the decision process. This is a rather time consuming process to compare hotel/flight/POI options.

H. Gao et al. (Eds.): WAIM 2012, LNCS 7418, pp. 399–404, 2012.

Our itinerary planning system intelligently provides a multi-city itinerary for tourists, including POIs, hotels, transportations, and the detailed visiting schedules for these destinations. In addition, our system allows travelers specifying constraints, e.g., the starting and ending location/time, budgets, the selected attractions, etc. Moreover, we decompose this multi-city itinerary planning problem into two sub-problems: inter-city planning and intra-city planning. This process can degrade complexity, reduce the solution space and satisfy different traveling purposes, business or tourism. To the best of our knowledge, this is the first approach which provides a general solution to build itinerary plans between and inside destinations.

To achieve these functionalities, in this paper, we propose an ACO-based itinerary planning algorithms to efficiently discover high utility itineraries. The experimental result confirms that ACO algorithm outperforms other trip planning algorithms [4, 5] in terms of effectiveness and efficiency. Moreover, we implement an itinerary planning system, gTravel, based on our proposed algorithms. A case study on gTravel is also provided to show the practicability of the proposed approach.

2   Itinerary Planning Model

In this section, we propose an automatic itinerary planning approach by using ACO algorithm-based algorithm and utility functions (measure the quality of itineraries). We first introduce the problem definition. The itinerary planning problem, in this paper, is formally defined as: given a set of constraints, such as budget, a starting/ending place/time, visiting cities, etc., system generates a personalized itinerary, which includes POIs, hotels, transportations between cities and routes between POIs.

2.1   Algorithm

Our approach divides multi-city planning into inter-city planning which considers the resource of transportations between cities and lodgings at each destination and intra-city planning of POIs and visiting sequences. We also estimate visiting time and transiting time for these POIs inside cities. Both inter-city and intra-city planning use ACO approach to discover proper traveling plans for users, so we take intra-city algorithm in this paper as an example to show how to develop itinerary algorithms.

The resources considered in our ACO algorithm include three types: POIs, hotels and transportations. In ACO algorithm, the domain of each resource consists of solution components. Each components of resource r_i is denoted by r_i^j if r_i being planned after r_j with its domain value of v_i^j. A pheromone model is used to probabilistically generate solutions from a finite set of solution components and iteratively update the pheromone values to search in regions which contain high quality solutions. The pheromone model consists of pheromone trail parameters. The pheromone trail value, denoted by τ_i^j, is associated with each component r_i^j. A number of artificial ants search for good solutions to the considered optimization problem. The ants construct solutions from a finite set of available solution components. At the beginning, solutions are initialized with an empty set of components. Then, at each construction step, the current partial solution is extended by adding a feasible solution component r_i^j. Each ant selects r_i^j in a probabilistic way by the following equation:

$$P_k(r_i^j) = \begin{cases} \dfrac{[\tau_i^j]^\alpha [\eta(r_i^j)]^\beta}{\sum_{j \neq tabu_k} [\tau_i^j]^\alpha [\eta(r_i^j)]^\beta} & if \ j \neq tabu_k \\ \quad 0 & otherwise \end{cases} \tag{1}$$

where $P_k(r_i^j)$ denotes the probability of ant k choosing the solution component r_i^j; $tabu_k$ stores the list of resources that ant k has passed which ensures that ant k does not choose resources repeatedly; and $\eta(r_i^j)$ is a function to compute the domain value v_i^j.

In addition, α and β represent the importance of the pheromone value and the heuristic information respectively. The higher the value of α is, the more importance of other ants guidance is. We note that it tends to be greedy algorithm if β set as a high value.

The ants construct the solutions as follows. Each ant selects tourism resources in a probabilistic way incrementally building a solution until it reaches the ending time. The algorithm then evaluates the quality of the solutions that ants found by computing the utility (using Equation 6 in the following part). After each round of the solution finding process, the system would estimate whether the termination criterion is satisfied. If the algorithm doesn't meet the termination condition, it continues to update the pheromone values by using the following equation:

$$\tau_i^j = (1 - \rho) * \tau_i^j + \rho \Delta \tau_i^j \tag{2}$$

where ρ is the volatile coefficient. $\Delta \tau_i^j$ expresses the increment of τ_i^j. The value of $\Delta \tau_i^j$ depends on the quality of the generated solutions. Subsequent ants utilize the pheromone information as a guide towards more promising regions of the searching space.

2.2   Utility Function

We note that in the ACO algorithm, we should define utility functions to calculate the utilities of generated solutions. There are many possible utility metrics and each of them is perhaps subjective to a certain extent. In the following parts, as a concrete example, we discuss how we define the utility metrics in this study.

We take the intra-city utility function as an example to show how to define utility functions. We assume that the higher is the rank of the POIs, the better is the itinerary; the shorter is the travel distances, the better is this itinerary, furthermore, the higher is the fraction of the time spent for visiting POIs instead of transportation, the better is the itinerary. Such, we propose the utility function for intra-city planning as follows:

$$f(I) = \sum_{i \in I} (k(r_i)) + e^{-(\sum_{i \in I} dist(r_i, r_{i+1}))} + pR \tag{6}$$

$$(x) = \frac{x - min(X)}{max(X) - min(X)} \tag{7}$$

where $(k(r_i))$, $e^{-(\Sigma_{i \in I} dist(r_i, r_{i+1}))}$ and pR normalize the rank of r_i, distances between r_i and r_{i+1}, and the fraction of visiting time into a given scale respectively. $k(r_i)$ denotes the rank of r_i .

We note that the inter-city utility function can be also defined similar to the intra-city utility function. When designing the inter-city utility functions, price, comfort level and distances should be considered and a similar function can also be defined.

3   Experimental

3.1   Experimental Results

We evaluate our approach on popular travel destinations in China using real and synthetic datasets extracted from DaoDao[1], elong[2], Flickr[3]. There are 26650 hotels, 9203 POIs and more than 3000 transportation choices. The goal of our experiments is to evaluate our itinerary planning approach in terms of the performance of the algorithm and the quality of generated itinerary by comparing with the greedy algorithm (Greedy) [4] and the guided local search meta-heuristic algorithm (GLS) [5]. Table 1 lists the means and variances of the utility and time of algorithms after running 100 times of each algorithm by the time budget and the number of POIs setting as 12 hours and 40 POIs.

Table 1. The construct of means and variances

	Utility		Running Time	
	Avg	Std	Avg	Std
ANT200	6.6792	0.00642	0.2664	0.006
ANT100	6.3042	0.00994	0.13345	0.00151
ANT50	6.225	0.01852	0.0816	$6.7425e^{-4}$
GLS	6.2083	----	0.3173	----
Greedy	5.1625	----	0.01	----

From the average value of utility and running time, we can see in this situation, our algorithm outperforms GLS and Greedy algorithm in terms of effectiveness. In the way of efficiency, our algorithm is better than GLS algorithm. Also, from the variance, we can see that our algorithm is relatively stable. We have also conducted few experiments to compare the performances between our model and other two commonly-used trip planning approaches, the results also confirm the effectiveness and efficiency of our approach. Due to the limitation of this paper, we do not show these results.

---

[1] http://www.daodao.com
[2] http://www.elong.com
[3] http://www.flickr.com

3.2   Case Study

We have built an itinerary planning system, gTravel, based on our proposed approach. In this section, we make use of an example to illustrate the practicability of gTravel: a traveler is going to visit Beijing, Shanghai and Guangzhou from San Francisco at 2012/3/15 and return at 2012/3/23; and he/she wants to spend about 2 days in each city; also, the budget of 20,000 Yuan and 2000 Yuan in each city are cost constraints. gTravel helps this potential traveler organize his/her itinerary by our proposed model. The detailed itinerary is listed in table 2.

Table 2. The itinerary that our system generated

Dest	Start Time		End Time		Name
1	**3/15**	**13:02**	**3/16**	**02:12**	**Flight CA8857**
2	3/16	03:30	3/16	08:00	Marriott Executive Apartment
3	3/16	08:15	3/16	09:30	Shanghai macrocosm
4	3/16	09:4	3/16	11:00	City God Temple
5	3/16	11:15	3/16	11:30	Yu Garden
6	3/16	11:45	3/16	15:00	Bund Sightseeing Tunnel
7	3/16	15:15	3/16	17:00	World Architecture Expo
8	3/16	17:15	3/16	19:00	Jin Mao Tower
9	3/16	19:15	3/16	20:30	Oriental Pearl TV Tower
10	3/16	20:45	3/17	08:00	Marriott Executive Apartment
11	**3/17**	**09:12**	**3/18**	**06:36**	**Train K511**
12	3/18	07:30	3/18	10:30	Southern Airlines Pearl Hotel
13	3/18	11:30	3/18	14:30	Six Banyan Temple
14	3/18	14:45	3/18	17:30	Commercial Pedestrian Street
15	3/18	18:30	3/19	08:00	Southern Airlines Pearl Hotel
16	3/19	09:00	3/19	11:30	One thousand sites of ancient road
17	3/19	11:45	3/19	14:15	Museum of the Nanyue King
18	3/19	14:30	3/19	16:45	Sun Yatsen Memorial Hall
19	3/19	17:00	3/19	19:00	Southern Theatre
20	3/19	20:00	3/20	08:00	Southern Airlines Pearl Hotel
21	3/20	09:00	3/20	11:00	Shangxiajiu pedestrian street
22	**3/20**	**14:55**	**3/20**	**19:55**	**Train K600**
23	3/20	21:00	3/21	08:00	Swissotel Beijing
24	3/21	08:15	3/21	09:45	Forbidden City
25	3/21	10:15	3/21	11:45	Temple of Heaven
26	3/21	12:15	3/21	13:45	National Grand Theatre
27	3/21	14:30	3/21	15:30	Tiananmen square
28	3/21	16:00	3/21	17:30	Jingshan Park
29	3/21	18:30	3/21	20:00	Bird Nest
30	3/21	21:00	3/22	08:00	Swissotel Beijing
31	3/22	09:00	3/22	10:30	Chinese Ethnic Culture Park
32	3/22	11:00	3/22	12:30	Beijing Aquarium
33	**3/22**	**13:15**	**3/23**	**00:45**	**Flight CA8888**

4   Conclusion

This work proposes an itinerary planning approach that incorporates ACO algorithm to find one of the best combinations of tourism resources, such as transportations, lodgings, and POIs. To the best of our knowledge, this is the first approach which is built to plan itineraries between and inside cities and provide a general solution for trip planning systems to generate itineraries. We have executed empirical studies on real data collected from the Web and the experimental results show that our model outperforms some existing itinerary planning approaches both in effectiveness and efficiency. In addition, we conduct a case study illustrating the versatility of our framework.

Acknowledgment. This work was supported partly by National Natural Science Foundation of China (No. 61103031), partly by China 863 program (No. 2012AA011203), partly by the State Key Lab for Software Development Environment (No. SKLSDE-2010-ZX-03), and partly by the Fundamental Research Funds for the Central Universities (No. YWF-12-RHRS-016).

References

[1] Xie, M., Lakshmanan, L.V.S., Wood, P.T.: Breaking out of the box of recommendations: From items to packages. In: 4th ACM Conference on Recommender Systems, New York, pp. 151–158 (2010)

[2] Vansteenwegen, P., Oudheusden, D.V.: The mobile tourist guide: An OR opportunity. J. OR Insights 20, 21–27 (2007)

[3] Xie, M., Lakshmanan, L.V., Wood, P.T.: CompRec-Trip: a Composite Recommendation System for Travel Planning. In: 27th International Conference on Data Engineering, pp. 1352–1355 (2011)

[4] Hagen, K.T., Kramer, R., Hermkes, M., Schumann, B., Mueller, P.: Semantic matching and heuristic search for a dynamic tour guide. In: Frew, A.J. (ed.) COMP SCI. 2005. LNCS, vol. 5, pp. 149–159. Springer, Austria (2005)

[5] Souffriau, W., Vansteenwegen, P., Vertommen, J., van den Berghe, G., van Oudheusden, D.: A personalized tourist trip design algorithm for mobile tourist guides. J. Appl. Artif. Intell. 22, 964–985 (2008)

[6] Roy, S.B., Amer-Yahia, S., Das, G., Yu, C.: Interactive Itinerary Planning. In: 27th International Conference on Data Engineering, Arlington, pp. 15–26 (2011)

A Transparent Approach for Database Schema Evolution Using View Mechanism

Jianxin Xue, Derong Shen, Tiezheng Nie, Yue Kou, and Ge Yu

College of Information Science and Engineering, Northeastern Unicersity, China
xuejianxin@research.neu.edu.cn,
{shenderong,nietiezheng,kouyue,yuge}@ise.neu.edu.cn

Abstract. Designing databases that evolve over time is still a major problem today. The database schema is assumed to be stable enough to remain valid even as the modeled environment changes. However, the database administrators are faced with the necessity of changing something in the overall configuration of the database schema. Even though some approaches proposed are provided in current database systems, schema evolution remains an error-prone and time-consuming undertaking. We propose an on-demand transparent solution to overcome the schema evolution, which usually impacts existing applications/queries that have been written against the schema. In order to improve the performance of our approach, we optimize our approach with mapping composition. To this end, we show that our approach has a better potential than traditional schema evolution technique. Our approach, as suggested by experimental evaluation, is much more efficient than the other schema evolution techniques.

Keywords: schema evolution, schema mapping, backward compatibility, schema composition.

1   Introduction

Database designers construct schemas with the goal of accurately reflecting the environment modeled by the database system. The resulting schema is assumed to be stable enough to remain valid even as the modeled environment changes. However, in practice, data models are not nearly as stable as commonly assumed by the database designers. As a result, modifying the database schema is a common, but often troublesome, occurrence in database administration. These are significant industrial concerns, both from the viewpoint of database system manufacturers and information system users. Schema evolution, and its stronger companion, schema versioning, have arisen in response to the need to retain data entered under schema definitions that have been amended. A more formal definition of schema evolution is the ability for a database schema to evolve without the loss of existing information [1].

Motivating Example. To better motivate the need for schema evolution support, we illustrate a schema evolution example in an employee database, which is used as a

H. Gao et al. (Eds.): WAIM 2012, LNCS 7418, pp. 405–418, 2012.

running example in the rest of the paper. Fig. 1 outlines our example, which has three schema versions, V_1 through V_3.

Due to a new government regulation, the company is now required to store more personal information about employees. At the same time, it was required to separate employees' personal profiles from their business-related information to ensure the privacy. For these reasons, the database layout was changed to the one in version V_2, where the information about the employees is enriched and divided into two tables: Personal, storing the personal information about the employees, and Employees, maintaining business-related information about the employees.

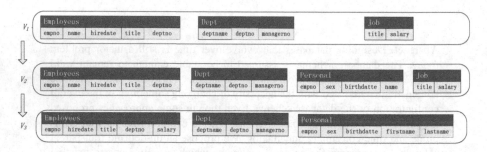

Fig. 1. Schema evolution in an employee database

The company chose to change its compensation policy: to achieve 'fair' compensation and to better motivate employees, the salaries are made dependent on their individual performance, rather than on their job titles. To support this, the salary attribute was moved to the table employees, and the table job was dropped. Another modification was also introduced that the first name and last name are now stored in two different columns to simplify the surname-based sorting of employees. These changes were represented as the last schema version, V_3.

Assume that a department works for a project, and view E_P is used to correlate the employees with the projects they work for. On top of this view, we consider that the report module contains an aggregate query that calculates the expenses of the project per month by summing up the salaries of all employees working for it and compare them with the budget of the project. Along with the evolution of the schema, the view E_P becomes unavailable. How to effectively keep the application available is our object.

Contribution. Our main contributions in this paper are summarized as follows.

1. A novel on-demand schema evolution approach based on virtual views (named as on-demand approach) is proposed, which only deals with the schemas effected by the schema evolution and preserves the new data instead of the version data.

2. Subsequently we construct view version with mapping inversion, which can recover all the source information with complementary approach.

3. Next, an optimization algorithm is presented to improve application/queries efficiency.

4. At last, we have performed comprehensive experiments to compare our approach with traditional database schema evolution techniques, which show that our approach achieves remarkable improvement in efficiency.

The remainder of this paper is organized as follows: Section 2 discusses related works, Section 3 introduces preliminary definitions, Section 4 discusses in details the design of our schema evolution method, Section 5 is dedicated to experimental results. We conclude in Section 6 where we also state our future work.

2   Related Works

Schema evolution means modifying schemas within a database without loss of existing data. With the acceleration of database schema modification frequency of Internet and enterprise, database schema evolution becomes a hotspot in current database research. However, current database schema evolution is mainly implemented manually.

Schema evolution has been extensively addressed in the past and a variety of techniques have been proposed to execute change in the least erratic way possible to avoid disruption of the operation of the database. A bibliography on schema evolution lists about 216 papers in the area [2]. The proposed solutions for schema evolution can be categorized mainly by following one of these approaches: modification, versioning and views.

Modification. The original schema and its corresponding data are replaced by a new schema and new data. This approach does not exactly adhere to the schema evolution definition, which makes the applications that use the original schemas inconsistent with the new database schemas [3]. This renders the approach unsuitable in most real cases, yet it remains the most popular with existing DBMS. In [4], the authors discuss extensions to the conventional relation algebra to support both aspects of evolution of a database's contents and evolution of a database's schema. In [5], authors present an approach to schema evolution through changes to the ER schema of a database. In [6], they describe an algorithm that is implemented in the O_2 object database system for automatically bringing the database to a consistent state after a schema update has been performed.

Versioning. The old schema and it corresponding data are preserved and continued to be used by existing applications, but a new version of the schemas is created, which incorporates the desired changes [7]. There are two most used versioning methods. The first is sequential revisions which consists of making each new version a modification of the most recent schema. This approach is adopted in Orion Database System [8]. The second method is a complex one, which is called parallel revisions. For instance, Encor Database System[9]. Generally, the versioning approach presents performance problems.

View. A view is a derived table. It makes possible to change the schema without stopping the database and destroying its coherence with existing applications.

Bellahsene [10] proposes a method that uses view to simulate the schema changes and the data migration is not needed, i.e., views are viewed as the target schema. However, this method has scalability limitations. In fact, after several evolution steps, the applications/queries may involve long chains of views and thus deliver poor performance. Moreover, it is difficult to change current schema.

In [11], the authors deal with the adaption of the view definition in the presence of changes in the underlying database schema. [12], [13] deal also with a specialized aspect of the view adaptation problem. The work of [14] employs a directed graph for representing the object dependencies in O-O database environments and finding the impact of changes in database objects towards application objects.

As of today, in [15], the authors tend to extend the work of [16]. They first consider a set of evolution changes occurring at the schema of a data warehouse and provide an informal algorithm for adapting affected queries and views to such changes. The most representative achievement is PRISM developed by Curino. One contribution of the work on PRISM is a language of Schema Modification Operator. Meanwhile, automatic query rewriting of queries specified against schema version N into semantically equivalent queries against schema Version $N+1$, and vice versa [17]. SMOS have a good semantic express for schema evolution, but still has some limitations.

3   Preliminaries

A schema R is a finite sequence $(R_1,\ldots,R_k)$ of relation symbols, where each R_i has a fixed arity. An instance I over R is a sequence $(R^i_1,\ldots,R^i_k)$, where each R^i_i is a finite relation of the same arity as R_i. We shall often use R_i to denote both the relation symbol and the relation R^I_i that instantiates it. We assume that we have a countable infinite set Const of constants and a countably infinite set Var of labeled nulls that is disjoint from Const. A fact of an instance I is an expression $R^i_i(v_1,\ldots,v_m)$ where R_i is a relation symbol of R and $v_1,\ldots,v_m$ are constants or labeled nulls such that $(v_1,\ldots,v_m)\in R^i_i$. The expression $v_1,\ldots,v_m$ is also sometimes referred to as a tuple of R_i. An instance is often identified with its set of facts.

A schema mapping is a triple $M=(S,T,\Sigma)$, where S is the source schema, T is a target schema, and Σ is a set of constraints that describe the relationship between S and T. M is semantically identified with the binary relation:

$$Inst(M)=\{(I,J)\mid I\in S_i, J\in T_i,(I,J)\models \Sigma\}. \tag{1}$$

Here, S_i is the instance of source schema and T_i is the instance of target schema. We will use the notation $(I, J)\models \Sigma$ to denote that the ordered pair (I, J) satisfies the constraints of Σ; furthermore, we will sometimes define schema mappings by simply defining the set of ordered pairs (I, J) that constitute M (instead of giving a set of constraints that specify M). If $(I, J)\in M$, we say that J is a solution of I (with respect to M).

In general, the constraints in Σ are formulas in some logical formalism. In this paper, we will focus on schema mappings specified by source-to-target tuple-generating dependencies.

An atom is an expression of the form $R(x_1,...,x_n)$. A source-to-target tuple-generating dependency (s-t tgd) is a first-order sentence of the form as follow:

$$\forall x(\varphi(x) \rightarrow \exists y\, \psi(x,y)). \tag{2}$$

where $\varphi(x)$ is a conjunction of atoms over S, each variable in x occurs in at least one atom in $\varphi(x)$, and $\psi(x,y)$ is a conjunction of atoms over T with variables in x and y. For simplicity, we will often suppress writing the universal quantifiers $\forall x$ in the above formula. Another name for S-T tgds is global-and-local-as-view (GLAV) constraints. They contain GAV and LAV constraints, which we now define, as important special cases.

A GAV (global-as-view) constraint is an S-T tgd in which the right-hand side is a single atom with no existentially quantified variables, that is, it is denoted by the following equation:

$$\forall x(\varphi(x) \rightarrow P(x)). \tag{3}$$

where $P(x)$ is an atom over the target schema. A LAV (local-as-view) constraint is an s-t tgd in which the left-hand side is a single atom, that is, it is denoted by the following equation:

$$\forall x(Q(x) \rightarrow \exists y\, \psi(x,y)). \tag{4}$$

where $Q(x)$ is an atom over the source schema.

4   Schema Evolution Management

In this section we will discuss the details of our on-demand approach (supporting backward compatibility) and the optimization of our approach. Here, the mappings between source schema and target schema are expressed by the S-T tgds. A main requirement for database schema evolution management is thus to propagate the schema changes to the instance, i.e., to execute instance migration correctly and efficiently. S-T tgds represent the semantics of conversion from source schema instance to target schema instance. The S-T tgds can express both simple changes, such as addition, modification or deletion of individual schema constructs, and complex changes refer to multiple simple changes. The converted semantics of S-T tgds can automatically execute instance migration from source schemas to target schemas with the change of schemas.

4.1   Our On-Demand Approach

In our approach, the original schema and its corresponding data are not preserved. To reuse the legacy applications/queries, we must support backward compatibility. Here, virtual versions (view version) are proposed to support backward compatibility, which can avoid the costly adaptation of applications. In schema evolution progress, many times the evolution operation only involves several tables. We create views not for all the tables but only for the tables evolved.

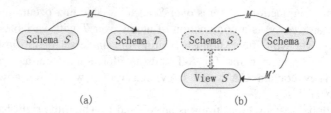

(a)   (b)

Fig. 2. Generation of virtual version

The functionality that our approach aims to achieve is illustrated in Fig. 2, where schema version S represents an initial (legacy) schema that goes through mapping M forming target schema version T. The instances migrate automatically through conversion (chase). In order to save memory space and improve performance, we delete the instances of schema version S and create view S that has the same name with the deleted schema to realize the backward compatibility. Applications deal with views the same way they deal with base tables. Supporting different explicit view versions for schemas realizes evolution transparency. The concept of view S can be created through mapping M', which is obviously the mapping inverse of M.

The most important thing is to calculate the inverse of mapping M. The ideal goal for schema mapping inversion is to be able to recover the instances of source schema. Concretely, if we apply the mapping M on some source instances and then the inverse on the result of mapping M is used to obtain the original source instance. Here, applying a schema mapping M to an instance means generating the instance by chase. However, a schema mapping may drop some of the source information, and hence it is not possible to recover the same amount of information.

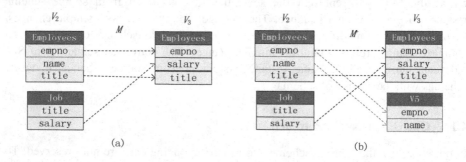

(a)   (b)

Fig. 3. Example of data loss

The example of the scenario described in Fig. 3(a) illustrates the schema evolution process with information loss. Consider the following two schema versions V_2 and V_3 in Fig. 1, where V_2 consists of two relation symbols Employees and Job, and schema V_3 consists of one ternary relation symbol Employees that associates each employee with salary and title. Given existing the schema mapping $M_{23}=(V_2, V_3, \Sigma_{23})$, where

$$\Sigma_{23}=\{(Employees(e,n,t) \wedge Job(t,s) \rightarrow Employees(e,s,t)\}.$$

The nature "inverse" that one would expect here is the following mapping:

$$\Sigma_{32}=\{(Employees(e,s,t) \rightarrow \exists n Employees(e,n,t), Employees(e,s,t) \rightarrow Job(t,s) \}$$

Here, we verify whether mapping Σ_{23} can recover instances of source schema versions V_2. If we start with a source instance I for schema V_2 where the source tuples contain some constant values, and then apply the chase with mapping Σ_{23} and then the reverse chase with Σ_{32}. Another source instance I', where the tuples have nulls in the name position, is obtained. Consequently, the resulting source instance I' cannot be equivalent to the original instance I. To give a concrete example, consider the source instance I over version V_2 that is shown in Fig. 3.

Schema version V_2 could not be created from V_3 by mapping Σ_{32}. In our approach we also could not get the views the same as V_2. In order to solve this problem, a separate table is established for the lost data. In Fig. 3(b), the red rectangle represents the loss data, which are stored in table V_3. The special table is name V_3, which works only when computing mapping inverstion and evolution rollback. The special table cannot be modified. The inversion of mapping Σ_{23} can be expressed as Σ^+_{32}.

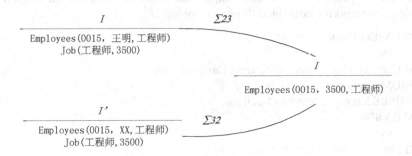

I

Employees(0015，王明, 工程师)
Job(工程师, 3500)

$\Sigma 23$

I

Employees(0015，3500, 工程师)

I'

Employees(0015，XX, 工程师)
Job(工程师, 3500)

$\Sigma 32$

Fig. 4. The verification of mapping inverstion

$$\Sigma^+_{32}=\{((Employees(e,s,t) \vee V_5(e,n) \rightarrow Employees(e,n,t)),$$

$$(Employees(e,s,t) \rightarrow Job(t,s))\}.$$

Mapping Σ^+_{32} can completely recover the instances of source schema versions. The structure of old schema versions can be correctly described by views that are created by mapping Σ^+_{32}. We present the algorithm for calculating the views of old schema in the following.

Algorithm 1. Virtual View Create
Input: source schema S, mapping M between source schema and target schema
1:   Let $M^+=\phi$, $M^{-1}=\phi$
2:   If M is full
3:       $T=chase_M(I_s)$
4:       $M^{-1}=INVERS(M)$
5:   Else
6:       $M^+=SUPPLEMENT(M)$
7:       $T=chase_{M^+}(I_s)$
8:       $M^{-1}=INVERS(M)$
9:       $V_S=VIEW\ CREATE(M^{-1})$

First, we determine whether there is information loss before schema evolving. If the mapping is not full, i.e., there is information loss. We will create a special table and modify the mapping M'. The operator of SUPPLEMENT() aims to modify the mapping M without information loss. Here, data migrate is automatically executed with chase algorithm. We compute inversion of full mapping M and generate the view concept with the operator VIEW CREATE().

Example of Fig. 3 is used to illustrate our algorithm. The inverse of mapping M^+ can be computed by operator INVERS().

$$\Sigma^+_{32}=\{\forall e,n,t,s(Employees(e,s,t)\vee V_5(e,n)\rightarrow Employees(e,n,t),$$
$$\forall e,n,t,s\ (\ Employees(e,s,t)\rightarrow Job(t,s)\}.$$

We get the mapping M^{-1}, which can compute the views that describe the old schema. The views concepts are computed through mapping M^{-1}.

```
CREATE VIEW Employees V₂
      AS
   SELECT Employees.empno, V₃.name, Employees.title
   FROM Employees, V₃
   WHERE Employees.empno= V₃.empno;
CREATE VIEW Job V₂
      AS
   SELECT Employees.title, Personal.salary
   FROM Employees
```

4.2   Optimized Approach

In on-demand approach, we support backward compatibility to reuse the legacy applications and queries by creating virtual version. But it has scalability limitations. In fact, after several evolution steps, each application execution may involve long chain of views and thus deliver poor performance. Fig. 5 shows the limit of our naïve approach. Schema version $S_1,\ldots,S_n$ represents each version in the progress of schema evolution. S_n is the current schema version, then the other schema versions are represented by views. The old schema versions connect with existing schema version through long views chains, e.g., schema version S_1 is mapped into schema version S_n with views chain $M_1\cup...\cup M_n$. To avoid the costly implementation of applications/queries, the chains of views should get shorter as much as possible. As Fig. 5 shows, we hope that each view version would have been directly mapped into schema version S_n, rather than through the intermediate steps. So the implementation of applications/queries could not operate physic data through long views chains. Mapping composition is a good choice.

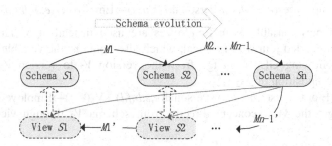

Fig. 5. The composition of views mapping

We do not preserve the data of old schema version. The mappings created by schema version matching are not accurate and cannot reflect the semantic of schema evolution process. In this paper, we compute the views concepts directly from existing schema version by mapping composition. The algorithm of view mapping composition is given in the following.

Algorithm 2. View Composition

Input: current schema version S_c, old schema version S_{o1} and S_{o2}, mapping M_1 between S_c and S_{o1} and mapping M_2 between S_{o1} and S_{o2}

1:   Let $M' =\phi$
2:   If M_1 is no the GAV s-t tuple tgds
3:       $M_1 =$ Splitup(M_1)
4:       $M' = M_2$
5:   While the atoms of the right side of M_1 appear on the left side of M_2
6:       $M' =$ REPLACE($M_1\square M'$)
7:   $M' =$ SIMIPLIFY(M')

Here, we illustrate the computed progress of our optimized method with an concrete example of schema evolution. We also use employees database as our example. Mapping Σ_{21} is the relation between version V_2 and V_1.

$$\Sigma_{21} =\{(\text{Employees}(e,h,t,d)\wedge\text{Personal}(e,s,b,n)\rightarrow\text{Employees}(e,n,h,t,d)\}.$$

The view mapping from V_3 to V_2 is Σ_{32}.

$\Sigma_{32}=\{(\text{Employees}(e,h,t,d) \rightarrow \text{Employees}(e,n,h,t,d), \text{Employees}(e,h,t,d) \rightarrow \text{Job}(t,s),$
$\text{Personal}(e,s,b,f,l) \wedge V_5(e,n) \rightarrow \text{Personal}(e,s,b,n)\}.$

Intuitively, the composition algorithm will replace each relation symbol from Employees and Personal in Σ_{21} by relation symbols from that use the GAV S-T tgds of Σ_{32}. In this case, the fact $\text{Personal}(e,s,b,f,l)$ that occurs on the left-hand side of Σ_{32} can be replaced by a Employees fact, according to the first GAV s-t tgd of Σ_{32}, we arrive at an intermediate tgd shown in the following.

$\text{Employees}(e,h,t,d,s') \wedge \text{Personal}(e,s,b,n) \rightarrow \text{Employees}(e,,n,h,t,d)$

Observe that new variable S' in Employees are used instead of S. This avoid an otherwise unintended join with Personal, which also contains the variable S. We then obtain the following GLAV s-t tgd from the version V_3 to version V_3. This tgds specify the composition of $\Sigma_{32} \circ \Sigma_{21}$.

$\Sigma_{32} \circ \Sigma_{21} = \{(\text{Employees}(e,h,t,d,s') \wedge \text{Personal}(e,s,b,f,l) \wedge V_5(e,n) \rightarrow \text{Employees}(e,,n,h,t,d)$

We can get the view concept directly from schema V_3 through view mapping $\Sigma_{32} \circ \Sigma_{21}$.

5   Experimental Evaluation

In this section, we report the results of a set of experiments designed to evaluate our proposed approach for schema evolution. Table 1 describes our experimental environment. The data-set used in these experiments is obtained from the schema evolution benchmark of [17] and consists of actual queries, schema and data derived from Wikipedia. We also do some experiments on the actual database data-set.

Table 1. Experimental Setting

Machine	RAM:	4Gb
	CPU(2x)	
	Disks:	500G
OS	Distribution	Linux Ubuntu server 11
MySQL	Version	5.022

Now approaches for schema evolution management mostly base on version management. We get the data of wikipadia from 2009-11-3 to 2012-03-7. The size of data grows from 31.2GB to 57.8GB. If we use versioning approach to manage the schema evolution, we must spent 1056.7GB memory space to store all versions of wikipadia from 2009-11-3 to 2012-03-7. However, our approach only needs to store the existing version, i.e., 57.8GB.

We evaluate the effectiveness of our approach with the following two metrics: (1) overall percentage of queries supported, and (2) the applications/queries running time. To make comparison with the PRISM, we use the same data-set obtained from the schema evolution of [17].

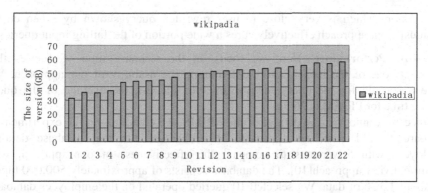

Fig. 6. The increase of wikipadia

Support for Backward Compatibility. An important measure of performance of our system is the percentage of queries supported by the old version. To this purpose we select the 66 most common query templates designed to run against version 28 of the Wikipedia schema and execute them against every subsequent schema version. The overall percentage of queries supported is computed by the following formulas.

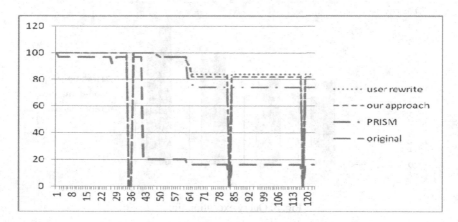

Fig. 7. Queries success rate

$$\text{Percent of queries supported} = \frac{\text{the queries number correctly implemented in version i}}{\text{Percent of queries supported}} \tag{5}$$

Fig.7 provides a graphical representation of the percent of queries supported for the various schema evolution approaches. Our approach can fully automatically realize the backward compatibility without rewriting queries. The original queries, failing when columns or tables are modified, highlight the portion of the schema affected by the evolution. The original query is run by the Modification. The black curve that represents the original queries shows how the schema evolution invalidates at most 82%of the schema. For PRISM, in the last version, about 26% of the queries fail due to the schema evolution. However, for our approach only 16% queries fail over the last version, which is very close to user rewritten query shown by green curve. Obviously, our approach effectively cures a wide portion of the failing input queries.

Run-time Performance. Here we focus on the response time of queries that represents one of the many factors determining the usability of our approach. We make a statistic for the query rewrite times of PRISM, the average of the query rewrite time for PRISM is 26.5s.

Since we cannot get the dataset of the PRIMS, we could not do the experiment to compare PRISM with our approach. Here, we use the real database data-set employees with 5 versions to compare the effectiveness of our approach with traditional view approach[10]. The database consists of approximately 500,000 tuples for about 1.3Gb of data. We selected 10 queried operated on the employees database, the response times of them are shown in Fig. 8: (1) traditional represents the approach with traditional view version; (2) on-demand is our on-demand approach; (3) optimized is our optimized approach. The testing results demonstrate that the response times of traditional approach and our naive approach grows with schema evolving, while our optimized method is close to constant. It is obvious that our optimized method has outperformed traditional approach and our naïve approach.

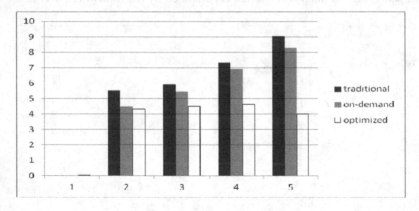

Fig. 8. Query execution time

6   Conclusion and Future Work

In this paper, we present a novel approach for schema evolution, which supports the backward compatibility. Traditionally, database schema evolution management is conducted by versioning, which generates high costs and requires much memory space. Such a time and space consuming approach severely limits the usability and convenience of the databases.

We exploit the virtual version approach which supports the applications/queries of old schemas. Our optimization for the virtual version makes our approach low time-consuming and high scalability. Both analysis and experiments verify the plausibleness of our approach and show that it consumes much less time and scales better that others schema evolution approach.

Despite recent progress we therefore see a need for substantially more research on schema evolution. For example, distributed architectures with many schemas and mappings need powerful mapping and evolution support, e.g., to propagate changes of a data source schema to merged schemas. New challenges are also posed by dynamic settings such a stream systems where the data to be analyzed may change its schema.

Acknowledgement. The National Natural Science Foundation of China (Grant No. 60973021, 61003060), and the Fundamental Research Funds for the Central Universities (N100704001).

References

1. Roddick, J.F.: A Survey of Schema Versioning Issues for Database Systems. J. Information and Software Technology 37(7), 383–393 (1995)
2. Roddick, J.F.: Schema Evolution in Database System–An Annotated Bibliography. SIGMOD RECORD 21(4) (1992)
3. Banerjee, J., Kim, W.: Semantics and Implementation of Schema Evolution in Object-Oriented Databases. SIGMOD RECORD 21, 311–322 (1987)
4. McKenzie, L.E., Snodgrass, R.T.: Schema Evolution and the Relational Algebra. J. Information System 15, 195–197 (1990)
5. Liu, C.T., Chrysanthis, P.K., Chang, S.K.: Database schema evolution through the specification and maintenance of the changes on entities and relationships. In: Loucopoulos, P. (ed.) ER 1994. LNCS, vol. 881, pp. 13–16. Springer, Heidelberg (1994)
6. Ferrandina, F., Meyer, T., Zicari, R., Ferran, G.: Schema and database Evolution in the O2 Object Dabase System. In: VLDB 1995, pp. 170–181. ACM, New York (1995)
7. Kim, W., Chou, H. T.: Versions of Schema for Object Oriented Databases. In: VLDB 1988, pp. 148–159. Morgan Kaufmann, San Francisco (1988)
8. Munch, B.P.: Versioning in a software Engineering Database–the Changes Oriented Way. Division of Computer Systems and Telematics. Norwegian Institute of Technology (1995)
9. Andany, J., Leonard, M., Palisser, C.: Management of schema evolution in databases. In: VLDB 1991, pp. 161–170. Morgan Kaufmann, San Francisco (1991)
10. Bellahsène, Z.: View mechanism for schema evolution. In: Morrison, R., Kennedy, J. (eds.) BNCOD 1996. LNCS, vol. 1094, pp. 18–35. Springer, Heidelberg (1996)
11. Bellahsene, Z.: Schema evolution in data warehouses. J. Knowledge and Information System, 283–304 (2002)
12. Nica, A., Lee, A.J., Rundensteiner, E.A.: The CVS Algorithm for View Synchronization in Evolvable Large-Scale Information Systems. In: Schek, H.-J., Saltor, F., Ramos, I., Alonso, G. (eds.) EDBT 1998. LNCS, vol. 1377, pp. 359–373. Springer, Heidelberg (1998)
13. Rundensteiner, E.A., Lee, A.J., Nica, N.: On preserving views in evolving environments. In: KRDB 1997, pp. 13.1–13.11 (1997)
14. Karahasanovic, A., Sjøberg, D.I.K.: Visualizing Impacts of Database schema changes-A Controlled Experiment. In: HCC 2001, p. 358. IEEE, Washington, DC (2001)

15. Favre, C., Bentayeb, F., Boussaid, O.: Evolution of Data Warehouses' Optimization: A Workload Perspective. In: Song, I.-Y., Eder, J., Nguyen, T.M. (eds.) DaWaK 2007. LNCS, vol. 4654, pp. 13–22. Springer, Heidelberg (2007)
16. Papastefanatos,G. Vassiliadis, P., Vassiliou, Y.: Adaptive Query Formulation to Handle Database Evolution (Extended Version), In: CAiSE 2006, pp.5-9, Springer, Heidelberg (2006)
17. Curino, C.A., Moon, H.J., Zaniolo, C.: Graceful database schema evolution: the prism work-bench. In: Proc. VLDB Conf., pp. 761–772. ACM, USA (2008)

WYSIWYE*: An Algebra for Expressing Spatial and Textual Rules for Information Extraction

Vijil Chenthamarakshan[1], Ramakrishna Varadarajan[2], Prasad M. Deshpande[1],
Raghuram Krishnapuram[1], and Knut Stolze[3]

[1] IBM Research
ecvijil@us.ibm.com, {prasdesh,kraghura}@in.ibm.com
[2] University of Wisconsin-Madison
ramkris@cs.wisc.edu
[3] IBM Germany Research & Development
stolze@de.ibm.com

Abstract. The visual layout of a webpage can provide valuable clues for certain types of Information Extraction (IE) tasks. In traditional rule based IE frameworks, these layout cues are mapped to rules that operate on the HTML source of the webpages. In contrast, we have developed a framework in which the rules can be specified directly at the layout level. This has many advantages, since the higher level of abstraction leads to simpler extraction rules that are largely independent of the source code of the page, and, therefore, more robust. It can also enable specification of new types of rules that are not otherwise possible. To the best of our knowledge, there is no general framework that allows declarative specification of information extraction rules based on spatial layout. Our framework is complementary to traditional text based rules framework and allows a seamless combination of spatial layout based rules with traditional text based rules. We describe the algebra that enables such a system and its efficient implementation using standard relational and text indexing features of a relational database. We demonstrate the simplicity and efficiency of this system for a task involving the extraction of software system requirements from software product pages.

1 Introduction

Information in web pages is laid out in a way that aids human perception using specification languages that can be understood by a web browser, such as HTML, CSS, and Javascript. The visual layout of elements in a page contain valuable clues that can be used for extracting information from the page. Indeed, there have been several efforts to use layout information for specific tasks such as web page segmentation [1] and table extraction [2]. There are two ways to use layout information:

1. **Source Based Approach:** Map the layout rule to equivalent rules based on the source code (html) of the page. For example, alignment of elements can be achieved in HTML by using a list (`<li>`) or a table row (`<tr>`) tag.

---

* What You See Is What You Extract.

H. Gao et al. (Eds.): WAIM 2012, LNCS 7418, pp. 419–433, 2012.
© Springer-Verlag Berlin Heidelberg 2012

2. **Layout Based Approach:** Use the layout information (coordinates) of various elements obtained by rendering the page to extract relevant information.

Both these approaches achieve the same end result, but the implementations are different as illustrated in the example below.

Example 1. Figure 1 shows the system requirements page for an IBM software product. The IE task is to extract the set of operating systems supported by the product (listed in a column in the table indicated by Q3). In the source based approach, the rules need to identify the table, its rows and columns, the row or column containing the word 'Operating Systems', and finally a list of entities, all based on the tags that can be used to implement them. In the layout based approach, the rule can be stated as: '*From each System Requirements page, extract a list of operating system names that appear strictly*[1] *to the south of the word 'Operating Systems' and are vertically aligned*'. The higher level layout based rule is simpler, and is more robust to future changes in these web pages.

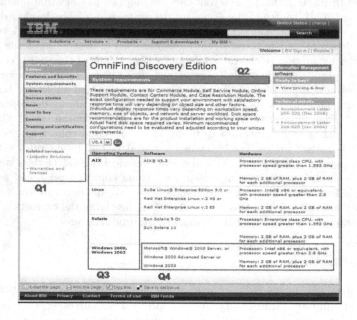

Fig. 1. System requirements page

Source based rules have several serious limitations, as listed below:

– An abstract visual pattern can be implemented in many different ways by the web designer. For example, a tabular structure can be implemented using any of `<table>`, `<div>` and `<li>` tags. Lerman et al [3] show that only

---

[1] See section 3.3 for a definition of strictness.

a fraction of tables are implemented using the `<table>` tag. Source-based rules that use layout cues need to cover all possible ways in which the layout can be achieved. Our experience with large scale IE tasks suggest that rules that depend on HTML tags and DOM trees work reasonably well on template based machine–generated pages, but become too complex and brittle to maintain when applied to manually authored web pages.

- Proximity of two entities in the HTML source code does not necessarily imply visual proximity [4], and so it may not be possible to encode visual proximity cues using simple source based rules.
- Specification languages are becoming more complex and difficult to analyze. Visualization logic is often embedded in CSS and Javascript, making the process of rule writing difficult.
- Rules based on HTML tags and DOM trees are often sensitive to even minor modifications of the web page, and rule maintenance becomes messy.

Layout based approaches overcome these limitations since they are at a higher level and independent of the page source code. Previous efforts at using layout based approaches were targeted at specific tasks such as page segmentation, wrapper extraction, table extraction, etc and are implemented using custom code. Existing rule based information extraction frameworks do not provide a mechanism to express rules based on the visual layout of a page. Our goal is to address this gap by augmenting a rule based information extraction framework to be able to express layout based rules. Rule based system can be either declarative [5,6] or procedural [7]. It has been shown that expressing information extraction (IE) tasks using an algebra, rather than procedural rules or custom code, enables systematic optimizations making the extraction process very efficient [5,6]. Hence, we focus on an algebraic information extraction framework described in [5] and extend its algebra with a visual operator algebra that can express rules based on spatial layout cues. One of the challenges is that not all rules can be expressed using layout cues alone. For some rules, it may be necessary to use traditional text–based matching such as regular expressions and dictionaries, and combine them with spatial layout based rules. The framework thus needs to support rules that use both traditional textual matching and high–level spatial layout cues. In summary, our contributions are as follows:

- We have developed an algebraic framework for rule–based information extraction that allows us to seamlessly combine traditional text–based rules with high–level rules based on the spatial layout of the page by extending an existing algebra for traditional text based information extraction [5], with a visual operator algebra. We would like to reiterate that our focus is not on developing spatial rules for a specific task, rather we want to develop an algebra using which spatial rules for many different tasks can be expressed.
- We implement the system using a relational database and demonstrate how the algebra enables optimizations by systematically mapping the algebra expressions to SQL. Thus, the system can benefit from the indexing and optimization features provided by relational databases.

– We demonstrate the simplicity of the visual rules compared to source based rules for the tasks we considered. We also conduct performance studies on a dataset with about 20 million regions and describe our experience with the optimizations using region and text indices.

2   Related Work

Information Extraction(IE): IE is a mature area of research that has received widespread attention in the NLP, AI, web and database communities [8]. Both rule based and machine learning based approaches have been proposed and widely used in real life settings. In this paper, we extend the operator algebra of System T [5] to support rules based on spatial layout.

Frameworks for Information Extraction: The NLP community has developed several software architectures for sharing annotators, such as GATE [9] and UIMA [10]. The motivation is to provide a reusable framework where annotators developed by different providers can be integrated and executed in a workflow.

Visual Information Extraction: There is a lot of work on using visual information for specific tasks. We list some representative work below. The VIPS algorithm described in [1] segments a DOM tree based on visual cues retrieved from browser's rendering. The VIPS algorithm complements our work as it can act as a good preprocessing tool performing task-independent page structure analysis before the actual visual extraction takes place - thereby improving extraction accuracy. A top-down approach to segment a web page and detect its content structure by dividing and merging blocks is given in [11]. [12] use visual information to build up a "M-tree", a concept similar to the DOM tree enhanced with screen coordinates. [2] describe a completely domain-independent method for IE from web tables, using visual information from Mozilla browser. All these approaches are implemented as monolithic programs that are meant for specific tasks. On the other hand, we are not targeting a specific task; rather our framework can be used for different tasks by allowing declarative specification of both textual and visual extraction rules.

Another body of work that is somewhat related is automatic and semi-automatic wrapper induction for information extraction [13].

These methods learn the a template expression for extracting information based on some training sets. The wrapper based methods work well on pages that have been generated using a template, but do not work well on human authored pages.

3   Visual Algebra

3.1   Overview of Algebraic Information Extraction

We start with a system proposed by Reiss et al [5] and extend it to support visual extraction rules. First, we give a quick summary of their algebra. For complete details, we request the reader to refer to the original paper.

Data Model. A document is considered to be a sequence of characters ignoring its layout and other visual information. The fundamental concept in the algebra is that of a span, an ordered pair $\langle begin, end \rangle$ that denotes a region or text within a document identified by its "begin" and "end" positions. Each annotator finds regions of the document that satisfy a set of rules, and marks each region with an object called a span.

The algebra operates over a simple relational data model with three data types: span, tuple, and relation. A tuple is an finite sequence of w spans $\langle s_1, ..., s_w \rangle$; where w is the width of the tuple. A relation is a multiset of tuples, with the constraint that every tuple in the relation must be of the same width. Each operator takes zero or more relations as input and produces a single output relation.

Operator Algebra. The set of operators in the algebra can be categorized broadly into relational operators, span extraction operators, and span aggregation operators as shown in Table 1. Relational operators include the standard operators such as select, project, join, union, etc. The span extraction operators identify segments of text that match some pattern and produce spans corresponding to these matches. The two common span extraction operators are the regular expression matcher ϵ_{re} and the dictionary matcher ϵ_d. The regular expression matcher takes a regular expression r, matches it to the input text and outputs spans corresponding to these matches. The dictionary matcher takes a dictionary $dict$, consisting of a set of words/phrases, matches these to the input text and outputs spans corresponding to each occurence of a dictionary item in the input text.

Table 1. Operator Algebra for Information Extraction

Operator class	Operators	Explanation
Relational	$\sigma, \pi, \times, \cup, \cap, \ldots$	
Span extraction	$\epsilon_{re}, \epsilon_d$	
Span aggregation	$\Omega_o, \Omega_c, \beta$	
	$s_1 \preceq_d s_2$	s_1 and s_2 do not overlap, s_1 precedes s_2, and there are at most d characters between them
Predicates	$s_1 \simeq s_2$	the spans overlap
	$s_1 \subset s_2$	s_1 is strictly contained within s_2
	$s_1 = s_2$	spans are identical

The span aggregation operators take in a set of input spans and produce a set of output spans by performing certain aggregate operations over the input spans. There are two main types of aggregation operators - *consolidation* and *block*. The consolidation operators are used to resolve overlapping matches of the same concept in the text. Consolidation can be done using different rules. Containment consolidation (Ω_c) is used to discard annotation spans that are wholly contained within other spans. Overlap consolidation (Ω_o) is used to produce new spans by

Table 2. Visual Operators

	Operator	Explanation
Span Generating	$\Re(d)$ Ancestors(vs) Descendants(vs)	Return all the visual spans for the document d Return all ancestor visual spans of vs Return all descendant visual spans of vs
Directional Predicate	NorthOf(vs_1, vs_2) StrictNorthOf(vs_1, vs_2)	Span vs_1 occurs above vs_2 in the page layout Span vs_1 occurs strictly above vs_2 in the page
Containment Predicate	Contains(vs_1, vs_2) Touches(vs_1, vs_2) Intersects(vs_1, vs_2)	vs_1 is contained within vs_2 vs_1 touches vs_2 on one of the four edges vs_1 and vs_2 intersect
Generalization, Specialization	MaximalRegion(vs)/ MinimalRegion(vs)	Returns the largest/smallest visual span vs_m that contains vs and the same text content as vs
Geometric	Area(vs) Centroid(vs)	Returns the area corresponding to vs Returns a visual span that has x and y coordinates corresponding to the centroid of vs and text span identical to vs
Grouping	(Horizontally/Vertically)Aligned (VS, consecutive, maxdist)	Returns groups of horizontally/vertically aligned visual spans from VS. If the *consecutive* flag is set, the visual spans have to be consecutive with no non-aligned span in between. The *maxdist* limits the maximum distance possible between two consecutive visual spans in a group
Aggregation	MinimalSuperRegion(VS) MinimalBoundingRegion(VS)	Returns the smallest visual span that contains all the visual spans in set VS Returns a minimum bounding rectangle of all visual spans in set VS

merging overlapping spans. The block aggregation operator (β) identifies spans of text enclosing a minimum number of input spans such that no two consecutive spans are more than a specified distance apart. It is useful in combining a set of consecutive input spans into bigger spans that represent aggregate concepts. The algebra also includes some new selection predicates that apply only to spans as shown in Table 1.

3.2   Extensions for Visual Information Extraction

We extend the algebra described in order to support information extraction based on visual rules. In addition to the span, we add two new types in our model – *Region* and *VisualSpan*. A *Region* represents a visual box in the layout of the page and has the attributes: $\langle x_l, y_l, x_h, y_h \rangle$. (x_l, y_l) and (x_h, y_h) denote the bounding box of the identified region in the visual layout of the document. We assume that the regions are rectangles, which applies to most markup languages such as HTML. A *VisualSpan* is a combination of a text based span and a visual region with the following attributes: $\langle s, r \rangle$, where s is a text span having attributes *begin* and *end* as before and r is the region corresponding to the span.

The operators are also modified to work with visual spans. The relational operators are unchanged. The span extraction operators are modified to return visual spans rather than spans. For example, the regular expression operator ϵ_{re} matches the regular expression r to the input text and for each matching text span s it returns its corresponding visual span. Similarly, the dictionary matcher ϵ_d outputs visual spans corresponding to occurences of dictionary items in the input text. The behavior of the span aggregation operators (Ω_c and Ω_o)

is also affected. Thus containment consolidation Ω_c will discard visual spans whose region and span are both contained in the region and span of some other visual span. Overlap consolidation (Ω_o) aggregates visual spans whose text spans overlap. It produces a new visual span whose text span is the merge of the overlapping text spans and bounding box is the region corresponding to the closest HTML element that contains the merged text span.

There are two flavors to the block aggregation operator (β). The text block operator (β_s) is identical to the earlier β operator. It identifies spans of text enclosing a minimum number of input spans such that no two consecutive spans are more than a specified distance apart. The region block operator (β_v) takes as input a X distance x and Y distance y. It finds visual spans whose region contains a minimum number of input visual spans that can be ordered such that the X distance between two consecutive spans is less than x and the Y distance is less than y. The text span of the output visual spans is the actual span of the text corresponding to its region.

The predicates described in Table 1 can still be applied to the text span part of the visual spans. To compare the region part of the visual spans, we need many new predicates, which are described in the next section.

3.3   Visual Operators

We introduce many new operators in the algebra to enable writing of rules based on visual regions. The operators can be classified as span generating, scalar or grouping operators and a subset has been listed in Table 2. Many of these operators are borrowed from spatial (GIS) databases. For example, the operators Contains, Touches and Intersects are available in a GIS database like DB2 Spatial Extender[2]. However, to our best knowledge this is the first application of using these constructs for Information Extraction.

Span Generating Operators. These operators produce a set of visual spans as output and include the $\Re(d)$, *Ancestors* and *Descendents* operators.

Scalar Operators. The scalar operators take as input one or more values from a single tuple and return a single value. Boolean scalar operators can be used in predicates and are further classified as directional or containment operators. The directional operators allow visual spans to be compared based on their positions in the layout. Due to lack of space, we have listed only *NorthOf*, however we have similar predicates for other directions. Other scalar operators include the generalization/specialization operators and the geometric operators.

Grouping Operators. The grouping operators are used to group multiple tuples based on some criteria and apply an aggregation function to each group, similar to the GROUP BY functionality in SQL.

---

[2] http://www-01.ibm.com/software/data/spatial/db2spatial/

3.4   Comparison with Source Based Approach

If visual algebra is not supported, we would have to impelement a given task using only source based rules. The visual algebra is a superset of the existing source based algebra. Expressing a visual rule using existing algebra as a source based rule can be categorized into one of the following cases:

1. **Identical Semantics:** Some of the visual operators can be mapped directly into source level rules keeping the semantics intact. For example, the operator *VerticallyAligned* can be mapped to an expression based on constructs in html that are used for alignment such as `<tr>`, `<li>` or `<p>`, depending on the exact task at hand.
2. **Approximate Semantics:** Mapping a visual rule to a source based rule with identical semantics may lead to very complex rules since there are many ways to achieve the same visual layout. It may be possible to get approximately similar results by simplifing the rules if we know that the layout for the pages in the dataset is achieved in one particular way. For example, in a particular template, alignment may always be implemented using rows of a table (the `<tr>` tag), so the source based rule can cover only this case.
3. **Alternate Semantics:** In some cases, it is not possible to obtain even similar semantics from the source based rules. For example, rules based on *Area, Centroid, Contains, Touches* and *Intersects* cannot be mapped to source based rules, since it is not possible to check these conditions without actually rendering of the page. In such cases, we have to use alternate source based rules for the same task.

4   System Architecture and Implementation

This section describes the architecture and our implementation of the visual extraction system. There are two models typically used for information extraction – document level processing, in which rules are applied to one document at a time and collection level processing, in which the rules are matched against the entire document collection at once. The document at a time processing is suitable in the scenario where the document collection is dynamic and new documents are added over time. The collection level processing is useful when the document collection is static and the rules are dynamic, i.e. new rules are being developed on the same collection over time. Previous work has demonstrated an order of magnitude improvement in performance by collection level processing compared to document level processing with the use of indices for evaluating regular expression rules [14]. The visual algebra can be implemented using either a document level processing model or a collection level processing model. We implemented a collection level processing approach using a relational database with extensions for inverted indices on text for efficient query processing. Figure 2 depicts the overall system architecture. Collection level processing has two

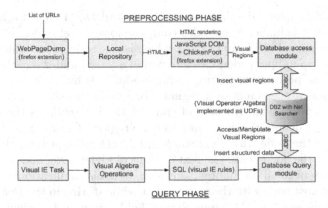

Fig. 2. System Architecture

phases: (a) Preprocessing phase comprising computations that can be done of-fline and (b) Query phase that includes the online computations done during interactive query time.

Preprocessing Phase. In the preprocessing phase, web pages from which in-formation is to be extracted are crawled and a local repository of the web pages is created. Along with the HTML source of the web page, all components that are required to render the page accurately, such as embedded images and stylesheets, should also be downloaded and appropriately linked from the local copy of the page. We use an open source Firefox extension called WebPageDump[3] specifi-cally designed for this purpose. Each page is then rendered in a browser and for each node in the DOM tree, its visual region and text is extracted (using the Chickenfoot Firefox extension[4]) and stored in a relational database (IBM DB2 UDB). We also use the indexing and text search capabilities of DB2 Net Search Extender[5] to speed up queries that can benefit from an inverted index.

Query Phase. During the interactive query phase, the user expresses the infor-mation extraction task as operations in the visual algebra. The visual algebraic operations are then translated to standard SQL queries and executed on the database.

4.1 Implementing Visual Algebra Queries Using a Database

Schema. The visual regions computed in the pre-processing stage are stored in table called *Regions* with the following schema:
$< Pageid, Regionid, x_l, y_l, x_h, y_h, TextStart, TextEnd,$
$Text, HtmlTag, MinimalRegion, MaximalRegion >.$

---

[3] http://www.dbai.tuwien.ac.at/user/pollak/webpagedump/
[4] http://groups.csail.mit.edu/uid/chickenfoot/
[5] http://www.ibm.com/software/data/db2/extenders/netsearch/

The *Pageid* uniquely identifies a page. The html DOM tree is a hierarchical structure where the higher level nodes comprise lower level nodes. For example, a `<td>` may be nested inside a `<tr>` tag, which is nested inside a `<table>`, and so on. The *Regionid* uniquely identifies a region in a page annd is a path expression that encodes the path to the corresponding node. This makes it easy to identify the parents and descendants of a region. For example, a node 1.2 indicates a node reached by following the second child of the first child of the root node. The x_l, y_l, x_h, y_h denote the coordinates of the region. The *Text* field stores the text content of the node with *TextStart* and *TextEnd* indicating the offsets of the text within the document. The text content of higher level nodes is the union of the text content of all its children. However, to avoid duplication, we associate only the innermost node with the text content while storing in the *Regions* table. The *MinimalRegion* and *MaximalRegion* fields are used to quickly identify a descendant or ancestor that has the identical text content as this node.

Implementation of Operators. The visual algebra is implemented using a combination of standard SQL and User Defined Functions (UDFs). Due to space constraints, we mention the mapping of only some representative operators without going into complete detail in Table 3. For simplicity, we have shown the SQL for each operator separately. Applying these rules for a general algebra expression will produce a nested SQL statement that can be flattened out into a single SQL using the regular transformation rules for SQL sub-queries. We also experimented with using a spatial database to implement our algebra, but found that it was not very efficient. Spatial databases can handle complex geometries, but are not optimized for the simple rectangular geometries that the visual regions have. Conditions arising from simple rectangular geometries can be easily mapped to simple conditions on the region coordinate columns in a regular relational database.

Table 3. Mapping to SQL

Operator	Mapping
$\Re$	`SELECT Pageid, Regionid FROM Regions`
$\epsilon_{re}(exp)$	`SELECT Pageid, Regionid FROM Regions R` `WHERE MatchesRegex(R.Text, exp)`
$\epsilon_d(dict)$	`SELECT Pageid, Regionid FROM Regions R` `WHERE MatchesDict(R.Text, dict)`
$Ancestors(v)$	`SELECT Pageid, Regionid FROM Regions R` `WHERE IsPrefix(R.Regionid, v.RegionId)`
$StrictNorthOf(v_1, v_2)$	`...WHERE` $v_1.y_h \leq v_2.y_l$ `AND` $v_1.x_l > v_2.x_l$ `AND` $v_1.x_h \leq v_2.x_h$
$MinimalRegion(v)$	`SELECT Pageid, MinimalRegion FROM Regions R` `WHERE R.Regionid = v.Regionid`
$HorizontallyAligned(R)$	`...FROM R GROUP BY` $R.x_l$
$MinimalBoundingRegion(V)$	`SELECT` $min(x_l), min(y_l), max(x_h), max(y_h)$ `FROM V`

Visual Span Producing Operators: The ϵ_{re} and ϵ_d operators are implemented using UDFs that implement regular expression and dictionary matching respectively. $Anscestors(v)$ and $Descendants(v)$ are implemented using the path expression in the region id of vs. Searching for all prefixes of the $Regionid$ returns the ancestors and searching for all extensions of $Regionid$ returns the descendants.

Span Aggregation Operators: The span aggregation operators (Ω_o, Ω_c and β_v) cannot be easily mapped to existing operators in SQL. We implement these in Java, external to the database.

Other Visual Operators: The scalar visual operators include the directional predicates, containment predicates, generalization/specialization operators and geometric operators. The predicates map to expressions in the $WHERE$ clause. The generalization/specialization predicates are implemented using the precomputed values in the columns $MinimalRegion$ and $MaximalRegion$. The grouping operators map to $GROUPBY$ clause in SQL and the aggregate functions can be mapped to SQL aggregate functions in a straightforward way as shown for $HorizontallyAligned$ and $MinimalBoundingRegion$.

Use of Indices. Indices can be used to speed up the text and region predicates. Instead of the $MatchesRegex$ UDF, we can use the $CONTAINS$ operation provided by the text index. We also build indices on x_l, x_l, x_h, y_h columns to speed up visual operators. Once the visual algebra query is mapped to a SQL query, the optimizer performs the task of deciding what indices to use for the query based on cost implications. Example of a mapping is shown in Table 4.

5   Experiments

The goal of the experiments is two fold - to demonstrate the simplicity of visual queries and to study the effectiveness of mapping the visual algebra queries to database queries. We describe the visual algebra queries for a representative set of tasks, map them to SQL queries in a database system and study the effect of indexing on the performance.

5.1   Experimental Setup

The document corpus for our experiments consists of software product information pages from IBM web site [6]. We crawled these pages resulting in a corpus of 44726 pages. Our goal was to extract the system requirements information for these products from their web pages (see Figure 1). Extracting the system requirements is a challenging task since the pages are manually created and don't have a standard format. This can be broken into sub-tasks that we use as representative queries for our experiments. The queries are listed below. The visual algebra expression and the equivalent SQL query over the spatial database are

---

[6] http://www.ibm.com/software/products/us/en?pgel=lnav

listed in Table 4. For ease of expression, the visual algebra queries are specified using a SQL like syntax. The functions $RegEx$ and $Dict$ represent the operators ϵ_r and ϵ_d respectively. For each of these sub-tasks, it is possible to write more precise queries. However, our goal here is to show how visual queries can be used for a variety of extraction tasks without focusing too much on the precision and recall of these queries.

- Filter the navigational bar at the left edge before extracting the system requirements.

  Q1: Retrieve vertically aligned regions with more than n regions such that the region bounding the group is contained within a virtual region $A(x_l, y_l, x_h, y_h)$. For our domain, we found that a virtual region of $A(0, 90, 500, \infty)$ works well.
- Identify whether a page is systems requirements page. We use the heuristic that system requirement pages have the term "system requirements" mentioned near the top of the page.

  Q2: Retrieve the region in the page containing the term 'system requirements' contained in a region A. In this case, we use a virtual region, $A(450, 0, \infty, 500)$
- To identify various operating systems that are supported, the following query can be used.

  Q3: Find all regions R, such that R contains one of the operating systems mentioned in a dictionary T and are to the strict south or to the strict east of a region containing the term "Operating Systems".
- To find the actual system requirements for a particular operating system, the following query can be used.

  Q4: Find a region that contains the term "Windows" that occurs to the strict south of a region containing the term "Operating Systems" and extract a region to the strict right of such a region.

Due to lack of space, we show the visual algebra expression and the equivalent SQL query (Section 4.1) for only query Q4 in Table 4. For ease of expression, the visual algebra queries are specified using a SQL like syntax.

Table 4. Queries

Q	Visual Query	SQL Query
4	select R3.VisualSpan from RegEx('operating system', D) as R1, RegEx('windows', D) as R2, $\Re(D)$ as R3 where StrictSouthOf(R2, R1) and StrictEastOf(R3, R2)	SELECT R3.pageid, R3.regionid FROM regions R1, regions R2, regions R3 WHERE r1.pageid = R2.pageid AND R2.pageid = R3.pageid AND contains(R1.text, '"Operating Systems"') = 1 AND contains(R2.text, '"Windows"') = 1 AND $R2.y_l \geq R1.y_h$ AND $R2.x_l \geq R1.x_l$ AND $R2.x_h \leq R1.x_h$ AND $R3.x_l \geq R2.x_h$ AND $R3.y_l \geq R2.y_l$ AND $R3.y_h \leq R2.y_h$

5.2   Accuracy of Spatial Rules

We measured the accuracy of our spatial rules using manually annotated data from a subset of pages in our corpus. The test set for Q2 and Q4 consists of 116 manually tagged pages. The test set for Q1 and Q3 contains 3310 regions from 10 pages with 525 positive examples for Q1 and 23 positive examples for Q3. Please note that for Q1 and Q3 we need to manually tag each region in a page. Since there are few hundred regions in a page, we manually tagged only 10 pages. The rules were developed by looking at different patterns that occur in a random sample of the entire corpus. The results are reported in Figure 4. Since our tasks were well suited for extraction using spatial rules, we were able to obtain a high level of accuracy using relatively simple rules.

5.3   Performance

We measured the performance of these queries on the document collection. Since the queries have selection predicates on the text column and the coordinates (x_l, y_l, x_h, y_h), we build indices to speed them up. We also index the text column using DB2 Net search extender. The running time for the queries are shown in Figure 3. We compare various options of using no indices, using only text index and using both text index and indices on the region coordinates. For $Q1$, the text index does not make a difference since there is no text predicate. The region index leads to big improvement in the time. $Q2$, $Q3$ and $Q4$ have both text and region predicates and thus benefit from the text index as well as the region indices. The benefit of the text index is found to be compartively larger. In all the cases, we can see that using indices leads to a three to fifteen times improvement in the query execution times.

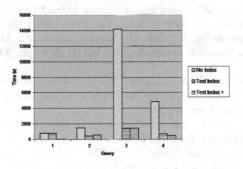

Query	Q1	Q2	Q3	Q4
Recall	100	96	100	100
Precision	84	85	88	100

Fig. 3. Effect of Indices   **Fig. 4.** Accuracy

6   Discussion

We have demonstrated an extension to the traditional rule based IE framework that allows the user to specify layout based rules. This framework can be used for many information extraction tasks that require spatial analysis without having to use custom code. The WYSIWYE algebra we propose allows the user to seamlessly combine traditional text based rules with high level rules based on spatial layout. The visual algebra can be systematically mapped to SQL statements, thus enabling optimization by the database. We have evaluated our system in terms of usability and performance for a task of extracting software system requirements from software web pages. The rules expressed using the visual algebra are much simpler than the corresponding source based rules and more robust to changes in the source code. The performance results show that by mapping the queries to SQL and using text and region indexes in the database, we can get significant improvement in the time required to apply the rules.

Layout based rules are useful for certain types of pages, where the layout information provides cues on the information to extract. A significant source of variation in web pages (different source code, same visual layout) can be addressed by rule based information extraction systems based on a visual algebra, leading to simpler rules. Visual rules are not always a replacement for the text based rules, rather they are complementary. In our system, we can write rules that combine both text based and layout based rules in one general framework.

References

1. Cai, D., Yu, S., Wen, J.R., Ma, W.Y.: Vips: a vision-based page segmentation algorithm. Technical report, Microsoft Research (2003)
2. Gatterbauer, W., Bohunsky, P., Herzog, M., Krüpl, B., Pollak, B.: Towards domain-independent information extraction from web tables. In: WWW 2007, Banff, Alberta, Canada, pp. 71–80. ACM (2007)
3. Lerman, K., Getoor, L., Minton, S., Knoblock, C.: Using the structure of web sites for automatic segmentation of tables. In: SIGMOD 2004, pp. 119–130. ACM, New York (2004)
4. Krüpl, B., Herzog, M., Gatterbauer, W.: Using visual cues for extraction of tabular data from arbitrary html documents. In: WWW 2005, pp. 1000–1001 (2005)
5. Reiss, F., Raghavan, S., Krishnamurthy, R., Zhu, H., Vaithyanathan, S.: An algebraic approach to rule-based information extraction. In: ICDE 2008, pp. 933–942 (2008)
6. Shen, W., Doan, A., Naughton, J.F., Ramakrishnan, R.: Declarative information extraction using datalog with embedded extraction predicates. In: VLDB 2007, pp. 1033–1044. VLDB Endowment, Vienna (2007)
7. Appelt, D.E., Onyshkevych, B.: The common pattern specification language. In: Proceedings of a Workshop on Held at Baltimore, Maryland, Morristown, NJ, USA, pp. 23–30. Association for Computational Linguistics (1996)
8. Sarawagi, S.: Information extraction. FnT Databases 1(3) (2008)
9. Cunningham, H., Wilks, Y., Gaizauskas, R.J.: Gate - a general architecture for text engineering (1996)
10. Ferrucci, D., Lally, A.: Uima: an architectural approach to unstructured information processing in the corporate research environment. Nat. Lang. Eng. 10(3-4), 327–348 (2004)
11. Gu, X.-D., Chen, J., Ma, W.-Y., Chen, G.-L.: Visual Based Content Understanding towards Web Adaptation. In: De Bra, P., Brusilovsky, P., Conejo, R. (eds.) AH 2002. LNCS, vol. 2347, pp. 164–173. Springer, Heidelberg (2002)

12. Kovacevic, M., Diligenti, M., Gori, M., Milutinovic, V.: Recognition of common areas in a web page using visual information: a possible application in a page classification. In: ICDM 2002, p. 250. IEEE Computer Society, Washington, DC (2002)
13. Arasu, A., Garcia-Molina, H.: Extracting structured data from web pages. In: SIGMOD Conference, pp. 337–348 (2003)
14. Ramakrishnan, G., Balakrishnan, S., Joshi, S.: Entity annotation based on inverse index operations. In: EMNLP 2006, pp. 492–500. Association for Computational Linguistics, Sydney (2006)

A Scalable Algorithm for Detecting Community Outliers in Social Networks

Tengfei Ji*, Jun Gao, and Dongqing Yang

School of Electronics Engineering and Computer Science
Peking University, Beijing, 100871 China
{tfji,gaojun,dqyang}@pku.edu.cn

Abstract. Outlier detection is an important problem that has been researched and applied in a myriad of domains ranging from fraudulent transactions to intrusion detection. Most existing methods have been specially developed for detecting global and (or) local outliers by using either content information or structure information. Unfortunately, these conventional algorithms have been facing with unprecedented challenges in social networks, where data and link information are tightly integrated.

In this paper, a novel measurement named Community Outlying Factor is put forward for community outlier, besides its descriptive definition. A scalable community outliers detection algorithm (SCODA), which fully considers both content and structure information of social networks, is proposed. Furthermore, SCODA takes effective measures to minimize the number of input parameters down to only one, the number of outliers. Experimental results demonstrate that the time complexity of SCODA is linear to the number of nodes, which means that our algorithm can easily deal with very large data sets.

Keywords: community outlier, outlier detection, social networks.

1 Introduction

Outlier detection is a task that seeks to determine and report such data objects which are grossly different from or inconsistent with other members of the sample [1,2]. The technique has the ability to potentially shed light on the unexpected knowledge with underlying value. Therefore, outlier detection has attracted much attention within diverse areas, ranging from fraudulent transactions to intrusion detection [3,4]. Recently, the advent of social networks exemplified by websites such as Facebook and MySpace has brought the following unprecedented challenges to outlier detection.

- **Content & Structure:** A *social network* is defined as a graph where the objects are represented by vertices, and the interactions between them are denoted by edges. In the social network representation, in addition to content

---

* Corresponding author.

H. Gao et al. (Eds.): WAIM 2012, LNCS 7418, pp. 434–445, 2012.

information associated with nodes, topological structure embedded in links is also available. Therefore, the new detecting approaches that fully take both content aspect and structure aspect into consideration are required.

- **Tremendous Amounts of Information:** Compared with average data sets, social networks are significantly larger in size where graphs may involve millions of nodes and billions of edges [7]. Typically, in the case of Facebook with more than 10^8 nodes, the potential number of edges could be of the order of 10^{11}. The availability of massive amounts of data has given a new impetus towards a robust study of outlier detection in social networks [8]. The issue of scalability should be at the top of the priority list because of storage and efficiency constraints.
- **Context of Community:** Communities, in the social network sense, are groups of entities that presumably share some common properties [6]. The profile of community provides the background for the specific contextual outlier detection. This kind of outlier may appear to be normal behavior compared to the entire data set while can present abnormal expression according to specific context. For instance, from a global point of view, normal as 1.8 meter in height seems, but it is an obvious deviation to a boy of 10 years old. Conventional outlier detection algorithms that focused solely on detecting global and(or) local outliers will fail to detect contextual outlier.

In this paper, *community outliers* are defined as those objects have a higher density of external links, compared with its internal links. As mentioned above, the community here is a group of entities that presumably share some common properties [6].

The example in figure 1 is adopted to illustrate directly the feature of community outlier. According to income, individuals in figure 1 are partitioned into three communities, namely low-income, middle-income and high-income community. The links between nodes disclose the friend relationship. In most cases, as a well-known proverb says, "A man is known by the company he keeps", one is supposed to make friends with those who have the same income level. Most of nodes in figure 1 correspond to prediction. However, node v_{13} is deviating widely from others. It is located in low-income group with 1 low-income friends, while its high-income friends and middle-income friends are 2 and 2, respectively. Community outlier node v_{13} is usually regard as a rising star in the social networks, for instance, a young and promising entrepreneur [9,10]. Obviously, different from global and local outliers, node v_{13} can hardly be detected by algorithms based on either content information or structure information.

The contribution of our work can be summarized as follows:

- Besides descriptive concept, we put forward a novel measurable definition of community outlier, namely *Community Outlying Factor*. To the best of our knowledge, this is the first straightforward concept of a community outlier which quantifies how outlying a community member is.
- We propose a scalable community outliers detection algorithm (SCODA), which fully considers the content information and the structure information of social networks. The communities that produced by the modified

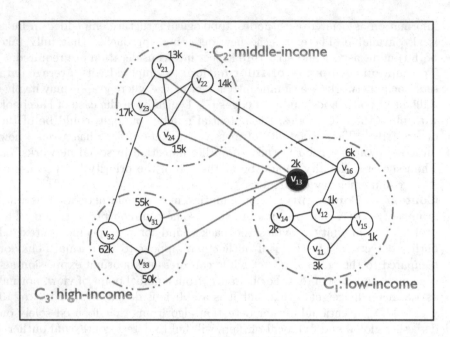

Fig. 1. An Example for community outlier

usmSqueezer algorithm(MSqueezer) according to the content information in phase I will be utilized as a context in phase II of SCODA for detecting community outliers with abnormal structure information.

– Our algorithm takes effective measures to minimize input parameters. Only the number of needed outliers is required as an input parameter.
– The extensive experiments demonstrate that the time complexity of SCODA is linear to the number of nodes. This property means that the algorithm can easily deal with very large data sets.

The rest part of this work is organized as follows: Section 2 discusses the recent related work; Section 3 gives the preliminaries about Squeezer algorithm; Section 4 proposes our scalable community outliers detection algorithm,SCODA; Section 5 gives experiments for our approach on both real and synthetic data sets, and shows the achieved results. Section 6 makes a conclusion about the whole work.

2   Related Work

To focus on the theme, the majority algorithms that aim at detecting global or local outliers by using either content or structure information will no more be introduced in this paper. We are eager to discuss some state-of-art algorithms that encompass both individual object and network information.

Some literatures[15,16] cluster data points according to the combination of data and link information. However, instead of outlier detection, they are proposed for community discovery, assuming that there are no outliers in social networks.

A contextual outlier detection method in [9] couples content and relationship information by modeling networked data as a mixture model composed of multiple normal communities and a set of randomly generated outliers. The probabilistic model characterizes both data and links simultaneously by defining their joint distribution based on hidden Markov random fields (HMRF). Maximizing the data likelihood and the posterior of the model gives the solution to the outlier inference problem. Unfortunately, the algorithm could not overcome the weakness of statistical outlier detection techniques which are limited in the sense that the data distribution and underlying parametric formulation are difficult to directly obtain in advance [11]. Moreover, most of the distribution models are not suitable for multi-dimensional space. Another limitation with these approaches is that they typically do not scale well to large or even moderately large datasets [14].

Some methods [17-19] solve the problem by first conducting graph partition using link information, and then adopting distance-based approaches or density-based approaches within each community. However, they suffer expensive computational cost as they require the calculation of the distances [5] or the analysis of the neighborhood density [12,13].

In summary, limitations of aforementioned approaches prompted us to look for a scalable community outlier detection algorithm that considers both content and structural information with low time cost and minimum input parameters.

3   Preliminaries

3.1   The usmSqueezer Algorithm

Dataset D is featured by m attributes and A_i is the i-th attribute. There is an assumption that the first p continuous attributes before the rest (m-p) categorical attributes. The domain of the i-th categorical attribute A_i denotes as $Dom(A_i)$, which has r_i different values. The usmSqueezer algorithm [20] reads each tuple t in sequence, either assigning t to an existing cluster, or creating t as a new cluster, which is determined by the *usmSimilarity* between t and clusters. *usmSimilarity*(C,t) is the sum of similarity measure between *categorical attributes* and similarity measure between *numerical attributes*. More details can be found in reference [20].

$$Summary = \{c_i \mid 1 \leq i \leq p\} \bigcup \{(A_{ij}, sup(A_{ij})) \mid p+1 \leq i \leq m, 1 \leq j \leq r\}$$

where sup() is a function to measure the frequency of categorical attribute a_i, and c_i represents the mean of first p numeric attributes.

Intuitively, the cluster stores its own summary information. The information contained in *Summary* is sufficient enough to compute the similarity between a tuple and Cluster.

4   SCODA Algorithm

Our SCODA algorithm involves two major phases. In the first phase, the modified usmSqueezer algorithm (MSqueezer for short) efficiently groups data objects into communities by their content information. The second phase is community outlier detection. SCODA identifies community outliers within communities generated in the first phase according to *Community Outlying Degree Factor*, which is a novel measurable standard considering the structure information of social networks.

4.1   Phase I: Content-Based Clustering

The first phase aims to efficiently obtain the partitions based on objects' information. The success of the usmSqueezer algorithm in rapidly producing high-quality clustering results in high-dimensional datasets mixed type of attributes motivates us to take advantage of it in our first phase. However, the number and the size of clusters obtained from the usmSqueezer algorithm are suffering the influence of the similarity threshold value, which is a static parameter predefined by users. Since the partitions generated in the first phase are principal to community outlier detection in the second phase, we design a similarity threshold *dynamic update* mechanism with *no personal interventions* for usmSqueezer algorithm. The modified usmSqueezer algorithm is named MSqueezer for short.

The definitions of *usmSimilarity* and *Summary* are still maintained just as in section 3.

Definition 1 (SS and SS^2). Given a Cluster C with $|C|$ tuples, the Sum of usmSimilarities (SS) and the Sum of squared usmSimilarities (SS^2) for C are respectively defined as

$$SS = \sum_{i=1}^{|C|} sumSmililarity(C, t_i), \quad t_i \in C$$

$$SS^2 = \sum_{i=1}^{|C|} sumSmililarity(C, t_i)^2, \quad t_i \in C$$

Definition 2 (Mortified Cluster Structure: *MCS*). Given a Cluster C with $|C|$ tuples, the Mortified Cluster Structure(MCS) for C is defined as

$$MCS = \{Cluster, Summary, |C|, SS, SS^2\}$$

The MCS is the main data structure stored by MSqueezer algorithm, which could be used to compute *usmSimilarity* and *Dynamic Similarity Threshold*.

Definition 3 (Dynamic Similarity Threshold: δ). According to Chebyshevs inequality:

$$Pr(\mid \delta.lower - \mu \mid \geq k\sigma) \leq \frac{1}{k^2}$$

Here the real number k is set to $\sqrt{2}$, we obtain the $\delta.lower$ for C $\mu - \sqrt{2}\sigma$. Alternatively, the $\delta.upper$ for C is set to be μ.

Where μ is the sample expected value and σ is sample standard deviation, which can be easily determined from SS and SS^2.

The multi-granularity threshold is adopted to determine whether to receive tuple t as a new member or not, because 1) our purpose is to produce communities of proper size without personal interventions; 2) $\delta.lower$ is able to avoid the phenomenon that clusters tend to decrease in size, if increasing $\delta.upper$ is used as the only threshold; 3) This can partly reduce the sensitivity towards data input order.

Our MSqueezer algorithm partitions n tuples into communities according to objects' information. Initially, the first tuple is read in and a MCS is constructed. Then, the rest tuples are read iteratively.

For each tuple, MSqueezer computes its usmSimilarity with all existing clusters. We examine the usmSimilarities in descending order, if the i-th largest $usmSimilarity(C_k, t)$ is larger than the $\delta.upper$ of cluster k, the tuple t will be put into cluster k. If all $\delta.uppers$ are unsatisfied, tuple t will be put into the cluster whose $\delta.lower$ is the first to be satisfied. The corresponding MCS is also updated with the new tuple. If the above condition does not hold, a new cluster must be created with this tuple. The algorithm continues until all the tuples have been traversed.

Algorithm: MSqueezer Algorithm (High level definition)

Input: Dataset D;
Output: A group of communities

Step 1: Read in a tuple t in sequential order;
Step 2: If t is the first tuple, add t to new MCS, else goto step3;
Step 3: For each existing cluster C, compute $usmSimilarity(C, t)$ and then sort usmSimilarities in descending order;
Step 4: For each usmSimilarity, get the corresponding threshold $\delta.upper$. If usmSimilarity $\geq \delta.upper$, add tuple t into cluster C, update the MSC, $\delta.upper$ and $\delta.lower$ of C and then goto step7, else goto step5;
Step 5: For each usmSimilarity, get the corresponding threshold $\delta.lower$. If usmSimilarity $\geq \delta.lower$, add tuple t into cluster C, update the MSC, $\delta.upper$ and $\delta.lower$ of C and then goto step7, else goto step6;
Step 6: add t to new MCS;
Step 7: Goto step1 until there is no unread tuple.

Time Complexity Analysis: The computational complexity of the MSqueezer algorithm has two parts: 1) the complexity for executing original Squeezer algorithm on the dataset; 2) the complexity for sorting and updating dynamic similarity thresholds; . The time complexity of the original Squeezer algorithm is $O(k * m * n)$, where n, m, k are the number of nodes, attributes and clusters, respectively. As for part 2), sorting and updating dynamic similarity thresholds are $O(k \log k)$ $(k \ll n)$ and $O(n)$, respectively. Therefore, the overall time complexity for the MSqueezer algorithm is $O(k * m * n)$.

4.2   Phase II: Structure-Based Community Outlier Detecting

In this section, we develop a formal measurement of community outliers by taking full account of the link information between communities, which are produced in phase I. We will first establish some notations and definitions.

The network $G = \langle V, E \rangle$ consists of a set of nodes V and a set of links E. The weights attached to links are stored in an adjacency matrix. For $\forall v_{ij} \in V$, we define its intra-community neighbor and inter-community neighbor as follows.

Definition 4 (Intra-Community Neighbor). $\forall$ node $v_{ij} \in$ community C_i, if node v_{ij} can communicate with node v_{ip} ($v_{ip} \in$ community C_i, and $p \neq j$), node v_{ip} is called a intra-community neighbor of node v_{ij}, all the intra-community neighbors of node v_{ij} constitute its intra-community neighbor set Intra-NS(v_{ij}), including v_{ip}.

Definition 5 (Inter-Community Neighbor). $\forall v_{ij} \in V$, if node v_{ij} can communicate with node v_{qp} ($v_{qp} \in$ community C_q, and $q \neq i$), node v_{qp} is called a inter-community neighbor of node v_{ij}, all v_{ij}'s inter-community neighbors in community C_q constitute its inter-community neighbor set with respect to community C_q, which is denoted as Inter-NS$^{C_q}(v_{ij})$.

Example 1: In figure 1, v_{16} is a intra-community neighbor of v_{13}, while v_{21} and v_{24} constitute v_{16}'s inter-community neighbor sets with respect to community C_2, namely Inter-NS$^{C_2}(v_{13})$. In the same way, Inter-NS$^{C_3}(v_{13})$ includes v_{31} and v_{33}.

Definition 6 (Link Density: *LD*). Let k be the number of communities in the network. $\forall v_{ij} \in V$, the Link Density of (v_{ij}) to community C_q ($1 \leq q \leq k$) is defined as

$$LD^{C_q}(v_{ij}) = \frac{W_{v_{ij}}^{C_q}}{|C_q|} \tag{1}$$

where $W_{v_{ij}}^{C_q}$ denotes the sum of edges' weights. These edges are between v_{ij} and its neighbors in community C_q. $|C_q|$ is the number of nodes in community C_q.

Intuitively, an object is supposed to mainly communicate (make friends) with the intra-community nodes that presumably share some common properties. In contrast, a community outlier owns a high external link density, whereas links within the community to which it belongs have a comparatively lower density. In other words, the object's community outlying degree is directly proportional to its inter-link density with respect to Inter-NS, and inversely proportional to intra-link density.

Definition 7 (Community Outlying Factor: *COF*). Let k be the number of communities in the network. The community outlying factor of v_{ij} ($v_{ij} \in C_i$) is defined as

$$COF(v_{ij}) = \frac{\sum\limits_{q \neq i, q=1}^{k} LD^{C_q}(v_{ij}) + \varepsilon}{LD^{C_i}(v_{ij}) + \varepsilon} \qquad (2)$$

Note that the link density can be ∞ if v_{ij} has no intra-community neighbor. For simplicity, we handle this case by adding the same infinitesimal positive number ε (e.g. 10^{-6})to both numerator and denominator.

Example 2: Continuing with *Example 1*, the community outlying factor of v_{13} can be computed using *Definition 7* as: $COF(v_{13}) = \dfrac{\dfrac{2}{3} + \dfrac{2}{4}}{\dfrac{1}{6}} = 7$

Algorithm: SCODA Algorithm (High level definition)
Input: Dataset D, the number of community outliers n;
Output: n community outliers
Step 1: Get a set of communities partitioned by MSqueezer (Phase I)
Step 2: Compute community outlying factor for each object;
Step 3: Select and output the objects with the first n-largest COF;

Obviously, the overall time complexity for the SCODA algorithm mainly depends on that of MSqueezer algorithm, which is O(k*m*n) as we discussed in Phase I. The above analysis demonstrates that the time complexity of SCODA algorithm is approximately linearly dependent on the number of nodes, which makes this algorithm more scalable.

5   Experiments

In this section, we illustrate the general behavior of the proposed SCODA algorithm. We compare the accuracy performance of SCODA with several baseline methods on synthetic datasets, and we examine the scalability of SCODA on real datasets.

5.1   Data Description and Evaluation Measure

Real Data Sets: We perform scalability experiments on 3 real data sets. These networks come from the same data set (DBLP, dblp.uni-trier.de/), but they vary in the number of nodes and edges (up to more than 400K nodes). DBLP bibliography is one of the best formatted and organized compute science community datasets. In our representation, we consider a undirected co-authorship network. The weighted graph W is constructed by extracting author-paper information: each author is denoted as a node in W; journal and conference papers are represented as links that connect the authors together; the edge weight is the number of joint publications by these two authors.

Synthetic Data Sets: Since "community outlier" is a new concept, the proper benchmark datasets are rare. Therefore, we attempt to convert a few classification datasets where each object consists of attribute values and a class label into datasets for community outlier detection. We generate synthetic data sets through two steps. First, we select two real UCI machine learning data sets, Adult and Yeast. The Adult dataset, which is based on census data, contains 45222 instances corresponding to 2 classes. The Yeast dataset is composed of 1484 instances with 8 attributes. Then we apply link generation procedure on these two data sets. The distribution of links follows Zipf's law, i.e. roughly 80% of the links come from 20% of the nodes. The self-links and the nodes without any links are removed from data sets. We denote the synthetic data sets based on Yeast and Adult as SYN1 and SYN2, respectively.

More detail information about real and synthetic datasets is shown in Table 1.

Table 1. Summary of the Data Sets

	Dataset	Node	Edge
Synthetic Datasets	SYN1	1,400	7,000
	SYN2	4,000	50,000
Real Datasets	DBLP1	5,089	9,106
	DBLP2	315,688	1,659,853
	DBLP3	404,892	1,422,263

We measured the performance of different algorithms using well-known metric F1 measure, which is defined as follows.

$$F1 = \frac{2 \times Recall \times Precision}{Recall + Precision}$$

where recall is ratio of the number of relevant records retrieved to the total number of relevant records in the dataset; precision is ratio of the number of relevant records retrieved to the total number of irrelevant and relevant records retrieved.

5.2   The Accuracy of SCODA Algorithm

To evaluate the clustering performance, we compared SCODA algorithm against three other algorithms. The first approach is a well-known single content-based outlier detection method, which identifies global outlier by its k-nearest neighbors' distance. Therefore we denote it as the Content Approach (CA). The second one (CODA) takes advantage of the probabilistic model, which characterizes both data and links simultaneously by defining their joint distribution based on hidden Markov random fields (HMRF). In fairness to all algorithms, we set the same number of outliers (from 5 to 20) for each method. We compared the F1 of the three algorithms on two simulated data sets. Table 2 illustrates the comparison results.

Table 2. The Accuracy Comparison on the Synthetic Datasets

Dataset	Num.outlier	CA	CODA	SCODA
SYN1	n=10	0.1	1.00	1.00
	n=20	0.05	0.90	1.00
SYN2	n=20	0.00	0.75	0.80
	n=40	0.03	0.68	0.83

Table 2 obviously indicates that CA algorithm which completely ignores the inherent structure information of datasets is far inferior to other algorithms. The experiments once again prove that solely using one type of information cannot accomplish accurate community outlier detection. The performance of CODA and SCODA are satisfactory in the situations because of considering both object and link information. The effectiveness of our proposed algorithm SCODA surpasses the state-of-the-art CODA approach.

5.3   The Scalability of SCODA Algorithm

To evaluate the scalability of SCODA, the next series of tests report the computation time as we vary the number of nodes. Figure 2 indicates the scalability of SCODA Algorithm with increasing number of nodes.

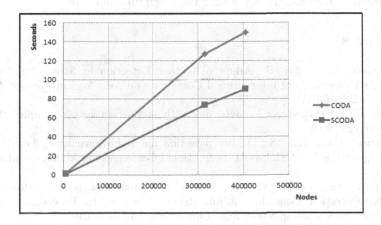

Fig. 2. Scalability Test of algorithms with increasing number of nodes

Figure 2 demonstrates that our method returns results that agree with our intuition, and there is a linear dependency of SCODAs processing time on the number of nodes in networks. Moreover, we can see that for the largest network, the computation time is less than one hundred seconds. This property means that the algorithm can easily deal with large data sets.

6 Conclusion

In this paper, we have investigated a novel outlier detection problem, namely community outlier detection, which springs from the advent of social network. Besides descriptive concept, we put forward a straightforward measurement named Community Outlying Factor, which quantifies how outlying a community member is. We propose a scalable community outliers detection algorithm (SCODA), which fully considers both content and structure information of social networks. Furthermore, we take effective measures to eliminate personal intervention by requiring a single input parameter. The experimental results on both real datasets and synthetic datasets clearly ascertain that SCODA algorithm is capable of detecting community outlier accurately and effectively. The scalability tests demonstrate SCODA algorithm is a scalable method that can efficiently work for large datasets.

Acknowledgment. This work was supported by the National High Technology Research and Development Program of China (Grant No. 2012AA011002), National Science and Technology Major Program (Grant No. 2010ZX01042-002-002-02, 2010ZX01042-001-003-05), National Science & Technology Pillar Program (Grant No. 2009BA H44B03), Natural Science Foundation of China 61073018, the Cultivation Fund of the Key Scientific and Technical Innovation Project, Ministry of Education of China (Grant No. 708001) and the Shenzhen-Hong Kong Innovation Cooperation Project (No. JSE201007160004A). We would like to thank anonymous reviewers for their helpful comments.

References

1. Toshniwal, D., Yadav, S.: Adaptive Outlier Detection in Streaming Time Series. In: Proceedings of International Conference on Asia Agriculture and Animal, ICAAA 2011 (2011)
2. Grubbs, F.E.: Procedures for detecting outlying observations in samples. Technometrics 11 (1969)
3. Aggarwal, C.C., Yu, P.S.: Outlier detection for high dimensional data. In: Proceedings of the ACM SIGMOD International Conference on Management of Data (2001)
4. Orair, G., Teixeira, C., Wang, Y., Meira, W., Parthasarathy, S.: Distance-Based Outlier Detection: Consolidation and Renewed Bearing. In: Proceedings of International Conference on Very Large Data Bases, VLDB (2010)
5. Hodge, V.J., Austin, J.: A Survey of Outlier Detection Methodologies. Artificial Intelligence Review 22 (2004)
6. Coscia, M., Giannotti, F., Pedreschi, D.: A Classification for Community Discovery Methods in Complex Networks. Statistical Analysis and Data Mining 4 (2011)
7. Parthasarathy, S., Ruan, Y., Satuluri, V.: Community Discovery in Social Networks: Applications, Methods and Emerging Trends. Social Network Data Analytics, 79–113 (2011)
8. Aggarwal, C.C.: An Introduction to Social Network Data Analytics. Social Network Data Analytics (2011)

9. Gao, J., Liang, F., Fan, W., Wang, C., Sun, Y., Han, J.: On Community Outliers and their Efficient Detection in Information Networks. In: Proceedings of the ACM SIGKDD International Conference on Knowledge Discovery and Data Mining (2010)
10. Aggarwal, C.C., Zhao, Y., Yu, P.S.: Outlier Detection in Graph Streams. In: Proceedings of the International Conference on Data Engineering, ICDE (2011)
11. Zhang, J.: Towards Outlier Detection for High-demential Data Streams using Projected Outlier Analysis Strategy. PhD thesis, Dalhousie University (2009)
12. Breunig, M.M., Kriegel, H.P., Ng, R.T., Sander, J.: Lof: Identifying density-based local outliers. In: Proceedings of the ACM SIGMOD International Conference on Management of Data (2000)
13. Papadimitriou, S., Kitagawa, H., Gibbons, P., Faloutsos, C.: LOCI: Fast outlier detection using the local correlation integral. In: Proceedings of the International Conference on Data Engineering, ICDE (2003)
14. Orair, G.H., Teixeira, C.H.C., Meira Jr., W., Wang, Y., Parthasarathy, S.: Distance-Based Outlier Detection: Consolidation and Renewed Bearing. In: Proceedings of the International Conference on Very Large Data Bases, VLDB (2010)
15. Moser, F., Ge, R., Ester, M.: Joint cluster analysis of attribute and relationship data withouta-priori specification of the number of clusters. In: Proceedings of the ACM SIGKDD International Conference on Knowledge Discovery and Data Mining, KDD (2007)
16. Yang, T., Jin, R., Chi, Y., Zhu, S.: Combining link and content for community detection: a discriminative approach. In: Proceedings of the ACM SIGKDD International Conference on Knowledge Discovery and Data Mining, KDD (2009)
17. Li, X., Li, Z., Han, J., Lee, J.-G.: Temporal Outlier Detection in Vehicle Traffic Data. In: Proceedings of the IEEE International Conference on Data Engineering, ICDE (2009)
18. Chakrabarti, D.: AutoPart: Parameter-Free Graph Partitioning and Outlier Detection. In: Boulicaut, J.-F., Esposito, F., Giannotti, F., Pedreschi, D. (eds.) PKDD 2004. LNCS (LNAI), vol. 3202, pp. 112–124. Springer, Heidelberg (2004)
19. Fortunato, S.: Community detection in graphs. Physics Reports 486 (2009)
20. He, Z., Xu, X., Deng, S.: Scalable Algorithms for Clustering Large Datasets with Mixed Type Attributes. International Journal of Intelligent Systems 20 (2005)

An Efficient Index for Top-k Keyword Search on Social Networks

Xudong Du

Department of Computer Science and Technology
Tsinghua University, Beijing 100084, China
andy2005cst@gmail.com

Abstract. Social networks (e.g., Facebook, Twitter) have attracted significant attention recently. Many users have a search requirement to find new or existing friendships with similar interests in social networks. A well-known computing model is keyword search, which provides a user-friendly interface to meet users search demands. However traditional keyword search techniques only consider the textual proximity and ignore the relationship closeness between different users. It is a big challenge to integrate social relationship and textual proximity and it calls for an effective method to support keyword search in social networks. To address these challenges, we present a tree decomposition based hierarchical keyword index structure (TDK-Index) to solve the problem. Our major contributions are: (1)TDK-Index which integrate keyword index and relationship closeness index as a whole; (2)Two-phase TA algorithm which narrows the threshold obviously compared to existing methods and speed up top-k query by a factor of two; and (3)flexible solution which adopts different application circumstances by parameter adjustment. Our experiments provide evidences of the efficiency and scalability of our solution.

1 Introduction

Social networks have attracted significant attention recently due to its fast user growth and high user stickiness. Many users have a search requirement to find new or existing friendships with similar interests in social networks. A well-known computing model is keyword search, which provides a user-friendly interface to meet users search demands. For example, Andy a Facebook user wants to invite several people to join a football match with him. He may search "football" and get some football fan candidates. There are tons of football fans, however Andy actually prefer to candidates who has a closer relationship with him. Another example, Yvonne a user of Linkedin is seeking for an IT job, and she wants to get some references through her social connections. She may search "Silicon Valley" or "Google" to find some people who are both highly related to the keywords and more familiar with her. Traditional keyword search techniques only consider the textual proximity and ignore the relationship closeness between users, So our user may not satisfied with the results.

H. Gao et al. (Eds.): WAIM 2012, LNCS 7418, pp. 446–458, 2012.

Although techniques of traditional keyword search are well studied [14], it is a big challenge to integrate social relationship and textual proximity to support keyword search on social networks. On social networks, users are connected by friendship or similar relationships. In this paper, relationship closeness of directly connected users can be identified by interactive activities such as replies, visits and so on. Relationship closeness of indirectly connected users are associated with the friendship chains between them. The idea is very intuitive from observation of real life. Two strangers in real life usually know each other by some mutual friends and the closer their friendship with the mutual friends the closer relationship they may have. Without an intergraded framework, we need to carry out keyword search and relationship closeness calculation separately. One possible solution is that we first carry out traditional keyword search to generate some candidates and then calculate relationship closeness for them to generate the final top-k. Or we may follow a breadth-first idea to calculate their textual proximities until top-k vacancies are full filled. Neither of the two solutions are efficient because candidates with high textual proximity may far away from the query user and candidates close to the query user may not related to the keywords at all.

In this paper, we map social networks to a graph structure and reduce the problem to a top-k keyword search over graph and focus on ranking functions constituted by textual proximity and relationship closeness. We make the following major contributions:

1. We propose TDK-Index which integrate keyword index and relationship closeness index as a whole. The index structure efficiently solve the challenges mentioned above.
2. We develop Two-phase TA algorithm which narrows the threshold obviously compared to existing methods and speed up top-k query by a factor of two.
3. Our solutions are flexible for parameter adjustment to adopt different application circumstances.

The rest of the paper is organized as follows. We formulate the problem in Section 2 and present TDK-Index and Two-phase algorithm in Section 3 and Section 4. Experiment results are reported in Section 5 and related works are discussed in Section 6. Finally, we make conclusions in Section 7.

2   Problem Formulation

Given an undirected positive weight keyword embedded graph $G(V, E, W, T)$, where vertex $v \in V$ edge $G(V, E, W, T)$, and $w_{ij} \in W$ present user, friendship and distance between users. Keyword set $t_i \in T$ belongs to user v_i. We define the top-k problem as bellow.

Definition 1 (UCK Query(User Closeness aware top-k Keyword query)). *A UCK Query $Q = (v, k, t_1, t_2, \ldots)$ is asked by user v and constituted by keywords $\{t_1, t_2, \ldots\}$ to retrieve other users with highest top-k ranking*

score which is monotonic increasing with the user closeness and keywords score.
v is called the owner of Q. User closeness scores are measured by a function
monotonic decreasing with the shortest distance between users over the graph.

In this paper, we use the following ranking function as an example. User closeness
score for u is determined by reciprocal of its shortest distance from query owner
v. Two parts of the score are normalized and weighted by parameters.

$$score_u(v, t_1, t_1, \ldots) = \alpha \cdot \frac{1/distance_{u,v}}{1/distance_{v,max}} + \beta \cdot \sum_i \frac{score_{t_i}}{maxScore_{t_i}} . \qquad (1)$$

3   TDK-Index Structure

The basic idea of TDK-Index is to integrate keyword and user closeness indexes
together and provide a standard sorted and random access interface for TA [5]
alike framework. Fig.1 gives a overview of the TDK-Index. The TDK-index is a
tree structure and each tree node are constituted by several vertexes of original
social graph. Offline calculated distance matrix and other useful information are
stored at each node to speed up shortest distance calculation of vertexes. And
we have two type of keyword indexes, a single global index located at the root
and multiple local indexes located at some selected nodes. Now, We will discuss
why and how to build the index structure in detail.

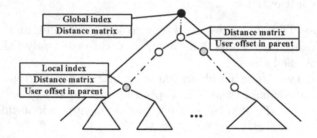

Fig. 1. Overview of TDK-Index structure

To solve the top-k query problem in a TA [5] alike framework, we need to provide sorted and random access for keywords rank lists and user closeness rank
list. For keywords, sorted and random access are available by building keyword
inverted lists and forward lists. The challenges are how to provide efficient random and sorted access to user closeness. One observation of sorted access to a
query owner's closeness rank list is that we visit the nearest(the nearest one has
the highest user closeness score) unvisited candidate one by one, so we do not
need to prepare a completed rank list. Actually we can do it in a lazy manner
by maintaining a priority queue. When a sorted access is needed, we pop out the
nearest candidate, push its unvisited neighbors and update distance if needed.
For the random access, we adopt a similar technique as TEDI [11] to do tree
decomposition on social graph and transform shortest distance calculation from
graph to tree.

3.1   Tree Decomposition Based Random Access of Relationship Closeness

In graph theory, tree decomposition is a methodology of mapping a graph into a tree to speed up problems on the original graph.

Definition 2 (Tree Decomposition[1]). *Given graph $G = (V, E)$, a tree decomposition is a pair (X, T) where $X = X_1, \ldots, X_n$ is a family of subsets of V, and T is a tree whose nodes are subsets X_i, satisfying the following properties:*

1. *$\bigcup_i X_i = V$.*
2. *For every edge $(u, v) \in E$, it exists at least one X_i satisfying $u \in X_i$ and $v \in X_i$.*
3. *If X_i and X_j both contain vertex u, then all nodes X_k between X_i and X_j contain u as well. That is all the nodes containing u form a connected subset of T.*

According to the definition, one graph may be mapped into multiple different trees. A trivial mapping is a tree with only one node which contains all the vertexes of original graph. However, we are interested in better trees whose nodes contain fewer vertexes. The vertex cardinality of a node is called the node size here. We try to generate some trees with the maximal node size smaller. Finding a best tree decomposition with smallest maximal node size is NP-hard problem. Inspired by TEDI [11], we adopt a similar technique to do tree decomposition on our social network graph and transform the graph into a tree like structure. The methodology deletes vertexes with smallest degree one by one until degree of the vertexes left is larger than parameter d or an empty graph remains. Different from TEDI [11], we also stop if there are less or equal than $d + 1$ vertexes remaining. when a vertex is deleted, the vertex and its neighbors are united to be a node. We push the node into a stack and new edges are added to make its neighbors left to be clique. After the deleting process, tree structure is built by vertexes left and nodes in the stack. We hang every node to a as low as possible place satisfying the parent node should contain all the users in the child node except the deleted user when the child node is generated in previous process. Fig.2 shows the process of our running example with parameter $d = 2$. At the beginning, vertex 1 is deleted because its degree is the smallest. We unite vertex 1 and its neighbors vertex 4 and 5 to be a node and push it into the stack. The same operations are carried out when deleting vertex 2 and 3. Once a vertex is deleted, we need to exam if its neighbors are clique. If not, we need to add necessary edges. In our example, when vertex 3 is deleted, new edge between 5 and 6 is created. When only $d + 1 = 3$ vertexes are left, we unit vertex 4, 5 and 6 into a node. Next, we will build the tree structure in a reversed order. Last generated node will be insert into the tree first. When node A with vertex 4, 5 and 6 is inserted into an empty tree, it becomes the root. Next one is node B, the lowest possible location is the child of A satisfying that node A contains

---

[1] http://en.wikipedia.org/wiki/Tree_decomposition

all vertexes in node B except vertex 3 which is the deleted vertex when node B is generated. After every nodes are inserted one by one, we construct a tree composition of original graph. Follow the method, every node size is smaller than or equal to $d + 1$ except the root, and the root size can be controlled by adjust parameter d. The good feature is very helpful to reduce shortest distance calculation complexities.

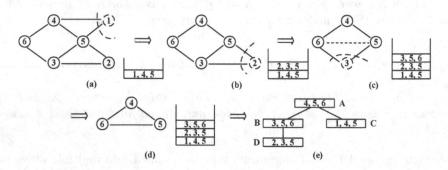

Fig. 2. Tree decomposition process of running example with parameter $d = 2$

Every vertex may exist in serval nodes, we call the highest node is the vertex's node or the vertex is the node's owner. In our running example, vertex 3's node is B. With the structure, shortest distance of two vertexes in original graph can be solved by a bottom up algorithm. The algorithm starts from the two vertexes' node to their common ancestor. The connection of two nodes of the tree is constituted by the same vertexes they share and there are at most d shared vertexes. The shortest path is formed by choosing the best shared vertex to step up until the common ancestor and the steps is at most equal to the height of the tree h. So, shortest path calculation complexity is $O(d^2 \cdot h)$. Proof of the algorithm correctness can be found in TEDI [11] which realizes very efficient shortest hop distance calculation on the structure. In our problem, shortest distance is not hop counts but weights sum along the path.

With the tree decomposition based index, random access of relationship closeness measured by shortest distance is well solved. Combined with inverted lists, forward lists and priority queue techniques, We have already make the top-k problem applicable for a TA [5] alike framework.

3.2   Hierarchical Keyword Index

When doing TA [5] algorithm over multiple rank lists, candidates with high score at one list may not able to rank into final top-k because they may rank very low at other lists. In our problem, relationship closeness score is decreasing with shortest distance. One observation is that many candidates far away from the query owner may get very high textual proximity score, but fail to rank into final top-k according to our ranking function. If we directly use previously discussed techniques to answer the top-k queries, lots of such useless candidates will be visited. Actually, we can reduce the useless visits by filtering candidates far from query owner. Our method is to create a hierarchical keyword index structure instead of a single global index.

Theorem 1. *Based on our tree decomposition method, the social graph has been transformed into a tree structure and every directly connected vertexes of v is guaranteed to be contained by the subtree whose root is v's node.*

Suppose N_v's owner is v and there is a u which is a neighbor of v, then there must be a node N' containing both u and v according to tree decomposition definition property 2. There must be a connected path from N' and N_v and every node along the path contains v according to property 3. If N' does not belong to N_v's subtree, there must be some other node N along the path with a higher position than N_v. That is to say N_v can not be v's node. The theorem is proved.

Every vertex's neighbors are contained by its node's subtree, and if we build some local keyword indexes on some selected nodes, we may do keyword search on some smaller index which locates next to the query owner's node to filter candidates far away. It is not wise to build local index on every node, because the results generated may not full fill top-k vacancies and the large space cost is incurred. Besides a global keyword index, our TDK-Index build an local keyword index every m un-indexed vertexes accumulated from leaf to root. And every local index will cover all the keywords of the subtree. To reduce the storage cost, only the offset positions of the global index are stored in local indexes. We will use our running example to show our local index strategies. Given the tree structure of Fig.2(e) and parameter $m = 4$, we start from leaf node D and 3 un-indexed vertexes here. We move up to node B and accumulated un-indexed number reaches 4 equals to m, so we create a local index at node B containing all the keywords of vertexes $2, 3, 5, 6$. We continue the process until the root. The same operations are carried out from the other leaf C and finally we need to build local index on node B and A. Because A is the root already with global index, we do not need to create an local index again. Fig.3 builds global and local keyword indexes for our running example.

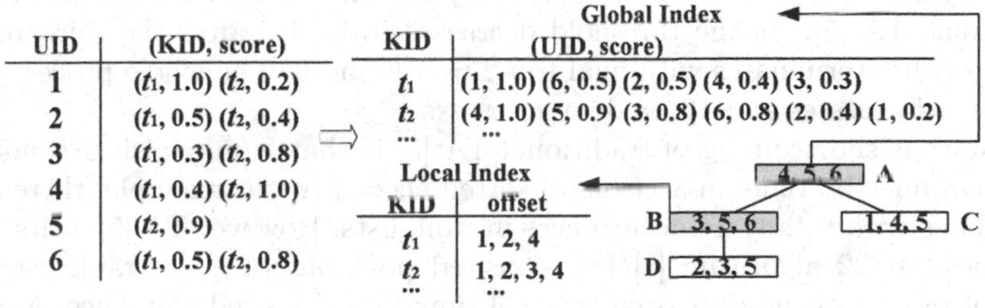

Fig. 3. Hierarchical index for running example

To sum up, based on tree decomposition, TDK-Index map original social network graph into tree structure and build both shortest path indexes and keyword indexes on the integrated system. To filter candidates far away from query owner, hierarchical local index structure is adopted. We will present Two-phase TA algorithm to efficiently answer top-k queries based on TDK-Index and discuss parameter optimizations in following section.

4   Two-Phase TA Algorithm for Top-k Query

For keyword search on social networks, many candidates far away from the query owner may get very high textual proximity score, but fail to rank into final top-k according to overall ranking function. Traditional TA [5] algorithm will visit candidates at every rank list from top to down. However, candidates far away from query owner actually should be skipped even if they have very high rank at textual proximity lists. It calls efficient algorithms to reduce the visits to candidates with low probability of top-k.

4.1   One Phase Solutions

Before discussion of Two-phase TA algorithm, Let's first take a look at how one phase TA [5] works. The algorithm make sorted access to rank lists from top to down and calculate candidates overall score by necessary random access. It terminates once enough results with score above threshold. We will use an example UCK query $Q = (2, 2, t_1, t_2)$ to show the process. The query owner user 2 is trying to find top-2 users(user 2 herself is excluded) with highest overall score. For simplicity, ranking function $Score(uc, t_1, t_2) = score_{uc} + score_{t_1} + score_{t_2}$ is used here, where uc is user closeness. As shown in Fig.4, user 3, 1 and 4 are visited in the first round by sorted access to uc, t_1 and t_1 rank lists separately. To calculate overall score for user 3, random accesses to rank lists of t_1 and t_1 are carried out. We calculate overall score for user 1 and 4 similarly. After the first round, the top-2 score are 2.1 and 1.5 and the calculations cost us 3 sorted accesses and 6 random accesses. A threshold is set to estimate the highest possible score of unseen candidates, and the algorithm can terminate once there are 2 candidates' score larger or equal than the threshold. Sum of currently lowest sorted access score of every rank list is used as the threshold in traditional TA [5]. As the threshold decreases to be 1.7 after the third round, the algorithm terminates with final top-2 user 6 and 3. The whole process costs us 9 sorted accesses and 16 random accesses.

A obvious shortcoming of traditional TA [5] is that it may visit a candidate multiple times by random access and sorted access. In our example, there are 5 candidates with totally 14 occurrences in rank lists. However TA [5] visits them 25 times. BPA2 algorithm [1] keeps visited positions of every rank lists and guarantees every candidate occurrence of rank lists is visited only once. And the algorithm use the sum of highest possible unseen score of every rank list as the threshold which is much tighter than TA [5]. We use the same query example

Rank list of uc		Rank list of t_1		Rank list of t_2		TA algorithm			
UID	score	UID	score	UID	score	threshold	sorted access #	random access #	(candidate, score)
3	1.0	1	1.0	4	1.0	3.0	3	6	(3, 2.1) (4, 1.5) ...
6	0.9	6	0.5	5	0.9	2.3	3	5	(6, 2.2) (3, 2.1) ...
5	0.5	4	0.4	3	0.8	1.7	3	5	(6, 2.2) (3, 2.1) ...
1	0.2	3	0.3	6	0.8				
4	0.1			1	0.2				
		(a)						(b)	

Fig. 4. Traditional TA algorithm solution for example Query $Q = (2, 2, t_1, t_2)$

to show its calculation process in Fig.5. The same as TA [5], sorted and random access to user 3, 1 and 4 are carried out in first round. After the second round, the top 4 positions of t_1 are all be visited by sorted access or random access, so the highest possible unseen score is equal to 0.3. Similarly, the highest possible unseen score for uc and t_2 are 0.1 and 0.2 respectively. The overall threshold reduces to 0.6 which is much narrower than TA [5] and the algorithm terminates immediately with totally sorted access cost 5 and random access cost 9.

Rank list of uc		Rank list of t_1		Rank list of t_2		BPA2 algorithm						
UID	score	UID	score	UID	score	threshold	sorted access #	random access #	(candidate, score)	uc	t_1	t_2
3	1.0	1	1.0	4	1.0	3.0	3	6	(3, 2.1) (4, 1.5) ...	2	2	2
6	0.9	6	0.5	5	0.9	0.6	2	3	(6, 2.2) (3, 2.1) ...	6	5	6
5	0.5	4	0.4	3	0.8							
1	0.2	3	0.3	6	0.8							
4	0.1			1	0.2							
		(a)							(b)			

Fig. 5. BPA2 algorithm solution for example Query $Q = (2, 2, t_1, t_2)$

4.2 Two-Phase TA Algorithm

It is obviously that user 1 and 4 have very high textual proximity score for t_1 and t_2. Both TA [5] and BPA2 [1] pay visiting costs for them. Our Two-phase TA algorithm will help to reduce the extra costs. Two-phase TA is constituted by two phase. It generates local top-k candidates(with high probability of final top-k) and tight the threshold in phase 1 and enrich and verify the final result in phase 2. Algorithm.1 shows the pseudo code of Two-phase TA algorithm. First, we need to determine a proper local index to run phase 1. Here we assume local index of the nearest ancestor of the query owner is selected and we will discuss local index selection later. In this paper, Very similar to BPA2 [1], we generate local top-k over the local rank lists. One difference is that we maintain 2 thresholds: one is called local threshold and the other is global threshold. They are calculated by the sum of highest possible unseen score of every local rank list and global rank list respectively. If local top-k with score higher than global threshold, the algorithm terminates. If local top-k with score higher than local threshold, the algorithm enters phase 2. Phase 2 is carried out over global rank lists and the process is similar to BPA2 [1].

Algorithm 1. Two-phase TA Algorithm

1: // Phase 1: sorted access local rank lists $(L_1, L_2, \ldots)$ to update global and local threshold T_g, T_l. Best unvisited candidate of global and local for keyword i is $B_{i,g}$ and $B_{i,l}$. Heap S_g and S_l storage and rank candidates with score larger than T_g and T_l.

2: **while** Size(S_g)+Size$(S_l) < k$ and (!Empty(Q) or !IsScannedOver$(L_1, L_2, \ldots)$) **do**

3:    u=Top(Q), Pop(Q), Push(neighbors of u)), score=Score(u), Update$(B_{i,g})$, Update$(B_{i,l})$, Update(T_g), Update(T_g)

4:    **if** score$>T_g$ **then**

5:       Push(S_g, u)

6:    **else if** score$>T_l$ **then**

7:       Push(S_l, u)

8:    **end if**

9:    **for** each in $L_1, L_2, \ldots$ **do**

10:       u=SortedAccess$(L_i, B_{i,l})$

11:       Update$(B_{i,g})$, Update$(B_{i,l})$

12:    **end for**

13:    Update(T_g), Update(T_g)

14:    early terminates if enough results with score above threshold.

15:    **for** each sorted accessed candidate u **do**

16:       score=Score(u), Update$(B_{i,g})$, Update$(B_{i,l})$, Update(T_g), Update(T_g)

17:       ... // The same as line 4-8

18:    **end for**

19: **end while**

20: // Phase 2: the only differences from phase 1 are only $B_{i,g}$ and T_g are maintained and we scan global rank lists $G_1, G_2, \ldots$.

21: **while** Size$(S_l)>0$ and TopScore$(S_l)>T_g$ **do**

22:    Add$(S_g,$ Top$(S_l))$, Pop(S_l)

23: **end while**

24: **while** Size$(S_g)< k$ and (!Empty(Q) or !IsScannedOver$(G_1, G_2, \ldots)$) **do**

25:    ... // Similar to phase 1, we omit code here for simplicity.

26: **end while**

Fig.6 shows how the algorithm works for the same example query $Q = (2, 2, t_1, t_2)$. Local index B is the nearest index for query owner user 2, so we run phase 1 over the local rank lists of keyword index B. User 3, 6 and 5 are visited by sorted and random access in first round and local threshold is reduces to 1.6 calculated by the sum of highest possible unseen score of local rank lists. However global threshold is 2.5 and early termination is not satisfied. The algorithm enters phase 2 running on the global keyword index A. After sorted access to user 1 and 4, the algorithm terminates without further random accesses as the threshold is reduced to be 1.8. Sorted access and random access costs are 10 in total which is much smaller compared to 14 of BPA2 [1] and 25 of TA [5].

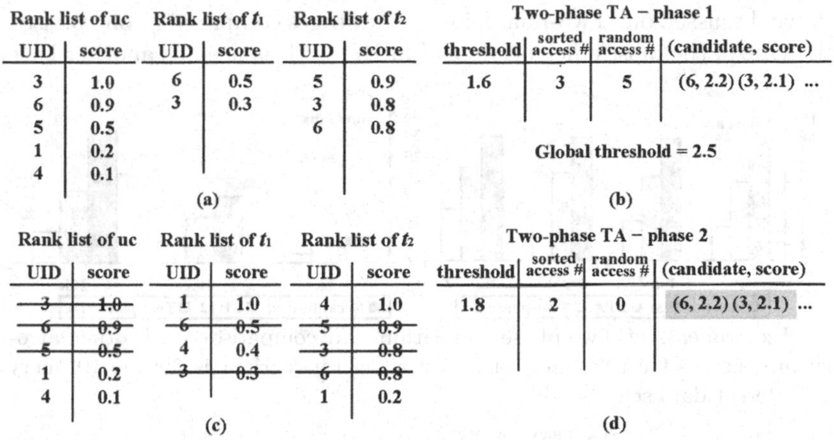

Rank list of uc

UID	score
3	1.0
6	0.9
5	0.5
1	0.2
4	0.1

Rank list of t₁

UID	score
6	0.5
3	0.3

Rank list of t₂

UID	score
5	0.9
3	0.8
6	0.8

(a)

Two-phase TA – phase 1

threshold	sorted access #	random access #	(candidate, score)
1.6	3	5	(6, 2.2) (3, 2.1) ...

Global threshold = 2.5

(b)

Rank list of uc

UID	score
~~3~~	~~1.0~~
~~6~~	~~0.9~~
~~5~~	~~0.5~~
1	0.2
4	0.1

Rank list of t₁

UID	score
1	1.0
~~6~~	~~0.5~~
4	0.4
~~3~~	~~0.3~~

Rank list of t₂

UID	score
4	1.0
~~5~~	~~0.9~~
~~3~~	~~0.8~~
~~6~~	~~0.8~~
1	0.2

(c)

Two-phase TA – phase 2

threshold	sorted access #	random access #	(candidate, score)
1.8	2	0	(6, 2.2) (3, 2.1) ...

(d)

Fig. 6. Two-phase TA algorithm solution for example Query $Q = (2, 2, t_1, t_2)$

4.3 Optimization and Parameters Selections

Along the path from the query owner' node to the root, there may be several nodes with local index. In previous section, we assume phase one of the Two-phase TA algorithm is carried out on the nearest ancestor with local index. Actually, we can do some optimizations to realize better performance under different application circumstances. Intuitively, in top-k query, the larger k is, the larger probability for candidate with far distance to be a final result. So, we need to pick up a proper local index with consideration of the result number needed. We use a parameter γ to help with selection of local index. In algorithm phase one, we prepare local rank list for every keyword with length at least $\gamma \cdot k$. That is to say we will keep on moving up along the path from the query owner's node until rank list with enough length is reached or root node is reached. And rank lists for different keywords may come from different nodes along the path. Experimental study will show the benefits of our strategies.

5   Experimental Study

The graph data is generated from random sample from DBLP dataset with vertexes number range from 5 thousand to 20 thousand and keywords number associated with a vertex follow a normal distribution with mean 100 and standard deviation 100. The keywords number of every test query follow a normal distribution with mean 2 and standard deviation 1. All the experiments are done on a computer with Four-core Intel(R) Xeon(R) CPU E5420 @ 2.50GHz and 16G memory.

To evaluate our solution, some existing solutions are selected as baseline. Without an integrated index structure, we may follow a breadth-first alike manner to calculate scores from near to far until enough results are found. In Section

4 we have discussed our Two-phase TA's advantages compared to one phase algorithms. One of the most efficient algorithm BPA2 [1] is used as another baseline.

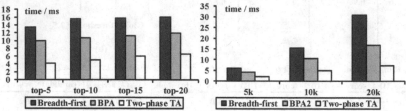

(a) Experiments of Two-phase Algorithm and comparison with other algorithms. Left is for 10K data set and varying top-k, right is for top-10 query in different data sets.

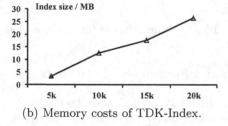

(b) Memory costs of TDK-Index.

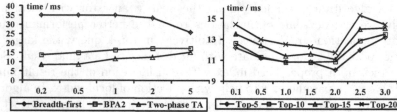

(c) Experiments of Parameter adjustments running on 20K data set. Left one presents time costs associated with different ratios of $\alpha : \beta$, right one presents time costs associated with different parameter γ.

Fig. 7. Experiment results

As shown in Fig.7(a), Two-phase TA algorithm is always better than breadth-first algorithm without TDK-Index and One phase BPA2 [1] algorithm with TDK-Index. Fig.7(b) presents the efficient memory costs of our TDK-Index(Global keyword index is not included here, as it is not changed by our solution) and Parameter adjustments under different application circumstance is presented in Fig.7(c).

6   Related Works

Recent years, search problems in social networks have attracted a significant attentions [4,3,7,9,2]. Some researches of top-k query over social networks focus on building ranking score model by shared tags [9]. Some papers design and evaluate new ranking functions to incorporate different properties of social networks [2,7]. Access control of keyword search over social networks is also attract research studies [4,3].

Although the problem of top-k keyword search with consideration of relationship closeness measured by shortest distance is not solved well by existing techniques, traditional keyword search and shortest distance calculation are well studied separately. Inverted list [14] is a state-of-the-art technique and there are many studies with a focus on inverted file compression [10,13] and management [8]. For shortest distance problems, there is a survey paper [6] covering different algorithms. Instead of online calculation, index based methods are studied [12,11]. Our study in this paper is different and our solution efficiently integrates top-k keyword search and shortest path index problems.

7   Conclusion

In this paper, we have proposed TDK-Index and Two-phase algorithm which solve the big challenge of integrating social relationship and textual proximity for keyword search over social networks. Experimental studies provide evidence of efficiency of our solutions and the flexibility to adopt different application circumstances.

References

1. Akbarinia, R., Pacitti, E., Valduriez, P.: Best position algorithms for top-k queries. In: VLDB, pp. 495–506 (2007)
2. Bao, S., Xue, G.-R., Wu, X., Yu, Y., Fei, B., Su, Z.: Optimizing web search using social annotations. In: WWW, pp. 501–510 (2007)
3. Bjørklund, T.A., Götz, M., Gehrke, J.: Search in social networks with access control. In: KEYS, p. 4 (2010)
4. Bjørklund, T.A., Götz, M., Gehrke, J., Grimsmo, N.: Workload-aware indexing for keyword search in social networks. In: CIKM, pp. 535–544 (2011)
5. Fagin, R., Lotem, A., Naor, M.: Optimal aggregation algorithms for middleware. In: PODS. ACM (2001)
6. Goldberg, A.V.: Point-to-Point Shortest Path Algorithms with Preprocessing. In: van Leeuwen, J., Italiano, G.F., van der Hoek, W., Meinel, C., Sack, H., Plášil, F. (eds.) SOFSEM 2007. LNCS, vol. 4362, pp. 88–102. Springer, Heidelberg (2007)
7. Hotho, A., Jäschke, R., Schmitz, C., Stumme, G.: Information Retrieval in Folksonomies: Search and Ranking. In: Sure, Y., Domingue, J. (eds.) ESWC 2006. LNCS, vol. 4011, pp. 411–426. Springer, Heidelberg (2006)
8. Margaritis, G., Anastasiadis, S.V.: Low-cost management of inverted files for online full-text search. In: CIKM, pp. 455–464 (2009)
9. Schenkel, R., Crecelius, T., Kacimi, M., Michel, S., Neumann, T., Parreira, J.X., Weikum, G.: Efficient top-k querying over social-tagging networks. In: SIGIR, pp. 523–530 (2008)
10. Scholer, F., Williams, H., Yiannis, J., Zobel, J.: Compression of inverted indexes for fast query evaluation (2002)

11. Wei, F.: Tedi: efficient shortest path query answering on graphs. In: SIGMOD Conference, pp. 99–110 (2010)
12. Xiao, Y., Wu, W., Pei, J., Wang, W., He, Z.: Efficiently indexing shortest paths by exploiting symmetry in graphs. In: EDBT, pp. 493–504 (2009)
13. Zhang, J., Long, X., Suel, T.: Performance of compressed inverted list caching in search engines. In: WWW, pp. 387–396 (2008)
14. Zobel, J., Moffat, A.: Inverted files for text search engines. ACM Comput. Surv. 38(2) (2006)

Engineering Pathway
for User Personal Knowledge Recommendation

Yunlu Zhang[1,2], Guofu Zhou[1,*], Jingxing Zhang[1], Ming Xie[1],
Wei Yu[1], and Shijun Li[1,2]

[1] State Key Laboratory of Software Engineering,
Wuhan University, 430072 Wuhan, China
[2] School of Computer, Wuhan University, 430072 Wuhan, China
gfzhou@whu.edu.cn

Abstract. K-Gray Engineering Pathway (EP) is a Digital library website that allows search and catalog of engineering education and computer
science education resources for higher education and k-12 educators and
students. In this paper, we propose a new EP that can give different
and personal search recommendation for users with different educational
background and accomplish this function automatically. For data, we
explore semantic relationships among knowledge, and then we classify
them and establish the knowledge relationships model. For users, we can
set up user profile by user log, then classify them and establish user
model. When a frequent user come to EP looking for something, we can
give information directly related and recommend knowledge not directly
related but can arouse their interest, based on these two models. The
experiments shows that we make EP a more excellent expert who know
users well enough to guide them, according to the statistic information
such as education background, and by improving the collaborative recommend results in EP, our users can make the best use of their time by
EP learning.

Keywords: search recommendation, user model, digital library, guide
user.

1 Introduction

Based on the users' education background, EP can give them different search
results when the input the same the query by users' active behavior, who know
their own education background and set the search area themselves, which make
EP search result with high correction, but also leads to potential information
loosing, for users have no idea of these knowledge exiting, for their limited educational. User also can get all search results from the homepage, independent
of education background, but it made low correction and lots of time wasting
to find what they really needed. There are mainly 6 branches is our home-
pages for users who needs some information: (1)Advanced Search (including

---

* Corresponding author.

H. Gao et al. (Eds.): WAIM 2012, LNCS 7418, pp. 459–470, 2012.
© Springer-Verlag Berlin Heidelberg 2012

K-12 search, Higher Education search, Geocentric search), (2) Browse Learning Resources(including Subject/Disciplinary Content Areas, Special Topics, Grade Levels, Learning Resource Types, ABET Outcomes (Students, Learning), Host Collections, PR2OVE-IT Interventions), (3)K-12 Community, (4)Higher Education Community, (5)Disciplinary Communities, (6)Broadening Participation. And 2 branches for introduction about us: Premier Award and About us, and 2 branches used for interaction between user and us: (1)Submit Resource and (2)My workspace. The homepage as shown in Figure 1.

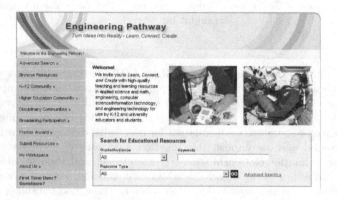

Fig. 1. Snapshot of EP User Interface

What EP can do now are: (1)Based on the users' education background, give them different search results when they input the same the query (K-12, and so on), this function is implemented by users' active behavior, they know their own education background, and set the search area by themselves. Pros: high correction; Cons: needs users active input; lose some information that may help them, they do not know the exiting of these knowledge, constrains by their limited educational level; (2)User can get all the search results in the homepage, independent of their education background, In this function, user's active behavior is no need, and the pros and cons are as follows: Pros are mostly, no useful information and knowledge will be missed, Cons are low correction and waste lots of time to find what they really needed.

So in this paper, we proposed a new EP that can give different and personal search recommendation for users with different educational level and accomplish this function automatically. For example, if the user is an undergraduate, he/she want to search some information about Advanced Mathematics, we can give his/her course ware, video and homework answers, or something can help them prepare for the test. What if we give him/her something in graduate, such as some papers and instructions about engineering application with advanced mathematics? If the user is a college teacher, and he/she also want to search some information about Advanced Mathematics, we can give him/her something the same as the former user, such as course ware and video, to help them improve teaching, but we can also give him/her some research material, such as papers with higher theoretical level –which is contrary to the former undergraduate users. So what else we can do here, since we already have users' log by powerful EP.

For data, firstly, we research semantic relationships among knowledge, then we classify or cluster them, finally we establish the knowledge relationships models. For users, firstly, based on user log, we can set up user profile, then classify or cluster the users type, finally we establish the users model, So, when a frequent user come to EP looking for something, we can give information directly related and recommend knowledge not directly related but can arouse their interest, based on the users model and knowledge relationships models. Based on users log mining, user model, data relationship analysis(semantic, classification, clustering), we will make EP a more excellent expert who know users well enough to guide them, according to the statistic information such as education background. And by improving the collaborative recommend results in EP, our users can make the best use of their time by EP learning.

2   Related Work

In our previous works, we had done lots work on Digital Library[1,3]. How can electronic course ware meet the diverse needs of curricula among a cross section of universities? How do educators adapt traditional teaching roles to fit new resources and delivery styles? What course ware access modes equally suit the needs of author, teacher, and student? Can an infrastructure designed for static course ware be adapted to dynamically changing information on the World Wide Web? In [4], for the experience of Synthesis/NEEDS(the National Engineering Education Delivery System) answered these questions while opening more issues in distance independent education by striving to integrate multidisciplinary, open-ended problem solving into the varied engineering curricula of its members.

The focus of this research [5] lies in ascertaining tacit knowledge to model the information needs of the users of an engineering information system. It is proposed that the combination of reading time and the semantics of documents accessed by users reflect their tacit knowledge. By combining the computational text analysis tool of Latent Semantic Analysis with analyzes of on-line user transaction logs, we introduce the technique of Latent Interest Analysis (LIA) to model information needs based on tacit knowledge through user's queries and prior documents downloaded; it was incorporated into our digital library to recommend engineering education materials to users.

There are lots of new trends in recently information ages, The digital library are improving with the change of information format as reported in the following research work. For many new services 2.0 are appearing everyday,eg., Facebook, Flickr, Jesus et.al[6] designs a system allows the reduction of the necessary time to find collaborators and information about digital resources depending on the user needs by using Google Wave to extend the concept of Library 2.0. Social tagging or collaborative tagging is also a new trend in digital age. So,by linking social tags to a controlled vocabulary, Yi [7] did a study to investigate ways of predicting relevant subject headings for resources from social tags assigned to the resources based on different similarity measuring techniques. Another trend is construction of semantic digital libraries. Jiang et al. [8] researched a clustering method based on normalized compression distance for "affiliation," its an important type of meta data in publications areas, and also a question hard to resolve when converts its meta data of digital resources into its semantic web data.

There are also many recent researches focus on the relationship among the retrieval system and users' search behavior. Such as Catherine L. Smith and Paul B. Kantor [9], through adapting a two-way street, they did a factorial experiment by manipulated a standard search system to produce degraded results and studied how people solve the problem of search failure.

Like our EP, Lin and Smucker[10] researched a content-similarity browsing tool can compensate for poor retrieval results to help their users. The goal of educational digital library applications is to provide both teachers and students enriching and motivational educational resources [11, 12, 13] Jing and Qingjun [14]and Ankem[15]provide teachers and students a virtual knowledge environment where students and teachers enjoy a high rate of participation. Alias et al.[16] describe an implementation of digital libraries that integrates semantic research. Mobile ad hoc networks are becoming an important part of the digital library ecology[17,13,18].

3   Users' Workspace

We have workspace for each register uses, when users using EP search find something interested in, they can add them to their own workspace, For example, user Lucy login EP, and do some research for "data mining," "information retrieval" in the "Advance Search" from the Discipline "Computer Science" and From grade "college Freshman" through "Continuing Education" as shown in Figure 2, and we can get lots of search result as shown in Figure 3, Lucy choice the one with title "Using the Data Warehouse", click the link under the title, and she can get the detail information about this learning resource, as shown in Figure 4, then, if she interested in it ,she can save it to her work space, after saving the interested learning resource to her workspace, you can edit it or remove it from her workspace, as shown in Figure 5, the last one with title "Using the Data Warehouse", which her added it just now.

Based on so many users' profiles in our data base, we can analyze the users' interest and calculate the similarity between their interested resources and the others, finally give them what resources they may interested in but they have no idea of their exiting. For doing the database SQL query on three tables in our database: smete_user , collection, and collection_member, we can find out there are 102 users have more than 10 learning resources in their workspace, 33 users have more than 50 learning resources in their workspace, 20 users have more than 100 learning resources in their workspace, 5 users have more than 500

Fig. 2. Search behavior of User Lucy

Fig. 3. Search results for Lucy's Search

Fig. 4. Detail information for "Using the Data Warehouse"

Fig. 5. Success to Save the "Using the data warehouse" to Lucy's workspace

learning resources in their workspace, and the top 3 users have 4535,1738,1122 learning resources.

4 CURE Cluster Algorithm

There are three main algorithms we can use for large database clustering algorithm setting: BIRCH[19], MST[20], and CURE[21], we take the CURE for short text clustering. In our experiment, the short text clustering are the titles in user's workspace. We still take Lucy's profile as an example; there are 26 learning resources in her workspace, we using their titles as input, in our cluster part of recommendation system, we can see the results from Figure 6. And we can figure out that Lucy's interest keywords maybe: "Design", "BPC-A", "Computing", "Education", "Collection" and "John Wiley Sons",according to these Lucy's potential interesting area, we can get her potential interesting learning resources are as shown in Figure 7.

In our system, user can decide the number of cluster we want to get, or we can decide the number of item in each cluster, or we do not decide everything, but leave the algorithm to select the best number of cluster based on different evaluations for the distances between clusters: such as Euclidian distance,

Jaccard similarity, Dice similarity, and Manhattan distance. The interface of our system as shown in Figure 8.

Fig. 6. Lucy's interest clustering

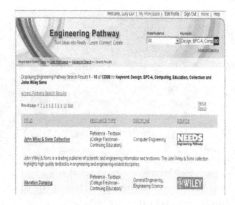

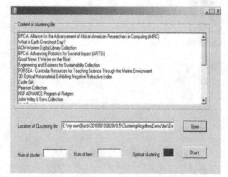

Fig. 7. Lucy's interesting keywords   **Fig. 8.** Interface of recommendation system by CURE

In the process of title clustering, we use cosine similarity to identify the correlated title pairs. The features of the first title to be clustered is $X = < x_1, x_2, ..., x_n >$, and the features of the second title to be clustered is $Y = < y_1, y_2, ..., y_n >$. The cosine similarity of titles to be clustered X and Y is
$C_{xy} = \frac{\sum_i \omega_i x_i y_i}{\sqrt{\sum_i x_i^2 \sum_i y_i^2}}$, where the ω_i is the weight of the edge between these two nodes in the term graph.

Given the dataset D and the number of clusters K, $C = < C_1, C_2, ..., C_k >$, we can evaluate this cluster results by $CH(Calinski - Harabasz)$ [28]index method, by doing Trial and error k, we can get the best cluster number when gets its maximum value. This index is computed as:

$$V_{CH}k = \frac{traceB/(k-1)}{traceW/(N-k)} \quad \exists traceB = \sum_{j=1}^{k} n_j \|z_j - z\| \bigwedge traceW = \sum_{j=1}^{k} \sum_{i=1}^{n_j} \|x_i - z_j\|$$

$$(1)$$

Algorithm 1. Title cluster based on CURE Algorithm

1: **Input**: S, k, α, γ ▷ dataset, impact factory, number of representation, iteration limit
2: **Output**: Best k
3: Initialize Tree T and heap Q
4: For every cluster u (each input point), in u.mean and u.rep store the mean of the points in the cluster and a set of c representative points of the cluster
5: initially c = 1   ▷ each cluster has one data point
6: u.closest = the cluster closest to u.
7: Trial and error(k)
8: While size(Q)> k
9: Remove the top element of Q(say u) and merge it with its closest cluster u.closest(say v) and compute
10: Create new merged cluster w = $< u, v >$
11: Use T to find best match for w
12: Update T, new representative points for w.
13: Remove u and v from T and Q.
14: Also for all the clusters x in Q, update x.closest and relocate x
15: insert w into Q
16: repeat to 8
17: index(k)=$\|interclass(k), intraclass(k)\|$
18: k ≡Max($index(k)$)
19: return k and k clusters representation point

Algorithm 2. EP Recom algorithm

1: Get the Learning resource titles in user's workspace
2: Calculate each element in word using TFIDF and get its weight
3: Change each element in tfidfVector to a CureCluster and add each CureCluster into CureClusterList
4: Calculate the distance between two CureCluster and fill the distance into the clos-estDistanceMatrix
5: Use the heap ordered by CureCluster.distance and
6: Use the MergeCureCluster to merge the two closest CureCluster until the CureCluster's number meets user's need
7: Based on the represetation of each cluster, get the learning resource keywords user may interested in
8: Go back to EP search get the learning resource user may interested in

---

The maximum hierarchy level is used to indicate the correct number of partitions in the data. $traceB$ is the trace of the between cluster scatter matrix B, and $traceW$ is the trace of the within cluster scatter matrix W. The meanings of parameters for them are: n_j is the number of points in cluster, $\|...\|$ means a certain distance calculation method, z is the centroid of the entire data set. z_j is the centroid of the C_j data set, x_i is the i-th item in C_j, and the number is x in each C_j is obviously n_j .

5   Experiment

In this section, we present the results of our experiment. The EPRecom dataset originally contained 120,000 ratings from 3125 users on 6250 learning resources. We gathered the top-20 recommendations for all users in the EPRecom dataset by using 4 methods, ItemRank, Tangent, PPTM, UPKR with c=0.01,0.001, c value is chosen experimentally. Since our method and Tangent require precomputed relevance scores, we run Tangent and PPTM based on the relevance scores of ItemRank.

5.1   Popularity

The Figure 9 shows comparison of recommendation from three algorithms in terms of popularity. In this figure, we present the difference between the distribution of rating and recommendation. From the results, our method outperformed competitors in terms of popularity matching. As analyzed in Section III, existing methods tend to recommend more items having high popularity. However, the result of UPKR is quite similar to that of rating so the differences is small for all degree of popularity. The comparison results for each individual are dropped from this paper because they are similar to the overall comparison results. All results for each individual are published on our website.

5.2   Diversity

We also measure the novelty of our method in terms of diversity in a way similar to the one we used in the analysis. The Figure 14 is the distribution of top-10 recommendation using UPKR with c = 0.002. The figure is generated in same way as we did in the investigation. Our method shows more diversified results and the over concentrated items set is smaller than that of others. In addition, the inner line indicating the coverage of UPKR is also located higher than that of other methods.To compare the results more clearly, we measure the diversity and coverage of each method (Figure 10)quantitatively. First, we measure EMD distance for the distribution of each method from that of rating (Figure 12). Our method always shows a smaller EMD distance,which indicates more diversified results, than other methods regardless of the number of recommendations. Note that,in this evaluation, EMD distance measures the distance between distribution of rating and recommendation. Thus, it's independent from EMD distance for PPTs. Furthermore,the coverage of our method is also higher than that of other competitors (Figure 13). When we recommend 50 items for each user, the coverage of UPKR is over 60% but that of ItemRank is near 45% and that of Tangent is around 25%.All these evaluation results show that our method is better than other methods in terms of diversity.

5.3   Accuracy

We evaluate the accuracy of our method and competitor. To compare the accuracy, we run 3-fold cross validation 30 times for each 942 users and aggregated the result. For more detail, we divide rating history of a user into 3 folds. After that, we use 2 folders to infer the taste of user,and measure the accuracy by comparing the recommendation result with remaining 1 fold, which is a relevant item set for test. We use recall to measure accuracy. It is because our method is a top-k recommendation method thus traditional accuracy measures, which are based on scores or the order of item, do not work. In this equation, the accuracy increases when the method recommends more items which are in the relevant item set for the test. We measure the accuracy change according to the change of the number of recommendations, and compare the accuracy of UPKR and Tangent to that of ItemRank (Figure 10). If the accuracy of a method is the same as ItemRank,it will be 1 for all k. It turns out that Tangent suffers from a performance degradation of around 15%-20%. On the other hand, our method shows better results than ItemRank when c is 0.001.Excessive engagement of PPT also results in a performance degradation of around 10 % but, still, it is better than Tangent.

5.4   Qualitative Evaluation

We conduct a case study for user 585 to evaluate the recommendation results also in a qualitative way. To the best of our knowledge, the user prefers films having practical value and reputation. In terms of popularity, the user also study many courses that gross satisfy rate is high. The satisfy rates of most courses that learned are less than. The average gross was 80%. However, in the recommendations of ItemRank, there are already 3 out of 10 courses that given less than 75% satisfy rate. The average satisfy rate of ItemRank recommendation is higher than that of the learned list. In the case of Tangent, it is more severe. Only 2 courses, Introduction to Computers and Descriptive Introduction to Physics, given more than 75% satisfy rate. The average satisfy rate of UPKR is 81% which is higher than the average the learned list. Furthermore, newly added items do not seem to be suitable for the interest of the user. After this we can use our recommendation system as shown in the bottom of Figure 13 "Create user Profile", for these users, there are two methods for them who want to get recommended learning resources with similar interests of they viewed before, the first one to tell

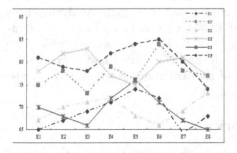
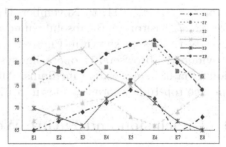

Fig. 9. c=0.001   **Fig. 10.** c=0.002

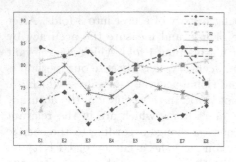
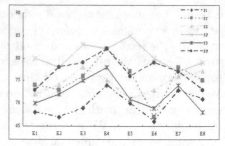

Fig. 11. c=0.003   **Fig. 12.** c=0.004

our EP about more detail about themselves,such as major, education backgroud
and so on, so we can get recommendation ,based on the similarity among profiles
of users and profiles of learning resources, (see Figure 13). The other method
is to cluster the learning resources in user's workspace,based on the profiles of
learning resources, we can infer a particular user's profile.

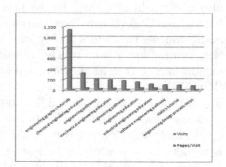

Fig. 13. Success to Save the "Using the   **Fig. 14.** Top-10 Visit Keywords
data warehouse" to Lucy's workspace

Then we do some demographics of the users during an extended time period
from January 1, 2011, to November 31, 2011 based on Google Analytic, there
are 52,572 visitors with absolute unique 42,951 visitors came from 166 coun-
tries/territories, as shown in Figure 6, they use 83 languages with 9,698 service
providers, 31 browsers, 19 operating systems, and 66 browser and OS combinations, via 685 sources and mediums. All of the search are via 12,821 keywords,
and the top-10 keywords as shown in Figure 14.

6   Conclusion and Future Work

We proposed the implementation of user recommendations to support deliver of
materials from the Engineering Pathways (NSDL) website. a clustering mecha-

nism is used as the basis for generating recommendations for users of the Engineering Pathway DL, and an experiment was conducted in order to assess the validity or potential of the proposed approach. By describing an optimal Cure clustering algorithms for short text are and summarizing web-log data from users, we can personalize search results based on users' profiles. Based on our system, we can implement much more personal and much more precise learning resource recommendation,in fact, the CURE algorithm could also be used for discovering the outlier learning resource information,which in our case, the outline learning resources must not relevant with user's major directly, but it may be another new research area this user may interested in, or maybe a potential information contributions to the user's major, in this case it will help mining the relationship among multiple and different majors and educational background in our future work.

Acknowledgments. This work was supported by a grant from the National Natural Science Foundation of China (No. 60970018).

References

1. Agogino, A.M.: Broadening Participation in Computing with the K-Gray Engineering Pathway Digital Library. In: Proceedings of the 8th ACM/IEEE-CS Joint Conference on Digital Libraries, JCDL 2008. ACM, New York (2008)
2. Agogino, A.M.: Engineering Pathway Education Digital Library. In: ABET Workshop, ASEE Meeting, June 24 (2007)
3. Dong, A., Agogino, A.M.: Design Principles for the Information Architecture of a SMET Education Digital Library. In: Proceedings of the 8th ACM/IEEE-CS Joint Conference on Digital Libraries, JCDL 2003. ACM, New York (2003)
4. Agogino, A.M.: Engineering Courseware Content and Delivery: The NEEDS Infrastructure for Distance Independent Education
5. Agogino, A.M.: Modeling Information Needs in Engineering Databases Using Tacit Knowledge
6. Jesus, S.-G., Enrique, H.-V., Olivas Jose, A., et al.: A google wave-based fuzzy recommender system to disseminate information in University Digital Libraries 2.0. Information Sciences 181(9), 1503–1516 (2011)
7. Yi, K.: A semantic similarity approach to predicting Library of Congress subject headings for social tags. Journal of the American Society for Information Science and Technology 61(8), 1658–1672 (2010)
8. Jiang, Y., Zheng, H.-T., Wang, X.: Affiliation disambiguation for constructing semantic digital libraries. Journal of the American Society for Information Science and Technology 62(6), 1029–1041 (2011)
9. Smith, C.L., Kantor, P.B.: User Adaptation: Good Results from Poor Systems. In: SIGIR 2008, Singapore, July 20-24 (2008)
10. Lin, J., Smucker, M.D.: How Do Users Find Things with PubMed? Towards Automatic Utility Evaluation with User Simulations. In: SIGIR 2008, Singapore, July 20-24 (2008)

11. Brusilovsky, P., Cassel, L., Delcambre, L., Fox, E., Furuta, R., Garcia, D.D., Shipman III, F.M., Bogen, P., Yudelson, M.: Enhancing Digital Libraries with Social Navigation: The Case of Ensemble. In: Lalmas, M., Jose, J., Rauber, A., Sebastiani, F., Frommholz, I. (eds.) ECDL 2010. LNCS, vol. 6273, pp. 116–123. Springer, Heidelberg (2010)
12. Fernandez-Villavicencio, N.G.: Helping students become literate in a digital, networking-based society: a literature review and discussion. International Information and Library Review 42(2), 124–136 (2010)
13. Hsu, K.-K., Tsai, D.-R.: Mobile Ad Hoc Network Applications in the Library. In: Proceedings of the 2010 Sixth International Conference on Intelligent Information Hiding and Multimedia Signal Processing (IIHMSP 2010), pp. 700–703 (2010)
14. Jing, H., Qingjun, G.: Prospect Application in the Library of SNS. In: 2011 Third Pacific-Asia Conference on Circuits Communications and System (PACCS), Wuhan, China, July 17-18 (2011)
15. Ankem, K.: The Extent of Adoption of Internet Resource-Based Value-Added Processes by Faculty in LIS Education. Canadian Journal of Information and Library Science-Revue 34(2), 213–232 (2010)
16. Alias, N.A.R., Noah, S.A., Abdullah, Z., et al.: Application of semantic technology in digital library. In: Proceedings of 2010 International Symposium on Information Technology (ITSim 2010), pp. 1514–1518 (2010)
17. Datta, E., Agogino, A.M.: Mobile Learning and Digital Libraries: Designing for Diversity. In: Proceedings of ASME Congress (2007) ISBN 0-7918-3812-9
18. Ryokai, K., Oehlberg, L., Agogino, A.M.: Green Hat: Exploring the Natural Environment Through Experts Perspectives. In: ACM CHI 2011: Proceedings of the 29th International Conference on Human Factors in Computing Systems, pp. 2149–2152 (2011)
19. Zhang, T., Ramakrishnan, R., Livny, M.: BIRCH: a new data clustering algorithm and its applications. Data Mining and Knowledge Discovery 1(2), 141–182 (1997)
20. Suk, M.S., Song, O.Y.: Curvilinear Feature-Extraction Using Minimum Spanning-Trees. Computer Vision Graphics and Image Processing 26(3), 400–411 (1984)
21. Guha, S., Rastogi, R., Shim, K.: Cure: An efficient clustering algorithm for large databases. Information Systems 26(1), 35–58 (2001)

Pick-Up Tree Based Route Recommendation from Taxi Trajectories

Haoran Hu[1], Zhiang Wu[2,*], Bo Mao[2], Yi Zhuang[3], Jie Cao[2], and Jingui Pan[1]

[1] State Key Lab. for Novel Software Technology, Nanjing University, Nanjing, China
[2] Jiangsu Provincial Key Laboratory of E-Business,
Nanjing University of Finance and Economics, Nanjing, China
[3] College of Computer and Information Engineering,
Zhejiang Gongshang University, Hangzhou, China
zawuster@gmail.com

Abstract. Recommending suitable routes to taxi drivers for picking up passengers is helpful to raise their incomes and reduce the gasoline consumption. In this paper, a pick-up tree based route recommender system is proposed to minimize the traveling distance without carrying passengers for a given taxis set. Firstly, we apply clustering approach to the GPS trajectory data of a large number of taxis that indicates state variance from "free" to "occupied", and take the centroids as potential pick-up points. Secondly, we propose a heuristic based on skyline computation to construct a pick-up tree in which current position is its root node that connects all centroids. Then, we present a probability model to estimate gasoline consumption of every route. By adopting the estimated gasoline consumption as the weight of every route, the weighted Round-Robin recommendation method for the set of taxis is proposed. Our experimental results on real-world taxi trajectories data set have shown that the proposed recommendation method effectively reduce the driving distance before carrying passengers, especially when the number of cabs becomes large. Meanwhile, the time-cost of our method is also lower than the existing methods.

Keywords: Taxi trajectories, pick-up tree, route recommendation, clustering, skyline.

1 Introduction

The rapid growth of wireless sensors and development of Global Positioning System (GPS) technologies [1] make it increasingly convenient to obtain the time-stamped trajectory data of taxis. In practice, cab drivers wish to be recommended a fastest route to pick up passengers. Such a large number of trajectories provide us unprecedented opportunity to mine useful knowledge and to recommend efficient routes for cab drivers. This recommender system not only helps cab drivers to raise the income, but also decreases the gasoline consumption which is good for environmental protection.

---

* Corresponding author.

H. Gao et al. (Eds.): WAIM 2012, LNCS 7418, pp. 471–483, 2012.

In the literature, a great deal of research has been devoted to mobile recommendations. These studies mainly focus on the following subareas: the mobile tourist recommender systems [2, 3, 4, 5], taxi driving fraud detection systems [6], driving direction recommender systems [7, 8], and routes for carrying passengers recommendation [9], which have greatly advanced the research to a higher level.

The scope of this paper belongs to the field of recommending routes for carrying passengers. The idea of this paper stems from the work in [9] which aims to recommend a travel route for a cab driver in a way such that the potential travel distance before carrying passengers is minimized. In their approach, for the length of suggested driving route $\mathcal{L}$, there are $C_K^{\mathcal{L}}$ candidate routes where K is the number of pick-up points. Then, routes which are not dominated by any other route are recommended to cab drivers. The main task in [9] is to search the best routes among these $C_K^{\mathcal{L}}$ candidate routes. Although a satisfying strategy called *SkyRoute* is proposed [9], it is still time-consuming as the increase of $\mathcal{L}$ and K. In the meanwhile, a multitude of routes are recommended to cab drivers in the equal possibilities without considering the number of passengers and driving distance of the routes, as shown in the left side of the Fig. 1. If we consider the scene that many taxis head for a narrow route with few passengers, conflict for carrying customers or even the traffic jam in that route may happen.

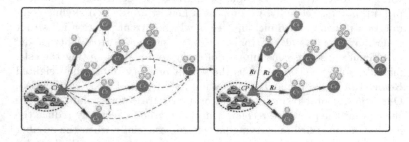

Fig. 1. The illustration for our research motivation

To meet this dilemma, a data structure named *pick-up tree* is presented which takes the current position as its root node and connects all pick-up points is presented. The better routes are recommended in a high probability as shown in thick line in right side of Fig. 1. Moreover, if we assume there are a set of cabs around CP, by adopting our recommendation, most of cabs will cruise along with the better routes and other relatively worse routes will also be covered by less cabs. There remain two problems to solve when we follow this idea:

- An objective function measuring the global profit of the set of taxis by adopting recommendation should be presented. The probabilities for picking up passengers at points should be estimated. Also, the expectation for measuring the weight of routes should be given.
- Finding the optimal value of objective function among all routes is a combinatorial optimization problem which is NP_hard. How to apply a heuristic to this NP_hard problem should be solved.

In this paper, we focus on the recommendation and minimization of the gasoline consumption which is close related to idle driving distance for *a set of cabs* around a position, rather than *one cab*. Historical pick-up points are extracted and clustered, and routes connecting the centroids are taken as recommendation choise. We then propose a heuristic to construct a *pick-up tree* to cover all centroids. The model for estimating oil consumption before carrying passengers of every route is presented. By adopting the oil consumption as the weight of every route, the recommendation method for the set of taxis is proposed.

2   Problem Definition

Let $\mathcal{C}$ be the set of K potential pick-up points $\mathcal{C} = \{C_1, C_2, \cdots, C_K\}$. We then assume there are W target taxis around a position CP, $\mathcal{T} = \{T_1, T_2, \cdots, T_W\}$. The probability that a taxi could carry passengers in the pick-up point C_i is written $P(C_i)$, and the set of mutually independence probability is written $\mathcal{P} = \{P(C_1), P(C_2), \cdots, P(C_K)\}$.

The route recommendation aims to generate W routes for every taxi, and minimize the global gasoline consumption before picking up passengers for all taxis. So, we have the objective function as follows:

Definition 1 (Global Gasoline Consumption Function). *Given W recommended routes $\mathcal{R} = \{R_1, R_2, \cdots, R_W\}$, the global oil consumption function:*

$$O_1 : \min_{\mathcal{R} \in \Omega} \sum_{r=1}^{W} \mathbb{C}(CP, R_r) \tag{1}$$

In the *definition* 1, CP is the current position of taxis, the set of all possible routes is known as Ω, and we try to find a subset of Ω with size of W for minimization. The problem is NP_hard due to the exponential scale of Ω. The oil consumption of a route is measured by the $\mathbb{C}(\bullet)$ function which will be addressed in Section 4. Note that the problem should not only minimize the global gasoline consumption, but also attempt to cover all pick-up points.

3   CabRec Design

In this section, we propose a route recommendation method, named CabRec, for cab drivers. To illustrate it, we first present a heuristic algorithm to generate a pick-up tree. Then, we show how to compute the gasoline consumption for each route. Finally,we describe the weighted Round-Robin recommendation process.

3.1   Pick-Up Tree Generation

The problem is to construct a tree which uses CP as the root and connects all points in $\mathcal{C}$. The straightforward way is to add an edge between any of two nodes, but the derived network will be too complex to process. In our solution, edges between a point and its skyline points are added. The set of skyline points consists of the points that are not dominated by any other point [10]. We should begin with the definition of the point dominance.

Definition 2 (Point Dominance). *Let CP' denote the current node, C_i and C_j are two different pick-up points. These two pick-up points can be described by two dimensions $C_i = (P(C_i), D(C_i, CP'))$ and $C_j = (P(C_j), D(C_j, CP'))$. We say that point C_i dominates point C_j iff one of the cases happens: (1) $P(C_i) = P(C_j)$ and $D(C_i, CP') < D(C_j, CP')$; (2) $P(C_i) > P(C_j)$ and $D(C_i, CP') = D(C_j, CP')$; (3) $P(C_i) > P(C_j)$ and $D(C_i, CP') < D(C_j, CP')$.*

The pseudocode of the pick-up tree generation algorithm is given as follows. Lines 2 and 3 are about initialization. The *FindSkyline* function in line 2 finds skyline points for the current position CP. Then, in line 5, we use a heuristic metric with maximum ratio of pick-up probability to distance to select the next expanding node. The algorithm stops when all points in $\mathcal{C}$ are added to the pick-up tree.

DISCUSSION. If we construct a rectangular coordinate system with CP' as its origin, all remaining pick-up points in $\mathcal{C}$ can be depicted by $C_i = (P(C_i), D(C_i, CP'))$. That there is at least a skyline point for CP'. So, in each *while* loop, at least a point is added to the pick-up tree. The pick-up tree guarantees to include all points in $\mathcal{C}$.

Algorithm 1. Pick-up Tree Generation Algorithm

1: **procedure** CREATEPTREE($CP, \mathcal{C}$)
2:      $V \leftarrow$ FindSkyline($CP, \mathcal{C}$)
3:      Add edges between CP and its skyline points
4:      **while** $(\mathcal{C} - V) \neq NULL$ **do**
5:          $CP' \leftarrow \arg\max_i \{ \frac{P(C_i)}{D(C_i, CP')}, i = 1, \cdots, |V| \}$
6:          $V \leftarrow V +$ FindSkyline($CP', \mathcal{C} - V$)
7:          Add edges between CP' and its skyline points
8:      **end while**
9: **end procedure**

It can be seen clearly that skyline computation encapsulated in the *FindSkyline* function can be done in polynomial time. The straightforward method is to compare each point with all the other points to check whether it can be dominated by some points. If so, remove it, otherwise mark this point as a skyline point. The time complexity of this method is $O(2n^2)$, but with the increase of the number of pick-up points, the performance *FindSkyline* will become the bottleneck. In this article, we use a sort-based skyline computation to implement *FindSkyline*.

Obviously, *FindSkyline* is much more faster than the straightforward way, because comparison is made just on the skyline points and all the points only need to visit once. Also, *Theorem* 1 is presented to guarantee the correctness of the process of *FindSkyline*.

Theorem 1. *After sorting on the data set $\mathcal{D}$, the one with the maximum value will be certainly the skyline point.*

PROOF: If the maximum point p_{max} is not a skyline point, it must be dominated by at least one point represented by q. According to the *definition* 1, $P(q) - D(q, CP') > P(p_{max}) - D(p_{max}, CP')$ is surely satisfied. So, p_{max} is not the maximum point and contradiction appears. Proof done.

Algorithm 2. Sort-based Skyline Computation Algorithm

1: **procedure** FINDSKYLINE($CP', \mathcal{D}$)
2:      Sort all points in $\mathcal{D}$ with $P(D_i) - D(D_i, CP')$,get $\mathcal{D}'$
3:      Maintain a set S for skyline points
4:      Move the first point into S from $\mathcal{D}'$
5:      **while** $\mathcal{D}' \neq NULL$ **do**
6:          Let p_{max} denote the current maximum in $\mathcal{D}'$
7:          Compare p_{max} with all the points in S
8:          If p_{max} is dominated by some points in S,
9:          Just remove p_{max} from $\mathcal{D}'$
10:          Else move p_{max} from $\mathcal{D}'$ to S
11:      **end while**
12: **end procedure**

3.2 Computational Issues

We assume the oil consumption increases proportionately with the driving distance before carrying passengers. Let k_{max} denote the length of the route $\boldsymbol{R_r} = \{C_1, C_2, \cdots, C_{k_{max}}\}$. Two cases may happen when a taxi selects the route: (1) the taxi carries a customer in one of the k_{max} nodes; (2) the taxi does not carry any customer in that route.

1. Assume the taxi carries a passenger in $C_k (k = 1, 2, \cdots, k_{max})$, the travel distance without customers is as follows:

$$\mathcal{F}_k = \prod_{i=1}^{k-1}(1 - P(C_i))P(C_k)\sum_{i=1}^{k} D_i \qquad (2)$$

In Eq. (2), D_i is the distance between $C_{(i-1)}$ and C_i, and D_1 is the distance between CP and C_1. The most common way for computing D_i is to use Euclidean distance. However, the earth is roughly a great circle, and the latitude and longitude are defined globally in respect to the earth surface instead of a plane. In this article, we employ Vincenty's formula [11] with the assumption of spherical earth. Let $C_{(i-1)} = (\phi_{(i-1)}, \lambda_{(i-1)})$ and

$C_i = (\phi_i, \lambda_i)$ denote the two points, where ϕ and λ are the latitude and longitude, respectively.

$$D_i = r \cdot \arctan(\frac{\sqrt{(\cos \phi_i \sin \triangle\lambda)^2 + (\cos \phi_{(i-1)} \sin \phi_i - \sin \phi_{(i-1)} \cos \phi_i \cos \triangle\lambda)^2}}{\sin \phi_{(i-1)} \sin \phi_i + \cos \phi_{(i-1)} \cos \phi_i \cos \triangle\lambda})$$

$$(3)$$

In Eq. (3), r is the radius of the Earth and is set to $6372.795km$, and $\triangle\lambda = |\lambda_i - \lambda_{i-1}|$. Eq. (4) measures the average oil consumption in the case of carrying a customer in the route. The smaller $\mathcal{F}(CP, R_r)$ is, the more valuable the route will be.

$$\mathcal{F}(CP, R_r) = \sum_{k=1}^{k_{max}} \mathcal{F}_k \qquad (4)$$

2. The event that a taxi does not carry a customer in a route may also happen. We have the probability of occurrence as follows:

$$P_\phi(R_r) = \prod_{i=1}^{k_{max}} (1 - P(C_i)) \qquad (5)$$

In Eq. (5), $P_\phi(R_r)$ decreases with the increase of the length of a route, and the smaller it is, the more valuable the route is.

We then define a cost function to combine above-mentioned two cases:

$$\mathbb{C}(CP, R_r) = \alpha^{P_\phi} \mathcal{F}(CP, R_r) \qquad (6)$$

In Eq. (6), an exponential function is employed to magnify P_ϕ. We set $\alpha > 1$ to make the cost monotone increasing with both P_ϕ and $\mathcal{F}(\bullet)$.

3.3   Recommendation Method

Assume the pick-up tree has $N(N \le K)$ leaf nodes, so there are N possible routes $\mathbf{R} = \{R_1, R_2, \cdots, R_N\}$. In this subsection, we introduce the method for recommending the N routes to W target taxis. We propose to use a standard weight to measure the importance of a route. The cost for each route is obtained by Eq. (6). We then define the weight of each route as follows:

$$\omega(R_r) = \frac{1}{N-1} \frac{\sum_{i=1}^{N} \mathbb{C}(CP, R_i) - \mathbb{C}(CP, R_r)}{\sum_{i=1}^{N} \mathbb{C}(CP, R_i)} \qquad (7)$$

Obviously, weights for all routes are normalized, $\sum_{i=1}^{N} \omega(R_i) = 1$. A Round-Robin method is used to make recommendation for multiple cabs in [9], and k optimal routes are used to generate the circle list. The Round-Robin approach recommends routes in the circle list to the coming empty cabs in turn. If there

Table 1. Description of the 10 centroids

No.	C_0	C_1	C_2	C_3	C_4	C_5	C_6	C_7	C_8	C_9
latitude	37.7901	.7773	.7752	.7472	.7990	.7971	.7641	.7869	.7705	.7800
longitude	-122.4199	.3949	.4169	.4198	.4344	.4043	.4277	.4062	.4657	.4391
Size	3075	1357	1838	1763	2154	3818	3506	4964	1253	2245
Dist.(km)	1.136	2.195	2.243	5.319	2.383	0.466	3.759	0.816	5.711	3.151
$P(C_i)$	0.120	0.082	0.120	0.093	0.131	0.090	0.097	0.097	0.101	0.117

Note: (1) Only the fractional part of latitude and longitude are kept.
(2) Size is the number of pick-up points in this cluster.

Table 2. The cost and weight of recommended routes

No.	Recommended routes	Cost	Weight
1	(CP, C_0)	0.195	0.192
2	(CP, C_4)	0.444	0.182
3	(CP, C_7)	0.114	0.195
4	(CP, C_5, C_1)	0.354	0.185
5	(CP, C_5, C_2, C_6)	1.006	0.158
6	$(CP, C_5, C_2, C_9, C_8, C_3)$	2.731	0.088

are more than k empty cabs, recommendations are repeated from the first route again after the kth empty cab.

In this paper, we firstly generate the circle list according to the weight of each route. For W empty cabs, we add $\lceil \omega(R_r) \cdot W \rceil$ times into the circle list for route R_r. Also, we randomly arrange routes in the circle list and employ the Round-Robin method to generate a recommended route for each cab. We benefits the proposed recommendation method from two aspects: (1) each route appears in the circle list at least once, so none pick-up point is ignored; (2) the route with low cost has more opportunities appearing in the circle list, so more cabs will cruise to that route.

DISCUSSION. In section 2, the green recommendation problem is formulated a combinatorial optimization problem which is NP_hard. So, our proposed CabRec is a heuristic in essence. However, the cost estimation for a route and the recommendation method are deterministic. In fact, we apply heuristic during construction of the pick-up tree, and once the pick-up tree is generated, the recommendation list can be produced. This strategy is more efficient The proposed CabRec does not always recommend the best route to cabs, but recommend better routes with high probability.

4   Experimental Results

In this section, we demonstrate the effectiveness of the proposed CabRec. Specifically, we will show: (1) An example of the pick-up tree and the cost and weight of every route; (2) The superior performance of CabRec compared with $\mathcal{LCP}$ on driving distance before carrying passengers and time-cost.

4.1 Experiment Setup

Data Sets. For our experiments, we use a real-world data set containing GPS location traces of approximately 500 taxis collected around 30 days in the San Francisco Bay Area, USA. To distinguish from other works, state variance points are deemed to be the potential pick-up points. The state of the pick-up point is $1(= occupied)$, and the state of its previous point is $0(= free)$. Then, we extract the pick-up points of all cab drivers (536 cabs) on two time periods: *2PM-3PM* and *6PM-7PM*. In total, 25973 points are obtained during *2PM-3PM*, and 6203 points are obtained during *6PM-7PM*. Simple K-means provided by WEKA[1] is used for clustering where "Euclidean Distance" function, "10" seed and "500" max iterations are set.

Methods. Two types of route recommendation methods are implemented for performance comparison. The first one is the proposed CabRec, and another one is $\mathcal{LCP}$ proposed in [9], coded by ourselves in Java. We use $\mathcal{LCP}_x$ to denote the $\mathcal{LCP}$ method of which the length of recommended route is x.

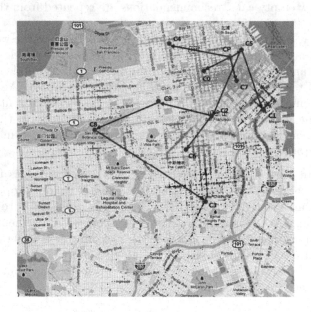

Fig. 2. An example of the pick-up tree

4.2 Results of Pick-Up Tree

In this subsection, we illustrate the output of the pick-up trees with different CP and the number of clusters K. To facilitate the exhibition, we limit the range of map with the latitude in the interval $[37.80654, 37.70846]$ and the longitude in the

---

[1] http://www.cs.waikato.ac.nz/ml/weka/

interval [-122.49979, -122.3794]. We then extract all state variance points of 536 cabs during *2PM-3PM* and obtain 25973 potential pick-up points. The current position CP is located at the point (37.7941, -122.4080) which is the China Town in the San Francisco Bay Area. All potential pick-up points are clustered into 10 clusters. The way to estimate $P(C_i)$ of the ith centroid is as follows: finding the adjacent points in the ith cluster for every taxi and computing average difference of time all taxis denoted as $t_{avg}(C_i)$. So, we have $P(C_i) = 1/t_{avg}(C_i)$. Note that two adjacent points implicate the taxi has carried passengers, since the point in each cluster represents state variance from *free* to *occupied*. Table 1 lists the information of the 10 centroids, and the derived pick-up tree is shown in Fig. 2.

In this example, there are 6 routes from CP to the leaf nodes. We set $\alpha = 1.5$ to get the values of cost function by Eq. (6), and then obtain the standard weights of every route by Eq. (7). Table 2 shows the cost and weight of these 6 recommended routes.

4.3   Performance Comparison

Here, we compare the proposed CabRec with $\mathcal{LCP}$ on the driving distance before carrying passengers and time-cost. We design the procedure of our experiment as follows:

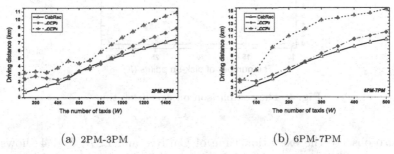

(a) 2PM-3PM   (b) 6PM-7PM

Fig. 3. A comparison on the average driving distance before carrying passengers

- **Step 1: Pick-Up Points Generation by Clustering.** The state variance points of all taxis during the given time period are clustered into K clusters, and the K centroids are returned as the pick-up points.
- **Step 2: Route Recommendation.** We employ CabRec and $\mathcal{LCP}$ for route recommendation. CabRec utilizes the weighted Round-Robin method to generate recommended routes, while $\mathcal{LCP}$ takes all routes equal and uses the simple Round-Robin method.
- **Step 3: Simulation of Taxis to Pick Up Passengers.** In this step, we assume there are W target taxis around a given current position. Since the time span of the experimental data set is about 30 days, the size of every cluster divided by 30 days is the average number of passengers of each pick-up point every day. After these W target taxis adopt recommended routes

generated by both CabRec and $\mathcal{LCP}$, once a taxi passes a pick-up point having non-zero remaining passengers, the taxi is deemed to be occupied and the number of remaining passenger of that pick-up point should subtract 1.

In the above-mentioned procedure, we are readily to compute the average driving distance of W taxis before carrying passengers. Note that if the taxi does not carry any passenger in the recommended route we extra add $10km$ to its driving distance as penalty. We set $K = 20$ and range W in different scale according to the number of state variance points of different time period. Fig. 3 show the comparison results on two time periods. As can be seen, CabRec works much more better than both $\mathcal{LCP}_3$ and $\mathcal{LCP}_4$, especially when W is large. The reason lies in that since CabRec has estimated the cost of every route and taken it as the weight, more cabs will cruise along with the high-weight routes with the increase of W. In contrast, $\mathcal{LCP}$ takes every route as equal and many taxis head for the routes with few passengers, which reduce the performance dramatically.

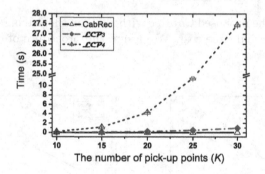

Fig. 4. The comparison on execution time

We also observe the execution time of CabRec and $\mathcal{LCP}$. Fig. 4 shows the execution time with the increase of the number of pick-up points (K centroids). As can be seen, the execution time of $\mathcal{LCP}_4$ goes up sharply. Although the time-cost of $\mathcal{LCP}_3$ is not very high, our CabRec performs even more fast than $\mathcal{LCP}_3$.

To sum up, we have two conclusions from the experiments: (1) The proposed CabRec can successfully decrease the average driving distance of the set of taxis before carrying passengers. (2) The time-cost of our CabRec is also lower than the existing methods.

5   Related Work

With the prevalent of the ubiquitous computing, it is more and more convenient to obtain the information about location, time, trajectory, surrounding, etc. Mobile recommender systems have attracted more attentions. Some research opportunities on mobile recommender systems were discussed in [12]. Quercia et al. have designed a social events recommender system using mobile phone location data [13]. Several kinds of mobile tourist recommender systems have been presented in [2, 3, 4, 5].

In addition, great research efforts have been taken to use GPS trajectories of cabs for analysis. Liu et al.focused on cabdriver operation patterns analysis rather than recommendation based on the large scale cabdrivers' behavior in Shenzhen, China [14]. A mobility-based clustering analysis method is used for identifying hot spots of moving vehicles [15]. Chang et al. have proposed a context-aware prediction model for taxi demand hotspots [16]. Wu et al. investigated service site selection using GIS data [17]. Zheng and Xie et al. have conducted many studies on knowledge discovery from GPS trajectory data in Beijing, China. For instance, the computation method of user similarity based on location history is proposed in [18], friends and locations recommendation method is then studied in [19], and the knowledge extracted from GPS data is also used to solve the congested traffic problems [20, 8]. Chen et al. studied a new problem of searching the k Best Connected Trajectories (k-BCT) by locations from a database, in which context the query is only a small set of locations [21]. Chen et al. also discovered the Most Popular Route(MPR) between two locations by observing the traveling behaviors of many previous users [22]. In this article, we focus on green recommendation for cab drivers, and our experiments utilize the cab traces data set in the San Francisco Bay Area, which is also used by [9, 6].

6   Conclusion and Future Work

Upon taxi trajectories data, this paper proposes a system called CabRec for route recommendation. In CabRec, the state variance points that imply the taxis have carried passengers are clustered, and the centroids are taken as potential pick-up points. Then, a heuristic is employed to construct the pick-up tree which takes the current position as its root node and connect all centroids. A probability model to estimate the weight of every route and the weighted Round-Robin recommendation method for the set of taxis is proposed. Our experimental results on real-world taxi trajectories data set have shown the effectiveness and efficiency of the CabRec.

There are a wealth of research directions that we are currently considering, such as implementing a cab recommender system with dynamic visualization, employing several large-scale data sets, and expanding other recommender applications in the area of intelligent transportation system, and more.

Acknowledgments. This research is supported by National Natural Science Foundation of China (No.61103229, 71072172, and 61003074), Industry Projects in the Jiangsu Science & Technology Pillar Program (No.BE2011198), Jiangsu Provincial Colleges and Universities Outstanding Science & Technology Innovation Team Fund(No.2001013), International Science & Technology Cooperation Program of China (No.2011DFA12910), National Key Technologies R&D

sub Program in 12th five-year-plan (No. SQ2011GX07E03990), Jiangsu Provincial Key Laboratory of Network and Information Security(Southeast University) (No.BM2003201), Natural Science Foundation of Jiangsu Province of China (BK2010373) and Postgraduate Cultivation and Innovation Foundation of Jiangsu Province (CXZZ11_0045).

References

[1] Wang, S., Wu, C.: Application of context-aware and personalized recommendation to implement an adaptive ubiquitous learning system. Expert Systems with Applications (2011)

[2] Abowd, G.D., Atkeson, C.G., Hong, J., Long, S., Kooper, R., Pinkerton, M.: Cyberguide: A mobile context-aware tour guide. Wireless Networks 3(5), 421–433 (1997)

[3] Staab, S., Werthner, H.: Intelligent systems for tourism. IEEE Intelligent Systems 17(6), 53–66 (2002)

[4] Tao, Y., Ding, L., Lin, X., Pei, J.: Distance-based representative skyline. In: Proceedings of the 2009 IEEE International Conference on Data Engineering (ICDE 2009), pp. 892–903 (2009)

[5] Liu, Q., Ge, Y., Li, Z., Chen, E., Xiong, H.: Personalized travel package recommendation. In: IEEE 11th International Conference on Data Mining (ICDM 2011), pp. 407–416 (2011)

[6] Ge, Y., Xiong, H., Liu, C., Zhou, Z.: A taxi driving fraud detection system. In: IEEE 11th International Conference on Data Mining(ICDM 2011), pp. 181–190 (2011)

[7] Yuan, J., Zheng, Y., Zhang, C., Xie, W., Xie, X., Sun, G., Huang, Y.: T-drive: Driving directions based on taxi trajectories. In: Proceedings of the 18th SIGSPATIAL International Conference on Advances in Geographic Information Systems, pp. 99–108 (2010)

[8] Yuan, J., Zheng, Y., Xie, X., Sun, G.: Driving with knowledge from the physical world. In: Proceedings of the 17th ACM SIGKDD International Conference on Knowledge Discovery and Data Mining, pp. 316–324 (2011)

[9] Ge, Y., Xiong, H., Tuzhilin, A., Xiao, K., Gruteser, M., Pazzani, M.: An energy-efficient mobile recommender system. In: Proceedings of the 16th ACM SIGKDD International Conference on Knowledge Discovery and Data Mining, pp. 899–908 (2010)

[10] Yuan, Y., Lin, X., Liu, Q., Wang, W., Yu, J., Zhang, Q.: Efficient computation of the skyline cube. In: Proceedings of the 31st International Conference on Very Large Data Bases, pp. 241–252 (2005)

[11] Vincenty, T.: Direct and inverse solutions of geodesics on the ellipsoid with application of nested equations. Survey Review 23(176), 88–93 (1975)

[12] van der Heijden, H., Kotsis, G., Kronsteiner, R.: Mobile recommendation systems for decision making. In: Proceedings of International Conference on Mobile Business (ICMB 2005), pp. 137–143 (2005)

[13] Quercia, D., Lathia, N., Calabrese, F., Lorenzo, G.D., Crowcroft, J.: Recommending social events from mobile phone location data. In: IEEE 10th International Conference on Data Mining (ICDM 2010) (2010)

[14] Liu, L., Andris, C., Ratti, C.: Uncovering cabdrivers' behavior patterns from their digital traces. Computers, Environment and Urban Systems 34(6), 541–548 (2010)

[15] Liu, S., Liu, Y., Ni, L.M., Fan, J., Li, M.: Towards mobility-based clustering. In: Proceedings of the 16th ACM SIGKDD International Conference on Knowledge Discovery and Data Mining, pp. 919–927 (2010)

[16] Chang, H., Tai, Y., Hsu, J.: Context-aware taxi demand hotspots prediction. International Journal of Business Intelligence and Data Mining 5(1), 3–18 (2010)

[17] Wu, J., Chen, J., Ren, Y.: GIS enabled service site selection: Environmental analysis and beyond. Information Systems Frontier (13), 337–348 (2011)

[18] Li, Q., Zheng, Y., Xie, X., Chen, Y., Liu, W., Ma, W.Y.: Mining user similarity based on location history. In: Proceedings of the 16th ACM SIGSPATIAL International Conference on Advances in Geographic Information Systems (2008)

[19] Zheng, Y., Zhang, L., Ma, Z., Xie, X., Ma, W.: Recommending friends and locations based on individual location history. ACM Transactions on the Web (TWEB) 5(1), 5 (2011)

[20] Liu, W., Zheng, Y., Chawla, S., Yuan, J., Xing, X.: Discovering spatio-temporal causal interactions in traffic data streams. In: Proceedings of the 17th ACM SIGKDD International Conference on Knowledge Discovery and Data Mining, pp. 1010–1018 (2011)

[21] Chen, Z., Shen, H., Zhou, X., Zheng, Y., Xie, X.: Searching trajectories by locations: an efficiency study. In: Proceedings of the 2010 International Conference on Management of Data (SIGMOD 2010), pp. 255–266 (2010)

[22] Chen, Z., Shen, H., Zhou, X.: Discovering popular routes from trajectories. In: Proceedings of the 2009 International Conference on Data Engineering (ICDE 2011), pp. 900–911 (2011)

Author Index